T0213205

# Lecture Notes in Computer Science 12043

More information about this series at http://www.springer.com/series/7407

Roman Wyrzykowski · Ewa Deelman ·
Jack Dongarra · Konrad Karczewski (Eds.)

# Parallel Processing
# and Applied Mathematics

13th International Conference, PPAM 2019
Bialystok, Poland, September 8–11, 2019
Revised Selected Papers, Part I

 Springer

*Editors*
Roman Wyrzykowski
Czestochowa University of Technology
Czestochowa, Poland

Ewa Deelman
University of Southern California
Marina del Rey, CA, USA

Jack Dongarra
University of Tennessee
Knoxville, TN, USA

Konrad Karczewski
Czestochowa University of Technology
Czestochowa, Poland

ISSN 0302-9743      ISSN 1611-3349  (electronic)
Lecture Notes in Computer Science
ISBN 978-3-030-43228-7      ISBN 978-3-030-43229-4  (eBook)
https://doi.org/10.1007/978-3-030-43229-4

LNCS Sublibrary: SL1 – Theoretical Computer Science and General Issues

This Springer imprint is published by the registered company Springer Nature Switzerland AG
The registered company address is: Gewerbestrasse 11, 6330 Cham, Switzerland

# Preface

This volume comprises the proceedings of the 13th International Conference on Parallel Processing and Applied Mathematics (PPAM 2019), which was held in Białystok, Poland, September 8–11, 2019. It was organized by the Department of Computer and Information Science of the Częstochowa University of Technology together with Białystok University of Technology, under the patronage of the Committee of Informatics of the Polish Academy of Sciences, in technical cooperation with the IEEE Computer Society and IEEE Computational Intelligence Society. The main organizer was Roman Wyrzykowski.

PPAM is a biennial conference. 12 previous events have been held in different places in Poland since 1994, when the first PPAM took place in Częstochowa. Thus, the event in Białystok was an opportunity to celebrate the 25th anniversary of PPAM. The proceedings of the last nine conferences have been published by Springer in the Lecture Notes in Computer Science series (Nałęczów, 2001, vol. 2328; Częstochowa, 2003, vol. 3019; Poznań, 2005, vol. 3911; Gdańsk, 2007, vol. 4967; Wrocław, 2009, vols. 6067 and 6068; Toruń, 2011, vols. 7203 and 7204; Warsaw, 2013, vols. 8384 and 8385; Kraków, 2015, vols. 9573 and 9574; and Lublin, 2017, vols. 10777 and 10778).

The PPAM conferences have become an international forum for the exchange of the ideas between researchers involved in parallel and distributed computing, including theory and applications, as well as applied and computational mathematics. The focus of PPAM 2019 was on models, algorithms, and software tools which facilitate the efficient and convenient utilization of modern parallel and distributed computing architectures, as well as on large-scale applications, including artificial intelligence and machine learning problems.

This meeting gathered more than 170 participants from 26 countries. A strict refereeing process resulted in the acceptance of 91 contributed papers for publication in these conference proceedings. For regular tracks of the conference, 41 papers were selected from 89 submissions, thus resulting in an acceptance rate of about 46%.

The regular tracks covered important fields of parallel/distributed/cloud computing and applied mathematics such as:

- Numerical algorithms and parallel scientific computing, including parallel matrix factorizations
- Emerging HPC architectures
- GPU computing
- Parallel non-numerical algorithms
- Performance analysis in HPC systems
- Environments and frameworks for parallel/distributed/cloud computing
- Applications of parallel computing
- Soft computing with applications

The invited talks were presented by:

- David A. Bader from the New Jersey Institute of Technology (USA)
- Fabio Baruffa from the Intel Corporation
- Anne Benoit from ENS Lyon (France)
- Jack Dongarra from the University of Tennessee and ORNL (USA)
- Lin Gan from the Tsinghua University and National Supercomputing Center in Wuxi (China)
- Mary Hall from the University of Utah (USA)
- Torsten Hoefler from the ETH Zurich (Switzerland)
- Kate Keahey from the University of Chicago and Argonne National Lab (USA)
- Alexey Lastovetsky from the University College Dublin (Ireland)
- Miron Livny from the University of Wisconsin (USA)
- Satoshi Matsuoka from the Tokyo Institute of Technology (Japan)
- Bernd Mohr from the Jülich Supercomputing Centre (Germany)
- Manish Parashar from the Rutgers University (USA)
- Javier Setoain from ARM (UK)
- Leonel Sousa from the Technical University of Lisbon (Portugal)
- Jon Summers from the University of Lulea (Sweden)
- Manuel Ujaldon from the University of Malaga (Spain) and Nvidia
- Jeffrey Vetter from the Oak Ridge National Laboratory and Georgia Institute of Technology (USA)
- Tobias Weinzierl from the Durham University (UK)

Important and integral parts of the PPAM 2019 conference were the workshops:

- The 8th Workshop on Language-Based Parallel Programming Models (WLPP 2019), organized by Ami Marowka from the Bar-Ilan University (Israel)
- Workshop on Models, Algorithms and Methodologies for Hierarchical Parallelism in New HPC Systems, organized by Giulliano Laccetti and Marco Lapegna from the University of Naples Federico II (Italy), and Raffaele Montella from the University of Naples Parthenope (Italy)
- Workshop on Power and Energy Aspects of Computation (PEAC 2019), organized by Ariel Oleksiak from the Poznan Supercomputing and Networking Center (Poland) and Laurent Lefevre from Inria (France)
- Special Session on Tools for Energy Efficient Computing, organized by Tomas Kozubek and Lubomir Riha from the Technical University of Ostrava (Czech Republic), and Andrea Bartolini from the University of Bologna (Italy)
- Workshop on Scheduling for Parallel Computing (SPC 2019), organized by Maciej Drozdowski from the Poznań University of Technology (Poland)
- The Third Workshop on Applied High-Performance Numerical Algorithms in PDEs, organized by Piotr Krzyżanowski and Leszek Marcinkowski from the Warsaw University (Poland) and Talal Rahman from the Bergen University College (Norway)
- Minisymposium on HPC Applications in Physical Sciences, organized by Grzegorz Kamieniarz and Wojciech Florek from the A. Mickiewicz University in Poznań (Poland)

- Minisymposium on High Performance Computing Interval Methods, organized by Bartłomiej J. Kubica from the Warsaw University of Technology (Poland)
- Workshop on Complex Collective Systems, organized by Paweł Topa and Jarosław Wąs from the AGH University of Science and Technology in Kraków (Poland)

The PPAM 2019 meeting began with two tutorials:

- Modern GPU Computing, by Dominik Göddeke from the Stuttgart University (Germany), Robert Strzodka from the Heidelberg University (Germany), and Manuel Ujaldon from the University of Malaga (Spain) and Nvidia.
- Object Detection with Deep Learning: Performance Optimization of Neural Network Inference using the Intel OpenVINO Toolkit, by Evgenii Vasilev and Iosif Meyerov from the Lobachevsky State University of Nizhni Novgorod (Russia), and Nadezhda Kogteva and Anna Belova from Intel Corporation.

The PPAM Best Paper Award is awarded in recognition of the research paper quality, originality, and significance of the work in high performance computing (HPC). The PPAM Best Paper was first awarded at PPAM 2019 upon recommendation of the PPAM Chairs and Program Committee. For the main track, the PPAM 2019 winners were Evgeny Kuznetsov, Nikolay Kondratyuk, Mikhail Logunov, Vsevolod Nikolskiy, and Vladimir Stegailov from the National Research State University High School of Economics in Moscow and Russian Academy of Sciences (Russia), who submitted the paper "Performance and portability of state-of-art molecular dynamics software on modern GPUs." For workshops, the PPAM 2019 winners were Dominik Ernst, Georg Hager, and Gerhard Wellein from the Erlangen Regional Computing Center and Jonas Thies from the German Aerospace Center (Germany), who presented the paper "Performance Engineering for a Tall & Skinny Matrix Multiplication Kernel on GPUs."

A New Topic at PPAM 2019 was the Special Session on Tools for Energy Efficient Computing, focused on tools designed to improve the energy-efficiency of HPC applications running at scale.

With the steaming out of Moore's law and the end of Dennard's scaling, the pace dictated on the performance increase of HPC systems among generations has led to power constrained architectures and systems. In addition, the power consumption represents a significant cost factor in the overall HPC system economy. For these reasons, it is important to develop new tools and methodologies to measure and optimize the energy consumption of large scale high performance system installation. Due to the link between the energy consumption, power consumption, and execution time of the application executed by the final user, it is important for these tools and methodologies to consider all these aspects empowering the final user and the system administrator with the capability to find the best configuration given different high level objectives.

This special session provided a forum to discuss and present innovative solutions in following topics: (i) tools for fine grained power measurements and monitoring of HPC infrastructures, (ii) tools for hardware and system parameter tuning and its challenges in the HPC environment, (iii) tools and libraries for dynamic tuning of HPC applications at runtime, (iv) tools and methodology for identification of potential dynamic tuning for performance and energy, and (v) evaluation of applications in terms of runtime and energy savings.

These topics were covered by four presentations:

- Overview of application instrumentation for performance analysis and tuning (by O. Vysocky, L. Riha, and A. Bartolini) focused on automatizing the process of an application instrumentation which is an essential step for future optimization that leads to time and energy savings.
- Energy-efficiency tuning of the Lattice Boltzmann simulation using MERIC (by E. Calore, et al.) presents the impact of CPU core and uncore frequency dynamic tuning on energy savings, that reaches up to 24 % for this specific application.
- Evaluating the advantage of reactive MPI-aware power control policies (by D. Cesarini, C. Cavazzoni, and A. Bartolini) shows the COUNTDOWN library, that automatically down-scales the CPU core frequency during long-enough communication phases, with neither any modifications of the code nor complex application profiling.
- Application-aware power capping using Nornir (by D. De Sensi and M. Danelutto) presents how to combine DVFS and thread packing approaches to keep power consumption under a specified limit. This work shows that the proposed solution performs better than the state-of-the-art Intel RAPL power-capping approach for very low power budgets.

The organizers are indebted to PPAM 2019 sponsors, whose support was vital to the success of the conference. The main sponsor was the Intel Corporation and the other sponsors were byteLAKE and Gambit. We thank all the members of the International Program Committee and additional reviewers for their diligent work in refereeing the submitted papers. Finally, we thank to all of the local organizers from the Częstochowa University of Technology and the Białystok University of Technology, who helped us to run the event very smoothly. We are especially indebted to Łukasz Kuczyński, Marcin Woźniak, Tomasz Chmiel, Piotr Dzierżak, Grażyna Kołakowska, Urszula Kroczewska, and Ewa Szymczyk from the Częstochowa University of Technology; and to Marek Krętowski and Krzysztof Jurczuk from the Białystok University of Technology.

We hope that this volume will be of use to you. We would like everyone who reads it to feel welcome to attend the next conference, PPAM 2021, which will be held during September 12–15, 2021, in Gdańsk, the thousand-year old city on the Baltic coast and one of the largest academic centers in Poland.

January 2020

Roman Wyrzykowski
Jack Dongarra
Ewa Deelman
Konrad Karczewski

# Organization

## Program Committee

| | |
|---|---|
| Jan Węglarz (Honorary Chair) | Poznań University of Technology, Poland |
| Roman Wyrzykowski (Chair of Program Committee) | Częstochowa University of Technology, Poland |
| Ewa Deelman (Vice-chair of Program Committee) | University of Southern California, USA |
| Robert Adamski | Intel Corporation, Poland |
| Francisco Almeida | Universidad de La Laguna, Spain |
| Pedro Alonso | Universidad Politecnica de Valencia, Spain |
| Hartwig Anzt | University of Tennessee, USA |
| Peter Arbenz | ETH Zurich, Switzerland |
| Cevdet Aykanat | Bilkent University, Turkey |
| Marc Baboulin | University of Paris-Sud, France |
| Michael Bader | TU Munchen, Germany |
| Piotr Bała | ICM, Warsaw University, Poland |
| Krzysztof Banaś | AGH University of Science and Technology, Poland |
| Olivier Beaumont | Inria Bordeaux, France |
| Włodzimierz Bielecki | West Pomeranian University of Technology, Poland |
| Paolo Bientinesi | RWTH Aachen, Germany |
| Radim Blaheta | Institute of Geonics, Czech Academy of Sciences, Czech Republic |
| Jacek Błażewicz | Poznań University of Technology, Poland |
| Pascal Bouvry | University of Luxembourg, Luxembourg |
| Jerzy Brzeziński | Poznań University of Technology, Poland |
| Marian Bubak | AGH Kraków, Poland, and University of Amsterdam, The Netherlands |
| Tadeusz Burczyński | Polish Academy of Sciences, Poland |
| Christopher Carothers | Rensselaer Polytechnic Institute, USA |
| Jesus Carretero | Universidad Carlos III de Madrid, Spain |
| Andrea Clematis | IMATI-CNR, Italy |
| Pawel Czarnul | Gdańsk University of Technology, Poland |
| Zbigniew Czech | Silesia University of Technology, Poland |
| Jack Dongarra | University of Tennessee and ORNL, USA |
| Maciej Drozdowski | Poznań University of Technology, Poland |
| Mariusz Flasiński | Jagiellonian University, Poland |
| Tomas Fryza | Brno University of Technology, Czech Republic |
| Jose Daniel Garcia | Universidad Carlos III de Madrid, Spain |

| | |
|---|---|
| Pawel Gepner | Intel Corporation |
| Shamsollah Ghanbari | Iranian Distributed Computing and Systems Society, Iran |
| Domingo Gimenez | University of Murcia, Spain |
| Jacek Gondzio | University of Edinburgh, UK |
| Andrzej Gościński | Deakin University, Australia |
| Inge Gutheil | Forschungszentrum Juelich, Germany |
| Georg Hager | University of Erlangen-Nuremberg, Germany |
| José R. Herrero | Universitat Politecnica de Catalunya, Spain |
| Ladislav Hluchy | Slovak Academy of Sciences, Slovakia |
| Sasha Hunold | Vienna University of Technology, Austria |
| Aleksandar Ilic | Technical University of Lisbon, Portugal |
| Heike Jagode | University of Tennessee, USA |
| Ondrej Jakl | Institute of Geonics, Czech Academy of Sciences, Czech Republic |
| Krzysztof Jurczuk | Białystok University of Technology, Poland |
| Bo Kagstrom | Umea University, Sweden |
| Grzegorz Kamieniarz | A. Mickiewicz University in Poznań, Poland |
| Eleni Karatza | Aristotle University of Thessaloniki, Greece |
| Ayse Kiper | Middle East Technical University, Turkey |
| Jacek Kitowski | Institute of Computer Science, AGH, Poland |
| Joanna Kołodziej | Cracow University of Technology, Poland |
| Jozef Korbicz | University of Zielona Góra, Poland |
| Stanislaw Kozielski | Silesia University of Technology, Poland |
| Tomas Kozubek | Technical University of Ostrava, Czech Republic |
| Dieter Kranzlmueller | Ludwig-Maximillian University and Leibniz Supercomputing Centre, Germany |
| Henryk Krawczyk | Gdańsk University of Technology, Poland |
| Piotr Krzyżanowski | University of Warsaw, Poland |
| Krzysztof Kurowski | PSNC, Poland |
| Jan Kwiatkowski | Wrocław University of Technology, Poland |
| Giulliano Laccetti | University of Naples Federico II, Italy |
| Daniel Langr | Czech Technical University, Czech Republic |
| Marco Lapegna | University of Naples Federico II, Italy |
| Alexey Lastovetsky | University College Dublin, Ireland |
| Laurent Lefevre | Inria and University of Lyon, France |
| Joao Lourenco | University Nova of Lisbon, Portugal |
| Tze Meng Low | Carnegie Mellon University, USA |
| Hatem Ltaief | KAUST, Saudi Arabia |
| Emilio Luque | Universitat Autonoma de Barcelona, Spain |
| Piotr Luszczek | University of Tennessee, USA |
| Victor E. Malyshkin | Siberian Branch, Russian Academy of Sciences, Russia |
| Pierre Manneback | University of Mons, Belgium |
| Tomas Margalef | Universitat Autonoma de Barcelona, Spain |
| Svetozar Margenov | Bulgarian Academy of Sciences, Bulgaria |
| Ami Marowka | Bar-Ilan University, Israel |

| Norbert Meyer | PSNC, Poland |
| Iosif Meyerov | Lobachevsky State University of Nizhni Novgorod, Russia |
| Marek Michalewicz | ICM, Warsaw University, Poland |
| Ricardo Morla | INESC Porto, Portugal |
| Jarek Nabrzyski | University of Notre Dame, USA |
| Raymond Namyst | University of Bordeaux and Inria, France |
| Edoardo Di Napoli | Forschungszentrum Juelich, Germany |
| Gabriel Oksa | Slovak Academy of Sciences, Slovakia |
| Tomasz Olas | Częstochowa University of Technology, Poland |
| Ariel Oleksiak | PSNC, Poland |
| Marcin Paprzycki | IBS PAN and SWPS, Poland |
| Dana Petcu | West University of Timisoara, Romania |
| Jean-Marc Pierson | University Paul Sabatier, France |
| Loic Pottier | University of Southern California, USA |
| Radu Prodan | University of Innsbruck, Austria |
| Enrique S. Quintana-Ortí | Universitat Politecnica de Valencia, Spain |
| Omer Rana | Cardiff University, UK |
| Thomas Rauber | University of Bayreuth, Germany |
| Krzysztof Rojek | Częstochowa University of Technology, Poland |
| Witold Rudnicki | University of Białystok, Poland |
| Gudula Rünger | Chemnitz University of Technology, Germany |
| Leszek Rutkowski | Częstochowa University of Technology, Poland |
| Emmanuelle Saillard | Inria, France |
| Robert Schaefer | Institute of Computer Science, AGH, Poland |
| Olaf Schenk | Universita della Svizzera Italiana, Switzerland |
| Stanislav Sedukhin | University of Aizu, Japan |
| Franciszek Seredyński | Cardinal Stefan Wyszyński University in Warsaw, Poland |
| Happy Sithole | Centre for High Performance Computing, South Africa |
| Jurij Silc | Jozef Stefan Institute, Slovenia |
| Karolj Skala | Ruder Boskovic Institute, Croatia |
| Renata Słota | Institute of Computer Science, AGH, Poland |
| Masha Sosonkina | Old Dominion University, USA |
| Leonel Sousa | Technical University of Lisabon, Portugal |
| Vladimir Stegailov | Joint Institute for High Temperatures of RAS and MIPT/HSE, Russia |
| Przemysław Stpiczyński | Maria Curie-Skłodowska University, Poland |
| Reiji Suda | University of Tokio, Japan |
| Lukasz Szustak | Częstochowa University of Technology, Poland |
| Boleslaw Szymanski | Rensselaer Polytechnic Institute, USA |
| Domenico Talia | University of Calabria, Italy |
| Christian Terboven | RWTH Aachen, Germany |
| Andrei Tchernykh | CICESE Research Center, Mexico |
| Parimala Thulasiraman | University of Manitoba, Canada |
| Sivan Toledo | Tel-Aviv University, Israel |

## Steering Committee

# Contents – Part I

**Numerical Algorithms and Parallel Scientific Computing**

Multi-workgroup Tiling to Improve the Locality of Explicit One-Step
Methods for ODE Systems with Limited Access Distance on GPUs . . . . . . .   3
  *Matthias Korch and Tim Werner*

Structure-Aware Calculation of Many-Electron Wave Function Overlaps
on Multicore Processors. . . . . . . . . . . . . . . . . . . . . . . . . . . . . . . . .   13
  *Davor Davidović and Enrique S. Quintana-Ortí*

Lazy Stencil Integration in Multigrid Algorithms. . . . . . . . . . . . . . . . . .   25
  *Charles D. Murray and Tobias Weinzierl*

High Performance Tensor–Vector Multiplication
on Shared-Memory Systems. . . . . . . . . . . . . . . . . . . . . . . . . . . . . . .   38
  *Filip Pawłowski, Bora Uçar, and Albert-Jan Yzelman*

Efficient Modular Squaring in Binary Fields on CPU Supporting AVX
and GPU . . . . . . . . . . . . . . . . . . . . . . . . . . . . . . . . . . . . . . . . . . .   49
  *Paweł Augustynowicz and Andrzej Paszkiewicz*

Parallel Robust Computation of Generalized Eigenvectors
of Matrix Pencils . . . . . . . . . . . . . . . . . . . . . . . . . . . . . . . . . . . . . .   58
  *Carl Christian Kjelgaard Mikkelsen and Mirko Myllykoski*

Introduction to StarNEig—A Task-Based Library for Solving
Nonsymmetric Eigenvalue Problems . . . . . . . . . . . . . . . . . . . . . . . . .   70
  *Mirko Myllykoski and Carl Christian Kjelgaard Mikkelsen*

Robust Task-Parallel Solution of the Triangular Sylvester Equation . . . . . . . .   82
  *Angelika Schwarz and Carl Christian Kjelgaard Mikkelsen*

Vectorized Parallel Solver for Tridiagonal Toeplitz Systems
of Linear Equations. . . . . . . . . . . . . . . . . . . . . . . . . . . . . . . . . . . . .   93
  *Beata Dmitruk and Przemysław Stpiczyński*

Parallel Performance of an Iterative Solver Based on the Golub-Kahan
Bidiagonalization . . . . . . . . . . . . . . . . . . . . . . . . . . . . . . . . . . . . . .   104
  *Carola Kruse, Masha Sosonkina, Mario Arioli, Nicolas Tardieu,
  and Ulrich Rüde*

A High-Performance Implementation of a Robust Preconditioner
for Heterogeneous Problems . . . . . . . . . . . . . . . . . . . . . . . . . . . . . . . . . . . 117
  Linus Seelinger, Anne Reinarz, and Robert Scheichl

Hybrid Solver for Quasi Block Diagonal Linear Systems . . . . . . . . . . . . . . 129
  Viviana Arrigoni and Annalisa Massini

Parallel Adaptive Cross Approximation for the Multi-trace Formulation
of Scattering Problems . . . . . . . . . . . . . . . . . . . . . . . . . . . . . . . . . . . . . . . . 141
  Michal Kravčenko, Jan Zapletal, Xavier Claeys, and Michal Merta

Implementation of Parallel 3-D Real FFT with 2-D Decomposition
on Intel Xeon Phi Clusters . . . . . . . . . . . . . . . . . . . . . . . . . . . . . . . . . . . . 151
  Daisuke Takahashi

Exploiting Symmetries of Small Prime-Sized DFTs . . . . . . . . . . . . . . . . . . 162
  Doru Thom Popovici, Devangi N. Parikh, Daniele G. Spampinato,
  and Tze Meng Low

Parallel Computations for Various Scalarization Schemes in Multicriteria
Optimization Problems . . . . . . . . . . . . . . . . . . . . . . . . . . . . . . . . . . . . . . . . 174
  Victor Gergel and Evgeny Kozinov

## Emerging HPC Architectures

Early Performance Assessment of the ThunderX2 Processor for Lattice
Based Simulations . . . . . . . . . . . . . . . . . . . . . . . . . . . . . . . . . . . . . . . . . . . 187
  Enrico Calore, Alessandro Gabbana, Fabio Rinaldi,
  Sebastiano Fabio Schifano, and Raffaele Tripiccione

An Area Efficient and Reusable HEVC 1D-DCT Hardware Accelerator . . . . . 199
  Mate Cobrnic, Alen Duspara, Leon Dragic, Igor Piljic, Hrvoje Mlinaric,
  and Mario Kovac

## Performance Analysis and Scheduling in HPC Systems

Improving Locality-Aware Scheduling with Acyclic Directed
Graph Partitioning . . . . . . . . . . . . . . . . . . . . . . . . . . . . . . . . . . . . . . . . . . 211
  M. Yusuf Özkaya, Anne Benoit, and Ümit V. Çatalyürek

Isoefficiency Maps for Divisible Computations in Hierarchical
Memory Systems . . . . . . . . . . . . . . . . . . . . . . . . . . . . . . . . . . . . . . . . . . . 224
  Maciej Drozdowski, Gaurav Singh, and Jędrzej M. Marszałkowski

## Environments and Frameworks for Parallel/Distributed/Cloud Computing

OpenMP Target Device Offloading for the SX-Aurora TSUBASA
Vector Engine .................................................... 237
   Tim Cramer, Manoel Römmer, Boris Kosmynin, Erich Focht,
   and Matthias S. Müller

On the Road to DiPOSH: Adventures in High-Performance OpenSHMEM. . .    250
   Camille Coti and Allen D. Malony

Click-Fraud Detection for Online Advertising .................... 261
   Roman Wiatr, Vladyslav Lyutenko, Miłosz Demczuk, Renata Słota,
   and Jacek Kitowski

Parallel Graph Partitioning Optimization Under PEGASUS DA Application
Global State Monitoring. ........................................ 272
   Adam Smyk, Marek Tudruj, and Lukasz Grochal

Cloud Infrastructure Automation for Scientific Workflows ............. 287
   Bartosz Balis, Michal Orzechowski, Krystian Pawlik, Maciej Pawlik,
   and Maciej Malawski

## Applications of Parallel Computing

Posit NPB: Assessing the Precision Improvement in HPC
Scientific Applications. ......................................... 301
   Steven W. D. Chien, Ivy B. Peng, and Stefano Markidis

A High-Order Discontinuous Galerkin Solver with Dynamic Adaptive
Mesh Refinement to Simulate Cloud Formation Processes ............. 311
   Lukas Krenz, Leonhard Rannabauer, and Michael Bader

Performance and Portability of State-of-Art Molecular Dynamics
Software on Modern GPUs ....................................... 324
   Evgeny Kuznetsov, Nikolay Kondratyuk, Mikhail Logunov,
   Vsevolod Nikolskiy, and Vladimir Stegailov

Exploiting Parallelism on Shared Memory in the QED Particle-in-Cell
Code PICADOR with Greedy Load Balancing ..................... 335
   Iosif Meyerov, Alexander Panov, Sergei Bastrakov, Aleksei Bashinov,
   Evgeny Efimenko, Elena Panova, Igor Surmin, Valentin Volokitin,
   and Arkady Gonoskov

Parallelized Construction of Extension Velocities
for the Level-Set Method. ....................................... 348
   Michael Quell, Paul Manstetten, Andreas Hössinger,
   Siegfried Selberherr, and Josef Weinbub

Relative Expression Classification Tree. A Preliminary
GPU-Based Implementation . . . . . . . . . . . . . . . . . . . . . . . . . . . . . . . . 359
    Marcin Czajkowski, Krzysztof Jurczuk, and Marek Kretowski

Performance Optimizations for Parallel Modeling of Solidification
with Dynamic Intensity of Computation . . . . . . . . . . . . . . . . . . . . . . . . . 370
    Kamil Halbiniak, Lukasz Szustak, Adam Kulawik, and Pawel Gepner

## Parallel Non-numerical Algorithms

SIMD-node Transformations for Non-blocking Data Structures . . . . . . . . . . 385
    Joel Fuentes, Wei-yu Chen, Guei-yuan Lueh, Arturo Garza,
    and Isaac D. Scherson

Stained Glass Image Generation Using Voronoi Diagram
and Its GPU Acceleration . . . . . . . . . . . . . . . . . . . . . . . . . . . . . . . . . . 396
    Hironobu Kobayashi, Yasuaki Ito, and Koji Nakano

Modifying Queries Strategy for Graph-Based Speculative Query Execution
for RDBMS . . . . . . . . . . . . . . . . . . . . . . . . . . . . . . . . . . . . . . . . . . . 408
    Anna Sasak-Okoń

## Soft Computing with Applications

Accelerating GPU-based Evolutionary Induction of Decision
Trees - Fitness Evaluation Reuse . . . . . . . . . . . . . . . . . . . . . . . . . . . . . 421
    Krzysztof Jurczuk, Marcin Czajkowski, and Marek Kretowski

A Distributed Modular Scalable and Generic Framework for Parallelizing
Population-Based Metaheuristics . . . . . . . . . . . . . . . . . . . . . . . . . . . . . . 432
    Hatem Khalloof, Phil Ostheimer, Wilfried Jakob, Shadi Shahoud,
    Clemens Duepmeier, and Veit Hagenmeyer

Parallel Processing of Images Represented by Linguistic Description
in Databases . . . . . . . . . . . . . . . . . . . . . . . . . . . . . . . . . . . . . . . . . . . 445
    Danuta Rutkowska and Krzysztof Wiaderek

An OpenMP Parallelization of the K-means Algorithm Accelerated
Using KD-trees . . . . . . . . . . . . . . . . . . . . . . . . . . . . . . . . . . . . . . . . . 457
    Wojciech Kwedlo and Michał Łubowicz

Evaluating the Use of Policy Gradient Optimization Approach
for Automatic Cloud Resource Provisioning . . . . . . . . . . . . . . . . . . . . . . 467
    Włodzimierz Funika and Paweł Koperek

Improving Efficiency of Automatic Labeling by Image Transformations
on CPU and GPU .......................................... 479
  Łukasz Karbowiak

## Special Session on GPU Computing

Efficient Triangular Matrix Vector Multiplication on the GPU ........... 493
  Takahiro Inoue, Hiroki Tokura, Koji Nakano, and Yasuaki Ito

Performance Engineering for a Tall & Skinny Matrix Multiplication
Kernels on GPUs ......................................... 505
  Dominik Ernst, Georg Hager, Jonas Thies, and Gerhard Wellein

Reproducible BLAS Routines with Tunable Accuracy Using Ozaki
Scheme for Many-Core Architectures ........................... 516
  Daichi Mukunoki, Takeshi Ogita, and Katsuhisa Ozaki

Portable Monte Carlo Transport Performance Evaluation
in the PATMOS Prototype .................................. 528
  Tao Chang, Emeric Brun, and Christophe Calvin

## Special Session on Parallel Matrix Factorizations

Multifrontal Non-negative Matrix Factorization ..................... 543
  Piyush Sao and Ramakrishnan Kannan

Preconditioned Jacobi SVD Algorithm Outperforms PDGESVD ........... 555
  Martin Bečka and Gabriel Okša

A Parallel Factorization for Generating Orthogonal Matrices ............. 567
  Marek Parfieniuk

**Author Index** ......................................... 579

# Contents – Part II

**Workshop on Language-Based Parallel Programming Models (WLPP 2019)**

Parallel Fully Vectorized *Marsa-LFIB4*: Algorithmic and Language-Based Optimization of Recursive Computations . . . . . . . . . . . . . . . . . . . . . . . . . 3
  Przemysław Stpiczyński

Studying the Performance of Vector-Based Quicksort Algorithm . . . . . . . . . . 13
  Ami Marowka

Parallel Tiled Cache and Energy Efficient Code for Zuker's RNA Folding . . . 25
  Marek Palkowski and Wlodzimierz Bielecki

Examining Performance Portability with Kokkos for an Ewald Sum Coulomb Solver . . . . . . . . . . . . . . . . . . . . . . . . . . . . . . . . . . . . . . . . . 35
  Rene Halver, Jan H. Meinke, and Godehard Sutmann

Efficient cuDNN-Compatible Convolution-Pooling on the GPU . . . . . . . . . . 46
  Shunsuke Suita, Takahiro Nishimura, Hiroki Tokura, Koji Nakano,
  Yasuaki Ito, Akihiko Kasagi, and Tsuguchika Tabaru

Reactive Task Migration for Hybrid MPI+OpenMP Applications . . . . . . . . . 59
  Jannis Klinkenberg, Philipp Samfass, Michael Bader,
  Christian Terboven, and Matthias S. Müller

**Workshop on Models Algorithms and Methodologies for Hybrid Parallelism in New HPC Systems**

Ab-initio Functional Decomposition of Kalman Filter: A Feasibility Analysis on Constrained Least Squares Problems . . . . . . . . . . . . . . . . . . . 75
  Luisa D'Amore, Rosalba Cacciapuoti, and Valeria Mele

Performance Analysis of a Parallel Denoising Algorithm on Intel Xeon Computer System . . . . . . . . . . . . . . . . . . . . . . . . . . . . . . . . . . . . . . . 93
  Ivan Lirkov

An Adaptive Strategy for Dynamic Data Clustering with the K-Means Algorithm . . . . . . . . . . . . . . . . . . . . . . . . . . . . . . . . . 101
  Marco Lapegna, Valeria Mele, and Diego Romano

Security and Storage Issues in Internet of Floating Things Edge-Cloud
Data Movement . . . . . . . . . . . . . . . . . . . . . . . . . . . . . . . . . . . . . . . . . . .    111
    *Raffaele Montella, Diana Di Luccio, Sokol Kosta, Aniello Castiglione,*
    *and Antonio Maratea*

**Workshop on Power and Energy Aspects of Computations
(PEAC 2019)**

Performance/Energy Aware Optimization of Parallel Applications on GPUs
Under Power Capping . . . . . . . . . . . . . . . . . . . . . . . . . . . . . . . . . . . . . . .    123
    *Adam Krzywaniak and Paweł Czarnul*

Improving Energy Consumption in Iterative Problems Using
Machine Learning . . . . . . . . . . . . . . . . . . . . . . . . . . . . . . . . . . . . . . . . . .    134
    *Alberto Cabrera, Francisco Almeida, Vicente Blanco,*
    *and Dagoberto Castellanos–Nieves*

Automatic Software Tuning of Parallel Programs for Energy-Aware
Executions . . . . . . . . . . . . . . . . . . . . . . . . . . . . . . . . . . . . . . . . . . . . . . .    144
    *Sébastien Varrette, Frédéric Pinel, Emmanuel Kieffer, Grégoire Danoy,*
    *and Pascal Bouvry*

**Special Session on Tools for Energy Efficient Computing**

Overview of Application Instrumentation for Performance Analysis
and Tuning. . . . . . . . . . . . . . . . . . . . . . . . . . . . . . . . . . . . . . . . . . . . . . .    159
    *Ondrej Vysocky, Lubomir Riha, and Andrea Bartolini*

Energy-Efficiency Tuning of a Lattice Boltzmann Simulation
Using MERIC . . . . . . . . . . . . . . . . . . . . . . . . . . . . . . . . . . . . . . . . . . . . .    169
    *Enrico Calore, Alessandro Gabbana, Sebastiano Fabio Schifano,*
    *and Raffaele Tripiccione*

Evaluating the Advantage of Reactive MPI-aware Power Control Policies . . .    181
    *Daniele Cesarini, Carlo Cavazzoni, and Andrea Bartolini*

Application-Aware Power Capping Using Nornir . . . . . . . . . . . . . . . . . . . .    191
    *Daniele De Sensi and Marco Danelutto*

**Workshop on Scheduling for Parallel Computing (SPC 2019)**

A New Hardware Counters Based Thread Migration Strategy
for NUMA Systems. . . . . . . . . . . . . . . . . . . . . . . . . . . . . . . . . . . . . . . . .    205
    *Oscar García Lorenzo, Rubén Laso Rodríguez, Tomás Fernández Pena,*
    *Jose Carlos Cabaleiro Domínguez, Francisco Fernández Rivera,*
    *and Juan Ángel Lorenzo del Castillo*

Alea – Complex Job Scheduling Simulator. . . . . . . . . . . . . . . . . . . . . . . .   217
   *Dalibor Klusáček, Mehmet Soysal, and Frédéric Suter*

Makespan Minimization in Data Gathering Networks with Dataset
Release Times . . . . . . . . . . . . . . . . . . . . . . . . . . . . . . . . . . . . . . . . . . . .   230
   *Joanna Berlińska*

## Workshop on Applied High Performance Numerical Algorithms for PDEs

Overlapping Schwarz Preconditioner for Fourth Order Multiscale
Elliptic Problems. . . . . . . . . . . . . . . . . . . . . . . . . . . . . . . . . . . . . . . . . .   245
   *Leszek Marcinkowski and Talal Rahman*

MATLAB Implementation of C1 Finite Elements:
Bogner-Fox-Schmit Rectangle. . . . . . . . . . . . . . . . . . . . . . . . . . . . . . . . .   256
   *Jan Valdman*

Simple Preconditioner for a Thin Membrane Diffusion Problem . . . . . . . . . .   267
   *Piotr Krzyżanowski*

A Numerical Scheme for Evacuation Dynamics . . . . . . . . . . . . . . . . . . . .   277
   *Maria Gokieli and Andrzej Szczepańczyk*

Additive Average Schwarz with Adaptive Coarse Space for Morley FE. . . . .   287
   *Salah Alrabeei, Mahmood Jokar, and Leszek Marcinkowski*

## Minisymposium on HPC Applications in Physical Sciences

Application of Multiscale Computational Techniques to the Study
of Magnetic Nanoparticle Systems. . . . . . . . . . . . . . . . . . . . . . . . . . . . . .   301
   *Marianna Vasilakaki, Nikolaos Ntallis, and Kalliopi N. Trohidou*

*clique*: A Parallel Tool for the Molecular Nanomagnets Simulation
and Modelling . . . . . . . . . . . . . . . . . . . . . . . . . . . . . . . . . . . . . . . . . . . . .   312
   *Michał Antkowiak, Łukasz Kucharski, and Monika Haglauer*

Modelling of Limitations of BHJ Architecture in Organic Solar Cells . . . . . .   323
   *Jacek Wojtkiewicz and Marek Pilch*

Monte Carlo Study of Spherical and Cylindrical Micelles in Multiblock
Copolymer Solutions. . . . . . . . . . . . . . . . . . . . . . . . . . . . . . . . . . . . . . . . .   333
   *Krzysztof Lewandowski, Karolina Gębicka, Anna Kotlarska,*
   *Agata Krzywicka, Aneta Łasoń, and Michał Banaszak*

Electronic and Optical Properties of Carbon Nanotubes Directed to Their
Applications in Solar Cells. . . . . . . . . . . . . . . . . . . . . . . . . . . . . . . . .     341
  Jacek Wojtkiewicz, Bartosz Brzostowski, and Marek Pilch

## Minisymposium on High Performance Computing Interval Methods

The MPFI Library: Towards IEEE 1788–2015 Compliance:
(*In Memoriam Dmitry Nadezhin*). . . . . . . . . . . . . . . . . . . . . . . . . . . .     353
  Nathalie Revol

Softmax and McFadden's Discrete Choice Under Interval
(and Other) Uncertainty. . . . . . . . . . . . . . . . . . . . . . . . . . . . . . . . . .     364
  Bartlomiej Jacek Kubica, Laxman Bokati, Olga Kosheleva,
  and Vladik Kreinovich

Improvements of Monotonicity Approach to Solve Interval Parametric
Linear Systems. . . . . . . . . . . . . . . . . . . . . . . . . . . . . . . . . . . . . . . .     374
  Iwona Skalna, Marcin Pietroń, and Milan Hladík

The First Approach to the Interval Generalized Finite Differences. . . . . . . . .     384
  Malgorzata A. Jankowska and Andrzej Marciniak

An Interval Calculus Based Approach to Determining the Area
of Integration of the Entropy Density Function. . . . . . . . . . . . . . . . . . . . .     395
  Bartłomiej Jacek Kubica and Arkadiusz Janusz Orłowski

An Interval Difference Method of Second Order for Solving
an Elliptical BVP . . . . . . . . . . . . . . . . . . . . . . . . . . . . . . . . . . . . . . .     407
  Andrzej Marciniak, Malgorzata A. Jankowska, and Tomasz Hoffmann

A Parallel Method of Verifying Solutions for Systems
of Two Nonlinear Equations. . . . . . . . . . . . . . . . . . . . . . . . . . . . . . . .     418
  Bartłomiej Jacek Kubica and Jarosław Kurek

## Workshop on Complex Collective Systems

Experiments with Heterogenous Automata-Based Multi-agent Systems . . . . .     433
  Franciszek Seredyński, Jakub Gąsior, Rolf Hoffmann,
  and Dominique Désérable

Cellular Automata Model for Crowd Behavior Management in Airports. . . . .     445
  Martha Mitsopoulou, Nikolaos Dourvas, Ioakeim G. Georgoudas,
  and Georgios Ch. Sirakoulis

A Conjunction of the Discrete-Continuous Pedestrian Dynamics Model
SigmaEva with Fundamental Diagrams . . . . . . . . . . . . . . . . . . . . . . . . .     457
  Ekaterina Kirik, Tatýana Vitova, Andrey Malyshev, and Egor Popel

Simulating Pedestrians' Motion in Different Scenarios with Modified Social
Force Model. . . . . . . . . . . . . . . . . . . . . . . . . . . . . . . . . . . . . . . . . .   467
    Karolina Tytko, Maria Mamica, Agnieszka Pękala, and Jarosław Wąs

Traffic Prediction Based on Modified Nagel-Schreckenberg Model.
Case Study for Traffic in the City of Darmstadt . . . . . . . . . . . . . . . . . . .   478
    Łukasz Gosek, Fryderyk Muras, Przemysław Michałek,
    and Jarosław Wąs

HPC Large-Scale Pedestrian Simulation Based on Proxemics Rules. . . . . . . .   489
    Paweł Renc, Maciej Bielech, Tomasz Pęcak, Piotr Morawiecki,
    Mateusz Paciorek, Wojciech Turek, Aleksander Byrski,
    and Jarosław Wąs

**Author Index** . . . . . . . . . . . . . . . . . . . . . . . . . . . . . . . . . . . . . . . .   501

# Numerical Algorithms and Parallel Scientific Computing

# Multi-workgroup Tiling to Improve the Locality of Explicit One-Step Methods for ODE Systems with Limited Access Distance on GPUs

Matthias Korch[✉] and Tim Werner

Department of Computer Science,
University of Bayreuth, Bayreuth, Germany
{korch,werner}@uni-bayreuth.de

**Abstract.** Solving an initial value problem of a large system of ordinary differential equations (ODEs) on a GPU is often memory bound, which makes optimizing the locality of memory references important. We exploit the limited access distance, which is a property of a large class of right-hand-side functions, to enable hexagonal or trapezoidal tiling across the stages of the ODE method. Since previous work showed that the traditional approach of launching one workgroup per tile is worthwhile only for small limited access distances, we introduce an approach where several workgroups cooperate on a tile (multi-workgroup tiling) and investigate several optimizations and variations. Finally, we show the superiority of the multi-workgroup tiling over the traditional single-workgroup tiling for large access distances by a detailed experimental evaluation using two different Runge–Kutta (RK) methods.

**Keywords:** ODE methods · Runge–Kutta methods · Parallel · GPU · Tiling · Multi-workgroup tiling · Limited access distance

## 1 Introduction

Many scientific simulations use systems of differential equations as mathematical models to approximate phenomena of the real world. This paper considers initial value problems (IVPs) of systems of ordinary differential equations (ODEs):

$$\mathbf{y}'(t) = \mathbf{f}(t, \mathbf{y}(t)), \quad \mathbf{y}(t_0) = \mathbf{y}_0, \quad t \in [t_0, t_e]. \tag{1}$$

The classical numerical solution approach, which is also used by the methods considered in this paper, applies a time-stepping procedure that starts the simulation at time $t_0$ with initial state $\mathbf{y}_0$ and performs a series of time steps $t_\kappa \to t_{\kappa+1}$ for $\kappa = 0, 1, 2, \ldots$ until the final simulation time $t_e$ is reached. At each time step $\kappa$, by applying the right-hand-side (RHS) function $\mathbf{f} : \mathbb{R} \times \mathbb{R}^n \to \mathbb{R}^n$, a new simulation state $\mathbf{y}_{\kappa+1}$ is computed which approximates the exact solution function $\mathbf{y}(t)$ at time $t_{\kappa+1}$. A detailed treatment of the subject is given in [3].

© Springer Nature Switzerland AG 2020
R. Wyrzykowski et al. (Eds.): PPAM 2019, LNCS 12043, pp. 3–12, 2020.
https://doi.org/10.1007/978-3-030-43229-4_1

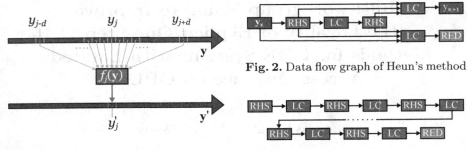

$y_{j-d}$    $y_j$    $y_{j+d}$

**Fig. 2.** Data flow graph of Heun's method

**Fig. 1.** Limited access distance $d$

**Fig. 3.** Dependency chain

We consider the parallel solution of IVPs by explicit one-step methods on GPUs, which – in contrast to implicit and multi-step methods – only use one input approximation $y_\kappa$ and a number of evaluations of the RHS function to compute the output approximation $y_{\kappa+1}$. For sparse and cheap RHS functions, the linear combinations of the stages dominate the runtime and make these methods memory bound. Therefore, optimizing the locality of memory references is important. To reduce the time-to-solution on GPUs, we use the well-known technique of *kernel fusion*, which exploits the on-die memories of the GPU (caches, scratchpad, registers) to increase data reuse. In [6], we focused on kernel fusion for problems with an arbitrary coupling of the RHS. Since this allowed the fusion of different stages of the method only if they were independent, we lifted this restriction in [7] by a specialization in problems with a limited access distance, which commonly arise, for example, from stencil-based problems (e.g., partial differential equations (PDEs) discretized by the method of lines) or block-structured electrical circuits. The access distance $d(\mathbf{f})$ is the smallest value $d$, such that each component function $f_j(t, \mathbf{y})$, $j = 1, \ldots, n$, accesses only the subset $\{y_{j-d}, \ldots, y_{j+d}\}$ of the components of the argument vector $\mathbf{y}$ (see Fig. 1). By exploiting this special structure, we could derive tilings across the stages where each tile is processed by a single workgroup. This lead to improved performance, but only for small access distances in the order of up $10^3$.

The contribution of this paper is the increase of the range of access distances for which a speedup can be obtained by two orders of magnitude, thus significantly increasing the practical usability, in particular for 2D stencils on quadratic or near-quadratic grids. This is achieved by an extension of the single-workgroup tiling approach across the stages to a multi-workgroup tiling where several workgroups and hence multiple GPU cores collaborate on each tile. The multi-workgroup tiling is implemented as part of an automatic framework that can generate CUDA or OpenCL kernels for different tiling parameters automatically. Finally, we contribute a detailed experimental evaluation which investigates the influencing factors on the performance of the approach.

# 2  Related and Preliminary Work

Modern GPUs from Intel, AMD and NVIDIA have similar basic architectures [4], but there is no common *GPU terminology*, and vendor specific terms are often on purpose misleading. In this paper, we continue to use the terminology of [6,7] (cf. Table 1), which is similar to the common CPU terminology and [4].

**Table 1.** Comparison of GPU and CPU terminologies (preferred terms emphasized).

| NVIDIA | OpenCL |
|---|---|
| Thread | *Workitem* |
| Thread block | *Workgroup* |

| NVIDIA | OpenCL | CPU |
|---|---|---|
| Streaming multiprocessor | Compute unit | *Core* |
| Warp | — | *Thread* |
| CUDA core | Processing element | *SIMD lane* |

*Kernel fusion* is one of the most important state-of-the-art techniques to optimize data reuse on GPUs, and several domain-specific approaches have been proposed, e.g., [1,9]. Kernel fusion with multi-workgroup barriers was considered in [10,11] for dynamic programming algorithms, bitonic sort, and FFT. However, the multi-workgroup barriers were only used to reduce the overhead of kernel launches and not to share data between different workgroups.

*Tiling* is a classical locality optimization technique used on CPUs and GPUs to break down the working sets of loops so that the resulting tiles fit into faster levels of the memory hierarchy. One important theoretical model that allows to determine possible tilings for loops with dependencies is the polyhedral model, which is employed, e.g., by the PPCG compiler [2]. However, up to now, PPCG can only generate single-workgroup tilings. The collaboration of several CPU cores on a single tile (*multi-core tiling*) is considered in [8].

*ODE methods*, in contrast to typical PDE and stencil approaches, compute several stages within one time step, and often step size control takes the effect of a barrier between time steps. Therefore, locality optimizations for ODE methods often focus on the loop structure inside one time step. If a limited access distance can be exploited, additional loop transformations are possible. For example, for explicit RK methods, a time-skewing strategy for Adams–Bashforth methods to provide a temporal tiling of time steps was investigated in [5].

The *application of kernel fusion to ODE methods on GPUs* was considered in [6]. To enable the automatic generation of fused kernel code, the computations of one time step of an explicit ODE method were abstracted by a data flow graph (see Fig. 2 for an example) consisting of the following three basic operations: evaluations of the RHS, linear combinations (LC) for combining system states and temporal derivatives of system states to a new system state and reduction operations (RED) for the stepsize control. Although the graph may have an arbitrary structure, for many methods, such as explicit Runge–Kutta (RK) methods, parallel iterated RK (PIRK) methods or extrapolation methods [3], its transitive reduction is or contains a dependency chain of RHS → LC links with an optional reduction operation in the end of the chain for step size control,

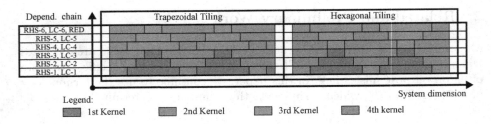

**Fig. 4.** Tiling schemes

as it is shown in Fig. 3. In addition to being dependent on the preceding RHS evaluation, linear combinations in this chain may also depend on other previous RHS evaluations and linear combinations. For general ODE systems, as considered in [6], it is only allowed to fuse RHS → LC, LC → LC and LC → RED dependencies, while a global barrier is required for each LC → RHS dependency.

For ODE systems with *limited access distance*, i.e., systems where the evaluation of component $j$ of the RHS, $f_j(t, \mathbf{y})$, only accesses components of $\mathbf{y}$ located nearby $j$ (Fig. 1), we showed in [7] that while a LC → RHS dependency is still not fusible to a single kernel, a distributed fusion to a pair of kernels is possible. Based on this, we derived two parameterized 2-dimensional strategies for *tiling across the stages* (trapezoidal and the hexagonal tiling, Fig. 4) with the parameters tile width (tile size along the system dimension) and tile height (tile size along the time dimension). However, in [7] each tile is processed by a single-workgroup only (*single-workgroup tiling*) and a tile is only able to use the private on-die memory resources and computational power of one GPU core. Consequently, many tiles are required to saturate the GPU, which is why even medium tile sizes cause massive register spilling and cache thrashing. Since larger access distances also require larger tile widths for the tiling to be efficient or even possible, single-workgroup tiling is inefficient for quite small access distances.

In this paper, we avoid this problem by extending our previous approach to use *multi-workgroup tiling*. There, several workgroups and hence multiple GPU cores collaborate on each tile, which in the process makes larger tile widths more efficient. While other works have already employed multi-core tiling for CPUs [8], to the best of our knowledge, nobody has yet applied multi-workgroup tiling on GPUs to increase data reuse of ODE methods with limited access distance.

## 3   Multi-workgroup Tiling of Explicit One-Step Methods

As the single-workgroup tiling [7], the multi-workgroup tiling is derived from a data flow graph of the ODE method by applying a distributed fusion to a pair of kernels along a dependency chain to generate trapezoidal or hexagonal tile shapes as illustrated in Fig. 4, where the tiles with the same color are processed in parallel by the same 1D kernel, while the tiles with different colors are processed sequentially by different kernels. In such a 1D kernel, each $n$ consecutive workgroups, which are assumed to be executed simultaneously, collaborate on

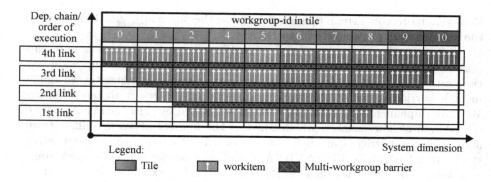

**Fig. 5.** Multi-workgroup tiling

the same tile. In such a tile, the components of a basic operation are evaluated by the workitems of the workgroups in parallel, while the basic operations in the dependency chain are processed one after another (Fig. 5).

To avoid deadlocks, we additionally require that, first, the workgroup scheduler dispatches the workgroups to the cores in an ascending order starting with the 0th workgroup and, second, a workgroup is only assigned to a core if a core has a free slot for this workgroup. Unfortunately, GPU vendors typically do not guarantee any concurrent execution or the order of execution for different workgroups of the same kernel on their GPUs. However, the restrictions above hold true on all NVIDIA GPUs considered.

Because we assign each workitem in a tile the same component of all basic operations it computes, the fusion of RHS $\rightarrow$ LC, RHS $\rightarrow$ RED, LC $\rightarrow$ LC, and LC $\rightarrow$ RED dependencies does not cause dependencies between the workitems of the fused kernel. Thus, they can be fused completely via the register file.

However, dependencies between different workitems and workgroups in a tile occur, if a LC $\rightarrow$ RHS dependency is fused. Hence, this fusion requires communication between the workgroups of the tile and the synchronization of communication steps to avoid data races. Because of that, a multi-workgroup tile requires three steps to evaluate a fused LC $\rightarrow$ RHS dependency: First, the workgroups of the tile write those components of the linear combination which the workgroups have computed beforehand and which the workgroups later require to evaluate the RHS from the registers to a global memory buffer which is private to this tile. Then the workgroups of the tile execute a costly barrier spanning across all workgroups of the tile to ensure that all writes to the buffer have finished and are visible. Finally, the workgroups of the tile evaluate the RHS and, for that purpose, read from the buffer. For fusing more than two links, one has the choice between double buffering (two global memory buffers and one barrier for each fused link) and single buffering (one global memory buffer and two barriers for each fused link). Experiments have shown that double buffering is faster due to the smaller number of costly barriers. To decrease barrier costs further, one can vary the number of workgroups cooperating on a tile, so that the GPU does not run a single but several tiles concurrently.

However, more concurrent tiles also decrease the tile width and thus increase the amount of data transferred via DRAM.

Unfortunately, on modern GPUs signal-based synchronization is only supported between the threads of a single-workgroup via a workgroup wide barrier instruction, but not across multiple workgroups. Consequently, to implement a barrier across several workgroups, one has to resort to busy waiting. We have implemented two barrier variants suggested by [11]: Barrier-A consists of a shared atomic counter, which contains the number of workgroups already arrived at the barrier. Barrier-B consists of a master workgroup that controls the barrier. On Volta GPUs, NVIDIA also offers a barrier across *all* workgroups of a kernel, to which we refer as Barrier-C. Benchmarks have shown that, as expected, Barrier-B has a higher overhead than Barrier-A, but it is also less sequential and therefore it scales better. However, even with many workgroups per core on a modern Volta GPU, the better scaling of Barrier-B cannot compensate its higher overhead. The benchmarks have also shown that Barrier-C has a very low performance, suggesting that it is a not optimally implemented software barrier.

As a conclusion of this section, multi-workgroup tiling is expected to exploit the on chip memories for data reuse by:

- using registers to transfer data along fused RHS → LC, RHS → RED, LC → LC and LC → RED dependencies,
- using caches to transfer data along a fused LC → RHS dependency,
- using a register to avoid redundant loads when several linear combinations sharing an argument vector that resides in DRAM are fused, or
- writing back only those vector components to DRAM which are actually read by a succeeding kernel.

## 4   Experimental Evaluation

### 4.1   Setup and Experiments

We have added the ability to generate multi-workgroup tilings for explicit one-step methods along a user defined dependency chain to our automatic prototype framework [6,7]. The framework as introduced in [6] allows a user to solve an arbitrary IVP by an arbitrary explicit ODE method of several supported classes (RK methods, PIRK methods, peer methods, Adams–Bashforth methods). This generality is achieved by an internal data flow representation of the ODE method from which, after several transformations, CUDA or OpenCL kernels with all memory optimizations from Sect. 3 can be generated and executed automatically with automatic memory management. To automatically tune performance, the framework performs an exhaustive online search over a predefined set of values to optimize several performance parameters, e.g., tile height and tile width. Single-workgroup tiling for explicit one-step methods was added previously [7].

Since double precision (DP) is usually desirable for scientific simulations, we consider the most recent NVIDIA GPU architecture that provides reasonable

DP performance: Volta (Titan V, 80 cores). For this GPU, we let our framework generate CUDA code.

As ODE methods we consider two embedded explicit RK methods of different order: Bogacki–Shampine 2(3) with 4 stages and Verner 5(6) with 8 stages.

The IVP we consider is BRUSS2D, a chemical reaction diffusion system of two chemical substances, discretized on a 2D grid by a 5-point diffusion stencil. Hence, BRUSS2D has a limited access distance $d$ of two times its $x$-dimension. For simulating the impact of the access distance, we set the $x$-dimension to $d/2$ and adjust the $y$-dimension so that the overall size of the ODE system $n$ remains almost constant ($n = 16 \cdot 1024 \cdot 1024$ on Volta and $n = 8 \cdot 1024 \cdot 1024$ on Kepler). Note that due to the problem statement, we only exploit the limited access distance of BRUSS2D resulting in a 2D iteration space (system dimension and time). However, we do not exploit the 2D spatial structure of the stencil, which would result in a 3D iteration space ($x$-, $y$-, and time dimension).

As reference for the speedup computation we use a non-tiled implementation which fuses all but the LC → RHS dependencies.

With this setup the following experiments have been performed:

- Figure 6: Impact of the access distance on the speedup of the two tiling strategies using the optimal choice of tile width, tile height, and number of concurrent tiles.
- Figure 7: Impact of the access distance on the speedup of the tiling strategies for different tile heights using Verner's method with optimal choice of the tile width, and number of concurrent tiles.
- Figure 8: Best tile width and best number of concurrent tiles as a function of the access distance for several fixed tile heights using Verner's method with hexagonal tiling.

## 4.2   Discussion

*For which access distances is the multi-workgroup tiling worthwhile? (see* Fig. 6*).* There is a wide range of access distances where multi-workgroup tiling yields a significant speedup over the reference implementation and single-workgroup tiling. For example, for Verner's method, multi-workgroup tiling becomes faster than single-workgroup tiling for access distances larger than 256, yields a speedup of 2.14 over the reference implementation for an access distance of 10 240, a speedup of 1.23 for an access distance of 245 760, and a speedup smaller than one only for access distances larger than 327 680. This wide range of access distances, for which multi-workgroup tiling yields a speedup over the reference implementation, is a remarkable improvement over single-workgroup tiling, which only yields a speedup for access distances smaller than 1 536. An access distance of 245 760 would correspond to a quadratic 2D grid size of 122 880 × 122 880 and a cubic 3D grid size of 350 × 350 × 350.

*How does the number of stages of the method impact the performance? (see* Fig. 6*).* The more stages a method has, the larger the speedup over the reference

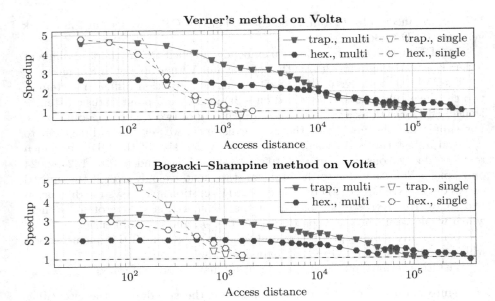

**Fig. 6.** Impact of the access distance on the speedup of the tiling strategies using the optimal choice (auto-tuned) of tile width, tile height, and number of concurrent tiles.

implementation for small access distances is. This may be explained by methods with more stages having a lower arithmetic intensity. For example, a limited access distance of 32 results in a speedup of 4.56 for Verner's method and a speedup of 3.25 for the Bogacki–Shampine method. Moreover, for larger access distances one might expect the tiling to become less efficient the more stages a method has, since those methods also have a higher working set. Yet, the measurements show that for both methods investigated the tiling offers a similar speedup if the access distances are large.

*Which tiling strategy is better?* (see Fig. 6). The trapezoidal tiling has a better speedup than the hexagonal tiling for small access distances, while for larger access distances (e.g., $d \geq 12\,288$ for Verner's method or $d \geq 40\,960$ for the Bogacki–Shampine method), the hexagonal tiling is better.

*How does the tile height influence the performance depending on the access distance?* (see Fig. 7). For smaller access distances, larger tile heights are more efficient because they reduce the DRAM data transfers further. However, larger tile heights also have a larger working set and hence scale worse when increasing the access distance. By contrast, smaller tile heights yield smaller speedups for smaller access distances, but also degrade far less for larger access distances.

*How does the tile width and the number of concurrent tiles impact the performance depending on the tile height and the access distance?* (see Fig. 8). For all tile heights, the best number of concurrent tiles decreases with the access

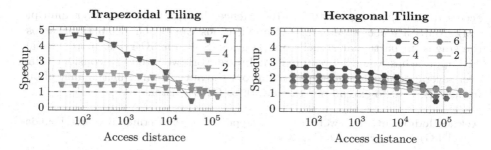

**Fig. 7.** Impact of the access distance on the speedup of the tiling strategies for different tile heights using Verner's method with optimal choice of the tile width on Volta.

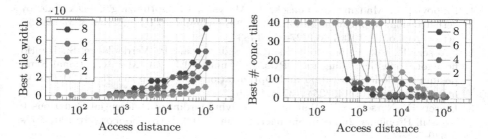

**Fig. 8.** Impact of the access distance on the best tile width (left) and the best number of concurrent tiles (right) for several fixed tile heights using Verner's method with hexagonal tiling on Volta.

distance while the best tile width increases. The larger the tile height, the larger the optimal tile width and the smaller the optimal number of concurrent tiles.

*How does the multi-workgroup tiling scale on older GPU architectures?* (measurements not displayed in this paper). Since older GPU architectures have less on chip memory and a less severe memory bottleneck, multi-workgroup tiling yields a significantly worse scaling than on newer GPU architectures, e.g. on a Kepler Titan Black only a speedup of 1.14 for a quite small access distance of 14336 and Verner's method. Consequently, it is expected that multi-workgroup tiling will scale even better on future GPU architectures.

## 5   Conclusions

We have seen that on modern GPU architectures such as NVIDIA Volta multi-workgroup tiling based on the generic notion of the access distance of an ODE system applied to the stages of an explicit one-step method could improve the performance for access distances of up to $\approx 2.5 \cdot 10^5$. This is a tremendous improvement over single-workgroup tiling, which only yielded a speedup for access distances of up to $\approx 1.5 \cdot 10^3$. Hence, by spanning tiles over multiple workgroups a much wider range of applications can benefit from tiling across the stages, in particular stencils on 2D grids. Still, for 3D grids, more specialized approaches would be required. As future work we intend to consider tiling over several adjacent time steps, to improve the auto-tuning functionality of our framework, and to extend our framework to GPU clusters.

**Acknowledgment.** This work has been supported by the German Research Foundation (DFG) under grant KO 2252/3-2.

# References

1. Filipovič, J., Madzin, M., Fousek, J., Matyska, L.: Optimizing CUDA code by kernel fusion: application on BLAS. J. Supercomput. **71**(10), 3934–3957 (2015). https://doi.org/10.1007/s11227-015-1483-z
2. Grosser, T., Cohen, A., Holewinski, J., Sadayappan, P., Verdoolaege, S.: Hybrid hexagonal/classical tiling for GPUs. In: Annual IEEE/ACM International Symposium on Code Generation and Optimization (CGO), pp. 66–75. ACM (2014). https://doi.org/10.1145/2544137.2544160
3. Hairer, E., Nørsett, S.P., Wanner, G.: Solving Ordinary Differential Equations I: Nonstiff Problems, 2nd edn. Springer, Berlin (2000). https://doi.org/10.1007/978-3-540-78862-1
4. Hennessy, J.L., Patterson, D.A.: Computer Architecture: A Quantitative Approach, 5th edn. Morgan Kaufmann, Amsterdam (2011)
5. Korch, M.: Locality improvement of data-parallel Adams–Bashforth methods through block-based pipelining of time steps. In: Kaklamanis, C., Papatheodorou, T., Spirakis, P.G. (eds.) Euro-Par 2012. LNCS, vol. 7484, pp. 563–574. Springer, Heidelberg (2012). https://doi.org/10.1007/978-3-642-32820-6_56
6. Korch, M., Werner, T.: Accelerating explicit ODE methods by kernel fusion. Concurr. Comput. Pract. Exp. **30**(18), e4470 (2018). https://doi.org/10.1002/cpe.4470
7. Korch, M., Werner, T.: Exploiting limited access distance for kernel fusion across the stages of explicit one-step methods on GPUs. In: 30th International Symposium on Computer Architecture and High Performance Computing (SBAC-PAD), pp. 148–157 (2018). https://doi.org/10.1109/CAHPC.2018.8645892
8. Malas, T., Hager, G., Ltaief, H., Stengel, H., Wellein, G., Keyes, D.: Multicore-optimized wavefront diamond blocking for optimizing stencil updates. SIAM J. Sci. Comput. **37**(4), C439–C464 (2015). https://doi.org/10.1137/140991133
9. Wahib, M., Maruyama, N.: Automated GPU kernel transformations in large-scale production stencil applications. In: 24th International Symposium on High-Performance Parallel and Distributed Computing (HPDC), pp. 259–270 (2015). https://doi.org/10.1145/2749246.2749255
10. Xiao, S., Aji, A.M., Feng, W.: On the robust mapping of dynamic programming onto a graphics processing unit. In: 15th International Conference on Parallel and Distributed Systems (ICPADS), pp. 26–33 (December 2009). https://doi.org/10.1109/ICPADS.2009.110
11. Xiao, S., Feng, W.: Inter-block GPU communication via fast barrier synchronization. In: IEEE International Symposium on Parallel Distributed Processing (IPDPS), pp. 1–12 (April 2010). https://doi.org/10.1109/IPDPS.2010.5470477

# Structure-Aware Calculation of Many-Electron Wave Function Overlaps on Multicore Processors

Davor Davidović[1]([✉])[iD] and Enrique S. Quintana-Ortí[2][iD]

[1] Centre for Informatics and Computing, Ruđer Bošković Institute,
Bijenička cesta 54, 10000 Zagreb, Croatia
davor.davidovic@irb.hr
[2] Depto. de Informática de Sistemas y Computadores,
Universitat Politècnica de València, 46022 València, Spain
quintana@disca.upv.es

**Abstract.** We introduce a new algorithm that exploits the relationship between the determinants of a sequence of matrices that appear in the calculation of many-electron wave function overlaps, yielding a considerable reduction of the theoretical cost. The resulting enhanced algorithm is embarrassingly parallel and our comparison against the (embarrassingly parallel version of) original algorithm, on a computer node with 40 physical cores, shows acceleration factors which are close to 7 for the largest problems, consistent with the theoretical difference.

**Keywords:** Wave functions · LU factorization · Multicore processors

## 1 Introduction

Many-electron wave function (MEWF) overlaps are extensively used in the nonadiabatic dynamics and have significant importance in photochemical studies. Concretely, the overlap functions provide a straightforward mechanism to record the electronic states along different nuclear geometries. Therefore, they can be leveraged for constructing multi-state, multi-dimensional potential energy surfaces for quantum dynamics [6]. In the context of nonadiabatic dynamics simulations, for example, MEWF overlaps yield an approximation of time-derivative couplings (TDCs) in fewest-switch surface hopping (FSSH) calculations [5,11]. The main drawback of the approaches based on MEWF overlaps lies in their high computational complexity and poor scaling with the system size. Thus, accelerating the calculation of the overlap functions can significantly augment the dimension of the systems to which the FSSH method can be applied.

In this work, we improve the algorithm presented in [10] for computing the overlaps between excited states using CIS-type wave functions. In particular, we optimize the algorithm denoted there as OL2M –an approach based on the level-2 minors obtained from Laplace's recursive formula, or in other words,

R. Wyrzykowski et al. (Eds.): PPAM 2019, LNCS 12043, pp. 13–24, 2020.
https://doi.org/10.1007/978-3-030-43229-4_2

the minors obtained by removing two rows and two columns from the input referent matrix. In rough detail, our optimization targets the part of the OL2M algorithm in which the determinants of all the level-2 minors are computed and stored, introducing a structure-aware variant of the method that reduces the theoretical cost of that part of the algorithm by an order of magnitude via an *update of the LU factorization*; see, e.g., [4,8]. This enhancement results in significantly shorter execution times for large problems solved in parallel on a multicore processor.

The rest of the paper is structured as follows. In Sect. 2 we offer a brief introduction to the computations of MEWF overlaps. The columnwise structure-aware algorithm and its parallelization are described in Sect. 3; and the parallel experimental results are presented in Sect. 4. Finally, a few concluding remarks and future research directions close the paper in Sect. 5.

## 2   Problem Definition

Given two many-electron wave-functions, denoted by $\Psi_I$ and $\Psi_J$, the overlap between these functions is expressed as:

$$S_{IJ} = \langle \Psi_I | \Psi_J \rangle, \tag{1}$$

using bra-ket $\langle * | * \rangle$ notation [2], a common notation used in quantum mechanics to describe quantum spaces. The $S_{IJ}$ is the $(I, J)$ element of the overlap matrix $S$, $N_A$ is the number of states, and the indices $I, J \in \{1, 2, \ldots, N_A\}$. The $N_A$ states are described by CIS-type wave functions and, therefore, they can be expanded using Slater determinants. In the Slater determinants expansion, the electrons are divided into $n_\sigma$ and $m_\sigma$ respectively, representing the number of occupied and virtual orbitals for each spin $\sigma$.

The mathematical-physics problem can be reformulated into a matrix representation as that in Fig. 1. The main computational problem consists in obtaining the determinants of all the matrices constructed such that a row $i$ and column $j$ from the referent matrix $A_{ref}$ are replaced with the contents of row $i_\beta$ and column $j_\alpha$, respectively, from matrices $WF_\beta$ and $WF_\alpha$, for all possible combinations of $i, j, i_\beta, j_\alpha$.

The total number of possible matrices is $n_\sigma^2 m_\sigma^2$ for each spin $\sigma \in \{\alpha, \beta\}$. Furthermore, if an LU factorization [4] is applied to compute these determinants, which requires $2/3 n_\sigma^3$ flops (floating-point operations), this operation becomes the main bottleneck of the entire MEWF, yielding a total cost of $O(n_\sigma^5 m_\sigma^2)$ flops.

It is worth to note that, for a fixed row and column $(i, j)$ of $A_{ref}$, numerous rows and columns from $WF_\beta$ and $WF_\alpha$ are to be tested, while all other rows and columns in the referent matrix remain unchanged. Thus, the matrices, for which we have to compute the determinants differ only in one row and one column. In order to decrease the computational cost, the challenge is to reuse the unchanged parts of referent matrix for computing other determinants.

The first approach to exploit this property was presented in [7]. There, the authors reported a significant speedup by exploiting similarities between the

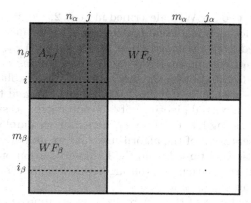

**Fig. 1.** Matrix representation of the computational problem. The referent matrix is square with dimension $n_\beta \times n_\alpha$. Rows/columns $i/j$ are replaced with rows/columns from the panels $WF_\beta$ and $WF_\alpha$, respectively.

consecutive matrices differ in only one row/column. Concretely, the referent matrix $A_{ref}$ was expanded by columns using the Laplace transformation. The determinants of the minors were then stored and reused to compute determinants of matrices which differ only in one column from the baseline determinant. With this strategy, replacing any column for $WF_\alpha$ is simply a linear combination of factors from $j_\alpha$ with pre-computed determinants of the minors.

The work presented in [10] improved the idea in [7] by computing the determinants of the minors obtained from the second-level recursive Laplace expansion. There, the minors were constructed by expanding the referent matrix along both rows and columns; then the LU factorization was repeatedly applied to obtain the determinants; and finally these results were reused to compute the overlap of the corresponding Slater determinants. Hereafter, we will refer to this algorithmic approach as DL2M. Although this strategy yields a considerable reduction in the computational complexity compared with the method in [7], obtaining the determinants of the minors remains the major computational bottleneck.

A related topic is that of computing the LU factorization of a large collection of small matrices. This problem has been tackled, on graphics processors, as part of an effort to develop a batched version of the Basic Linear Algebra Subprograms as well as in the framework of computing a block-Jacobi preconditioner for the iterative solution of sparse linear systems [1,3]. However, in those applications the matrices are independent and no structure-aware exploitation of the problem is possible.

## 3    Algorithms for DL2M Calculations

### 3.1    Original

Consider the matrix $A \in \mathbb{R}^{n \times n}$ representing the referent matrix $A_{ref}$ and assume, for simplicity, that $n = n_\alpha = n_\beta = m_\alpha = m_\beta$. The calculation of the overlap between any two MEWFs, described in Sect. 2, requires the computation of the determinants for all possible submatrices of $A$ where any two rows/columns have been eliminated from the matrix. Let us denote by $A_{r_1,r_2||c_1,c_2}$ the submatrix that results from eliminating rows $r_1, r_2$ and columns $c_1, c_2$ from $A$. The straightforward solution to obtain the determinants is to explicitly construct all possible submatrices with $m = n - 2$ rows/columns of $A$, and then compute the LU factorization (with partial pivoting) of each submatrix, as shown in the naive algorithm (NA) in Listing 1.1. For brevity, hereafter we employ pseudo-Matlab notation in the presentation of the algorithms, and we do not consider the effect of the row permutations obtained from the LU factorization on the determinant sign. All our actual realizations of the algorithms include partial pivoting to ensure numerical stability in practice.

The computational cost of the NA realization is, approximately,

$$\sum_{r_1=1}^{n} \sum_{r_2=1}^{r_1-1} \sum_{c_1=1}^{n} \sum_{c_2=1}^{c_1-1} 2n^3/3 \approx n^7/6 \text{ flops.}$$

```
1  for r1=1:n
2    for r2=1:r1-1
3      for c1=1:n
4        for c2=1:c1-1
5          [L,U,P] = lu(A_{r_1,r_2||c_1,c_2});
6          d[r1][r2][c1][c2] = prod(diag(U));
```

**Listing 1.1.** Naive algorithm for DL2M calculations.

## 3.2   Columwise Re-utilization

The naive algorithm in Listing 1.1 exposes that, between any two iterations of the two inner loops (that is, those indexed by c1, c2), the matrix that needs to be factorized only differs in two columns. A natural question is thus how to exploit the fact that all other matrix columns remain the same between these two iterations. To illustrate the response, let us define $A_r = A_{r_1,r_2||-} \in \mathbb{R}^{m \times n}$ as the submatrix with $m = n - 2$ rows and $n$ columns that results from eliminating only rows $r_1, r_2$ from $A$. Next, consider the LU factorization of this submatrix:

$$L^{-1}PA_r = L^{-1}P[a_1, a_2, \ldots, a_n] = [u_1, u_2, \ldots, u_n] = U, \qquad (2)$$

where $L \in \mathbb{R}^{m \times m}$ is a unit lower triangular comprising the Gauss transforms that are applied to annihilate the subdiagonal entries of the matrix, $P$ is the $m \times m$ permutation matrix due to the application of partial pivoting during the factorization, and $U \in \mathbb{R}^{m \times n}$ is the resulting upper triangular factor [4]. The answer we are searching for should state the relationship between the factorization of $A_r$ in (2) and that of the submatrix

$$A_{r_1,r_2||c_1,c_2} = [a_1, a_2, \ldots, a_{c_2-1}, a_{c_2+1}, \ldots, a_{c_1-1}, a_{c_1+1}, \ldots, a_n], \qquad (3)$$

for any two column values $c_1, c_2$. Since a Gauss transform, applied to a matrix from the left, simply performs an independent linear combination of each matrix column, the application of the factors $L$ and $P$ from the LU factorization in (2) to $A_{r_1, r_2 || c_1, c_2}$ results in:

$$L^{-1}PA_{r_1, r_2 || c_1, c_2} = [u_1, u_2, \ldots, u_{c_2-1}, u_{c_2+1}, \ldots, u_{c_1-1}, u_{c_1+1}, \ldots, u_n]; \quad (4)$$

which corresponds to the columns of $U$ in (2), except for $u_{c_1}$ and $u_{c_2}$, which have disappeared. The result is thus already upper triangular in the leftmost $c_2 - 1$ columns, but it contains zeros only below the first and second subdiagonals in columns $[u_{c_2+1}, u_{c_2+2}, \ldots, u_{c_1-1}]$ and $[u_{c_1+1}, \ldots, u_n]$, respectively; see the example in Fig. 2.

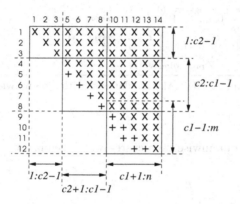

**Fig. 2.** Structure of the upper triangular matrix in (4), with $n = 14$, $m = n - 2 = 12$, $c_2 = 4$ and $c_1 = 9$. Nonzero entries below the main diagonal are identified with the symbol '+'. Columns are numbered taking into account that $c_1, c_2$ were eliminated.

In consequence, in order to obtain the desired upper triangular matrix, which yields the determinant for $A_{r_1, r_2 || c_1, c_2}$ we need to apply Gauss transforms (or, alternatively, any type of orthogonal transform [4]) to eliminate the nonzero entries below the main diagonal of this matrix. Let us consider the partitioning of the quasi-upper triangular factor in (4) as follows:

$$L^{-1}PA_{r_1, r_2 || c_1, c_2} =
\begin{array}{|c|c|c|}
\hline
U_{11} & U_{12} & U_{13} \\
\hline
 & U_{22} & U_{23} \\
\hline
 & & U_{33} \\
\hline
\end{array}
\begin{array}{l}
\left.\right\} 1 : c_2 - 1 \\
\left.\right\} c_2 : c_1 - 1 \\
\left.\right\} c_1 : m
\end{array} \quad (5)$$

$$\underbrace{1 : c_2 - 1}\ \underbrace{c_2 + 1 : c_1 - 1}\ \underbrace{c_1 + 1 : n}$$

where $U_{11}$ is upper triangular; and $U_{22}/U_{33}$ contain zeros below the first/second subdiagonal. Furthermore, consider $U_{23} = \begin{bmatrix} U_T \\ \hline u_B \end{bmatrix}$, where $u_B$ corresponds to the bottom row of $U_{23}$. The algorithm that exploits the relationship between the matrices factorized in the inner two loops is shown in Listing 1.2. Naturally, the LU factorizations involving the blocks $[U_{22}, U_{23}]$ and $U_{33}$ leverage the special quasi-upper triangular structure of these submatrices to reduce the cost of this alternative method.

```
1  for r1=1:n
2    for r2=1:r1-1
3      [L,U,P] = lu(A_{r1,r2||-});
4      for c1=1:n
5        for c2=1:c1-1
6          Partition U as in equation (5)  →  U11,U22,U23,U33
7          [L2,U2,P2] = lu([U22, U23]);% Exploit zeros below first subdiagonal
8          [L3,U3,P3] = lu([uB;
9                          U33]);     % Exploit zeros below second subdiagonal
10         d[r1][r2][c1][c2] = prod(diag(U11))
11                           * prod(diag(U2))
12                           * prod(diag(U3));
```

**Listing 1.2.** Algorithm for DL2M calculations with columnwise structure-aware reutilization.

The cost of the columnwise "structure-aware" realization is approximately given by

$$\sum_{r_1=1}^{n} \sum_{r_2=1}^{r_1-1} \underbrace{2m^3/3}_{\text{LU of } A_r} + \sum_{c_1=1}^{n} \sum_{c_2=1}^{c_1-1} \left( \underbrace{\sum_{j_1=c_2+1}^{c_1-1} 6(n-j_1)}_{\text{LU of } [U_{22},U_{23}]} + \underbrace{\sum_{j_2=c_1+1}^{n-1} 12(n-j_2)}_{\text{LU of } [u_B;U_{33}]} \right) \approx n^6/2 \text{ flops,}$$

where the sum for $j_1$ corresponds to the cost of the LU factorization for $[U_{22}, U_{23}]$ and the sum for $j_2$ to that for $U_{33}$. Taking into account that $m \approx n$, and neglecting the lower order terms, the cost for this approach is $n^6/2$ flops. Compared with NA, this represents a reduction in the cost of one order of magnitude.

### 3.3   Parallel Implementation

Both the original NA and the columnwise structure-aware algorithm (CSA), described in the previous subsections, are embarrassingly parallel. An analysis of dependencies and concurrency is, therefore, trivial. From Listing 1.1, we observe that the LU factorizations of the minors (and the accumulation of the diagonal elements of the resulting triangular factors), in the inner-most loop, are completely independent. In other words, each minor can be constructed independently of other minors, and the corresponding calculations can proceed in parallel. Although this approach exhibits a much larger memory footprint, because the minors are explicitly constructed, it accommodates a straight-forward parallelization. In contrast, in real-world use cases, the dimension of the initial matrix is up to $n = 200$, and exploiting only the parallelism intrinsic to the operations involved in a single LU factorization will surely exhibit very low performance. Therefore, our approach computes single-threaded (i.e. sequential) LU factorizations, but combines this with a parallelization of the outer loops around these calculations to compute several decompositions concurrently.

For the parallelization on multicore processors, in this work we leverage OpenMP. Concretely, in the DL2M NA case, in Listing 1.1, an OpenMP parallel for pragma is applied to the outer-most loop of the algorithm – that is, the loop over index $r_1$ – which corresponds to the first row to be removed from the initial matrix. The work corresponding to the three inner loops (over $r_2$, $c_1$, and $c_2$) is then executed in parallel for each iteration of the row index $r_1$.

For the CSA variant, in Listing 1.2, the inner-most LU factorizations and the products of diagonal elements of the factors $U_{\{11,2,3\}}$ are independent for each combination of the columns $c_1$ and $c_2$. They require only the $U$ factor obtained from the two outer-most loops (a unique combination of $r_1$ and $r_2$), as in Line 3 of Listing 1.2. In consequence, the parallelization of DL2M with columnwise reuse is done by distributing the iteration space across loop $c_1$; that is over the first column to be removed from the minor (without rows $r_1$ and $r_2$). For this variant, it is also possible to parallelize across the outer-most loop $r_1$ like it was done for L2M NA. That approach can be followed, for example, to parallelize the outer-most loop across multiple nodes, while the parallelization over $c_1$ can be applied at the node level. This multi-level parallel alternative for clusters is part of ongoing work.

## 4    Experimental Results

In this section we illustrate the gains in performance attained by the DL2M structure-aware algorithm, with columnwise re-utilization, for computing the determinants of the level-2 minors of the referent matrix. For this purpose, the new algorithm, CSA, is compared against the original NA realization, described in Subsect. 3.1. In addition, in the final part of this section, we compare the overall performance of the MEWF algorithms for computing the overlap of the wave functions, using our DL2M algorithm with columnwise re-utilization, with the algorithm OL2M, described in [10].

*Set Up.* The experiments in this section were obtained on the Juwels cluster from Juelich Supercomputing Center. Each node consists of two Intel Xeon Platinum 8168 processors, running at 2.7 GHz, with 24 cores each and 96 GB of RAM. The code was written in Fortran and C programming languages, compiled with GCC 8.3.0 and linked against the Intel MKL 2019.3.199 (sequential) library. All experiments are run on a single node and employ double precision arithmetic.

*Algorithmic Improvements.* The first experiment assesses the speed up gains achieved only by introducing algorithmic changes, without any parallelisation strategies. The columnwise reutilization strategy yields a significant performance gain up to 2.5× and 5× compared with the OL2M and NA variants, respectively, even when executed sequentially, as illustrated in Fig. 3. Note that the difference between CSA and other variants increases with the problem size. That is because of much lower computational cost of CSA compared to NA and OL2M variants. As an example, for $n = 100$ the flops count for NA is $10^{1}4/6$ while CSA exhibits $10^{1}2/2$ flops, that is approximately 33× less flops for CSA variant. However, in NA and OL2M variants more flops are "fast" flops based on LU and BLAS-3 operations, while in CSA, a part of flops, due to columnwise reutilization strategy, are based on BLAS-1 operations (the update step).

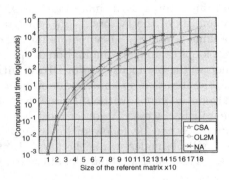

**Fig. 3.** Execution time of the sequential NA, OL2M and CSA algorithms.

*NA vs CSA.* This experiment compares the DL2M CSA variant (Listing 1.2) and the DL2M NA implementation (Listing 1.1). The input matrix for this test was generated with random entries and a number of columns/rows ranging from $n = 10$ to 180. The lower theoretical cost of the new CSA approach (of an order of magnitude) becomes evident in Fig. 4(left), which shows a reduction of the execution time by approximately one order of magnitude for sufficiently large matrices. When increasing the matrix size though, the speedup compared to NA significantly increases, because of the much lower computational cost of the CSA variant. Figure 4(right) shows that the speed up of CSA (with respect to NA) is consistent, independently of the number of cores in the test, and roughly grows by a factor of up to 7 for the largest test matrices.

We can observe that the CSA algorithm is superior in performance to the naive version for all configurations and matrix sizes, as presented in Fig. 4(right). It can been seen that the speedup is similar, no matter how many cores we used in test configuration, and is increasing up to more than 7× for larger test matrix sizes.

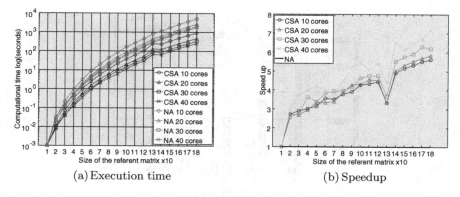

(a) Execution time                    (b) Speedup

**Fig. 4.** Execution time and speed-up (left and right plots, respectively) for the parallel DL2M CSA and NA algorithms for varying #cores and matrix size.

*OL2M vs CSA.* In Fig. 5(left) we compare the new DL2M CSA with the OL2M algorithm [10] (considering only the part that computes the determinants of the minors). In OL2M, the complete LU is computed inside the $c_1$ loop, and the accumulated $U$ factors are reused only inside the $c_2$ loop. By computing the LU before the start of the $c_1$ loop, the CSA variant offers a speed up of up to 3.3× for the largest problems, Fig. 5(right).

Note that the speedup of CSA w.r.t. OL2M is also independent on the number of cores (as in the case of NA) and that is almost the same for different test configurations and larger test matrix sizes.

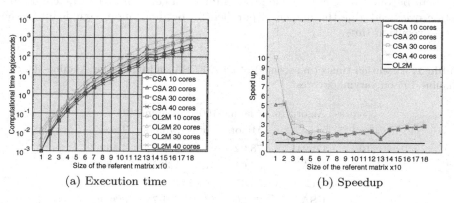

(a) Execution time                    (b) Speedup

**Fig. 5.** Execution time and speed-up (left and right plots, respectively) for the parallel DL2M CSA and OL2M algorithms for varying #cores and matrix size.

*Scalability.* Although CSA re-uses the columns from the $U$ factor for computing determinants of the subsequent minors, the algorithm achieves a good scalability with increasing number of cores, as presented in Fig. 6.

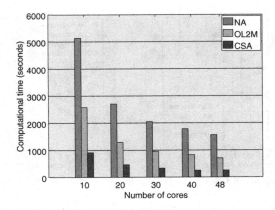

**Fig. 6.** Strong scaling of CSA variant with the number of processor for $n = 180$ (number of rows/columns of the referent matrix).

*Real Use-Case.* The final experiment compares DL2M CSA and the OL2M algorithm on real test cases corresponding to the excited state wave functions of poly-Alanine systems (with 100 and 195 occupied orbitals, i.e. the sizes of the referent matrices) obtained using different basis sets. The results of this test in Table 1 show that DL2M CSA offers higher scalability and achieves a speed up factor above 5× over DL2M, on 48 cores, which is aligned with the results reported in Fig. 5(right). For the Alanine-100 use-case, by increasing the number of cores over 40 a drop in performance is occured. That is expected since the problem becomes to small for the given number of cores in which the communication overhead, due the increased parallelism, becomes more significant in the total execution time.

**Table 1.** Execution times (in seconds) of DL2M CSA and OL2M for Alanine-100 and Alanine-195 on varying #cores.

| #cores | Algorithm | Alanine-100 | | Alanine-195 | |
|--------|-----------|--------|--------|----------|---------|
| | | Minors | Total | Minors | Total |
| 10 | OL2M | 78.02 | 92.57 | 4,233.16 | 4,432.2 |
| | CSA | 37.61 | 51.96 | 1,364.06 | 1,568.2 |
| 20 | OL2M | 39.47 | 53.00 | 2,117.29 | 2,303.6 |
| | CSA | 20.26 | 33.89 | 691.78 | 896.39 |
| 30 | OL2M | 29.29 | 43.74 | 1,557.32 | 1,755.0 |
| | CSA | 14.13 | 28.79 | 478.16 | 674.98 |
| 40 | OL2M | 28.33 | 44.11 | 1,306.56 | 1,508.5 |
| | CSA | 11.61 | 26.83 | 410.27 | 628.87 |
| 48 | OL2M | 29.41 | 44.92 | 1,142.5 | 1,365.7 |
| | CSA | 10.72 | 26.46 | 364.56 | 581.14 |

# 5    Conclusion and Future Work

In this work, we leverage the connection between the level-2 minors of the referent matrix appearing in MEWFs to save a considerable part of the computations required to obtain the corresponding determinants. For that purpose, we use an updating technique for the LU factorization, in order to reduce the cost by an order of magnitude. Our tests show that the new approach considerably accelerates the computation of the MEWF overlaps, by a factor of up to 7, in principle allowing the solution of larger problems.

As part of future work, we plan to explore the parallelization of this algorithm using different approaches and/or tools (e.g., to exploit task-level parallelism, or to combine a cluster-level parallelization with a finer grain concurrent execution). In addition, we will investigate how to exploit the connection between the level-2 minors across other dimensions to further reduce the theoretical cost of this type of computations.

The source code is available at [9] and is part of the *cto-nto* library for computing natural transition orbitals for CIS type wave functions.

**Acknowledgement.** This research was performed under project HPC-EUROPA3 (INFRAIA-2016-1-730897) and supported by Croatian Science Fundation under grant HRZZ IP-2016-06-1142, the Foundation of the Croatian Academy of Science and Arts, and the European Regional Development Fund under grant KK.01.1.1.01.0009 (DAT-ACROSS). Enrique S. Quintana-Ortí was supported by project TIN2017-82972-R of the MINECO and FEDER. The authors gratefully acknowledge the computer resources provided by the Juelich Supercomputing center, and to BSC where the initial testings and the code development were performed.

# References

1. Anzt, H., Dongarra, J., Flegar, G., Quintana-Ortí, E.S.: Variable-size batched LU for small matrices and its integration into block-Jacobi preconditioning. In: 2017 46th International Conference on Parallel Processing (ICPP), pp. 91–100 (2017). https://doi.org/10.1109/ICPP.2017.18
2. Dirac, P.A.M.: A new notation for quantum mechanics. Math. Proc. Cambridge Philos. Soc. **35**(3), 416–418 (1939). https://doi.org/10.1017/S0305004100021162
3. Dong, T., Haidar, A., Luszczek, P., Harris, J.A., Tomov, S., Dongarra, J.: Lu factorization of small matrices: accelerating batched DGETRF on the GPU. In: 2014 IEEE International Conference on High Performance Computing and Communications, 2014 IEEE 6th Internatinal Symposium on Cyberspace Safety and Security, 2014 IEEE 11th International Conference on Embedded Software and Systems (HPCC, CSS, ICESS), pp. 157–160, August 2014. https://doi.org/10.1109/HPCC.2014.30
4. Golub, G.H., Loan, C.F.V.: Matrix Computations, 3rd edn. The Johns Hopkins University Press, Baltimore (1996)
5. Hammes-Schiffer, S., Tully, J.C.: Proton transfer in solution: molecular dynamics with quantum transitions. J. Chem. Phys. **101**(6), 4657–4667 (1994). https://doi.org/10.1063/1.467455

6. Li, S.L., Truhlar, D.G., Schmidt, M.W., Gordon, M.S.: Model space diabatization for quantum photochemistry. J. Chem. Phys. **142**(6), 064106 (2015)
7. Plasser, F., Ruckenbauer, M., Mai, S., Oppel, M., Marquetand, P., González, L.: Efficient and flexible computation of many-electron wave function overlaps. J. Chem. Theory Comput. **12**(3), 1207–1219 (2016)
8. Quintana-Ortí, E.S., Van De Geijn, R.A.: Updating an LU factorization with pivoting. ACM Trans. Math. Softw. **35**(2), 11:1–11:16 (2008)
9. Sapunar, M.: Natural transition orbitals for CIS type wave functions. https://github.com/marin-sapunar/cis_nto. Accessed 24 Oct 2019
10. Sapunar, M., Piteša, T., Davidović, D., Došlić, N.: Highly efficient algorithms for CIS type excited state wave function overlaps. J. Chem. Theory Comput. **15**, 3461–3469 (2019)
11. Tully, J.C.: Molecular dynamics with electronic transitions. J. Chem. Phys. **93**(2), 1061–1071 (1990). https://doi.org/10.1063/1.459170

# Lazy Stencil Integration in Multigrid Algorithms

Charles D. Murray[✉] and Tobias Weinzierl

Department of Computer Science, Durham University, Durham, UK
{c.d.murray,tobias.weinzierl}@durham.ac.uk

**Abstract.** Multigrid algorithms are among the most efficient solvers for elliptic partial differential equations. However, we have to invest into an expensive matrix setup phase before we kick off the actual solve. This assembly effort is non-negligible; particularly if the fine grid stencil integration is laborious. Our manuscript proposes to start multigrid solves with very inaccurate, geometric fine grid stencils which are then updated and improved in parallel to the actual solve. This update can be realised greedily and adaptively. We furthermore propose that any operator update propagates at most one level at a time, which ensures that multiscale information propagation does not hold back the actual solve. The increased asynchronicity, i.e. the laziness improves the runtime without a loss of stability if we make the grid update sequence take into account that multiscale operator information propagates at finite speed.

**Keywords:** Additive multigrid · Matrix assembly · Asynchronous stencils

## 1 Introduction

Modern linear algebra dramatically reduces the total number of iterations required by iterative equation system solvers to reduce the error by a fixed amount. The most sophisticated solvers such as multigrid or multilevel preconditioners bring down this count to a constant—rather than increasing commensurately with the problem size. This holds for sufficiently well-behaved problems

$$- \nabla \left( \epsilon \nabla \right) u = f \quad \text{with} \quad \Omega = (0,1)^d, \ d \in \{2,3\}, \tag{1}$$

where $u : \Omega \mapsto \mathbb{R}$ is unknown, $f : \Omega \mapsto \mathbb{R}$ is any source or sink and $\epsilon : \Omega \mapsto \mathbb{R}^+$.

Equation (1) is not a toy problem. It is a fundamental building block in many applications, and it is difficult to solve: Its ellipticity implies that a change of any weight of a discretisation induces changes everywhere else. Multigrid solvers anticipate this global behaviour. They have a built-in separation of scales.

The work was funded by an EPSRC DTA PhD scholarship (award no. 1764342). It made use of the facilities of the Hamilton HPC Service of Durham University. The underlying project has received funding from the European Union's Horizon 2020 research and innovation programme under grant agreement No 671698 (ExaHyPE).

© Springer Nature Switzerland AG 2020
R. Wyrzykowski et al. (Eds.): PPAM 2019, LNCS 12043, pp. 25–37, 2020.
https://doi.org/10.1007/978-3-030-43229-4_3

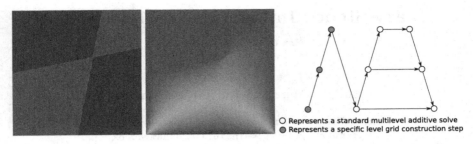

O Represents a standard multilevel additive solve
● Represents a specific level grid construction step

**Fig. 1.** $\epsilon$ value distribution used in our experiment (left) and a corresponding solution (middle). Right: Schematic display of standard additive multigrid where all operators are constructed fine-to-coarse prior to the multigrid cycles.

Tackling a problem on multiple scales is a seminal idea and yields optimal complexity for some problems. Yet, it does not yield solves for free. We still have to numerically integrate all stencils on the fine grid. On top of that, multigrid requires us to construct appropriate coarse grid stencils and inter-grid transfer operators. True geometric solvers work with operator rediscretisation and $d$-linear inter-grid transfer operators, where $d$ is the dimension (we test for $d = 2$, i.e. refer to bilinear functions), and thus avoid the latter overhead. Yet, their performance and robustness deteriorates when we face non-constant $\epsilon$ [17,18].

We have to accept that the operator construction (assembly) in robust multigrid makes up for a non-negligible part of the total runtime [20]. Many efforts have been made to parallelise and tune the assembly [3,8]. Structured grids [1], particularly octrees [9,11,13,19], are a popular solution. Their tensor product structure and built-in multiscality simplifies the assembly. Yet, assembly remains painful. It is the character of the assembly that complicates the situation further: Often, we are constrained by bandwidth. As the gap between compute power and memory bandwidth is widening [12], the assembly becomes hard to scale.

Our work proposes a lazy element-wise assembly: It starts from a trivial integration of the weak formulation of (1) where we sample $\epsilon$ in the center of the cell once. In parallel to the actual solve, we then improve the quadrature per stencil and compute new stencil entries with better and better accuracy. Whenever we need a stencil and its next higher accuracy is not yet available, we continue with the old "low-accuracy" one. It is a greedy process not delaying the solve. As soon as two subsequent integrations do not yield a large difference anymore, we stop it for this particular element. It is an adaptive process. As new stencils become available, we restrict their influence to the next coarser level in the next multigrid cycle. It is an iterative process where the algebraic coarse-grid computation and Ritz-Galerkin construction incrementally push the operator information up the resolutions. The operators ripple through the hierarchies.

Our experiments reveal that it is totally sufficient to kick off a solve with inaccurate stencils, as long as we aggressively improve them afterwards. There are however two surprising pitfalls: Publications around Bootstrap AMG [4] or Adaptive AMG [5,6], e.g., which use information gained during the solve to directly improve coarse grids suggest that we could also kick off with inaccurate coarse grid stencils and improve them iteratively. This hypothesis is not supported by our data. It seems that too crude coarse grid operators lead the solver initially into the wrong direction, which destroys the overall multigrid, i.e. mesh-independent convergence. It is better to incrementally develop from a 2-grid into a 3-grid into a 4-grid algorithm and so forth. Our second surprise relates to the scalability: Though the stencil integration is deployed to the background and thus increases the overall concurrency level, the scalability tends to suffer. It improves the time to solution though.

The remainder of the paper is organised as follows: We start our manuscript with a revision of the multigrid workflow including its setup steps (Sect. 2). A short description of the used multigrid realisation leads over to the introduction of our lazy approach (Sect. 3) which is studied in Sect. 4. The conclusion summarises the key observations and provides a brief outlook.

## 2 The Multigrid Workflow

Classic multigrid can be read as a sequence of activities

$$(\mathcal{C} \circ \ldots \circ \mathcal{C}) \circ \ldots \circ \mathcal{A}_{k^3 h} \circ \mathcal{A}_{k^2 h} \circ \mathcal{A}_{kh} \circ \mathcal{A}_h^*(\epsilon). \tag{2}$$

We start with the integration of a discretisation of (1) on a fine grid identified by mesh size $h$. $\mathcal{A}_h^*$ yields this discretisation matrix $\mathbf{A}_h$. This is the assembly in a finite element jargon. The material $\epsilon$ directly enters the arising linear equation system $\mathbf{A}_h$.

With a fine grid matrix to hand, we construct the multigrid operators. Per level, this is a two-step process. We use $\mathcal{A}$ in a multigrid sense compared to $\mathcal{A}^*$ on the finest grid. We furthermore assume that all coarse grid meshes are known. $\mathcal{A}$ first constructs the inter-grid transfer prolongation $\mathbf{P}_{kh}^h$ and restriction $\mathbf{R}_h^{kh}$. Here $k$ is the increase of the mesh size. Most literature uses $k = 2$ or $k = 3$. We employ the latter. $\mathbf{P}_{kh}^h$ and $\mathbf{R}_h^{kh}$ propagate a solution on the mesh with spacing $kh$ to the fine grid with mesh size $h$ and back. Their construction—we rely on BoxMG [10, 20]—takes the fine grid operator's effective local null space into account. After that, $\mathcal{A}$ determines the matrix $\mathbf{A}_{kh} = \mathbf{R}_h^{kh} \mathbf{A}_h \mathbf{P}_{kh}^h$ through the Ritz-Galerkin formulation. $\mathcal{A}_{kh}$ in total yields three matrices $\mathbf{A}_{kh}$, $\mathbf{R}_h^{kh}$ and $\mathbf{P}_{kh}^h$.

Once all operators are set up, each multigrid cycle $\mathcal{C}$ decomposes into a series of actions $\mathcal{C}_{k^n h}$ on their respective grids. Depending on how we arrange and design these sub-actions, the overall scheme denotes an additive or multiplicative cycle. They run through the grid hierarchy differently. While the number of resolution levels determines the length of the $\mathcal{A}$ sequence as well as level updates, the solve itself is iterative: The total number of $\mathcal{C}$ applications is typically steered by the residual. We terminate when the residual normalised by its initial value falls below a threshold, i.e. when the error has been diminished by this factor.

*Causal Dependencies.* The above discussion of (2) incorporates two assumptions: On the one hand, we do not explicitly include a coarse grid identification into our presentation. Indeed, finding a proper coarse grid is a time-consuming preamble to $\mathcal{A}_{kh}$ in pure algebraic multigrid. We assume that the multiresolution hierarchy is known [16]. On the other hand, we assume that we realise an algebraic multigrid operator construction. Alternatively, codes can realise a geometric approach: They choose $\mathbf{P} = C\mathbf{R}^T$ with a constant $C$ and make $\mathbf{P}$ a $d$-linear interpolation. For the coarse grid operator, they drop Ritz-Galerkin and construct $\mathbf{A}_{kh}$ through re-discretisation: On every level, they discretise (1) through the same finite element approach used on the fine grid, and use the arising equation system as coarse grid correction operator. Both schemes yield the same operators if $\epsilon$ is constant and regular grids are used which are embedded into each other.

For an algebraic operator construction, neither (1) nor its material parameter $\epsilon$ enter the outcome matrices of $\mathcal{A}_{kh}$ directly. They enter through $\mathbf{A}_h$. As we associate the construction of inter-grid transfer operators with the coarser mesh, $\mathcal{A}_{kh}$ depends solely on the outcome of $\mathcal{A}_h$. In line with machine learning terminology, the operators are meta arguments towards other operators.

We intentionally write (2) in a multiplicative way. One step (multiplication) has to finish before the subsequent operation may start. Yet, there is concurrency hidden within this formalism. Each assembly $\mathcal{A}_h$ constructs stencils tied to the level. This also holds for $\mathbf{P}_{kh}^h$ and $\mathbf{R}_h^{kh}$ as we can associate each prolongation or restriction operation to a single vertex on level $kh$. The global result matrices of $\mathcal{A}_h$ thus can be constructed concurrently as a sum of components over the individual stencils tied to the vertices $v$: $\mathcal{A}_h = \sum_v \mathcal{A}_h(v)$. An analogous reasoning holds for the updates within a $\mathcal{C}_h$ since we study point smoothers only.

Multiplicative multigrid relies on a fine-to-coarse paradigm: A given discretisation of (1) is smoothed on a fine mesh, i.e. its high frequency errors are eliminated. Its residual equation then is projected onto the next coarser mesh where the remaining low frequency errors exhibit high frequency components again. We recursively return to the first step. Additive multigrid realises the same fine-to-coarse idea, but the individual levels are not treated one after another. Instead, it uses the discretisation of (1) on the fine grid to derive a fine grid correction equation, restricts the associated residual immediately, i.e. without smoothing, to all resolution levels, and then updates all levels in one rush.

*Full Multigrid Cycles and Dynamic Adaptivity.* The full multigrid cycle extends the fine-to-coarse idea of multigrid by a coarse-to-fine flavour: We start with a rather coarse fine grid mesh and run our multigrid solve there. This initial mesh then is unfolded into the next finer resolution and we continue. As the unfolding is combined with a (higher order) prolongation—the higher order can be dropped if we are willing to invest additional smoothing steps—a solve on a level $kh$ serves as initial guess to the solve on level $h$. We solve

$$\ldots \circ \mathcal{A}_h \circ \mathcal{A}_{h/k} \circ \mathcal{A}_{h/k^2}^*(\epsilon) \circ \mathcal{U} \circ \mathcal{C} \circ \mathcal{A}_h \circ \mathcal{A}_{h/k}^*(\epsilon) \circ \mathcal{U} \circ \mathcal{C} \circ \mathcal{A}_h^*(\epsilon)$$

where $\mathcal{U}$ is the mesh unfolding operator.

The term "unfolding" technically describes mesh refinement. Whenever we equip our multigrid solver with dynamic adaptivity, i.e. the solver may add more degrees of freedom to the mesh throughout the solve, we technically inject (localised) $\mathcal{U}$ operators into the mesh and must trigger some reassembly. From an implementation point of view, full multigrid cycles and dynamic adaptivity are similar. All properties, including full multigrid mesh unfolding, thus hold for additive and multiplicative as well as both geometric and algebraic multigrid.

*Multigrid on Spacetrees.* Our code solves (1) for the artificial $\epsilon$ arrangement shown in Fig. 1. We compute all stencils using a Finite Element formulation with bilinear shape functions. The underlying grid is an adaptive Cartesian mesh as we obtain it from a two-dimensional spacetree [20, 21]. We embed the computational domain into a square and then cut each square equidistantly into $k$ pieces along each coordinate axis. This process continues recursively yet independently for each of the $k^2$ resulting squares. We end up with meshes of squares which are axis-aligned. No $\epsilon$ jump is therefore guaranteed to align with the mesh. A "simpler" equation to solve may arise if we constructed the mesh to exactly encode the discontinuities. However there is no geometric grid scheme that works for all discontinuities—curves in material data could never be exactly encoded in a mesh. We instead favour the simpler construction of equations geometric schemes offer.

It is the spacetree which identifies the coarse grids. We rely on a geometric coarse grid identification. The handling of the adaptivity boundaries is realised through FAS as discussed in [16, 20]. It follows the idea of FAC: Technically, we treat the individual resolution levels as completely independent data structures. A vertex or a cell, respectively, thus becomes unique through it spatial position plus its level. Multiple vertices from multiple levels can coincide spatially.

Our code works without any matrix data structure [20]. Each vertex directly holds the corresponding line from the assembly matrix, as well as the lines from the inter-grid transfer operators. The matrix is embedded into the mesh. Thus, we can traverse the meshes to compute the matrix-vector products (mat-vecs).

## 3   Lazy Operator Assembly

The entries of $\mathbf{A}_h$ stem from a weak formulation of (1). We express $u$ as linear combination over the basis functions and pick test functions $\phi(x, y)$ from the same function space:

$$\int_{\Omega_h} \epsilon\,(\nabla u, \nabla \phi)\ dx = \sum_{c \in \Omega_h} \int_c \epsilon\,(\nabla u, \nabla \phi)\ dx, \tag{3}$$

with cells $c \in \Omega_h$ where $\Omega_h$ is our chosen discretisation of $\Omega$. As a cell $c$ is aware of its adjacent vertices, $c$ "knows" to which stencil entries its integral contributes to. For non-trivial $\epsilon$, they require an accurate integration rule. We use piece-wise constant integration of $\epsilon$ but sample over each grid cell on a $p^d$ grid. $p \in \mathbf{N}^+$.

For analytic $\epsilon$, we can a priori prescribe a $p$ per mesh cell which makes the numerical evaluation of (3) fall below a prescribed error $dp > 0$. Otherwise, this is not straightforward.

*Adaptive Stencil Integration.* Let each vertex hold its $\mathbf{A}, \mathbf{P}$ and $\mathbf{R}$ contributions twice. $\mathbf{A}, \mathbf{P}, \mathbf{R}$ are the actual values used by the solver. $\mathbf{A}^{new}, \mathbf{P}^{new}, \mathbf{R}^{new}$ are helper variables. We make $p(c) \in \mathbb{N}^+ \cup \{\bot, \top\}$ with the following semantics: Initially, all cells carry $p(c) = \bot$. We have not yet computed the cell's contribution towards any stencil. If $p(c) = \top$, the cell's contributions towards $\mathbf{A}, \mathbf{P}, \mathbf{R}$ are sufficiently accurate. If neither condition is met, $\mathbf{A}, \mathbf{P}, \mathbf{R}$ hold an operator representation which results from a subsampling of $c$ with a $p(c) \times p(c)$ quadrature.

It is straightforward to phrase the assembly of the stencils, i.e. the entries of $\mathbf{A}$, iteratively, and to fuse it with the mesh traversal evaluating the multigrid's mat-vecs. Such a fusion ensures that the assembly does not introduce additional data reads and writes compared to the mat-vecs. It releases pressure from the memory interconnect: If a cell holds $\bot$, the grid has just been created due to $\mathcal{U}$ or this is the initial iterate. The cell belongs to the fine grid. In this case, we set $p = 1$, compute the cell's contribution towards $\mathbf{A}$ and continue. This is the initial instantiation of one task of $\mathcal{A}$ for one cell. If we encounter a cell with $p \neq \top$, we increase $p$ by one and recompute the cell's contribution towards $\mathbf{A}$. The recomputed cell contributions are stored in $\mathbf{A}^{new}$. We can then compare the relevant entries of $\mathbf{A}$ and $\mathbf{A}^{new}$. If the entries' maximum difference is below $dp$, we set $p = \top$. Finally, we roll over $\mathbf{A}^{new}$ into $\mathbf{A}$ to make it available to the solver. With this scheme, $\mathcal{A}_h(v)$ becomes a set of tasks parameterised though $p$.

We augment each cell with one further atomic boolean which is unset initially. Our greedy stencil integration then is formalised as follows. If we encounter a fine grid cell with

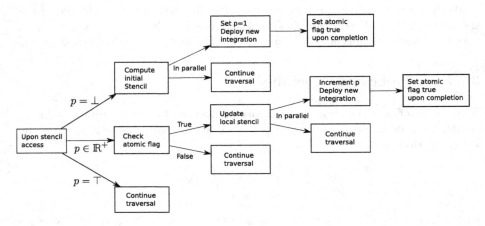

**Fig. 2.** Illustrative diagram of how we perform the lazy integration. All cells carry a $p$ that holds the number of samples per dimension of the quadrature.

1. $p = \bot$, we run the integration as sketched above. This yields a low accuracy "kick-off" integration of the system matrix.
2. $p = \top$, we do not alter the stencil.
3. $p \neq \bot$ and $p \neq \top$ whose atomic boolean is not set, we set the marker and spawn a new integration task running one adaptive integration as described above. It has low priority. Upon completion, the task unsets the marker.
4. $p \neq \bot$ and $p \neq \top$ whose atomic boolean is set, we continue.

These steps integrate easily into any multigrid code. The mat-vecs kick off with a low accuracy integration of (3) but create additional tasks that improve this accuracy. We postpone the accurate integration. Standard multicore work stealing ensures that idling cores take over the integration, i.e. the matrix "assembly" branches out (Fig. 2).

*Greedy Stencil Integration.* The overall scheme is anarchic, as we never explicitly synchronise the integration with the solves. Whenever an improved integration of a cell's stencil contribution becomes available, we use it in the mat-vecs. Technically, it would be convenient to protect the roll over of $\mathbf{A}^{new}$ into $\mathbf{A}$ by a semaphore, too. In practice, we found the code working nicely without such a protection.

*Multiscale Stencil Rippling.* Algebraic multigrid is sensitive to such an asynchronous integration, as any update of the fine grid operators propagates to all coarser levels. With an anarchic scheduling—we neither control the scheduling of integration improvements nor do we know whether a particular re-integration improves a stencil—it is unreasonable to make every stencil update propagate through the full hierarchy instantly. Instead, we update all inter-grid transfer and coarse grid operators throughout the grid traversals' mat-vecs and thus ensure that the re-assembly does not induce any additional data reads or writes.

In line with the fine grid stencils, the multigrid operators are held redundantly. Every cycle relies on the current value. At the same time, it computes an updated version in temporary buffers. At the end of the traversal, these buffers are rolled over. Such a scheme implies that any improved stencil information propagates fine-to-coarse by exactly one level per grid traversal. The information ripples through the system. We note that this does not impose any problems for multiplicative multigrid. Here, the propagation of the improved coarse grid operators follows the smoothing order. For additive multigrid, the operator propagation however lags behind. If there are problematic effects induced by the laziness, then such additive schemes are more vulnerable. Therefore we concentrate on such setups.

Geometric multigrid hardcodes the inter-grid transfer operators, but can apply the incremental integration on each and every level. The integral in (3) weights $\epsilon$ fluctuations relative to the mesh size automatically and thus yields appropriate stencil accuracy. We concentrate on algebraic operator construction here as it yields more robust solvers. In this context, we emphasise that not only the lazy integration, but also dynamic mesh refinement cause rippling: When we refine the mesh, a former fine grid stencil becomes a correction stencil subject to a Ritz-Galerkin construction and will hence be updated incremementally; an update which propagates through fine-to-coarse.

## 4   Results

Our results stem from an Intel Xeon E5-2650V4 (Broadwell) cluster in a two socket configuration hosting 24 cores. The cores have a nominal baseline frequency of 2.4 GHz. Yet, AVX2 instructions can reduce this frequency to 1.8 GHz [14,15]. We may assume that (3) yields many AVX operations once $p(c)$ grows reasonably big. The assembly of the Ritz-Galerkin operator and BoxMG require vector operations, too. If any coarse grid equations proves to be singular, that is the stencil is not diagonally dominant, it's contribution is removed, similar to the work in [21]. Coarse grid problems are solved using a jacobi smoothing step, in the same fashion as the fine grids. Each node is equipped with 64 GB of TruDDR4 memory. The parallelisation relies on Intel's Threading Building Blocks (TBB).

All experiments study plain additive multigrid [2,16] plus two variations of this additive scheme: adAFAC-PI [15,16] damps the additive corrections through an additional BPX-like term. Before the solver adds the correction of a level, it reduces this correction by an injection followed by a prolongation. This effectively damps any overshooting. adAFAC-Jac [15] also damps the additive correction, but uses an auxiliary coarse grid equation for this. Before the solver adds the

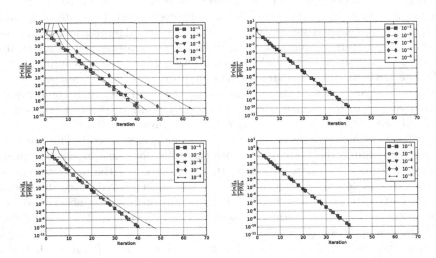

**Fig. 3.** Number of iterations until convergence is reached for the damped adAFAC-PI setup. with checkerboard material parameter as the size of the jump increases. We either start from a geometric coarse operator guess (left) or delay the switching on of a level's contribution until a meaningful operator has rippled through (right). adAFAC-PI (top) vs. adAFAC-Jac (bottom).

correction of a level, it studies this correction effect on the next level via a smoothed operator and subtracts this additional term. Both adAFAC-PI and adAFAC-Jac employ the standard multigrid operators, but are slightly more robust. Our studies navigate along the following questions: Does the laziness increase the iteration count and affect the robustness? How does the lazy initialisation within the solves affect the runtime? Does the lazy integration improve our codes' scalability?

*Robustness and Iteration Counts.* To test whether additional instabilities are introduced due to our laziness, we test a grid setup with a total of 531,441 degrees of freedom on the fine grid. With the geometric parameter distribution from 1 we fix $\epsilon = 1$ in the bottom left and top right subdomain and make $\epsilon = 10^{-k}$, $k \in \{1, 2, \ldots, 5\}$ otherwise. This results in $p(c)$ of 2 away from the discontinuity and most cells exhibiting $p(c)$ of around 20 when they hold a discontinuity. All data report normalised residuals, i.e. the residual development in the discrete $L_2$ norm relative to the initial one. We stop when the initial residual is reduced by ten orders of magnitude (Fig. 3).

Our benchmarks compare three methods how to initialise the stencils against each other: (i) An exact computation of all fine grid and coarse grid operators is done prior to the first multigrid cycle. (ii) An incremental, lazy update of the fine grid stencils plus an initial geometric guess for all coarse operators is used while updates ripple through the hierarchies. (iii) An incremental, lazy update of the fine grid stencils is used. Coarse grid updates are disabled until the first algebraic expression has rippled through. In the latter case, the algorithm develops from a 2-grid into a 3-grid into a 4-grid algorithm and so forth.

Our first tests vary $k$ (3). As soon as $k$ exceeds 2, both modified additive solver variants outperform their plain additive cousin as plain additive multigrid becomes unstable (not shown). Our modified additive solvers exhibit constant residual reduction as long as $k \leq 3$. If the parameter jump becomes bigger, they start to suffer from some instabilities in the first few iterations. This offsets the convergence curve. It is due to the rippling. Once we disable the rippling, i.e. involve only levels with reasonably valid operators, we retain the convergence of accurate precomputation (not shown as indistinguishable) (Fig. 4).

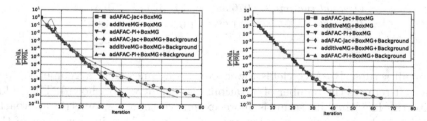

**Fig. 4.** Number of iterations until convergence for $k = 3$. We either start from a geometric coarse operator guess (left) or delay the switching on of a level's contribution until a meaningful operator has rippled through (right).

The offset is bigger for the additive solver that solves a real additional coarse grid problem (adAFAC-Jac). Previous studies [15] suggest that additive multigrid which employs a solver that tests permanently on an auxiliary grid whether it overshoots is superior to a BPX-like code which does not employ a coarse grid operator but solely **P** and **R**. With rippling, the coin switches. Now, solvers alike BPX become superior as they do not require Ritz-Galerkin operators (4). It is the slow migration from two-grid to three-grid and so forth that eliminates all penalties of the rippling again. Multiplicative multigrid exhibits similarities to the resulting scheme. It first solves on the finest grid, then on the next coarser, then on the next coarser. In our additive mindset, the individual multiplicative steps are replaced by an additive cycle. However, if we omit this gradual switching on of coarser and coarser meshes, it seems that too wrong coarse grid operators can lead the solver into the wrong direction.

*Runtime.* We continue with timings where we compare the solvers with an explicit a priori setup of all operators to the lazy variants (1). Different to the previous setup, we also benchmark the geometric multigrid variant against its algebraic counterpart employing BoxMG. The latter is computationally significantly more intense. All experiments fix $k = 3$ (Table 1).

**Table 1.** Total solver timings for BoxMG including all assembly time. The first row in each denotes the time-to-solution with a precise a priori assembly, the second the speedup obtained through lazy integration.

| Solver type | Run time [s]/Speedup | | | |
|---|---|---|---|---|
| | 2 | 4 | 8 | 12 |
| AdAFAC-Jac | | | | |
| *Original runtime* | 847.90 | 491.44 | 297.63 | 273.34 |
| *Speedup* | 1.54 | 1.50 | 1.50 | 1.604 |
| AdAFAC-PI | | | | |
| *Original runtime* | 864.16 | 503.31 | 293.83 | 291.64 |
| *Speedup* | 1.49 | 1.48 | 1.40 | 1.49 |
| Additive | | | | |
| *Original runtime* | 976.39 | 562.21 | 323.51 | 279.31 |
| *Speedup* | 1.68 | 1.62 | 1.61 | 1.63 |

For all setups, BoxMG yields lower execution times than a geometric approach as it reduces the number of iterations signficiantly (not shown). This holds despite the fact that its setup is more expensive than a geometric approach. Switching to background stencil integration yields a robust improvement of the total run time. Without it, the grid setup took roughly one third of the total execution time. adAFAC-Jac solves an additional equation compared to the sole evaluation of additional inter-grid transfer operators of adAFAC-PI. This allows it to solve the setup with fewer iterations. However, each additive cycle is slightly more expensive. In some cases, this nullifies its advantage and it is impossible to say which solver is superior. We reiterate however that sole additive multigrid becomes unstable for larger $k$.

*Scalability.* As our background integration lacks any NUMA-awareness, our scalability studies restrict to one socket only. To avoid the impact of the actual problem setup, we also study only the first 5 multigrid cycles. Despite the fact that our setup is almost trivial by means of arithmetic load, we see some scalability. It is limited however (5). Overall, the scaling behaviour is almost agnostic of the solver type (Fig. 5).

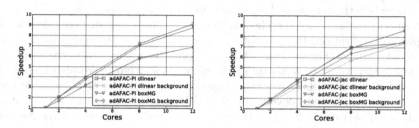

**Fig. 5.** Speedups for adAFAC-PI (left) and adAFAC-Jac (right).

The laziness has no advantageous impact on the scalability. It is counterproductive. We do increase the concurrency level, but the initial integration tasks are all extremely lightweight and impose a large overhead. At the same time, the lack of an (expensive) initial assembly implies that the first few sweeps are cheap. Later throughout the solve when the integration tasks are more expensive, there are only very few of them left. Our background integration eliminates the advantageous concurrency with high arithmetic intensity of the first sweep. This is an algorithmic character.

The lack of overall scalability is due to two implementation effects however: On the one hand, the integration relying heavily on AVX downclocks the chip [7]. On the other hand, our algebraic coarse grid assembly at the moment is change-agnostic. We recompute always. As restriction and prolongation operators gather information from many fine grid points and as we want to avoid data races, we apply a very aggressive $6^3$-colouring [20]. This effectively reduces the concurrency.

## 5    Conclusion and Outlook

We have introduced a lazy matrix assembly strategy for multigrid which is of practical relevance as it does not blend out the solver's initialisation cost. Despite the fact that we study something alike a worst case scenario—very low arithmetic cost and an additive solver family—and despite the fact that we employ a very rudimentary implementation—pessimistic a priori colouring avoids race conditions even though our approach would offer the opportunity to derive aggressive concurrency patterns; we know which entries change or do not change anymore—we robustly can improve the runtime.

The proposed scheme can be read as a multigrid realisation which develops from almost geometric multigrid (very inaccurate integration rules based upon few geometric samples) into an algebraic variant. It is thus an obvious idea to couple the present scheme with the operator compression introduced in [20]. Such a combination renders the doubling of the memory footprint in the present approach negligible. This will be subject of future work. The studied integration rule has to be studied further, too. We employ resampling with regular grids. On the long term, hierarchical approaches, i.e. adaptive integration, and higher order shape functions have to be plugged into our lazy approach.

An open problem only sketched is the exact scheduling of tasks tied to the background integration. In our prototype, we just spawn the integration as low priority TBB tasks, i.e. we leave it to this generic runtime to derive a proper scheduling, and we work agnostic of memory associativity. We also ignore the fact that low $p$ tasks likely are too cheap to map to tasks of reasonable cost. Future studies will have to develop proper scheduling and task strategies that get the balancing between solve and assembly right and furthermore take memory associativity into account.

# References

1. Bader, M., Schraufstetter, S., Vigh, C.A., Behrens, J.: Memory efficient adaptive mesh generation and implementation of multigrid algorithms using Sierpinski curves. Int. J. Comput. Sci. Eng. **4**(1), 12–21 (2008)
2. Bastian, P., Wittum, G., Hackbusch, W.: Additive and multiplicative multi-grid–a comparison. Computing **60**(4), 345–364 (1998)
3. Bell, N., Dalton, S., Olson, L.N.: Exposing fine-grained parallelism in algebraic multigrid methods. SIAM J. Sci. Comput. **34**(4), C123–C152 (2012)
4. Brandt, A., Brannick, J., Kahl, K., Livshits, I.: Bootstrap AMG. SIAM J. Sci. Comput. **33**(2), 612–632 (2011)
5. Brezina, M., Falgout, R., MacLachlan, S., Manteuffel, T., McCormick, S., Ruge, J.: Adaptive smoothed aggregation ($\alpha$ sa) multigrid. SIAM Rev. **47**(2), 317–346 (2005)
6. Brezina, M., Falgout, R., MacLachlan, S., Manteuffel, T., McCormick, S., Ruge, J.: Adaptive algebraic multigrid. SIAM J. Sci. Com. **27**(4), 1261–1286 (2006)
7. Charrier, D.E., Hazelwood, B., Kudryavtsev, A., Moskovsky, A., Tutlyaeva, E., Weinzierl, T.: Studies on the energy and deep memory behaviour of a cache-oblivious, task-based hyperbolic PDE solver. arXiv preprint arXiv:1810.03940 (2018)
8. Cleary, A.J., Falgout, R.D., Henson, V.E., Jones, J.E.: Coarse-grid selection for parallel algebraic multigrid. In: Ferreira, A., Rolim, J., Simon, H., Teng, S.-H. (eds.) IRREGULAR 1998. LNCS, vol. 1457, pp. 104–115. Springer, Heidelberg (1998). https://doi.org/10.1007/BFb0018531

9. Clevenger, T.C., Heister, T., Kanschat, G., Kronbichler, M.: A flexible, parallel, adaptive geometric multigrid method for fem. arXiv preprint arXiv:1904.03317 (2019)
10. Dendy, J.E.: Black box multigrid. J. Comput. Phys. **48**(3), 366–386 (1982)
11. Dick, C., Georgii, J., Westermann, R.: A hexahedral multigrid approach for simulating cuts in deformable objects. IEEE Trans. Visual. Comput. Graph. **17**(11), 1663–1675 (2011)
12. Dongarra, J., Hittinger, J., et al.: Applied Mathematics Research for Exascale Computing. http://www.netlib.org/utk/people/JackDongarra/PAPERS/doe-exascale-math-report.pdf. Accessed 14 Oct 2019
13. Haber, E., Heldmann, S.: An octree multigrid method for quasi-static maxwell's equations with highly discontinuous coefficients. J. Comput. Phys. **223**(2), 783–796 (2007)
14. Microway: Detailed specifications of the intel xeon e5–2600v4 broadwell-ep processors. https://www.microway.com/knowledge-center-articles/detailed-specifications-of-the-intel-xeon-e5-2600v4-broadwell-ep-processors/. Accessed 14 Oct 2019
15. Murray, C.D., Weinzierl, T.: Dynamically adaptive FAS for an additively damped AFAC variant. arXiv:1903.10367 (2019)
16. Reps, B., Weinzierl, T.: Complex additive geometric multilevel solvers for Helmholtz equations on spacetrees. ACM Trans. Math. Softw. (TOMS) **44**(1), 2 (2017)
17. Sampath, R.S., Adavani, S.S., Sundar, H., Lashuk, I., Biros, G.: Dendro: parallel algorithms for multigrid and AMR methods on 2:1 balanced octrees. In: Proceedings of the 2008 ACM/IEEE conference on Supercomputing, p. 18. IEEE Press (2008)
18. Sampath, R.S., Biros, G.: A parallel geometric multigrid method for finite elements on octree meshes. SIAM J. Sci. Comput. **32**(3), 1361–1392 (2010)
19. Sundar, H., Biros, G., Burstedde, C., Rudi, J., Ghattas, O., Stadler, G.: Parallel geometric-algebraic multigrid on unstructured forests of octrees. In: Proceedings of the International Conference on High Performance Computing, Networking, Storage and Analysis, p. 43. IEEE Computer Society Press (2012)
20. Weinzierl, M., Weinzierl, T.: Quasi-matrix-free hybrid multigrid on dynamically adaptive Cartesian grids. ACM Trans. Math. Softw. **44**(3), 32:1–32:44 (2018)
21. Weinzierl, T.: The peano software–parallel, automaton-based, dynamically adaptive grid traversals. ACM Trans. Math. Softw. (TOMS) **45**(2), 14 (2019)

# High Performance Tensor–Vector Multiplication on Shared-Memory Systems

Filip Pawłowski[1,2(✉)], Bora Uçar[2,3]🆔, and Albert-Jan Yzelman[1]

[1] Huawei Technologies France,
20 Quai du Point du Jour, 92100 Boulogne-Billancourt, France
{filip.pawlowski1,albertjan.yzelman}@huawei.com
[2] ENS Lyon, Lyon, France
filip.pawlowski@ens-lyon.fr
[3] CNRS and LIP (UMR5668, CNRS - ENS Lyon - UCB Lyon 1 - INRIA),
Lyon, France
bora.ucar@ens-lyon.fr

**Abstract.** Tensor–vector multiplication is one of the core components in tensor computations. We have recently investigated high performance, single core implementation of this bandwidth-bound operation. Here, we investigate its efficient, shared-memory implementations. Upon carefully analyzing the design space, we implement a number of alternatives using OpenMP and compare them experimentally. Experimental results on up to 8 socket systems show near peak performance for the proposed algorithms.

**Keywords:** Tensor computations · Tensor–vector multiplication · Shared-memory systems

## 1 Introduction

Tensor–vector multiply (*TVM*) operation, along with its higher level analogues tensor–matrix (*TMM*) and tensor–tensor multiplies (*TTM*) are the building blocks of many algorithms [1]. These operations are applied to a given mode (or dimension), or to given modes (in the case of *TTM*). Among these, *TVM* is the most bandwidth-bound. Recently, we have investigated this operation on single core systems, and proposed data structures and algorithms to achieve high performance and mode-oblivious behavior [10]. While high performance is a common term in the close by area of matrix computations, mode-obliviousness is mostly related to tensor computations. It requires that a given algorithm for a core operation (e.g., *TVM*) should have more or less the same performance no matter which mode it is applied to. In matrix terms, this corresponds to having the same performance in computing matrix–vector and matrix–transpose–vector multiplies. Our aim in this work is to develop high performance and mode oblivious parallel *TVM* algorithms on shared-memory systems.

© Springer Nature Switzerland AG 2020
R. Wyrzykowski et al. (Eds.): PPAM 2019, LNCS 12043, pp. 38–48, 2020.
https://doi.org/10.1007/978-3-030-43229-4_4

Let $\mathcal{A}$ be a tensor with $d$ modes, or for our purposes in this paper, a $d$-dimensional array. The $k$-mode tensor–vector multiplication produces another tensor whose $k$th mode is of size one. More formally, for $\mathcal{A} \in \mathbb{R}^{n_1 \times n_2 \times \cdots \times n_d}$ and $\mathbf{x} \in \mathbb{R}^{n_k}$, the $k$-mode $TVM$ operation $\mathcal{Y} = \mathcal{A} \times_k \mathbf{x}$ is defined as

$$y_{i_1,\ldots,i_{k-1},1,i_{k+1},\ldots,i_d} = \sum_{i_k=1}^{n_k} a_{i_1,\ldots,i_{k-1},i_k,i_{k+1},\ldots,i_d} x_{i_k},$$

for all $i_j \in \{1,\ldots,n_j\}$ with $j \in \{1,\ldots,d\}$, where $y_{i_1,\ldots,i_{k-1},1,i_{k+1},\ldots,i_d}$ is an element of $\mathcal{Y}$, and $a_{i_1,\ldots,i_{k-1},i_k,i_{k+1},\ldots,i_d}$ is an element of $\mathcal{A}$. The output tensor $\mathcal{Y} \in \mathbb{R}^{n_1 \times \cdots \times n_{k-1} \times 1 \times n_{k+1} \times \cdots \times n_d}$ is $d-1$-dimensional. That is why one can also state that the $k$-mode $TVM$ contracts a $d$-dimensional tensor along mode $k$ and forms a $d-1$-dimensional tensor. Let $n = \prod_{i=1}^{d} n_i$. Then, a $k$-mode $TVM$ performs $2n$ flops on $n + n/n_k + n_k$ data elements, and thus has arithmetic intensity of $\frac{2n}{n+n/n_k+n_k}$ flop per word, which is between 1 and 2. This amounts to a heavily bandwidth-bound computation even for sequential execution [10]. The multi-threaded case is even more challenging, as cores on a single socket compete for the local memory bandwidth.

We proposed [10] a blocking approach for obtaining efficient, mode-oblivious tensor computations by investigating the case of tensor–vector multiplication in the single core setting. Earlier approaches to related operations unfold the tensor (reorganize the whole tensor in the memory), and carry out the overall operation using a single matrix–matrix multiplication [5]. Recent alternatives [6] instead propose a parallel loop-based algorithm: a loop of the BLAS3 kernels which operate in-place on parts of the tensor such that no unfold is required, which we adopt for $TVM$. Other related work targets more complex operations [2] (called MTTKRP), and tensor–tensor multiplication [7,11,12]. Our $TVM$ routines address a special case of $TMM$, which is a special case of $TTM$, based on our earlier work [10]. Apart from not explicitly considering $TVM$, these do not adapt the tensor layout. Kjolstad et al. [4] propose The Tensor Algebra Compiler (taco) for tensor computations, which generates straightforward for-loops in our case.

We list the notation in Sect. 2, and provide a background on blocking algorithms we proposed earlier for sequential high performance. Section 3 contains $TVM$ algorithms whose analyses are presented in Sect. 3.3. Section 4 contains experiments on up to 8-socket 120 core systems. A deeper analysis of algorithms and experiments appears in the accompanying technical report [9], which we refer to for the sake of brevity.

## 2 Notation and Background

### 2.1 Notation

We use mostly the standard notation [5] (the full list of symbols is given in Table 1 in the technical report [9]). $\mathcal{A}$ is an order-$d$, or a $d$-dimensional tensor.

$\mathcal{A} \in \mathbb{R}^{n_1 \times n_2 \times \cdots \times n_d}$ has size $n_k$ in mode $k \in \{1, \ldots, d\}$. $\mathcal{Y}$ is a $(d-1)$-dimensional tensor obtained by multiplying $\mathcal{A}$ along a given mode $k$ with a suitably sized vector $\mathbf{x}$. Matrices are represented using boldface capital letters; vectors using boldface lowercase letters; and elements in them are represented by lowercase letters with subscripts for each dimension. When a subtensor, matrix, vector, or an element of a higher order object is referred, we retain the name of the parent object. For example $a_{i,j,k}$ is an element of the tensor $\mathcal{A}$. We use Matlab column notation for denoting all indices in a mode. For $k \in \{1, \ldots, d\}$, we use $I_k = \{1, \ldots, n_k\}$ to denote the index set for the mode $k$. We also use $n = \Pi_{i=1}^{d} n_i$ to refer to the total number of elements in $\mathcal{A}$. Likewise, $I = I_1 \times I_2 \times \cdots \times I_d$ is the Cartesian product of all index sets, whose elements are marked with boldface letters $\mathbf{i}$ and $\mathbf{j}$. A mode-$k$ fiber $\mathbf{a}_{i_1, \ldots, i_{k-1}, :, i_{k+1}, \ldots, i_d}$ in a tensor is obtained by fixing the indices in all modes except mode $k$. A hyperslice is obtained by fixing one of the indices, and varying all others. In third order tensors, a hyperslice become a slice, and therefore, a matrix. For example, $\mathbf{A}_{i,:,:}$ is the $i$th mode-1 slice of $\mathcal{A}$.

## 2.2   Sequential *TVM* and Dense Tensor Memory Layouts

We parallelize the *TVM* by distributing the input tensor between the physical cores of a shared-memory machine, while adopting the tensor layouts and *TVM* kernels from our earlier work [10], summarized below.

A layout $\rho$ maps tensor elements onto an array of size $n = \Pi_{i=1}^{d} n_i$. Let $\rho_\pi(\mathcal{A})$ be a layout, and $\pi$ an ordering (permutation) of $(1, \ldots, d)$ such that $\rho_\pi(\mathcal{A}) : (i_1, \ldots, i_d) \mapsto \sum_{k=1}^{d} \left( (i_{\pi_k} - 1) \prod_{j=k+1}^{d} n_{\pi_j} \right) + 1$, with the convention that $\prod_{j=k+1}^{d} \cdot = 1$ for $k = d$. The regularity of this layout allows such tensors be processed using BLAS in a loop without explicit tensor unfolds. Let $\rho_Z(\mathcal{A})$ be a Morton layout defined by the space-filling Morton order [8]. Such layout improves performance on systems with multi-level caches due to the locality preserving properties of the Morton order. However, $\rho_Z(\mathcal{A})$ is an irregular layout, and thus unsuitable for processing with BLAS routines.

Blocking is a well-known technique for improving data locality. A blocked tensor consists of blocks $\mathcal{A}_j \in \mathbb{R}^{b_1 \times \cdots \times b_d}$, where $j \in \{1, \ldots, \prod_{i=1}^{d} a_i\}$, and $n_k = a_k b_k$ for all modes $k$. We previously introduced a $\rho_Z \rho_\pi$ blocked layout which organizes elements into blocks, and uses $\rho_Z$ to order the blocks in memory, and $\rho_\pi$ to order the elements in individual blocks [10]. By using the regular layout at the lower level, we can use BLAS routines for processing the individual blocks, while benefiting from the properties of the Morton order (increased data reuse between blocks, and mode-oblivious performance).

## 3   Shared-Memory Parallel *TVM* Algorithms and Analysis

We assume a shared-memory architecture consisting of $p_s$ connected processors. Each processor supports running $p_t$ threads for a total of $p = p_s p_t$ threads. The set of all possible thread IDs is $P = \{1, \ldots, p\}$. Each processor has local memory which can be accessed faster than remote memory areas. We assume threads taking part in a parallel $TVM$ computation are *pinned* to a specific core, meaning that threads will not move from one core to another while a $TVM$ is executed. The pinning of the threads entails the notion of *explicit* versus *interleaved* memory use (see Section 3.1 of the accompanying technical report [9]).

A distribution of an order-$d$ tensor of size $n_1 \times \cdots \times n_d$ over $p$ threads is given by a map $\pi : I \to \{1, \ldots, p\}$. Let $\pi_{1D}$ be a regular 1D *block distribution* such that $\pi_{1D}(\mathcal{A}) : (i_1, i_2 \ldots, i_d) \mapsto \lfloor (i_1 - 1)/b_{1D} \rfloor + 1$, where *block size* $b_{1D} = \lceil n_1/p \rceil$ refers to the number of hyperslices. Let $m_s = |\pi_{1D}^{-1}(s)|$ count the number of elements local to thread $s$. We demand that a 1D distribution be *load-balanced*, $\max_{s \in P} m_s - \min_{s \in P} m_s \le n/n_1$. The choices to distribute over the first mode and to use a block distribution are without loss of generality (see Section 3.1 in the report [9]).

### 3.1  Baseline: LoopedBLAS

We assume $\mathcal{A}$ and $\mathcal{Y}$ have the default unfold layout. The $TVM$ operation could naively be written using $d$ nested for-loops, where the outermost loop that does not equal the mode $k$ of the $TVM$ is executed concurrently using OpenMP; such code is generated by taco. For a better performing parallel baseline, however, we observe that the $d - k$ inner for-loops correspond to a dense matrix–vector multiplication if $k < d$; we can thus write the parallel $TVM$ as a loop over BLAS-2 calls, and use highly optimized libraries for their execution. For $k = d$, the naively nested for-loops actually correspond to a dense matrix–transpose–vector multiplication, which is a standard BLAS-2 call as well.

We execute the loop over the BLAS-2 calls in parallel using OpenMP. For $k = d$, and for smaller tensors, this may not expose enough parallelism to make use of all available threads; we use any such left-over threads to parallelize the BLAS-2 calls themselves, while taking care that threads collaborating on the same BLAS-2 call are pinned close to each other to exploit shared caches as much as possible. Since all threads access both the input tensor and input vector, and since it cannot be predicted which thread accesses which part of the output tensor, all memory areas corresponding to $\mathcal{A}$, $\mathcal{Y}$, and $\mathbf{x}$ must be interleaved. We refer to the described algorithm as *loopedBLAS*.

### 3.2  Proposed 1D $TVM$ Algorithms

We explore a family of algorithms assuming the $\pi_{1D}$ distribution of the input and output tensors, thus resulting in $p$ disjoint input tensors $\mathcal{A}_s$ and $p$ disjoint output tensors $\mathcal{Y}_s$ where each of their unions correspond to $\mathcal{A}$ and $\mathcal{Y}$, respectively. For all but $k = 1$, a parallel $TVM$ amounts to a thread-local call to a sequential $TVM$ computing $\mathcal{Y}_s = \mathcal{A}_s \times_k \mathbf{x}$; each thread reads from its own part of $\mathcal{A}$ while writing to its own part of $\mathcal{Y}$. We may thus employ the $\rho_Z \rho_\pi$ layout for $\mathcal{A}_s$ and $\mathcal{Y}_s$ and use its high-performance sequential mode-oblivious kernel [10]; here, $\mathbf{x}$ is allocated interleaved while $\mathcal{A}_s$ and $\mathcal{Y}_s$ are explicit. The global tensors $\mathcal{A}$ and

$\mathcal{Y}$ are never materialized in shared-memory—only their distributed variants are required. We expect the explicit allocation of these two largest data entities involved with the $TVM$ computation to induce much better parallel efficiency compared to the *loopedBLAS* baseline where all data is interleaved.

For $k = 1$, the output tensor $\mathcal{Y}$ cannot be distributed. We define that $\mathcal{Y}$ is then instead subject to a 1D block distribution over mode 2, and assume $n_2 \geq p$. Since the distributions of $\mathcal{A}$ and $\mathcal{Y}$ then do not match, communication ensues. We suggest three variants that minimize data movement, characterized by the number of synchronization barriers they require: zero, one, or $p-1$. Before describing these variants, we first motivate why it is sufficient to only consider one-dimensional partitionings of $\mathcal{A}$.

Assume a tensor of size $n = \prod_{k=1}^{d} n_k$, with $n_i \geq n_{i+1}$ for $i = 1, \ldots, d-1$, and $n_1 \geq p > 1$. Consider a series of $d$ $TVMs$, $\mathcal{Y}_k = \mathcal{A} \times_k \mathbf{v}_k$, for all modes $k \in \{1, \ldots, d\}$. Assume *any* load-balanced distribution $\pi$ of $\mathcal{A}$ and $\mathcal{Y}$ such that thread $s$ has at most $2d\lceil n_1/p \rceil n/n_1$ work. For any $\mathbf{i} \in I$, the distribution $\pi$ defines which thread multiplies the input tensor element $a_\mathbf{i}$ with its corresponding input vector element $x_{i_k}$. The thread(s) in $\pi(i_1, \ldots, i_{k-1}, I_k, i_{k+1}, \ldots, i_d)$ are said to *contribute* to the reduction of $y_\mathbf{j}$, where $\mathbf{j} = (i_1, \ldots, i_{k-1}, 1, i_{k+1}, \ldots, i_d)$, as they perform local reductions of multiplicands to the same element $y_\mathbf{j}$. We do not assume a specific reduction algorithm and count the minimal work involved.

For any $\mathbf{i} \in I$, let $J_\mathbf{i} = \{\mathbf{j} \in I \mid \vee_{k=1}^{d} i_k = j_k\}$ be the set of elements lying on $d$ different axes which all go through $\mathbf{i}$, as illustrated in Fig. 1 (left). Let $X_\mathbf{i} = \pi(J_\mathbf{i})$, where $\pi$ is any distribution, describe the set of threads to which elements in $J_\mathbf{i}$ are mapped. Should $|X_\mathbf{i}| > 1$ for all $\mathbf{i} \in I$, then there is at least one $TVM$ for which all elements of $\mathcal{Y}$ are involved in a reduction, as at least two threads contribute to $y_\mathbf{j}$. For a 1D distribution, this amounts to $n/n_1$ reductions, occurring only for mode 1, which shows that this lower bound on communication complexity for a series of $TVMs$ is attainable. We will now consider if we can do better by allowing $\mathbf{i}$ for which $|X_\mathbf{i}| = 1$, and if so, by how much.

Suppose there exist $r = \prod_{k=1}^{d} r_k$ coordinates $\mathbf{i} \in I$ such that $X_\mathbf{i} = \{s\}$, which form a hyper-rectangular subtensor $\mathcal{B}$ of side length $r_k < n_k$ contained in $\mathcal{A}$, as in Fig. 1 (right). We choose a hyper-rectangular shape, so that the $r$ elements create the minimum amount of redundant work. Since $|X_\mathbf{i}| = 1$, the number of coordinates which must then also lie on thread $s$ is $r(\sum_{k=1}^{d} n_k/r_k - (d-1))$. If $r_k = 2^{1/(d-1)} n_k / p^{1/(d-1)}$, this already corresponds to a load exceeding the assumed load balance $(2n - n/n_1)/p$. Furthermore, with $r = 2^{d/(d-1)} n / p^{d/(d-1)}$ such coordinates, the lower bound on communication complexity may only be reduced to $n/n_1(1 - 2/p)$, where $r/r_1 = 2n/pn_1$ is the projection of the cube $r$ onto the $d-1$-dimensional output tensor. The data movement on the input vector is at most $(d-1)n_1$, which typically is significantly less than the data movement associated with the output tensor. Thus, the $\pi_{1D}$ distribution is asymptotically optimal when $n/n_1 \gg (d-1)n_1$ and $d > 2$.

In the following, we discuss two 1D algorithms, while our accompanying technical report [9, Section 3.3] contains three more.

**Fig. 1.** Illustrations of elements in $J_{\mathbf{i}}$, indicated via thick gray lines, for an arbitrarily chosen $\mathbf{i}$ depicted by a filled dot (left), and for a cube of $r$ elements $\mathbf{i}$ (right).

**0-sync.** We avoid performing a reduction on $\mathcal{Y}$ for $k = 1$ by storing $\mathcal{A}$ twice; once with a 1D distribution over mode 1, another time using a 1D distribution over mode $d$. Although the storage requirement is doubled, data movement remains minimal while explicit reduction for $k = 1$ is completely eliminated, since the copy with the 1D distribution over mode $d$ can then be used without penalty. In either case, the parallel *TVM* computation completes after a sequential thread-local *TVM*; this variant requires no barriers to resolve data dependencies.

**q-sync.** This variant stores $\mathcal{A}$ with a 1D distribution over mode 1. It also stores two versions of the output tensor, one interleaved $\mathcal{Y}$ and one thread-local $\mathcal{Y}_s$. The vector $\mathbf{x}$ is interleaved. Both $\mathcal{A}_s$ and $\mathcal{Y}_s$ are split into $q = \prod_{i=2}^{d} q_i \geq p$ parts, by splitting each object into $q_i$ parts across mode $i$. We index the resulting objects as $\mathcal{A}_{s,t}$, which are explicitly allocated to thread $s$, and $\mathcal{Y}_{s,t}$, which are both allocated as explicit and interleaved. The input vector $\mathbf{x}$ remains interleaved. The algorithm is seen below.

```
1: if k = 1 then
2:     Y = A_{s,s} ×_k x
3:     for t = 2 to q do
4:         barrier
5:         Y += A_{s,(t+s−1) mod q+1} ×_k x
6: else
7:     for t = 1 to q do
8:         Y_{s,t} += A_{s,t} ×_k x
```

If this algorithm is to re-use output of mode-0 *TVM*, then, similarly to the 0-sync variant each thread must re-synchronize its local $\mathcal{Y}_{s,t}$ with $\mathcal{Y}$. Thus, unless the need explicitly arises, implementations need not distribute $\mathcal{Y}$ over $n_2$ as part of a mode-1 *TVM* (at the cost of interleaved data movement on $\mathcal{Y}$).

### 3.3  Analysis

We investigate the amount of data moved during a *TVM* computation, mode-obliviousness, memory, and work. We divide data movement into intra-socket data movement (where cores contend for resources) and inter-socket data movement (where data is moved over a communication bus, instead of only to and from local memory). For quantifying data movement we assume perfect caching, meaning that all required data elements are touched exactly once. Since *TVM* is bandwidth bound, we consider memory overhead and efficiency versus the sequential memory requirement. Once we quantify algorithm properties in each of these five dimensions, we consider their *iso-efficiencies* [3]. Table 1 gives the summary, while the technical report contains an in-depth analysis [9, Section 4].

The *loopedBLAS* algorithm, thanks to interleaving, is both memory- and work-optimal. It does not include any cache-oblivious nor mode-oblivious optimizations, and has no barriers. Since all memory used is interleaved, the effective bandwidth is spread over intra and inter-socket bandwidth proportional to the number of CPU sockets $p_s$. Thus, assuming a balanced work distribution, its overhead $\mathcal{O}((p_s - 1)/p_s n(h - g))$ becomes $\Theta(n(h - g))$ as $p_s$ increases. For $p_s = 1$ the overhead is $\Theta(p_t n_k g)$, which excludes any underlying overhead of its parallel implementation. The 0-sync algorithm is work optimal, incurs $n$ words of extra storage (not memory optimal), and has no barriers. It fully exploits the cache- and mode-oblivious optimizations from our earlier work. The overhead of 0-sync is bounded by $\Theta(p n_k h)$ for $p_s > 1$, a significant improvement over *loopedBLAS*. The $q$-sync algorithm is work optimal, but not memory optimal as it stores $\mathcal{Y}$ twice. However, it improves upon 0-sync's overhead.

**Table 1.** Overheads of different *TVM* algorithms, and the allocation mode of $\mathcal{A}, \mathcal{Y}$, and $\mathbf{x}$. Optimal overheads are in **bold**. We display the worst-case asymptotics, i.e., assuming $p_s > 1$ and the worst-case $k$ for non mode-oblivious algorithms. Intra-socket throughput $g$ and the inter-socket throughput $h$ are in seconds per word (per socket), threads' compute speed is in $r$ seconds per flop, and a barrier completes in $L$ seconds.

| Method | Work | Memory | Movement | Barrier | Oblivious | Implicit | Explicit | $k$ |
|---|---|---|---|---|---|---|---|---|
| *loopedBLAS* | **0** | **0** | $n(h - g)$ | **0** | None | $\mathbf{x}, \mathcal{A}, \mathcal{Y}$ | - | - |
| 0-sync | **0** | $n$ | $\mathbf{p n_1 h + p_t n_1 g}$ | **0** | Full | $\mathbf{x}$ | $\mathcal{A}, \mathcal{A}, \mathcal{Y}$ | - |
| $q$-sync | **0** | $n/n_d$ | $pn/n_d h + p_t n/n_d g$ | $p^2 L$ | Good | $\mathbf{x}, \mathcal{Y}$ | $\mathcal{A}, \mathcal{Y}$ | 1 |

The *loopedBLAS* algorithm is highly sensitive to the mode $k$ of the *TVM*, while the algorithms based on the $\rho_Z \rho_\pi$ tensor layout are not [10]. The 0-sync variant exploits the $\rho_Z \rho_\pi$ maximally; thus, it is fully mode-oblivious. In the $q$-sync variant, $\mathcal{A}_s$ are split into $q$ parts, and each part is stored using a $\rho_Z \rho_\pi$ layout, which implies an overhead of $q - 1$ space-filling curves. Furthermore, in the worst-case, each element of $\mathcal{Y}$ is touched $p - 1$ times more than in a 0-sync variant, which hurts both cache efficiency and mode-obliviousness.

For *loopedBLAS*, efficiency is constant if $g/(h-g)$ decreases while $p_s$ increases, which does not scale. The 0-sync attains efficiency when $p$ grows linearly with $n/n_k$. The $q$-sync algorithm attains iso-efficiency when $p$ grows linearly with $n_k$.

## 4    Experiments

We run our experiments on three Intel Ivy Bridge nodes, described in Table 2. In the paper, the terms KB, MB, and GB denote $2^{10}$, $2^{20}$, and $2^{30}$ bytes, respectively. We do not use hyperthreading and limit the tests to at most $p/2$ threads equal the number of cores (each core supports 2 hyperthreads). We measure the maximum bandwidth of the systems with the STREAM benchmark, and report the maximum measured performance. The system uses CentOS 7 with Linux kernel 3.10.0 and software is compiled with GCC version 6.1. We use Intel MKL version 2018.2.199 for *loopedBLAS*. For algorithms based on blocked layouts (0- and q-sync), we run with LIBXSMM version 1.9-864 and Intel MKL, and retain the faster result. We conduct 10 experiments for each combination of dimension, mode, and algorithm and report the average performance (the effective bandwidth, GB/s) and its standard deviation among the modes.

We compare the synchronization-optimal *loopedBLAS*, the work- and communication-optimal 0-sync, and the work-optimal q-sync. We benchmark tensors of order 2 up to 5. We choose $n$ such that the combined input and output memory areas during a single *TVM* call have a total size of at least 10 GBs. The exact tensor sizes and block sizes are given in Tables 3 and 4, respectively. The block sizes selected ensure that computing a *TVM* on a block fits the L3 cache. This combination of tensor and block sizes ensures all algorithms run with perfect load balance and without requiring any padding of blocks. We additionally kept the sizes of tensors equal through all pairs of $(d, p_s)$, which enables comparison of different algorithms within the same $d$ and $p_s$.

**Table 2.** Machine configurations used. Nodes 1 and 2 use a quad-channel and node 3 uses an octa-channel memory configuration. Each processor has 32 KB of L1 and 256 KB of L2 cache per core, and $1.25p_t$ MB of L3 cache shared amongst the cores.

| | | | | | | Bandwidth | |
| Node | CPU (clock speed) | $p_s$ | $p_t$ | $p$ | Memory size (clock speed) | STREAM | Theoretical |
|---|---|---|---|---|---|---|---|
| 1 | E5-2690 v2 (3 GHz) | 2 | 20 | 40 | 256 GB (1600 MHz) | 76.7 GB/s | 95.37 GB/s |
| 2 | E7-4890 v2 (2.8 GHz) | 4 | 30 | 120 | 512 GB (1333 MHz) | 133.6 GB/s | 158.91 GB/s |
| 3 | E7-8890 v2 (2.8 GHz) | 8 | 30 | 240 | 2048 GB (1333 MHz) | 441.9 GB/s | 635.62 GB/s |

**Table 3.** Tensor sizes $n_1 \times \cdots \times n_d$ per tensor-order $d$ and node. The exact size in GBs is given in parentheses.

| $d$ | Node 1 | Node 2 | Node 3 |
|---|---|---|---|
| 2 | $45600 \times 45600$ (15.49) | $68400 \times 68400$ (34.86) | $136800 \times 136800$ (139.43) |
| 3 | $1360 \times 1360 \times 1360$ (18.74) | $4080 \times 680 \times 4080$ (84.34) | $4080 \times 680 \times 4080$ (84.34) |
| 4 | $440 \times 110 \times 88 \times 440$ (13.96) | $1320 \times 110 \times 132 \times 720$ (102.81) | $1440 \times 110 \times 66 \times 1440$ (112.16) |
| 5 | $240 \times 60 \times 36 \times 24 \times 240$ (22.25) | $720 \times 60 \times 36 \times 24 \times 360$ (100.11) | $720 \times 50 \times 36 \times 20 \times 720$ (139.05) |

## 4.1  Single-Socket Results

Table 5 shows the results for the single-socket of node 1. Here, all memory regions are allocated locally. As *loopedBLAS* relies on the unfold storage and requires a loop over subtensors for modes 1 and $d$, no for-loop parallelization is possible for these modes, and MKL parallelization is used instead. Its performance is

**Table 4.** Block sizes $b_1 \times \cdots \times b_d$ per tensor-order $d$ and node. Sizes are chosen such that all elements of a single block can be stored in L3 cache.

| $d$ | Node 1 | Node 2 | Node 3 |
|---|---|---|---|
| 2 | $570 \times 570$ | $570 \times 570$ | $570 \times 570$ |
| 3 | $68 \times 68 \times 68$ | $68 \times 68 \times 68$ | $34 \times 68 \times 34$ |
| 4 | $22 \times 22 \times 22 \times 22$ | $22 \times 22 \times 22 \times 12$ | $12 \times 22 \times 22 \times 12$ |
| 5 | $12 \times 12 \times 12 \times 12 \times 12$ | $12 \times 12 \times 12 \times 12 \times 6$ | $6 \times 10 \times 12 \times 10 \times 6$ |

highly mode-dependent, and thus it is outperformed by the algorithms based on $\rho_Z \rho_\pi$-storage. The block Morton order storage transfers the mode-obliviousness to parallel *TVMs* (the standard deviation oscillates within 1%).

**Table 5.** Average effective bandwidth (in GB/s) and relative standard deviation (in % of the average) over all possible $k \in \{1, \ldots, d\}$ of algorithms on a single-socket of node 1. The highest bandwidth and lowest standard deviation for each $d$ are in **bold**.

| | Average performance | | | Sample stddev. | | |
|---|---|---|---|---|---|---|
| $d$ | *loopedBLAS* | 0-sync | $q$-sync | *loopedBLAS* | 0-sync | $q$-sync |
| 2 | 40.23 | 42.28 | **42.54** | 0.63 | **0.55** | 0.65 |
| 3 | 36.43 | 39.34 | **39.87** | 24.93 | 2.55 | **2.50** |
| 4 | 37.63 | 39.02 | **39.05** | 21.29 | **4.35** | 4.40 |
| 5 | 34.56 | 36.53 | **36.65** | 22.43 | 5.14 | **4.26** |

## 4.2   Inter-socket Results

Table 6 shows results on the compute nodes for tensors of order-3 and 5 (the accompanying report [9] contains results for order-2 and 4 as well). These runtime results show a lack of scalability of *loopedBLAS*. This is due to the data structures being interleaved instead of making use of a 1D distribution. Interleaving or not only matters for multi-socket results, but since Table 5 conclusively shows that approaches based on our $\rho_Z \rho_\pi$-storage remain superior on single sockets, we conclude that our approach is superior at all scales. The performance drops slightly with the increasing $d$ for all variants. This is inherent to the BLAS libraries handling matrices with a lower row-to-column ratio better than tall-skinny or short-wide matrices [10].

As 0-sync does not require any synchronization for $k = 1$, it achieves the lowest standard deviation. Thus, for *TVMs* of mode 1, the 0-sync algorithm slightly outperforms the $q$-sync, while they achieve almost identical performance for all the other modes. Some results are faster than STREAM benchmark, as the output tensor fits the combined L3 size and enables super-linear speedup.

**Table 6.** Average effective bandwidth (in GB/s) and relative standard deviation (in % of average) over all possible $k \in \{1, \ldots, d\}$ of order-3 and -5 tensors of algorithms executed on different nodes (2 socket node 1, 4 socket node 2, and 8-socket node 3). The highest bandwidth and lowest standard deviation for different $d$ are stated in **bold**.

| | Order-3 | | | | | | Order-5 | | | | | |
|---|---|---|---|---|---|---|---|---|---|---|---|---|
| | Sample stddev. | | | Avg. performance | | | Sample stddev. | | | Avg. performance | | |
| Algorithm/$p_s$ | 2 | 4 | 8 | 2 | 4 | 8 | 2 | 4 | 8 | 2 | 4 | 8 |
| *loopedBLAS* | 9.57 | 16.52 | 23.05 | 63.89 | 55.68 | 13.66 | 15.37 | 19.70 | 32.03 | 56.11 | 54.04 | 12.43 |
| 0-sync | 2.80 | **1.38** | **3.42** | **77.06** | **145.07** | **467.31** | 3.47 | **5.01** | **5.02** | **71.71** | **129.80** | **421.98** |
| *q*-sync | **1.90** | 3.86 | 6.56 | 76.27 | 143.17 | 441.65 | 4.17 | 9.37 | 14.83 | 71.65 | 129.60 | 397.25 |

When passing from 4 to 8 processors the increase may be higher than twofold, due to the octa-channel memory on node 3. Overall, our measured performances are within the impressive range of 75–88%, 81–95%, and 66–77% of theoretical peak performance for node 1, 2, and 3, respectively.

Table 7 displays the parallel efficiency on three different nodes versus the performance of the *q*-sync on a single socket. Each node takes its own baseline since the tensor sizes differ between nodes as per Table 3; one can thus only compare parallel efficiencies over the *columns* of these tables. We compare algorithms, and do not investigate inter-socket scalability. The efficiencies larger than one are commonly due to cache-effects; here, output tensors fit in the combined cache of eight CPUs, but did not fit in cache of a single CPU. These tests conclusively show that both 0- and *q*-sync algorithms scale significantly better than looped-BLAS for $p_s > 1$, resulting in up to 35x higher efficiencies (for order-4 tensors on node 3).

**Table 7.** Parallel efficiency of algorithms on order-2 up to 5 tensors executed on 3 different nodes (2 socket node 1, 4 socket node 2, and 8-socket node 3), calculated against the single-socket runtime on a given node of *q*-sync algorithm on the same problem size and tensor order.

| | Order-2 | | | Order-3 | | | Order-4 | | | Order-5 | | |
|---|---|---|---|---|---|---|---|---|---|---|---|---|
| Algorithm/$p_s$ | 2 | 4 | 8 | 2 | 4 | 8 | 2 | 4 | 8 | 2 | 4 | 8 |
| *loopedBLAS* | 0.81 | 0.31 | 0.02 | 0.80 | 0.34 | 0.03 | 0.79 | 0.28 | 0.03 | 0.77 | 0.32 | 0.05 |
| 0-sync | 0.99 | 0.93 | 0.98 | 0.97 | 0.88 | 0.96 | 0.99 | 0.83 | 1.05 | 0.98 | 0.76 | 1.53 |
| *q*-sync | 0.99 | 0.93 | 0.97 | 0.96 | 0.87 | 0.91 | 0.98 | 0.82 | 1.00 | 0.98 | 0.76 | 1.44 |

## 5 Conclusions

We investigate the tensor–vector multiplication operation on shared-memory systems. Building on an earlier work, where we developed blocked and mode-oblivious layouts for tensors, we explore the design space of parallel shared-memory algorithms based on this same mode-oblivious layout, and propose several parallel algorithms. After analyzing those for work, memory, intra- and inter-socket data movement, the number of barriers, and mode obliviousness, we choose to implement two of them. These algorithms, called 0-sync and $q$-sync, deliver close to peak performance on three different systems, using 2, 4, and 8 sockets, and surpass a baseline algorithm based on looped BLAS calls that we optimized. For future work, we plan to investigate the use of the proposed algorithms in distributed memory systems.

# References

1. Bader, B.W., Kolda, T.G.: Algorithm 862: MATLAB tensor classes for fast algorithm prototyping. ACM TOMS **32**(4), 635–653 (2006)
2. Ballard, G., Knight, N., Rouse, K.: Communication lower bounds for matricized tensor times Khatri-Rao product. In: 2018 IPDPS, pp. 557–567. IEEE (2018)
3. Grama, A.Y., Gupta, A., Kumar, V.: Isoefficiency: measuring the scalability of parallel algorithms and architectures. IEEE Parallel Distrib. Technol. Syst. Appl. **1**(3), 12–21 (1993)
4. Kjolstad, F., Kamil, S., Chou, S., Lugato, D., Amarasinghe, S.: The tensor algebra compiler. Proc. ACM Program. Lang. **1**(OOPSLA), 77:1–77:29 (2017)
5. Kolda, T.G., Bader, B.W.: Tensor decompositions and applications. SIAM Rev. **51**(3), 455–500 (2009)
6. Li, J., Battaglino, C., Perros, I., Sun, J., Vuduc, R.: An input-adaptive and in-place approach to dense tensor-times-matrix multiply. In: SC 2015, pp. 76:1–76:12 (2015)
7. Matthews, D.: High-performance tensor contraction without transposition. SIAM J. Sci. Comput. **40**(1), C1–C24 (2018)
8. Morton, G.M.: A computer oriented geodetic data base and a new technique in file sequencing (1966)
9. Pawłowski, F., Uçar, B., Yzelman, A.J.N.: High performance tensor-vector multiples on shared memory systems. Technical report 9274, Inria, Grenoble-Rhône-Alpes (2019)
10. Pawłowski, F., Uçar, B., Yzelman, A.N.: A multi-dimensional Morton-ordered block storage for mode-oblivious tensor computations. J. Comput. Sci. (2019). https://doi.org/10.1016/j.jocs.2019.02.007
11. Solomonik, E., Matthews, D., Hammond, J.R., Stanton, J.F., Demmel, J.: A massively parallel tensor contraction framework for coupled-cluster computations. J. Parallel Distrib. Comput. **74**(12), 3176–3190 (2014)
12. Springer, P., Bientinesi, P.: Design of a high-performance GEMM-like tensor-tensor multiplication. ACM TOMS **44**(3), 1–29 (2018)

# Efficient Modular Squaring in Binary Fields on CPU Supporting AVX and GPU

Paweł Augustynowicz[(✉)] and Andrzej Paszkiewicz

Faculty of Cybernetics, Military University of Technology,
ul. gen. Sylwestra Kaliskiego 2, 00-908 Warsaw, Poland
{pawel.augustynowicz,andrzej.paszkiewicz}@wat.edu.pl

**Abstract.** This paper deals with the acceleration of modular squaring operation in binary fields on both modern CPUs and GPUs. The key idea is based on applying bit-slicing methodology with a view to maximizing the advantage of *Single Instruction Multiple Data* (SIMD) and *Single Instruction Multiple Threads* (SIMT) execution patterns. The developed implementation of modular squaring was adjusted to testing for the irreducibility of binary polynomials of some particular forms.

**Keywords:** GPU · SIMD · Parallel algorithms

## 1 Introduction

The finite field over GF(2), also called binary field has numerous applications in cryptography, coding theory and mathematics [5,9,10]. It is an essential part of various correcting codes and therefore plays a crucial role in reliable storage systems. Currently, software implementations of binary field arithmetic are typically performed transparently by carry-less multiplication instruction. Nevertheless, there are computing platforms, such as Graphical Processing Units (GPUs), which do not offer this advanced instruction. Furthermore, carry-less multiplication instruction on modern processors cannot be executed in *Single Instruction Multiple Data* (SIMD) pattern [7]. Consequently, there exist the cases, such as the construction of a binary field, where more specialized and advanced implementations of the arithmetic are required.

The construction of a binary field involves the choice of an irreducible polynomial over GF(2), which will be a reduction polynomial for the field. A search of an irreducible polynomial of a given degree can be realized by usage of one of the irreducibility tests. The most popular ones are the Ben-Or irreducibility test [3] and the Rabin irreducibility test [12]. Both algorithms are based on two primary operations: modular exponentiation and computing greatest common divisor of two polynomials. There is a requirement to accelerate modular exponentiation on the modern computing platforms owing to the fact, that it is a time-consuming operation. This paper deals with the problem of efficient

© Springer Nature Switzerland AG 2020
R. Wyrzykowski et al. (Eds.): PPAM 2019, LNCS 12043, pp. 49–57, 2020.
https://doi.org/10.1007/978-3-030-43229-4_5

modular squaring on both CPU and GPU platforms. In both cases, a bit-slicing method was applied similarly with a view to taking advantage of highly parallel computational platforms. Developed implementations were practically used in the computational experiment of a search of high degree irreducible sedimentary polynomials [1] and trinomials [2].

## 2    Related Work

The idea of applying the SIMD execution model to binary field arithmetic was examined both on CPU [6,11] and GPU [4] platforms. In both cases, a significant gain of performance was observed. Nonetheless, performed optimizations were restricted only for individual sizes of binary fields or certain forms of modular polynomials. It is insufficient for the wide range of possible usage scenarios. There are applications such as irreducibility testing which involves a lot of modular squaring or modular exponentiation operations.

The previous fast SIMD implementations of Galois field arithmetic [6,11] were restricted to multiplication in $GF(2^w)$ for $w \in \{4, 8, 16, 32\}$ and consequently cannot be directly used for the search of irreducible binary polynomials. Even more restrictive approach is presented in [8], where the list of available fields is limited to $GF(2^{16})$. Notwithstanding the latter method can be applied for both CPU and GPU implementations. Another idea to increase the efficiency of multiplication in binary fields was to combine SIMD methodology with register cache optimization [4]. It resulted in very fast GPU implementation for $GF(2^w)$, for $w \in \{32, 64, 96, \ldots, 2048\}$. Those as mentioned earlier, optimized implementations was significantly faster than the highly popular Number Theory Library (NTL) [13].

This paper is focused on accelerating finite field modular squaring for large binary polynomials of high degree. An innovative massively parallel algorithm was developed which outperforms the previous implementations, especially the NTL library, in case of both large and small finite fields. The algorithm was highly optimized for the search of irreducible trinomials and sedimentary polynomials via the Ben-Or irreducibility test.

## 3    Bit-Slicing Modular Squaring Algorithm

The presented algorithm is based on the following main assumptions:

1. Application of bit-slicing methodology allows a single thread to perform 32/64/128/256/512 modular squaring operations in parallel.
2. Particular forms of polynomials like trinomials, pentatomials or sedimentary polynomials ought to be much more effectively processed then the dense polynomials.
3. Both CPU and GPU implementations ought to be very fast.
4. In case of GPU as many threads as possible should be involved in computations, and all threads must execute the same operations.

The first step of the algorithm is the reorganization of the computations in such a way that single thread batches $\mathcal{B}$ operations together, where $\mathcal{B} \in \{32, 64, 128, 256, 512\}$. This methodology, referred to as bit-slicing packs multiple bits for parallel execution. For the sake of completeness, a data structure to represent multiple polynomials called chunk will be introduced.

**Definition 1.** *A chunk can be defined is an $\mathcal{B} \times n$ matrix $\mathbf{M}$ of bits, which is representation of a set of $\mathcal{B}$ binary polynomials $P_j$, $i \in \{0, 1, \ldots, \mathcal{B} - 1\}$, where $\deg(P_j) \leq n - 1$. Particularly every polynomial is represented in the following manner: $P_j = \sum_{i=0}^{n-1} \mathbf{M}_i^j x^i$.*

An example representation of a chunk of polynomials in the memory is depicted at Fig. 1. Note that in the following notation operations are performed on columns, not rows.

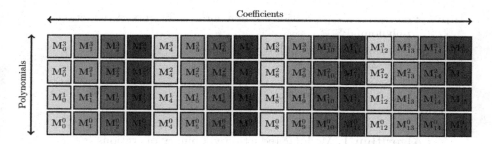

**Fig. 1.** A chunk of 4 binary polynomials of degree 15.

After the reorganization is completed, the second step of the algorithm, which is actual squaring operation can be performed. It can be effectively realized by memory transactions. Basically, the squaring is placing value 0 between every two coefficients of the binary polynomial (see Fig. 2). In the case of GPUs, the best way to handle memory efficiently is to take advantage of register cache and local memory, which are both very fast comparing to constant memory. It requires a lot of programming effort and adjusting, although results in several times faster implementation. In the case of CPUs, the operation is straightforward and does not require any further optimizations.

The last and the most time-consuming step of the modular squaring is a modular reduction (see Algorithm 1). This mathematical operation can be easily transformed into the language of binary, SIMD operations, which process all the columns at once. In the case of GPUs, the whole reduction step can be executed concurrently by different threads, whereas for CPU it is performed sequentially. As a consequence, the CPU implementation of the algorithm achieves the best results for very sparse binary polynomials such as trinomials and pentatomials, whereas the GPU one is dedicated for dense polynomials, such as sedimentary polynomials. Recall, that sedimentary polynomials are irreducible polynomials

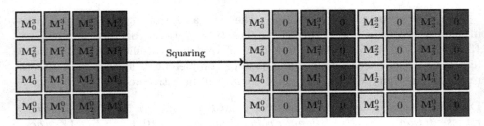

**Fig. 2.** Squaring of 4 binary polynomials of degree 3. Columns can be processed together using SIMD instructions.

of the form $x^n + g(x)$, where $g(x)$ is minimal in the sense of lexicographic order. Owing to the low degree of $g(x)$ it can be processed concurrently by the warp of threads on GPU, which results in efficiency advantage over CPU for polynomials with numerous non-zero coefficients.

**Input** : $a_0, a_1, \ldots, a_N$ - positions of non-zero coefficients of reduction
polynomial;
$\mathbf{M}_{\mathcal{B} \times 2 \cdot n}$ - chunk of polynomials;
**for** $i = 2 \cdot n - 2, \ldots, n$ **do**
$\quad$ **for** $j = 0, \ldots, N$ **do**
$\quad\quad |$ $\mathbf{M}_{i-a_N+a_j} = \mathbf{M}_i$ **xor** $\mathbf{M}_{i-a_N+a_j}$;
$\quad$ **end**
**end**

**Algorithm 1.** Modular reduction via binary operations.

## 4    Implementation and Optimization

### 4.1    CPU

The CPU implementation of the previously presented algorithm was made in two basic versions: 128-bit based on Streaming SIMD Extensions (SSE) and 256-bit based on Advanced Vector Extensions (AVX). It allows processing 128/256 coefficients of binary polynomials simultaneously. The following instructions were leveraged in the implementations:

**_mm_storeu_si128** - store 128-bits of data into memory;
**_mm_xor_si128** - compute the bitwise xor of two 128-bits integers;
**_mm256_store_si256** - store 256-bits of data into memory;
**_mm256_xor_si256** - compute the bitwise xor of two 256-bits integers.

It is possible to further extend the algorithm to process 512-bits of data at once with the use of AVX-512 instruction set. Nevertheless, the instruction set mentioned above is supported only in some Intel's processors, for instance, Xeon Phi family, which were not involved in conducted computational experiments.

Furthermore, it is strongly recommended to avoid the allocation of redundant memory due to limited cache size. The position of highest non-zero coefficients indicates the amount of required memory. It can be effortlessly determined by the **_mm_test_all_zeros** instruction. The avoidance is most advantageous in case of high degree polynomials that are not fitted into the cache.

## 4.2  GPU

The GPU implementation of the considered algorithm exploits a slightly different model of parallel execution than the CPU one. It is called *Single Instruction Multiple Threads* (SIMT) and envisages that threads are organised in warps. For the sake of clarity and completeness, the detailed algorithms modular squaring and modular reduction for GPU are depicted in Algorithms 2 and 3. Every warp within a block consist of 32 threads working simultaneously and every thread handle 32-bit instructions. In result, the $32 \times 32 = 1024$ bits are processed at once within every block. Nevertheless, the threads in the warp work with maximum efficiency if they process continuous chunks of memory. Therefore, the GPU implementation is suited only to particular forms of polynomials such as sedimentary polynomials or all one polynomials.

**Input** : $j$ - index of thread within the block;
$\qquad$ $\mathbf{M}_{\mathcal{B} \times 2 \cdot n}$ - chunk of polynomials;
**for** $i = 2 \cdot n - 2, \ldots, n$ **do**
$\quad \mid \quad \mathbf{M}_{2 \cdot j} = \mathbf{M}_j$;
**end**

**Algorithm 2.** Squaring on GPU.

Moreover, in the case of GPU implementation, it is vital to benefit from the efficiency of shared and local memory. Performing operations and manipulations on global memory are overly time-consuming and can undercut the gain from parallel processing.

**Input** : $j$ - index of thread within the block;
$\qquad$ $a_0, a_1, \ldots, a_i, \ldots a_N$ - position of non-zero coefficients of reduction polynomial;
$\qquad$ $\mathbf{M}_{\mathcal{B} \times 2 \cdot n}$ - chunk of polynomials;
**for** $i = 2 \cdot n - 2, \ldots, n$ **do**
$\quad \mid \quad \mathbf{M}_{i - a_N + a_j} = \mathbf{M}_i \ \mathbf{xor} \ \mathbf{M}_{i - a_N + a_j}$;
**end**

**Algorithm 3.** Modular reduction via binary operations on GPU.

Our experience reveals that the optimal strategy of GPU computations is to assign every block one chunk of polynomials. Applying another assignment can lead to bank conflicts which negatively influences the performance.

To summarize, GPU implementation of considered in article squaring algorithm has numerous limitations. However, there are certain forms of polynomials which are adjusted to these limitations. What is more, these forms are the worst case scenario for the CPU implementation of the algorithm mentioned above.

## 5  Performance Evaluation

The performance of the developed implementations for the most wide-spread binary fields $GF(2^w)$, for $w \in \{32, 64, 128, \ldots, 4096\}$ and particular forms of reduction polynomials was evaluated. The performance of $2^{23}$ modular squaring operations for above listed binary fields was measured and the results are presented in Tables 1, 2, 3, 4 and in Fig. 3. Each experiment used random chunks of polynomials for input and one chosen reduction polynomial of a special form. NTL version 11.3.2 was used as a reference implementation and baseline to compare achievements. It contains dedicated instruction for modular squaring operation that uses carry-less multiplication. All the computation were conducted on Intel Core i7 6700K, 4.0 GHz CPU, MSI GeForce GTX 960 4 GB GDDR5 with 8 GB of RAM.

**Table 1.** Performance evaluation for septanomials.

| Field size | NTL [s] | SSE [s] | AVX [s] | NTL/SSE | NTL/AVX | SSE/AVX |
|---|---|---|---|---|---|---|
| $2^{32}$ | 0,25 | 0,0060 | 0,0019 | 36,18 | 111 | 3,07 |
| $2^{64}$ | 0,49 | 0,0121 | 0,0039 | 36,01 | 110 | 3,07 |
| $2^{128}$ | 0,89 | 0,2458 | 0,0075 | 35,66 | 116 | 3,27 |
| $2^{256}$ | 1,81 | 0,0488 | 0,0150 | 35,88 | 116 | 3,25 |
| $2^{512}$ | 3,62 | 0,0975 | 0,0303 | 35,64 | 114 | 3,21 |
| $2^{1024}$ | 7,86 | 0,1949 | 0,0597 | 36,06 | 117 | 3,26 |
| $2^{2048}$ | 14,44 | 0,3928 | 0,1215 | 35,76 | 115 | 3,23 |
| $2^{4096}$ | 29,16 | 0,7791 | 0,2403 | 36,06 | 116 | 3,24 |

**Table 2.** Performance evaluation for pentanomials.

| Field size | NTL | SSE | AVX | NTL/SSE | NTL/AVX | SSE/AVX |
|---|---|---|---|---|---|---|
| $2^{32}$ | 0,22 | 0,0047 | 0,0008 | 46,7 | 254 | 5,45 |
| $2^{64}$ | 0,43 | 0,0095 | 0,0015 | 46,1 | 275 | 5,96 |
| $2^{128}$ | 0,87 | 0,0190 | 0,0033 | 46,0 | 264 | 5,74 |
| $2^{256}$ | 1,75 | 0,0379 | 0,0063 | 46,2 | 274 | 5,93 |
| $2^{512}$ | 3,47 | 0,0754 | 0,0132 | 46,0 | 263 | 5,71 |
| $2^{1024}$ | 7,03 | 0,1507 | 0,0259 | 46,6 | 271 | 5,81 |
| $2^{2048}$ | 14,05 | 0,3014 | 0,0517 | 46,6 | 271 | 5,82 |
| $2^{4096}$ | 28,10 | 0,6031 | 0,1040 | 46,5 | 269 | 5,79 |

**Table 3.** Performance evaluation for trinomials

| Field size | NTL [s] | SSE [s] | AVX [s] | NTL/SSE | NTL/AVX | SSE/AVX |
|---|---|---|---|---|---|---|
| $2^{32}$ | 0,28 | 0,003 | 0,0005 | 90,70 | 487 | 5,3 |
| $2^{64}$ | 0,56 | 0,006 | 0,0010 | 89,71 | 516 | 5,7 |
| $2^{128}$ | 1,13 | 0,012 | 0,0020 | 89,92 | 558 | 6,2 |
| $2^{256}$ | 2,26 | 0,025 | 0,0041 | 89,76 | 543 | 6,0 |
| $2^{512}$ | 4,52 | 0,050 | 0,0082 | 89,70 | 548 | 6,1 |
| $2^{1024}$ | 8,97 | 0,096 | 0,0163 | 92,99 | 547 | 5,8 |
| $2^{2048}$ | 17,79 | 0,201 | 0,0328 | 88,51 | 540 | 6,1 |
| $2^{4096}$ | 36,15 | 0,404 | 0,0657 | 89,48 | 549 | 6,1 |

**Table 4.** Performance evaluation for septanomials, CPU vs GPU.

| Field size | NTL [s] | SSE [s] | AVX [s] | CUDA [s] | NTL/CUDA | AVX/CUDA |
|---|---|---|---|---|---|---|
| $2^{32}$ | 0,22 | 0,0060 | 0,00197 | 0,00813 | 27,0 | 0,10 |
| $2^{64}$ | 0,438 | 0,0121 | 0,00395 | 0,01448 | 30,2 | 0,10 |
| $2^{128}$ | 0,877 | 0,2458 | 0,00750 | 0,02712 | 32,3 | 0,12 |
| $2^{256}$ | 1,753 | 0,0488 | 0,01502 | 0,05249 | 33,3 | 0,12 |
| $2^{512}$ | 3,477 | 0,0975 | 0,03038 | 0,10348 | 33,5 | 0,12 |
| $2^{1024}$ | 7,03 | 0,1949 | 0,05971 | 0,20481 | 34,3 | 0,12 |
| $2^{2048}$ | 14,052 | 0,3928 | 0,12153 | 0,40783 | 34,4 | 0,12 |
| $2^{4096}$ | 28,102 | 0,7791 | 0,24038 | 0,81518 | 34,4 | 0,12 |

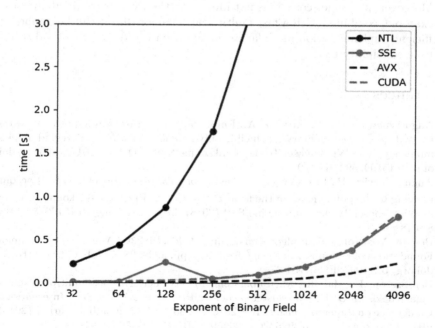

**Fig. 3.** Performance evaluation figure for septanomials, CPU vs GPU.

Tables 1, 2, 3, 4 and Fig. 3 reveals almost linear dependency between the field size and time of computations for implemented algorithms. Furthermore, their speedup over reference NTL implementation is significant and almost independent of the field size. Contrarily, the speedup factor over NTL implementation is strictly dependent on the number of nonzero coefficients in modular polynomial for CPU case.

# 6    Conclusion and Future Work

In this paper, the performance of modular squaring in the binary field was evaluated. Up to 500 times speedup for polynomials of degree $n = 4096$ was achieved relative to the NTL by Victor Shoup for the best case when the reduction polynomial is trinomial. Speedups for other forms of modular polynomials were also observed. Nevertheless, the efficiency of the implemented algorithm strictly depends on the number of non-zero coefficients of the modular polynomial.

The developed algorithm can also be applied to the GPU implementation of modular squaring in the binary field. The GPU implementation outperforms the reference implementation but for sparse polynomials has a considerably inferior efficiency than the CPU one.

The most significant advantage of the implemented squaring algorithm is its flexibility and universalism. It can be adjusted to heterogeneous computing platforms, different forms of reduction polynomials and various degrees of polynomials. Moreover, its usage scenario is not limited to the Ben-Or irreducibility test. For example, combined with a fast multiplication algorithm it can be applied in an efficient implementation modular exponentiation by squaring method or the Rabin irreducibility test.

# References

1. Augustynowicz, P., Paszkiewicz, A.: Empirical verification of a hypothesis on the inner degree of sedimentary irreducible polynomials over $GF(2)$. Przegląd Telekomunikacyjny + Wiadomości Telekomunikacyjne **8–9**, 799–802 (2017). https://doi.org/10.15199/59.2017.8-9.33
2. Augustynowicz, P., Paszkiewicz, A.: On trinomials irreducible over $GF(2)$ accompanying to the polynomials of the form $x^{2 \cdot 3^l} + x^{3^l} + 1$. Przegląd Telekomunikacyjny + Wiadomości Telekomunikacyjne **8–9** (2018). https://doi.org/10.15199/59.2018.8-9.48
3. Ben-Or, M.: Probabilistic algorithms in finite fields. In: 22nd Annual Symposium on Foundations of Computer Science (sfcs 1981), pp. 394–398, October 1981. https://doi.org/10.1109/SFCS.1981.37
4. Ben-Sasson, E., Hamilis, M., Silberstein, M., Tromer, E.: Fast multiplication in binary fields on GPUs via register cache. In: Proceedings of the 2016 International Conference on Supercomputing, ICS 2016, pp. 35:1–35:12, Istanbul, Turkey (2016). https://doi.org/10.1145/2925426.2926259, ISBN 978-1-4503-4361-9

5. Cohen, H., et al.: Handbook of Elliptic and Hyperelliptic Curve Cryptography, 2nd edn. Chapman & Hall/CRC (2012). ISBN 9781439840009

6. Feng, K., Ma, W., Huang, W., Zhang, Q., Gong, Y.: Speeding up Galois field arithmetic on Intel MIC architecture. In: Hsu, C.-H., Li, X., Shi, X., Zheng, R. (eds.) NPC 2013. LNCS, vol. 8147, pp. 143–154. Springer, Heidelberg (2013). https://doi.org/10.1007/978-3-642-40820-5_13

7. Flynn, M.J.: Some computer organizations and their effectiveness. IEEE Trans. Comput. **21**(9), 948–960 (1972). https://doi.org/10.1109/TC.1972.5009071, ISSN 0018–9340

8. Kalcher, S., Lindenstruth, V.: Accelerating Galois field arithmetic for Reed-Solomon erasure codes in storage applications. In: 2011 IEEE International Conference on Cluster Computing, pp. 290–298, September 2011. https://doi.org/10.1109/CLUSTER.2011.40

9. Lidl, R., Niederreiter, H.: Introduction to Finite Fields and Their Applications. Cambridge University Press, New York (1986)

10. Lidl, R., et al.: Finite Fields. EBL-Schweitzer t. 20, pkt 1. Cambridge University Press (1997). https://books.google.pl/books?id=xqMqxQTFUkMC, ISBN 9780521392310

11. Plank, J.S., Greenan, K.M., Miller, E.L.: Screaming fast Galois field arithmetic using intel SIMD instructions. In: Proceedings of the 11th USENIX Conference on File and Storage Technologies, FAST 2013, San Jose, CA, USA, 12–15 February 2013, pp. 299–306 (2013). https://www.usenix.org/conference/fast13/technicalsessions/presentation/plank%5C_james%5C_simd

12. Rabin, M.O.: Probabilistic algorithms in finite fields. SIAM J. Comput. **9**, 273–280 (1979)

13. Shoup,V.: NTL: A library for doing number theory (2003). http://www.shoup.net/ntl

# Parallel Robust Computation
# of Generalized Eigenvectors
# of Matrix Pencils

Carl Christian Kjelgaard Mikkelsen[(✉)] [iD] and Mirko Myllykoski[iD]

Department of Computing Science and HPC2N, Umeå University,
90187 Umeå, Sweden
{spock,mirkom}@cs.umu.se

**Abstract.** In this paper we consider the problem of computing generalized eigenvectors of a matrix pencil in real Schur form. In exact arithmetic, this problem can be solved using substitution. In practice, substitution is vulnerable to floating-point overflow. The robust solvers xtgevc in LAPACK prevent overflow by dynamically scaling the eigenvectors. These subroutines are scalar and sequential codes which compute the eigenvectors one by one. In this paper, we discuss how to derive robust algorithms which are blocked and parallel. The new StarNEig library contains a robust task-parallel solver Zazamoukh which runs on top of StarPU. Our numerical experiments show that Zazamoukh achieves a super-linear speedup compared with dtgevc for sufficiently large matrices.

**Keywords:** Generalized eigenvectors · Overflow protection · Task-parallelism

## 1  Introduction

Let $A \in \mathbb{R}^{m \times m}$ and let $B \in \mathbb{R}^{m \times m}$. The matrix pencil $(A, B)$ consists of all matrices of the form $A - \lambda B$ where $\lambda \in \mathbb{C}$. The set of (generalized) eigenvalues of the matrix pencil $(A, B)$ is given by

$$\lambda(A, B) = \{\lambda \in \mathbb{C} : \det(A - \lambda B) = 0\}.$$

We say that $x \in \mathbb{C}^m$ is a (generalized) eigenvector of the matrix pencil $(A, B)$ if and only if $x \neq 0$ and

$$Ax = \lambda Bx.$$

The eigenvalues of $(A, B)$ can be computed by first reducing $(A, B)$ to real Schur form $(S, T)$. Specifically, there exist orthogonal matrices $Q$ and $Z$ such that $S = Q^T A Z$ is quasi-upper triangular and $T = Q^T B Z$ is upper triangular. It is clear that

$$\lambda(A, B) = \lambda(S, T).$$

© Springer Nature Switzerland AG 2020
R. Wyrzykowski et al. (Eds.): PPAM 2019, LNCS 12043, pp. 58–69, 2020.
https://doi.org/10.1007/978-3-030-43229-4_6

Moreover, $y \in \mathbb{C}^m$ is a generalized eigenvector of $(S, T)$ corresponding to the eigenvalue $\lambda$, if and only if $x = Zy$ is a generalized eigenvector of $(A, B)$ corresponding to the eigenvalue $\lambda$.

In this paper, we consider the parallel computation of eigenvectors of a matrix pencil in real Schur form. In exact arithmetic, this problem can be solved using substitution. However, substitution is very vulnerable to floating-point overflow.

In LAPACK [3] there exists a family xtgevc of subroutines which compute the generalized eigenvectors of a matrix pencil in Schur form. They prevent overflow by dynamically scaling the eigenvectors. These subroutines are scalar codes which compute the eigenvectors one by one. In this paper we discuss the construction of algorithms which are not only robust, but blocked and parallel.

Our paper is organized as follows. In Sect. 2 we briefly review past work on robust algorithms for solving equations of triangular type. In Sect. 3 we consider the problem of computing the eigenvectors of a matrix pencil in real Schur form using real arithmetic. This problem is equivalent to solving a homogeneous matrix equation of the form

$$SVD - TVE = 0, \tag{1}$$

which respect to $V$. The matrix $D$ is diagonal and the matrix $E$ is block diagonal with diagonal blocks which are either 1-by-1 or 2-by-2. In Sect. 4 we present a blocked algorithm for solving this matrix equation. In Sect. 5 we discuss how to prevent overflow in this algorithm. The concept of an augmented matrix is central to this discussion. A robust task-parallel solver Zazamoukh has been developed and integrated into the new StarNEig library for solving non-symmetric eigenvalue problems [1,9]. The performance of Zazamoukh is compared to LAPACK in Sect. 7. We suggest directions for future work in Sect. 8.

## 2   Related Work on Robust Algorithms

LAPACK contains several robust routines for solving equations involving triangular matrices. These routines include xtrevc (standard eigenvectors), xtgevc (generalized eigenvectors), and xtrsyl (Sylvester matrix equations). They prevent overflow by scaling the right hand-side dynamically. The underlying principles were originally derived by Anderson and implemented in xlatrs [2]. This family of subroutines apply to triangular linear systems

$$Tx = b \tag{2}$$

with a single right-hand side. Mikkelsen and Karlsson [7] formalized the work of Anderson and derived a robust blocked algorithm for solving Eq. (2). In particular, Mikkelsen and Karlsson isolated two functions ProtectDivision and ProtectUpdate which can be used to prevent overflow in scalar divisions $y \leftarrow b/t$ and general linear updates of the form $Y \leftarrow Y - TX$. These two functions have the following key properties:

1. If $t \neq 0$ and $|b| \leq \Omega$, and

$$\xi = \texttt{ProtectDivision}(|b|, |t|)$$

then $\xi \in (0, 1]$ and $|\xi b| \leq |t|\Omega$. It follows that the scaled division

$$y \leftarrow \frac{(\xi b)}{t}$$

cannot exceed $\Omega$.

2. If $Z = Y - TX$ is defined, with

$$\|Y\|_\infty \leq \Omega, \quad \|T\|_\infty \leq \Omega, \quad \|X\|_\infty \leq \Omega,$$

and

$$\xi = \texttt{ProtectUpdate}(\|Y\|_\infty, \|T\|_\infty, \|X\|_\infty)$$

then $\xi \in (0, 1]$ and

$$\xi(\|Y\|_\infty + \|T\|_\infty \|X\|_\infty) \leq \Omega.$$

It follows that

$$Z \leftarrow (\xi Y) - T(\xi X) = (\xi Y) - (\xi T)X.$$

can be computed without any component of any intermediate or final result exceeding $\Omega$.

Mikkelsen, Schwarz, and Karlsson have derived a robust blocked algorithm for solving triangular linear systems

$$TX = B \tag{3}$$

with multiple right-hand sides. Their task-parallel implementation (Kiya) is significantly faster than dlatrs when numerical scaling is necessary and not significantly slower than dtrsm when numerical scaling is not required [8]. This paper also contains a formal proof of the correctness of ProtectUpdate and ProtectDivision.

## 3   Real Arithmetic

In this section we show that the problem of computing generalized eigenvectors is equivalent to solving a real homogeneous matrix equation of the type given by Eq. (1).

Let $A \in \mathbb{R}^{m \times m}$ and $B \in \mathbb{R}^{m \times m}$ be given. The set of generalized eigenvalues of the matrix pencil $(A, B)$ can be computed by first reducing $(A, B)$ to generalized real Schur form. Specifically, there exist orthogonal matrices $Q \in \mathbb{R}^{m \times m}$ and $Z \in \mathbb{R}^{m \times m}$ such that

$$S = Q^T A Z = \begin{bmatrix} S_{11} & S_{12} & \dots & S_{1p} \\ & S_{22} & \dots & S_{2p} \\ & & \ddots & \vdots \\ & & & S_{pp} \end{bmatrix}, \quad T = Q^T B Z = \begin{bmatrix} T_{11} & T_{12} & \dots & T_{1p} \\ & T_{22} & \dots & T_{2p} \\ & & \ddots & \vdots \\ & & & T_{pp} \end{bmatrix}$$

are upper block-triangular and $\dim(S_{jj}) = \dim(T_{jj}) \in \{1, 2\}$. It is clear that

$$\lambda(S, T) = \cup_{j=1}^{p} \lambda(S_{jj}, T_{jj}).$$

We follow the standard convention and represent eigenvalues $\lambda$ using an ordered pair $(\alpha, \beta)$ where $\alpha \in \mathbb{C}$ and $\beta \geq 0$. If $\beta > 0$, then $\lambda = \alpha/\beta$ is a finite eigenvalue. The case of $\alpha \in \mathbb{R}\backslash\{0\}$ and $\beta = 0$, corresponds to an infinite eigenvalue. The case of $\alpha = \beta = 0$ corresponds to an indefinite eigenvalue problem.

Let $n_j = \dim(S_{jj}) = \dim(T_{jj})$. In order to simplify the current discussion, we will make the following assumptions:

1. If $n_j = 1$, then $(S_{jj}, T_{jj})$ has an eigenvalue with is either real or infinite.
2. If $n_j = 2$, then $(S_{jj}, T_{jj})$ has two complex conjugate eigenvalues.
3. All eigenvalues are distinct.

By eliminating the possibility of multiple eigenvalues and indefinite problems we are free to formulate robust algorithms for well-defined problems. The question of how to handle problems which are not well-defined is certainly important but outside the scope of this paper.

## 3.1  Computing a Single Eigenvector

It this subsection, we note that the problem of computing a single generalized eigenvector of $(S, T)$ is equivalent to solving a tall homogeneous matrix equation involving real matrices. Let $\lambda \in \lambda(S_{jj}, T_{jj})$ and let $\lambda = \frac{\alpha_j}{\beta_j}$ where $\beta_j > 0$ and

$$\alpha_j = a_j + ib_j \in \mathbb{C}.$$

Let $n_j = \dim(S_{jj})$ and let $D_{jj} \in \mathbb{R}^{n_j \times n_j}$ and $E_{jj} \in \mathbb{R}^{n_j \times n_j}$ be given by

$$D_{jj} = [\beta_j], \quad E_{jj} = [a_j], \tag{4}$$

when $n_j = 1$ (or equivalently $b_j = 0$) and

$$D_{jj} = \begin{bmatrix} \beta_j & 0 \\ 0 & \beta_j \end{bmatrix}, \quad E_{jj} = \begin{bmatrix} a_j & b_j \\ -b_j & a_j \end{bmatrix}, \tag{5}$$

when $n_j = 2$ (or equivalently $b_j \neq 0$). With this notation, the problem of computing an eigenvector is equivalent to solving the homogeneous linear equation

$$SVD_{jj} - TVE_{jj} = 0 \tag{6}$$

with respect to $V \in \mathbb{R}^{m \times n_j}$. This follows from the following lemma.

**Lemma 1.** *Let $\lambda \in \lambda(S_{jj}, T_{jj})$ and let $\lambda = (a_j + ib_j)/\beta$ where $\beta > 0$. Then the following statements are true:*

1. *If $n_j = 1$, then $x \in \mathbb{R}^m$ is a real eigenvector of $(S, T)$ corresponding to the real eigenvalue $\lambda \in \mathbb{R}$ if and only if $V = [x]$ has rank 1 and solves (6).*
2. *If $n_j = 2$, then $z = x + iy \in \mathbb{C}^m$ is a complex eigenvector of $(S, T)$ corresponding to the complex eigenvalue $\lambda \in \mathbb{C}$ if and only if $V = [x\ y]$ has rank 2 and solves (6).*

## 3.2   Computing All Eigenvectors

In this subsection, we note that the problem of computing all generalized eigenvectors of $(S, T)$ is equivalent to solving a homogeneous matrix equation involving real matrices. Specifically, let $p$ denote the number of 1-by-1 or 2-by-2 blocks on the diagonal of $S$, let $D \in \mathbb{R}^{m \times m}$ and $E \in \mathbb{R}^{m \times m}$ be given by

$$D = \text{diag}\{D_{11}, D_{22}, \ldots, D_{pp}\}, \quad B = \text{diag}\{E_{11}, E_{22}, \ldots, E_{pp}\},$$

where $D_{jj}$ and $E_{jj}$ are given by Eqs. (4) and (5). Then $V = \begin{bmatrix} V_1 & V_2 & \ldots & V_p \end{bmatrix} \in \mathbb{R}^{m \times m}$ solves the homogeneous matrix equation

$$SVD - TVE = 0, \tag{7}$$

if and only if $V_j \in \mathbb{R}^{m \times n_j}$ solves Eq. (6).

## 4   A Blocked Algorithm

In this section we present a blocked algorithm for solving the homogeneous matrix Eq. (7). We begin by *redefining* the partitioning of $S$. Let

$$S = \begin{bmatrix} S_{ij} \end{bmatrix}, \quad i, j \in \{1, 2, \ldots, M\}$$

denote *any* partitioning of $S$ into an $M$ by $M$ block matrix which does not split any of the 2-by-2 blocks along the diagonal of $S$. Apply the same partitioning to $T$, $D$, $B$, and $V$. The homogeneous matrix Eq. (7) can now be written as

$$
\begin{bmatrix} S_{11} & S_{12} & \ldots & S_{1M} \\ & S_{22} & \ldots & S_{2M} \\ & & \ddots & \vdots \\ & & & S_{MM} \end{bmatrix}
\begin{bmatrix} V_{11} & V_{12} & \ldots & V_{1M} \\ & V_{22} & \ldots & V_{2M} \\ & & \ddots & \vdots \\ & & & V_{MM} \end{bmatrix}
\begin{bmatrix} D_{11} & & & \\ & D_{22} & & \\ & & \ddots & \\ & & & D_{MM} \end{bmatrix}
$$
$$
= \begin{bmatrix} T_{11} & T_{12} & \ldots & T_{1M} \\ & T_{22} & \ldots & T_{2M} \\ & & \ddots & \vdots \\ & & & T_{MM} \end{bmatrix}
\begin{bmatrix} V_{11} & V_{12} & \ldots & V_{1M} \\ & V_{22} & \ldots & V_{2M} \\ & & \ddots & \vdots \\ & & & V_{MM} \end{bmatrix}
\begin{bmatrix} E_{11} & & & \\ & E_{22} & & \\ & & \ddots & \\ & & & E_{MM} \end{bmatrix} \tag{8}
$$

The block columns of $V$ can be computed concurrently, because $D$ and $E$ are block diagonal. It is straightforward to verify that Eq. (8) can be solved using Algorithm 1. Algorithm 1 can be implemented using three distinct kernels.

1. Kernel 1 solves small homogeneous matrix equations of the form

$$S_{jj} X D_{jj} = T_{jj} X E_{jj} \tag{9}$$

   with respect to $X$. This is equivalent to finding eigenvectors for the pencil $(S_{jj}, T_{jj})$.

---

**Algorithm 1.** Blocked computation of all generalized eigenvectors

---

1 **for** $j \leftarrow 1, 2, \ldots, M$ **do**
2     **for** $i \leftarrow 1, 2, \ldots, j$ **do**
3        $V_{ij} \leftarrow 0$;

4 **for** $j \leftarrow 1, 2, \ldots, M$ **do**
5     Solve
$$S_{jj} X D_{jj} = T_{jj} X E_{jj}$$
    with respect to $X$ and set $V_{jj} \leftarrow X$;
6     **for** $i \leftarrow j - 1, \ldots, 1$ **do**
7        **for** $k \leftarrow 1, 2, \ldots, i$ **do**
8           Perform the linear update
$$V_{kj} \leftarrow V_{kj} - (S_{k,i+1} V_{i+1,j} D_{jj} - T_{k,i+1} V_{i+1,j} E_{j,j})$$
9     Solve
$$S_{ii} X D_{jj} - T_{ii} X E_{jj} = V_{ij}$$
    with respect to $X$ and set $V_{ij} \leftarrow X$;

---

2. Kernel 2 performs specialized linear updates of the form

$$Y \leftarrow Y - (S_{k,i+1} X D_{jj} - T_{k,i+1} X E_{jj}) \tag{10}$$

3. Kernel 3 solves small matrix equations of the form

$$S_{ii} X D_{jj} - T_{ii} X E_{jj} = Y \tag{11}$$

with respect to $X$.

Once these kernels have been implemented, it is straightforward to parallelize Algorithm 1 using a task-based runtime system such as StarPU [4].

## 5    Constructing Robust Kernels

Algorithm 1 is not robust. Each of the three kernels are vulnerable to floating point overflow. The kernels needed for Algorithm 1 can be implemented using nested loops, divisions and linear updates. Therefore, it is not surprising that robust kernels can be implemented using the functions `ProtectDivision` and `ProtectUpdate` given in Sect. 2. We will now explain how this can be done without sacrificing the potential for level-3 BLAS operations. We will concentrate on Kernel 2 which executes the vast majority of the arithmetic operations needed for Algorithm 1.

We will use the concept of an *augmented* matrix introduced by Mikkelsen, Schwarz and Karlsson [8].

**Definition 1.** *Let $X \in \mathbb{R}^{m \times n}$ be partitioned into $k$ block columns*

$$X = \begin{bmatrix} X_1 \, X_2 \, \dots \, X_k \end{bmatrix}, \quad X_j \in \mathbb{R}^{m \times n_j},$$

*and let $\alpha = (\alpha_1, \alpha_2, \dots, \alpha_k) \in \mathbb{R}^k$ satisfy $\alpha_j \in (0, 1]$. The augmented matrix $\langle \alpha, X \rangle$ represents the real matrix $Y \in \mathbb{R}^{m \times n}$ given by*

$$Y = \begin{bmatrix} Y_1 \, Y_2 \, \dots \, Y_k \end{bmatrix}, \quad Y_j \in \mathbb{R}^{m \times n_j}, \quad Y_j = \alpha_j^{-1} X_j.$$

This is a trivial extension of the original definition which considered the case of $k = n$. The purpose of the scaling factors $\alpha_j$ is to extend the normal representational range of our floating-point numbers.

Now consider the problem of computing without overflow a specialized linear update of the form needed for Kernel 2, i.e., an update of the form

$$Y \leftarrow (Y - S(XD)) + T(XE). \tag{12}$$

where $S$ and $T$ are general dense matrices, $D$ is diagonal and $E$ is block diagonal with diagonal blocks that are 1-by-1 or 2-by-2. The parentheses are used indicated the order of evaluation. A representation of the matrices $Z_1 = XD$ and $Z_2 = XE$ can be obtained using augmented matrices and Algorithm 2. Similarly, it is possible to obtain a representation of the matrices $Z_3 = Y - SZ_1$ and $Y = Z_3 + TZ_2$ using augmented matrices and Algorithm 3.

---

**Algorithm 2.** Right updates with block diagonal matrix

---

**Data:** An augmented matrix $\langle \alpha, X \rangle$ where

$$X = \begin{bmatrix} X_1 \, X_2 \, \dots \, X_k \end{bmatrix}, \quad X_j \in \mathbb{R}^{m \times n_j}, \quad \|X_j\|_\infty \le \Omega,$$

and a block diagonal matrix matrix

$$F = \mathrm{diag}(F_1, F_2, \dots, F_k), \quad F_j \in \mathbb{R}^{n_j \times n_j}, \quad \|F_j\|_\infty \le \Omega.$$

**Result:** An augmented matrix $\langle \beta, Y \rangle$ where

$$Y = \begin{bmatrix} Y_1 \, Y_2 \, \dots \, Y_k \end{bmatrix}, \quad Y_j \in \mathbb{R}^{m \times n_j}, \quad \|Y_j\|_\infty \le \Omega,$$

such that
$$\beta_j^{-1} Y_j = (\alpha_j^{-1} X_j) F_j,$$

and $Y$ can be computed without exceeding $\Omega$.

1  **for** $j = 1, 2 \dots, k$ **do**
2  $\quad \gamma_j = \texttt{ProtectUpdate}(0, \|X_j\|_\infty, \|F_j\|_\infty);$
3  $\quad Y_j = (\gamma_j X_j) F_j;$
4  $\quad \beta_j = \alpha_j \gamma_j;$
5  **return** $\langle \beta, Y \rangle;$

---

---

**Algorithm 3.** Left update with dense matrix

---

**Data:** A dense matrix $T \in \mathbb{R}^{m \times l}$ and augmented matrices $\langle \alpha, X \rangle$ and $\langle \beta, Y \rangle$ where

$$X = \begin{bmatrix} X_1 \ X_2 \ldots X_k \end{bmatrix}, \quad X_j \in \mathbb{R}^{l \times n_j}, \quad \|X_j\|_\infty \leq \Omega,$$
$$Y = \begin{bmatrix} Y_1 \ Y_2 \ldots Y_k \end{bmatrix}, \quad Y_j \in \mathbb{R}^{m \times n_j}, \quad \|Y_j\|_\infty \leq \Omega.$$

**Result:** An augmented matrix $\langle \zeta, Z \rangle$ where

$$Z = \begin{bmatrix} Z_1 \ Z_2 \ldots Z_k \end{bmatrix}, \quad Z_j \in \mathbb{R}^{m \times n_j}, \quad \|Z_j\|_\infty \leq \Omega,$$

such that
$$\zeta_j^{-1} Z_j = \beta_j^{-1} Y_j - T(\alpha_j^{-1} X_j),$$
and $Z$ can be computed without exceeding $\Omega$.

1 **for** $j = 1, \ldots, k$ **do**
2     $\gamma_j = \min\{\alpha_j, \beta_j\}$;
3     $\delta_j = \texttt{ProtectUpdate}((\gamma_j/\beta_j)\|Y_j\|_\infty, \|T\|_\infty, (\gamma_j/\alpha_j)\|X_j\|_\infty)$;
4     $X_j \leftarrow \delta_j(\gamma_j/\alpha_j)X_j$;
5     $Y_j \leftarrow \delta_j(\gamma_j/\beta_j)Y_j$;
6     $\zeta_j = \delta_j \gamma_j$;
7 $Z \leftarrow Y - TX$;
8 **return** $\langle \zeta, Z \rangle$

---

We cannot escape the fact that the right updates, i.e., the calls to Algorithm 2 have low arithmetic intensity because we are essentially scaling the columns of the input matrix. However, each right update can be followed by a left update, i.e., a call to Algorithm 3 acting on the *same* data. Algorithm 3 consists of some light pre-processing (the for-loop spanning lines 1–6) which has low arithmetic intensity, but concludes with a regular level-3 BLAS operation (line 7) which contains the overwhelming majority of the necessary arithmetic operations. This explains why robustness can be combined with good performance.

In order to execute all linear updates needed for a robust variant of Algorithm 1 we require norms of certain submatrices of $S$ and $T$. In particular, we need the infinity norms of all super-diagonal blocks of $S$ and $T$. Moreover, we require the infinity norm of certain submatrices of $V$. These submatrices consists of either a single column (segment of real eigenvector) or two adjacent columns (segment of complex eigenvector). The infinity norm must be computed whenever a submatrix has been initialized or updated. ProtectUpdate requires that the input arguments are bounded by $\Omega$ and failure is possible if they are not.

# 6 Zazamoukh - A Task-Parallel Robust Solver

The new StarNEig library runs on top of StarPU and can be used to solve dense non-symmetric eigenvalue problems [1,9]. A robust variant of Algorithm 1 has been implemented in StarNEig. This implementation (Zazamoukh) uses augmented matrices and scaling factors which are integer powers of 2. Zazamoukh can compute eigenvectors corresponding to a subset of $\lambda(S, T)$ which is closed under complex conjugation. Zazamoukh is not subject to the assumptions made in Sect. 3. In particular, Zazamoukh can handle the case of multiple eigenvalues. Zazamoukh is currently limited to shared memory, but an extension to distributed memory is under development.

## 6.1   Memory Layout

Given block sizes $mb$ and $nb$ Zazamoukh partitions $S$, $T$ and $V$ conformally by rows and columns. In the absence of any 2-by-2 diagonal blocks on the diagonal of $S$ the tiles of $S$ and $T$ are $mb$ by $mb$ and the tiles of $V$ are $mb$ by $nb$. The only exceptions can be found along the right and lower boundaries of the matrices. This default configuration is adjusted minimally to prevent splitting any 2-by-2 block of $S$ or separating the real part and the imaginary part of a complex eigenvector into separate tile columns.

## 6.2   Tasks

Zazamoukh relies on four types of tasks:

1. Pre-processing tasks which compute all quantities needed for robustness. This includes the infinity norm of all super-diagonal tiles of $S$ and $T$ as well as all norms needed for the robust solution of equations of the type (11). If necessary, the matrices $S$ and $T$ are scaled minimally.
2. Robust solve tasks which use dtgevc to compute the lower *tips* of eigenvectors, i.e., Eq. (9)) and a robust solver based on dlaln2 to solve equations of the type given by Eq. (11).
3. Robust update tasks which execute updates of the type given by (10).
4. Post-processing tasks which enforce a consistent scaling on all segments of all eigenvectors.

## 6.3   Task Insertion Order and Priorities

Zazamoukh is closely related to Kiya which solves triangular linear systems with multiple right-hand sides. Apart from the pre-processing and post-processing tasks, the main task graph is the disjoint union of $p$ task-graphs, one for each block column of the matrix of eigenvectors. Zazamoukh uses the same task insertion order and priorities as Kiya to process each of the $p$ disjoint sub-graphs. Specifically, tasks corresponding to blocks of $(S, T)$ on the main block diagonal are assigned the highest possible priority. Tasks corresponding to blocks of $(S, T)$ on the $j$th superdiagonal are assigned priority $q - j$ until the number of distinct priorities are exhausted. The rationale behind this choice is to guide the scheduler towards rapid progress on the critical path.

**Table 1.** Comparison between sequential `dtgevc` and task-parallel `Zazamoukh` using 28 cores. The run-times are given in milli-seconds (ms). The last column gives the speedup of `Zazamoukh` over LAPACK. Values above 28 correspond to super-linear speedup. All eigenvectors were computed with a relative residual less than $2u$, where $u$ denotes the double precision unit roundoff.

| Dimension | Eigenvalue analysis | | | Run time (ms) | | SpeedUp |
|---|---|---|---|---|---|---|
| m | zeros | inf. | indef. | LAPACK | StarNEig | |
| 1000 | 11 | 13 | 0 | 295 | 175 | 1.6857 |
| 2000 | 25 | 16 | 0 | 1598 | 409 | 3.9071 |
| 3000 | 24 | 30 | 0 | 6182 | 929 | 6.6545 |
| 4000 | 42 | 49 | 0 | 15476 | 1796 | 8.6169 |
| 5000 | 54 | 37 | 0 | 30730 | 2113 | 14.5433 |
| 6000 | 61 | 64 | 0 | 53700 | 2637 | 20.3641 |
| 7000 | 67 | 64 | 0 | 84330 | 3541 | 23.8153 |
| 8000 | 56 | 69 | 0 | 122527 | 4769 | 25.6924 |
| 9000 | 91 | 91 | 0 | 171800 | 6189 | 27.7589 |
| 10000 | 108 | 94 | 0 | 242466 | 7821 | 31.0019 |
| 20000 | 175 | 197 | 0 | 2034664 | 49823 | 40.8378 |
| 30000 | 306 | 306 | 0 | 7183746 | 162747 | 44.1406 |
| 40000 | 366 | 382 | 0 | 17713267 | 380856 | 46.5091 |

# 7   Numerical Experiments

In this section we give the result of a set of experiments involving tiny ($m \leq 10\,000$) and small ($m \leq 40\,000$) matrices. Each experiment consisted of computing all eigenvectors of the matrix pencil. The run time was measured for `dtgevc` from LAPACK and `Zazamoukh`. Results related to somewhat larger matrices ($m \leq 80\,000$) can be found in the NLAFET Deliverable 2.7 [10].

The experiments were executed on an Intel Xeon E5-2690v4 (Broadwell) node with 28 cores arranged in two NUMA islands with 14 cores each. StarNEig was compiled with OpenBLAS 0.3.2 (includes LAPACK) and StarPU 1.2.8. We used the StarNEig test-program `starneig-test` to generate reproducible experiments. The default parameters produce matrix pencils where approximately 1% of all eigenvalues are zeros, 1% of all eigenvalues are infinities and there are no indefinite eigenvalues. `Zazamoukh` used the default tile size $mb = nb$ which is 1.6% of the matrix dimension for matrix pencils with dimension $m \geq 1000$.

All experiments were executed with exclusive access to a complete node (28 cores). LAPACK was run in sequential mode, while `Zazamoukh` used 28 StarPU workers and 1 master thread. The summary of our results are given in Table 1. The timings include all overhead needed to achieve robustness. The speedup of `Zazamoukh` over LAPACK is initially very modest as there is not enough tasks to keep 28 workers busy, but it picks up rapidly and `Zazamoukh` achieves a super-linear speedup over `dtgevc` when $m \geq 10\,000$. This is an expression of the fact that `Zazamoukh` uses a blocked algorithm, whereas `dtgevc` computes the eigenvectors one by one.

## 8   Conclusion

Previous work by Mikkelsen, Schwarz and Karlsson has shown that triangular linear systems can be solved without overflow and in a blocked and parallel manner using augmented matrices. In this paper we have shown that the eigenvectors of a matrix pencil can be computed without overflow and in a blocked and parallel manner using augmented matrices. Certainly, robust algorithms are slower than non-robust algorithms when numerical scaling is not required, but robust algorithms will always return a result which can be evaluated in the context of the user's application. To the best of our knowledge StarNEig is the only library which contains a parallel robust solver for computing the generalized eigenvectors of a dense non-symmetric matrix pencil. The StarNEig solver (`Zazamoukh`) runs on top of StarPU and uses augmented matrices and scaling factors with are integer powers of 2 to prevent overflow. It achieves super-linear speedup compared with `dtgevc` from LAPACK (OpenBLAS 0.3.2). In the immediate future we expect to pursue the following work:

1. Extend `Zazamoukh` to also compute left eigenvectors. Here the layout of the loops is different and we must use the 1-norm instead of the infinity norm when executing the overflow protection logic.
2. Extend `Zazamoukh` to distributed memory machines.
3. Extend `Zazamoukh`'s solver to use recursive blocking to reduce the run-time. The solve tasks all lie on the critical path of the task graph.
4. Extend `Zazamoukh` to complex data types. This case is simpler than real arithmetic because there are no 2-by-2 blocks on the main diagonal of $S$.
5. Revisit the complex division routine `dladiv` [6] which is the foundation for the `dlaln2` routine used by `Zazamoukh`'s solve tasks. In particular, the failure modes of `xladiv` have not been characterized [5].

**Acknowledgments.** This work is part of a project (NLAFET) that has received funding from the European Union's Horizon 2020 research and innovation programme under grant agreement No 671633. This work was supported by the Swedish strategic research programme eSSENCE. We thank the High Performance Computing Center North (HPC2N) at Umeå University for providing computational resources and valuable support during test and performance runs.

## References

1. StarNEig. https://nlafet.github.io/StarNEig/
2. Anderson, E.: LAPACK Working Note No. 36: Robust Triangular Solves for Use in Condition Estimation. Technical report. CS-UT-CS-91-142, University of Tennessee, Knoxville, TN, USA, August 1991

3. Anderson, E., et al.: LAPACK Users' Guide, 3rd edn. SIAM, Philadelphia (1999). https://doi.org/10.1137/1.9780898719604

4. Augonnet, C., Thibault, S., Namyst, R., Wacrenier, P.A.: StarPU: a unified platform for task scheduling on heterogeneous multicore architectures. CCPE - Spec. Issue: Euro-Par **2009**(23), 187–198 (2011). https://doi.org/10.1002/cpe.1631

5. Baudin, M.: Personal correspondance to C. C. Kjelgaard Mikkelsen (2019)

6. Baudin, M., Smith, R.L.: Robust Complex Division in Scilab. CoRR abs/1210.4539 (2012). http://arxiv.org/abs/1210.4539

7. Kjelgaard Mikkelsen, C.C., Karlsson, L.: Blocked algorithms for robust solution of triangular linear systems. In: Wyrzykowski, R., Dongarra, J., Deelman, E., Karczewski, K. (eds.) PPAM 2017. LNCS, vol. 10777, pp. 68–78. Springer, Cham (2018). https://doi.org/10.1007/978-3-319-78024-5_7

8. Kjelgaard Mikkelsen, C.C., Schwarz, A.B., Karlsson, L.: Parallel robust solution of triangular linear systems. CCPE (2018). https://doi.org/10.1002/cpe.5064

9. Myllykoski, M., Kjelgaard Mikkelsen, C.C.: Introduction to StarNEig – a task-based library for solving nonsymmetric eigenvalue problems. In: Wyrzykowski, R., et al. (eds.) PPAM 2019, LNCS, vol. 12043, pp. 70–81. Springer, Cham (2020). https://doi.org/10.1007/978-3-030-43229-4_7

10. Myllykoski, M., Kjelgaard Mikkelsen, C.C., Schwarz, A., Kågström, B.: D2.7 eigenvalue solvers for nonsymmetric problems. Technical report, Umeå University (2019). http://www.nlafet.eu/wp-content/uploads/2019/04/D2.7-EVP-solvers-evaluation-final.pdf

# Introduction to StarNEig—A Task-Based Library for Solving Nonsymmetric Eigenvalue Problems

Mirko Myllykoski$^{(\boxtimes)}$ (ID) and Carl Christian Kjelgaard Mikkelsen (ID)

Department of Computing Science and HPC2N, Umeå University,
901 87 Umeå, Sweden
{mirkom,spock}@cs.umu.se

**Abstract.** In this paper, we present the StarNEig library for solving dense nonsymmetric (generalized) eigenvalue problems. The library is built on top of the StarPU runtime system and targets both shared and distributed memory machines. Some components of the library support GPUs. The library is currently in an early beta state and only real arithmetic is supported. Support for complex data types is planned for a future release. This paper is aimed at potential users of the library. We describe the design choices and capabilities of the library, and contrast them to existing software such as ScaLAPACK. StarNEig implements a ScaLAPACK compatibility layer that should make it easy for new users to transition to StarNEig. We demonstrate the performance of the library with a small set of computational experiments.

**Keywords:** Eigenvalue problem · Task-based · Library

## 1 Introduction

In this paper, we present the StarNEig library [5] for solving dense nonsymmetric (generalized) eigenvalue problems. StarNEig differs from the existing libraries such as LAPACK [1] and ScaLAPACK [3] in that it relies on a modern task-based approach (see, e.g., [21] and references therein). More specifically, StarNEig is built on top of the StarPU runtime system [6]. This allows StarNEig to target both shared memory and distributed memory machines. Furthermore, some components of StarNEig support GPUs. The library is currently in an early beta state and under continuous development.

This paper targets potential users of the library. We hope that readers, who are already familiar with ScaLAPACK, will be able to decide if StarNEig is suitable for them. In particular, we want to communicate what type of changes are necessary to make their software work with StarNEig. We will explain, through an example, why the task-based approach can potentially lead to superior performance when compared to older, well-established, approaches. We also present

© Springer Nature Switzerland AG 2020
R. Wyrzykowski et al. (Eds.): PPAM 2019, LNCS 12043, pp. 70–81, 2020.
https://doi.org/10.1007/978-3-030-43229-4_7

a small sample of computational results which demonstrate the expected performance of StarNEig. We refer the reader to [20] for more comprehensive performance and accuracy evaluations.

The rest of this paper is organized as follows: Sect. 2 provides a brief introduction to the solution of dense nonsymmetric eigenvalue problems. Section 3 explains the task-based approach and Sect. 4 introduces the reader to some of the inner workings of StarNEig. Section 5 presents a small set of computational results and, finally, Sect. 6 concludes the paper.

## 2  Solution of Dense Nonsymmetric Eigenvalue Problems

Given a matrix $A \in \mathbb{R}^{n \times n}$, the standard eigenvalue problem consists of computing eigenvalues $\lambda_i \in \mathbb{C}$ and matching eigenvectors $x_i \in \mathbb{C}^n$, $x_i \neq 0$, such that

$$Ax_i = \lambda_i x_i. \tag{1}$$

Similarly, given matrices $A \in \mathbb{R}^{n \times n}$ and $B \in \mathbb{R}^{n \times n}$ the generalized eigenvalue problem for the matrix pair $(A, B)$ consists of computing generalized eigenvalues $\lambda_i \in \mathbb{C}$ and matching generalized eigenvectors $x_i \in \mathbb{C}^n$, $x_i \neq 0$, such that

$$Ax_i = \lambda_i B x_i. \tag{2}$$

If the matrices $A$ and $B$ are sparse, then the well-known SLEPc library [4] is one the better tools for solving the eigenvalue problems (1) and (2). Similarly, if the matrices $A$ and $B$ are is symmetric, then algorithms and software that take advantage of the symmetry are preferred (see, e.g., [2,9,14,17,18]). Otherwise, if the matrices are *dense* and *nonsymmetric*, then route of acquiring the (generalized) eigenvalues and the (generalized) eigenvectors usually includes the following three steps:

**Hessenberg(-triangular) reduction:** The matrix $A$ or the matrix pair $(A, B)$ is reduced to Hessenberg or Hessenberg-triangular form by an orthogonal similarity transformation

$$A = Q_1 H Q_1^T \text{ or } (A, B) = Q_1(H, R) Z_1^T, \tag{3}$$

where $H$ is upper Hessenberg, $R$ is a upper triangular, and $Q_1$ and $Z_1$ are orthogonal matrices.

**Schur reduction:** The Hessenberg matrix $H$ or the Hessenberg-triangular matrix pair $(H, R)$ is reduced to real Schur or generalized real Schur form by an orthogonal similarity transformation

$$H = Q_2 S Q_2^T \text{ or } (H, R) = Q_2(S, T) Z_2^T, \tag{4}$$

where $S$ is upper quasi-triangular with $1 \times 1$ and $2 \times 2$ blocks on the diagonal, $T$ is a upper triangular, and $Q_2$ and $Z_2$ are orthogonal matrices. The eigenvalues or generalized eigenvalues can be determined from the diagonal blocks of $S$ or $(S, T)$. In particular, the $2 \times 2$ blocks on the diagonal of $S$ correspond to the complex conjugate pairs of (generalized) eigenvalues.

**Eigenvectors:** Finally, we solve for vectors $y_i \in \mathbb{C}^n$ from

$$(S - \lambda_i I)y_i = 0 \quad \text{or} \quad (S - \lambda_i T)y_i = 0 \tag{5}$$

and backtransform to the original basis by

$$x_i = Q_1 Q_2 y_i \quad \text{or} \quad x_i = Z_1 Z_2 y_i. \tag{6}$$

Additionally, a fourth step can be performed to acquire an invariant subspace of $A$ or $(A, B)$ that is associated with a given subset of eigenvalues or a given subset of generalized eigenvalues:

**Eigenvalue reordering:** The real Schur form $S$ or the generalized real Schur form $(S, T)$ is reordered, such that a selected set of eigenvalues or generalized eigenvalues appears in the leading diagonal blocks of an updated real Schur form $\hat{S}$ or an updated generalized real Schur form $(\hat{S}, \hat{T})$, by an orthogonal similarity transformation

$$S = Q_3 \hat{S} Q_3^T \quad \text{or} \quad (S, T) = Q_3 (\hat{S}, \hat{T}) Z_3^T, \tag{7}$$

where $Q_3$ and $Z_3$ are orthogonal matrices.

See [11] for a detailed explanation of the underlying mathematical theory.

## 3   A Case for the Task-Based Approach

A task-based algorithm functions by cutting the computational work into self-contained tasks that all have a well defined set of inputs and outputs. In particular, StarNEig divides the matrices into (square) tiles and each task takes a set of tiles as its input and produces/modifies a set of tiles as its output. The main difference between tasks and regular function/subroutine calls is that a task-based algorithm does not call the associated computation kernels directly. Instead, the tasks are inserted into a runtime system that derives the task dependences and schedules the tasks to computational resources in a sequentially consistent order. The main benefit of this is that as long as the cutting is carefully done, the underlying parallelism is exposed automatically as the runtime system traverses the resulting task graph.

### 3.1   Novelty in StarNEig

The first main source of novelty in StarNEig comes from the way in which the computational work is cut into tasks and how the task are inserted into the runtime system. The Hessenberg reduction, Schur reduction and eigenvalue reordering steps are based on two-sided transformation algorithms. These algorithms lead to task graphs that are significantly more complicated than the task graphs arising from one-sided transformation algorithm such as the LU factorization. Designing such a task graph, that also leads to high performance, is a non-trivial task as the left and right hand side updates can easily interfere with each other.

Furthermore, the Schur reduction and eigenvalue reordering steps apply a series of overlapping local transformations to the matrices. Due to this overlap, there cannot exists a clear one-to-one mapping between the tasks and the (output) tiles since the local transformations must at some point cross between two or more tiles. Instead, most tasks end up modifying several tiles and this can introduce spurious task dependences that limit the concurrency. By a spurious task dependency, we mean a dependency that is created when two (or more) tasks modify non-overlapping parts of the same tile (i.e., the tasks are independent from each other) but the runtime system interprets this as a task dependency.

The second main source of novelty in StarNEig is related to the eigenvector computation step. Here, the task dependences are comparatively simple but the computations must be protected against floating-point overflow. This is a nontrivial issue to address in a parallel setting as explained in [15,16]. The implementation in StarNEig is both robust and tiled. The former means that the computed eigenvectors are always in the representable range of double precision floating-point numbers and the latter leads to level 3 BLAS performance. The same combination of robustness and performance does not exists neither in LAPACK nor ScaLAPACK since the the corresponding routines is LAPACK are scalar codes and the corresponding ScaLAPACK routines are not robust.

## 3.2 Bulge Chasing and Eigenvalue Reordering

We will now use the Schur reduction and eigenvalue reordering steps to illustrate some benefits of the task-based approach. The modern approach for obtaining a Schur form $S$ of $A$ is to apply the multishift QR algorithm with Aggressive Early Deflation (AED) to the upper Hessenberg form $H$ (see [7,8,13] and references therein). The algorithm is a sequence of steps of two types: AED and bulge chasing. The bulge chasing step creates a set of $3 \times 3$ bulges to the upper left corner of the matrix and the bulges are chased down the diagonal to complete one pipelined QR iteration. This is accomplished by applying sequences of overlapping $3 \times 3$ Householder reflectors to $H$. Similarly, the eigenvalue reordering step is based on applying sequences of overlapping Givens rotations and $3 \times 3$ Householder reflectors to $S$.

If the local transformations are applied one by one, then memory is accessed as shown in Fig. 1a. This is grossly inefficient for two reasons: (i) the transformations are so localized that parallelizing them would not produce any significant speedup and (ii) the matrix elements are touched only once thus leading to very low arithmetic intensity. The modern approach groups together a set of local transformation and initially applies them to a relatively small diagonal window as shown in Fig. 1b. The local transformations are accumulated into an accumulator matrix and later propagated as level 3 BLAS operations acting on the off-diagonal sections of the matrix. This leads to much higher arithmetic intensity and enables proper parallel implementations as *multiple* diagonal windows can be processed simultaneously.

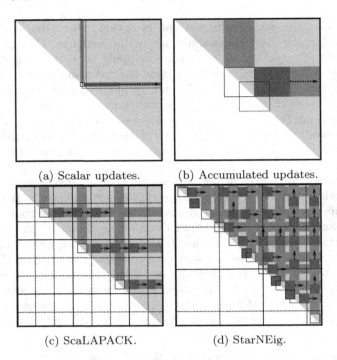

(a) Scalar updates.          (b) Accumulated updates.

(c) ScaLAPACK.               (d) StarNEig.

**Fig. 1.** Hypothetical snapshots taken during the computations. The currently active regions are highlighted with darker shade and the propagation directions of the off-diagonal updates are marked with arrows. In (a), the overlap between two overlapping transformations is highlighted with dashed lines. In (b), the overlap between two diagonal windows is highlighted with dashed lines. In (c) and (d), the dashed lines illustrate how the matrix is divided into distributed blocks and the solid lines illustrate the MPI process mesh.

In particular, the Schur reduction and eigenvalue reordering steps are implemented in ScaLAPACK as PDHSEQR [13] and PDTRSEN [12] subroutines, respectively. Following the ScaLAPACK convention, the matrices are distributed in two-dimensional block cyclic fashion [10]. The resulting memory access pattern is illustrated in Fig. 1c for a $3 \times 3$ MPI process mesh. In this example, three diagonal windows can be processed simultaneously. The related level 3 BLAS updates require careful coordination since the left and right hand side updates must be performed in a sequentially consistent order. In practice, this means (global or broadcast) synchronization after each set of updates have been applied.

In a task-based approach, this can be done using the following task types:

**Window task** applies a set of local transformations inside the diagonal window. Takes the intersecting tiles as input, and produces updated tiles and an accumulator matrix as output.

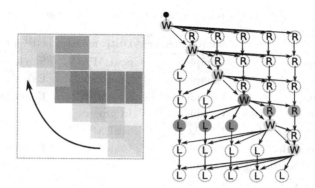

**Fig. 2.** A hypothetical task graph arising from a situation where a Schur form is reordered with a single chain of overlapping diagonal windows. We have simplified the graph by omitting dependences between the right (R) and left (L) update tasks as these dependences are enforced through the window tasks (W).

**Right update task** applies accumulated right-hand side updates using level 3 BLAS operations. Takes the intersecting tiles and an accumulator matrix as input, and produces updated tiles as output.

**Left update task** applies accumulated left-hand side updates using level 3 BLAS operations. Takes the intersecting tiles and an accumulator matrix as input, and produces updated tiles as output.

The tasks are inserted into the runtime system in a sequentially consistent order and each chain of overlapping diagonal windows leads to a task graph like the one shown in Fig. 2. Note that real live task graphs are significantly more complex that shown here, but also enclose more opportunities for parallelism. It is also critical to realize that the runtime system *guarantees* that the tasks are executed in a sequentially consistent order. In particular, there is no need for synchronization and different stages are allowed to overlap and merge together as illustrated in Fig. 1d. This can lead to a much higher concurrency since idle time can be reduced by delaying low priority tasks until computational resources start becoming idle. The AED step in the QR algorithm can also be overlapped with the bulge chasing steps and this improves the concurrency significantly. Other benefits of the task-based approach include, for example, better load balancing, task priorities, accelerators support and implicit MPI communication. See [19,20] for further information.

## 4    StarNEig Library

StarNEig is a C-library that runs on top of the StarPU task-based runtime system. StarPU handles low-level operations such as heterogeneous scheduling; data transfers and replication between various memory spaces; and MPI communication between compute nodes. In particular, StarPU is responsible for managing

**Table 1.** Current status of the StarNEig library.

| Step | Shared memory | Distr. memory | GPUs (CUDA) |
|------|---------------|---------------|-------------|
| Hessenberg | Complete | ScaLAPACK | Single GPU |
| Schur | Complete | Complete | Experimental |
| Reordering | Complete | Complete | Experimental |
| Eigenvectors | Complete | In progress | — |
| Hessenberg-triangular | LAPACK | ScaLAPACK | — |
| Generalized Schur | Complete | Complete | Experimental |
| Generalized reordering | Complete | Complete | Experimental |
| Generalized eigenvectors | Complete | In progress | — |

the various computational resources such as CPU cores and GPUs. The support for GPUs and distributed memory were the main reasons why StarPU was chosen as the runtime system.

StarPU manages a set of worker threads; usually one thread per computational resource. In addition, one thread is responsible for inserting the tasks into StarPU and tracking the state of the machine. If necessary, one additional thread is allocated for MPI communication. For these reasons, StarNEig must be used in a *one process per node* (1ppn) configuration, i.e., several CPU cores should be allocated for each MPI process (a node can be a full node, a NUMA island or some other reasonably large collection of CPU cores).

The current status of StarNEig is summarized in Table 1. The library is currently in an early beta state. The *Experimental* status indicates that the software component has not been tested as extensively as those software components that are considered *Complete*. In particular, the GPU functionality requires some additional involvement from the user (performance model calibration). At the time of writing this paper, only real arithmetic is supported and certain interface functions are implemented as LAPACK and ScaLAPACK wrapper functions. However, we emphasize that StarNEig supports real valued matrices that have complex eigenvalues and eigenvectors. Additional distributed memory functionality and support for complex data types are planned for a future release.

## 4.1 Distributed Memory

StarNEig distributes the matrices in rectangular blocks of a uniform size (excluding the last block row and column) as illustrated in Fig. 3a. The data distribution, i.e., the mapping from the distributed blocks to the MPI process rank space, can be arbitrary as illustrated in Fig. 3b. A user has three options:

1. Use the default data distribution. This is recommended for most users and leads to reasonable performance in most situations.
2. Use a two-dimensional block cyclic distribution (see Fig. 3c). In this case, the user may select the MPI process mesh dimensions and the rank ordering.

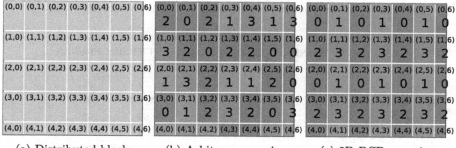

(a) Distributed blocks.      (b) Arbitrary mapping.      (c) 2D-BCD mapping.

**Fig. 3.** Examples of various data distributions supported by StarNEig, including two-dimensional block cyclic distribution (2D-BCD).

3. Define a data distribution function $d : \mathbb{Z}^+ \times \mathbb{Z}^+ \to \mathbb{Z}^+$ that maps the block row and column indices to the MPI rank space. For example, in Fig. 3b, the rank 0 owns the blocks $(0, 1)$, $(1, 2)$, $(1, 5)$, $(1, 6)$, $(2, 6)$, $(3, 0)$ and $(3, 5)$.

The library implements distribution agnostic copy, scatter and gather operations.

Users who are familiar with ScaLAPACK are likely accustomed to using relatively small distributed block sizes (between 64–256). In contrast, StarNEig functions optimally only if the distributed blocks are relatively large (at least 1000 but preferably must larger). This is due to the fact that StarNEig further divides the distributed blocks into tiles and a tiny tile size leads to excessive task scheduling overhead because the tile size is closely connected to the task granularity. Furthermore, as mentioned in the preceding section, StarNEig should be used in 1ppn configuration as opposed to a *one process per core* (1ppc) configuration which is common with ScaLAPACK.

### 4.2 ScaLAPACK Compatibility

StarNEig is fully compatible with ScaLAPACK and provides a ScaLAPACK compatibility layer that encapsulates BLACS contexts and descriptors [10] inside transparent objects, and implements a set of bidirectional conversion functions. The conversions are performed in-place and do not modify any of the underlying data structures. Thus, users can mix StarNEig interface functions with ScaLAPACK subroutines without intermediate conversions. The use of the ScaLAPACK compatibility layer requires the use of either the default data distribution or the two-dimensional block cyclic data distribution.

## 5 Performance Evaluation

Computational experiments were performed on the Kebnekaise system, located at the High Performance Computing Center North (HPC2N), Umeå University. Each regular compute node contains 28 Intel Xeon E5-2690v4 cores (2 NUMA

**Table 2.** A run time comparison between ScaLAPACK and StarNEig.

| $n$ | CPU cores | | Schur reduction *(secs)* | | Eigenvalue reordering *(secs)* | |
|---|---|---|---|---|---|---|
| | ScaLAPACK | StarNEig | PDHSEQR | StarNEig | PDTRSEN | StarNEig |
| 10 000 | 36 | 28 | 38 | 18 | 12 | 3 |
| 20 000 | 36 | 28 | 158 | 85 | 72 | 25 |
| 40 000 | 36 | 28 | 708 | 431 | 512 | 180 |
| 60 000 | 121 | 112 | 992 | 563 | 669 | 168 |
| 80 000 | 121 | 112 | 1667 | 904 | 1709 | 391 |
| 100 000 | 121 | 112 | 3319 | 1168 | 3285 | 737 |
| 120 000 | 256 | 252 | 3268 | 1111 | 2902 | 581 |

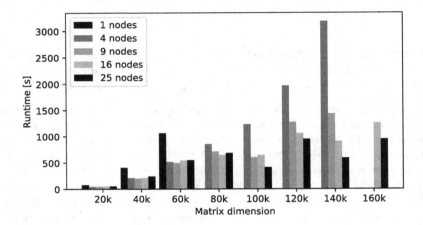

**Fig. 4.** Distributed memory scalability of StarNEig when computing a Schur form. Each node contains 28 CPU cores.

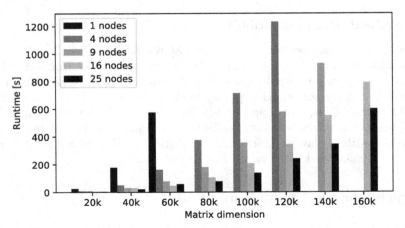

**Fig. 5.** Distributed memory scalability of StarNEig when reordering a Schur form. Each node contains 28 CPU cores.

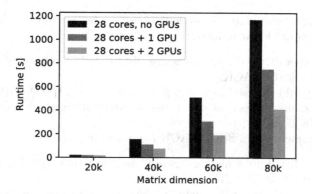

**Fig. 6.** GPU performance of StarNEig when reordering a Schur form. Each socket (14 cores) is connected to one NVidia Tesla V100 GPU.

islands) and 128 GB memory. The nodes are connected with FDR Infiniband. Each GPU compute node contains 28 Intel Xeon Gold 6132 cores (2 NUMA islands), 192 GB memory and two NVidia Tesla V100 GPUs.

The software was compiled with GCC 7.3.0 and linked to OpenMPI 3.1.3, OpenBLAS 0.3.2, ScaLAPACK 2.0.2, CUDA 9.2.88, and StarPU 1.2.8. All experiments were performed using a square MPI process grid. We always map each StarNEig process to a full node (28 cores) and each ScaLAPACK process to a single CPU core. The number of CPU cores in each ScaLAPACK experiment is always equal or larger than the number of CPU cores in the corresponding StarNEig experiment. The upper Hessenberg matrices for the Schur reduction experiments were computed from random matrices (entries uniformly distributed over the interval $[-1, 1]$).

Table 2 shows a comparison between ScaLAPACK and StarNEig[1]. We note that StarNEig is between 1.6 and 2.9 times faster than PDHSEQR and between 2.8 and 5.0 times faster than PDTRSEN. Figures 4 and 5 give some idea of how well the library is expected to scale in distributed memory. We note that StarNEig scales reasonably when computing the Schur form and almost linearly when reordering the Schur form. The iterative nature of the QR algorithm makes the Schur reduction results less predictable as different matrices require different number of bulge chasing steps. Figure 6 demonstrates that StarNEig can indeed take advantage of the available GPUs. See [20] for more comprehensive comparisons.

## 6   Summary

This paper presented a new library called StarNEig. The paper is aimed for potential users of the library. Various design choices were explained and con-

---

[1] StarNEig was compared against an updated version of PDHSEQR; see [13].

trasted to existing software. In particular, users who are already familiar with ScaLAPACK should know following:

- StarNEig expect that the matrices are distributed in relatively large blocks compared to ScaLAPACK.
- StarNEig should be used in a *one process per node* (1ppn) configuration as opposed to a *one process per core* (1ppc) configuration which is common with ScaLAPACK.
- StarNEig implements a ScaLAPACK compatibility layer.

The presented distributed memory results indicate that the library is highly competitive with ScaLAPACK.

Future work with StarNEig includes the implementation and integration of the missing software components. Support for complex valued matrices is also planned. The GPU support, and the multi-GPU support in particular, are still under active development. The authors hope to start a discussion which would help guide and prioritize the future development of the library.

**Acknowledgements.** StarNEig has been developed by the authors, Angelika Schwarz (who has written the standard eigenvector solver), Lars Karlsson, and Bo Kågström. This work is part of a project (NLAFET) that has received funding from the European Union's Horizon 2020 research and innovation programme under grant agreement No 671633. This work was supported by the Swedish strategic research programme eSSENCE. We thank the High Performance Computing Center North (HPC2N) at Umeå University for providing computational resources and valuable support during test and performance runs. Finally, the author thanks the anonymous reviewers for their valuable feedback.

# References

1. LAPACK: Linear Algebra PACKage. http://www.netlib.org/lapack
2. PLASMA: Parallel Linear Algebra Software for Multicore Architectures. http://icl.cs.utk.edu/plasma/software
3. ScaLAPACK: Scalable Linear Algebra PACKage. http://www.netlib.org/scalapack
4. SLEPc: The Scalable Library for Eigenvalue Problem Computations. http://slepc.upv.es
5. StarNEig: A task-based library for solving nonsymmetric eigenvalue problems. https://github.com/NLAFET/StarNEig
6. StarPU: A unified runtime system for heterogeneous multicore architectures. http://starpu.gforge.inria.fr
7. Braman, K., Byers, R., Mathias, R.: The multishift $QR$ algorithm. I. Maintaining well-focused shifts and level 3 performance. SIAM J. Matrix Anal. Appl. **23**(4), 929–947 (2002). https://doi.org/10.1137/S0895479801384573
8. Braman, K., Byers, R., Mathias, R.: The multishift $QR$ algorithm. II. Aggressive early deflation. SIAM J. Matrix Anal. Appl. **23**(4), 948–973 (2002). https://doi.org/10.1137/S0895479801384585

9. Buttari, A., Langou, J., Kurzak, J., Dongarra, J.: A class of parallel tiled linear algebra algorithms for multicore architectures. Parallel Comput. **35**(1), 38–53 (2009). https://doi.org/10.1016/j.parco.2008.10.002

10. Dongarra, J., Whaley, R.C.: A User's Guide to the BLACS, LAWN 94 (1997)

11. Golub, G.H., Van Loan, C.F.: Matrix Computations, 4th edn. Johns Hopkins University Press, Baltimore (2012)

12. Granat, R., Kågström, B., Kressner, D.: Parallel eigenvalue reordering in real Schur forms. Concurr. Comput.: Pract. Exp. **21**(9), 1225–1250 (2009), http://dx.doi.org/10.1002/cpe.1386

13. Granat, R., Kågström, B., Kressner, D., Shao, M.: ALGORITHM 953: parallel library software for the multishift QR algorithm with aggressive early deflation. ACM Trans. Math. Softw. **41**(4), 1–23 (2015). https://doi.org/10.1145/2699471

14. Imachi, H., Hoshi, T.: Hybrid numerical solvers for massively parallel eigenvalue computations and their benchmark with electronic structure calculations. J. Inf. Process. **24**(1), 164–172 (2016). https://doi.org/10.2197/ipsjjip.24.164

15. Kjelgaard Mikkelsen, C.C., Myllykoski, M.: Parallel robust computation of generalized eigenvectors of matrix pencils in real Schur form. Accepted to PPAM 2019 (2019)

16. Kjelgaard Mikkelsen, C.C., Schwarz, A.B., Karlsson, L.: Parallel robust solution of triangular linear systems. Concurr. Comput.: Pract. Exp. 1–19 (2018). https://doi.org/10.1002/cpe.5064

17. Luszczek, P., Ltaief, H., Dongarra, J.: Two-stage tridiagonal reduction for dense symmetric matrices using tile algorithms on multicore architectures. In: 2011 IEEE International Parallel & Distributed Processing Symposium (IPDPS), pp. 944–955. IEEE (2011). https://doi.org/10.1109/IPDPS.2011.91

18. Marek, A., et al.: The ELPA library: scalable parallel eigenvalue solutions for electronic structure theory and computational science. J. Phys.: Condens. Matter **26**(21), 201–213 (2014). https://doi.org/10.1088/0953-8984/26/21/213201

19. Myllykoski, M.: A task-based algorithm for reordering the eigenvalues of a matrix in real Schur form. In: Wyrzykowski, R., Dongarra, J., Deelman, E., Karczewski, K. (eds.) PPAM 2017. LNCS, vol. 10777, pp. 207–216. Springer, Cham (2018). https://doi.org/10.1007/978-3-319-78024-5_19

20. Myllykoski, M., Kjelgaard Mikkelsen, C.C., Schwarz, A., Kågström, B.: D2.7: eigenvalue solvers for nonsymmetric problems. Technical report, Umeå University (2019). http://www.nlafet.eu/wp-content/uploads/2019/04/D2.7-EVP-solvers-evaluation-final.pdf

21. Thibault, S.: On runtime systems for task-based programming on heterogeneous platforms, Habilitation à diriger des recherches, Université de Bordeaux (2018)

# Robust Task-Parallel Solution
# of the Triangular Sylvester Equation

Angelika Schwarz[(✉)] [ID] and Carl Christian Kjelgaard Mikkelsen [ID]

Department of Computing Science, Umeå University, Umeå, Sweden
{angies,spock}@cs.umu.se

**Abstract.** The Bartels-Stewart algorithm is a standard approach to solving the dense Sylvester equation. It reduces the problem to the solution of the triangular Sylvester equation. The triangular Sylvester equation is solved with a variant of backward substitution. Backward substitution is prone to overflow. Overflow can be avoided by dynamic scaling of the solution matrix. An algorithm which prevents overflow is said to be robust. The standard library LAPACK contains the robust scalar sequential solver `dtrsyl`. This paper derives a robust, level-3 BLAS-based task-parallel solver. By adding overflow protection, our robust solver closes the gap between problems solvable by LAPACK and problems solvable by existing non-robust task-parallel solvers. We demonstrate that our robust solver achieves a performance similar to non-robust solvers.

**Keywords:** Overflow protection · Task parallelism · Triangular Sylvester equation · Real Schur form

## 1 Introduction

The Bartels-Stewart algorithm is a standard approach to solving the general Sylvester equation

$$AX + XB = C \tag{1}$$

where $A \in \mathbb{R}^{m \times m}$, $B \in \mathbb{R}^{n \times n}$ and $C \in \mathbb{R}^{m \times n}$ are dense. It is well-known that (1) has a unique solution $X \in \mathbb{R}^{m \times n}$ if and only if the eigenvalues $\lambda_i^A$ of $A$ and $\lambda_j^B$ of $B$ satisfy $\lambda_i^A + \lambda_j^B \neq 0$ for all $i = 1, \ldots, m$ and $j = 1, \ldots, n$.

Sylvester equations occur in numerous applications including control and systems theory, signal processing and condition number estimation, see [13] for a summary. The case of $B = A^T$ corresponds to the continuous-time Lyapunov matrix equation, which is central in the analysis of linear time-invariant dynamical systems.

The Bartels-Stewart algorithm solves (1) by reducing $A$ and $B$ to upper quasi-triangular form $\tilde{A}$ and $\tilde{B}$. This reduces the problem to the solution of the triangular Sylvester equation

$$\tilde{A}Y + Y\tilde{B} = \tilde{C}. \tag{2}$$

© Springer Nature Switzerland AG 2020
R. Wyrzykowski et al. (Eds.): PPAM 2019, LNCS 12043, pp. 82–92, 2020.
https://doi.org/10.1007/978-3-030-43229-4_8

---

**Algorithm 1.** Triangular Sylvester Equation Solver

---

**Input:** $\tilde{A}$, $\tilde{B}$ as in (4), conformally partitioned $\tilde{C}$ as in (5).
**Output:** $Y \in \mathbb{R}^{m \times n}$ such that $\tilde{A}Y + Y\tilde{B} = \tilde{C}$.

1 $Y \leftarrow \tilde{C}$;
2 **for** $\ell \leftarrow 1, 2, \ldots, q$ **do**
3      **for** $k \leftarrow p, p-1, \ldots, 1$ **do**
4          Solve $\tilde{A}_{kk}Z + Z\tilde{B}_{\ell\ell} = Y_{k\ell}$ for $Z$;
5          $Y_{k\ell} \leftarrow Z$;
6          $Y_{1:k-1,\ell} \leftarrow Y_{1:k-1,\ell} - \tilde{A}_{1:k-1,k}Y_{k\ell}$;
7          $Y_{k,\ell+1:q} \leftarrow Y_{k,\ell+1:q} - Y_{k\ell}\tilde{B}_{\ell,\ell+1:q}$;
8 **return** $Y$;

---

During the solution of (2) through a variant of backward substitution, the entries of $Y$ can exhibit growth, possibly exceeding the representable floating-point range. To avoid such an overflow, the LAPACK 3.7.0 routine dtrsyl uses a scaling factor $\alpha \in (0, 1]$ to dynamically scale the solution. We say that an algorithm is *robust* if it cannot exceed the overflow threshold $\Omega$. With the scaling factor $\alpha$, the triangular Sylvester equation reads $\tilde{A}Y + Y\tilde{B} = \alpha\tilde{C}$. This paper focuses on the robust solution of the triangular Sylvester equation and improves existing non-robust task-parallel implementations for (2) by adding protection against overflow. This closes the gap between the class of problems solvable by existing task-parallel solvers and the class of problems solvable by LAPACK. Consequently, more problems can be solved efficiently in parallel through the Bartels-Stewart method.

We now describe the Bartels-Stewart algorithm for solving (1), see [1,6,13]. The algorithm computes the real Schur decompositions of $A$ and $B$

$$A = U\tilde{A}U^T, \quad B = V\tilde{B}V^T \tag{3}$$

using orthogonal transformations $U$ and $V$. Using (3), the general Sylvester Eq. (1) is transformed into the triangular Sylvester equation

$$\tilde{A}Y + Y\tilde{B} = \tilde{C}, \quad \tilde{C} = U^T CV, \quad Y = U^T XV.$$

The solution of the original system (1) is given by $X = UYV^T$. The real Schur forms $\tilde{A}$ and $\tilde{B}$ attain the shapes

$$\tilde{A} = \begin{bmatrix} \tilde{A}_{11} & \tilde{A}_{12} & \ldots & \tilde{A}_{1p} \\ & \tilde{A}_{22} & & \tilde{A}_{2p} \\ & & \ddots & \vdots \\ & & & \tilde{A}_{pp} \end{bmatrix} \in \mathbb{R}^{m \times m}, \quad \tilde{B} = \begin{bmatrix} \tilde{B}_{11} & \tilde{B}_{12} & \ldots & \tilde{B}_{1q} \\ & \tilde{B}_{22} & \ldots & \tilde{B}_{2q} \\ & & \ddots & \vdots \\ & & & \tilde{B}_{qq} \end{bmatrix} \in \mathbb{R}^{n \times n}, \quad (4)$$

where the diagonal blocks $\tilde{A}_{kk}$ and $\tilde{B}_{\ell\ell}$ are either 1-by-1 or 2-by-2. The right-hand side $\tilde{C} = U^T CV$ is partitioned conformally

$$\tilde{C} = U^T CV = \begin{bmatrix} \tilde{C}_{11} & \dots & \tilde{C}_{1q} \\ \vdots & \ddots & \vdots \\ \tilde{C}_{p1} & \dots & \tilde{C}_{pq} \end{bmatrix} \in \mathbb{R}^{m \times n}. \tag{5}$$

Adopting a block perspective, (2) reads

$$\tilde{A}_{kk} Y_{k\ell} + Y_{k\ell} \tilde{B}_{\ell\ell} = \tilde{C}_{k\ell} - \left( \sum_{i=k+1}^{p} \tilde{A}_{ki} Y_{i\ell} + \sum_{j=1}^{\ell-1} Y_{kj} \tilde{B}_{j\ell} \right) \tag{6}$$

for all blocks $k = 1, \dots, p$ and $\ell = 1, \dots q$. A straight-forward implementation of (6) is Algorithm 1. The algorithm starts at the bottom left corner ($k = p, \ell = 1$) and processes the block columns bottom-up from left to right. The flop count of the algorithm approximately corresponds to two backward substitutions and amounts to $\mathcal{O}(m^2 n + mn^2)$ flops.

The stability of Algorithm 1 and the Bartels-Stewart algorithm has been summarized in Higham [7]. In general, the computed solution $\hat{X}$ satisfies the bound

$$\|C - (A\hat{X} + \hat{X}B)\|_F \leq c_{m,n} u(\|A\|_F + \|B\|_F)\|X\|_F.$$

Hence, the relative residual is bounded by a small constant $c_{m,n}$ times the unit roundoff $u$. However, for the highly structured Sylvester equation, a small relative residual, does not imply a small backward error. Additional details can be found in Higham [7].

The rest of this paper is structured as follows. In Sect. 2 we formalize the definition of a robust algorithm and derive a new robust algorithm for solving the triangular Sylvester equation. Section 3 describes the execution environment used in the numerical experiments in Sect. 4. Section 5 summarizes the results.

## 2    Robust Algorithms for Triangular Sylvester Equations

In this section, we address the robust solution of the triangular Sylvester equation. The goal is to dynamically compute a scaling factor $\alpha \in (0, 1]$ such that the solution $Y$ of the scaled triangular Sylvester equation

$$\tilde{A} Y + Y \tilde{B} = \alpha \tilde{C} \tag{7}$$

can be obtained without ever exceeding the overflow threshold $\Omega > 0$. We derive two robust algorithms for solving (7). The first scalar algorithm can be viewed as an enhancement of LAPACK's dtrsyl by adding overflow protection to the linear updates. The second tiled algorithm redesigns the first algorithm such that most of the computation is executed as matrix-matrix multiplications.

## 2.1  Scalar Robust Algorithm

The central building block for adding robustness is PROTECTUPDATE, introduced by Mikkelsen and Karlsson in [9]. PROTECTUPDATE computes a scaling factor $\zeta \in (0, 1]$ such that the matrix update $\zeta C - A(\zeta Y)$ cannot overflow. PROTECTUPDATE uses the upper bounds $||C||_\infty$, $||A||_\infty$ and $||Y||_\infty$ to evaluate the maximum growth possible in the update. Provided that $||C||_\infty \leq \Omega$, $||A||_\infty \leq \Omega$ and $||Y||_\infty \leq \Omega$, PROTECTUPDATE computes a scaling factor $\zeta$ such that $\zeta(||C||_\infty + ||A||_\infty ||Y||_\infty) \leq \Omega$.

We use PROTECTUPDATE to protect the left and the right updates in the triangular Sylvester equation. We protect right updates by applying the scaling factor as $\zeta C - \tilde{A}(\zeta Y)$. We protect left updates by appling the scaling factor as $\zeta C - (\zeta Y)\tilde{B}$.

A solver for the triangular Sylvester equation requires the solution of small Sylvester equations $\tilde{A}_{kk} Z + Z\tilde{B}_{\ell\ell} = \beta Y_{k\ell}$, $\beta \in (0, 1]$. Since $\tilde{A}_{kk}$ and $\tilde{B}_{\ell\ell}$ are at most 2-by-2, these small Sylvester equations can be converted into linear systems of size at most 4-by-4, see [1]. We solve these linear systems robustly through Gaussian elimination with complete pivoting. This process requires linear updates and divisions to be executed robustly. We use PROTECTUPDATE for the small linear updates and guard divisions with PROTECTDIVISION from [9].

---

**Algorithm 2.** Robust Triangular Sylvester Equation Solver

---

**Input:** $\tilde{A}$, $\tilde{B}$ as in (4) with $||\tilde{A}_{ij}||_\infty \leq \Omega$, $||\tilde{B}_{k\ell}||_\infty \leq \Omega$, conformally partitioned $\tilde{C}$ as in (5) with $||\tilde{C}_{i\ell}||_\infty \leq \Omega$.
**Output:** $\alpha \in (0, 1]$, $Y \in \mathbb{R}^{m \times n}$ such that $\tilde{A}Y + Y\tilde{B} = \alpha\tilde{C}$.
**Ensure:** $||Y_{k\ell}||_\infty \leq \Omega$ for $Y$ partitioned analogously to $\tilde{C}$ as in (5).

1  ROBUSTSYL($\tilde{A}$, $\tilde{B}$, $\tilde{C}$)
2     $Y \leftarrow \tilde{C}$; $\alpha \leftarrow 1$;
3     for $\ell \leftarrow 1, 2, \ldots, q$ do
4        for $k \leftarrow p, p - 1, \ldots, 1$ do
5           Solve robustly $\tilde{A}_{kk} Z + Z\tilde{B}_{\ell\ell} = \beta Y_{k\ell}$ for $\beta$, $Z$;
6           $Y \leftarrow \beta Y$;
7           $Y_{k\ell} \leftarrow Z$;
8           $\gamma_1 \leftarrow$ PROTECTUPDATE($||Y_{1:k-1,\ell}||_\infty$, $||\tilde{A}_{1:k-1,k}||_\infty$, $||Y_{k\ell}||_\infty$);
9           $Y \leftarrow \gamma_1 Y$;
10          $Y_{1:k-1,\ell} \leftarrow Y_{1:k-1,\ell} - \tilde{A}_{1:k-1,k} Y_{k\ell}$;
11          $\gamma_2 \leftarrow$ PROTECTUPDATE(

$$\max_{\ell+1 \leq j \leq q}\{||Y_{k,j}||_\infty\}, ||Y_{k\ell}||_\infty, \max_{\ell+1 \leq j \leq q}\{||\tilde{B}_{\ell,j}||_\infty\});$$

12          $Y \leftarrow \gamma_2 Y$;
13          $Y_{k,\ell+1:q} \leftarrow Y_{k,\ell+1:q} - Y_{k\ell}\tilde{B}_{\ell,j+1:q}$;
14          $\alpha \leftarrow \alpha\beta\gamma_1\gamma_2$;
15    return $\alpha$, $Y$;

---

Algorithm 2 ROBUSTSYL adds overflow protection to Algorithm 1. ROBUSTSYL is dominated by level-2 BLAS-like thin linear updates. The next section develops a solver, which relies on efficient level-3 BLAS operations and uses RobustSyl as a basic building block.

## 2.2 Tiled Robust Algorithm

Solvers for the triangular Sylvester equation can be expressed as tiled algorithms [12] such that most of the computation corresponds to matrix-matrix multiplications. For this purpose, the matrices are partitioned into conforming, contiguous submatrices, so-called *tiles*. In order to decouple the tiles, the global scaling factor $\alpha$ is replaced with local scaling factors, one per tile of $Y$. The association of a tile with a scaling factor leads to *augmented tiles*.

**Definition 1.** *An augmented tile $\langle \alpha, X \rangle$ consists of a scalar $\alpha \in (0, 1]$ and a matrix $X \in \mathbb{R}^{m \times n}$ and represents the scaled matrix $Y = \alpha^{-1}X$. We say that two augmented tiles $\langle \alpha, X \rangle$ and $\langle \beta, Y \rangle$ are equivalent and we write $\langle \alpha, X \rangle = \langle \beta, Y \rangle$ if and only if $\alpha^{-1}X = \beta^{-1}Y$.*

**Definition 2.** *We say that two augmented tiles $\langle \alpha, X \rangle$ and $\langle \beta, Y \rangle$ are consistently scaled if $\alpha = \beta$.*

The idea of associating a scaling factor with a vector was introduced by Mikkelsen and Karlsson [9,10] who use augmented vectors to represent scaled vectors. We generalize their definition and associate a scaling factor with a tile. Their definition of consistently scaled vectors generalizes likewise to consistently scaled tiles.

Let $\tilde{A}$ and $\tilde{B}$ be as in (4). A partitioning of $\tilde{A}$ into M-by-M tiles and $\tilde{B}$ into N-by-N tiles such that 2-by-2 blocks on the diagonals are not split and the diagonal tiles are square induces a partitioning of $Y$ and $\tilde{C}$. We then solve the tiled equations

$$\tilde{A}_{kk}(\alpha_{k\ell}^{-1}Y_{k\ell}) + (\alpha_{k\ell}^{-1}Y_{k\ell})\tilde{B}_{\ell\ell} =$$
$$\tilde{C}_{k\ell} - \left( \sum_{i=k+1}^{M} \tilde{A}_{ki}(\alpha_{i\ell}^{-1}Y_{i\ell}) + \sum_{j=1}^{\ell-1}(\alpha_{kj}^{-1}Y_{kj})\tilde{B}_{j\ell} \right), \tag{8}$$

$k = 1, \ldots, M, \ell = 1, \ldots, N$ without explicitly forming any of the products $\alpha_{k\ell}^{-1}Y_{k\ell}$, $\alpha_{i\ell}^{-1}Y_{i\ell}$ and $\alpha_{ik}^{-1}Y_{ik}$. Note that (8) is structurally identical to (6). The solution of (8) requires augmented tiles to be updated. Algorithm 3 RobustUpdate executes such an update robustly. Combining RobustUpdate and RobustSyl leads to Algorithm 4, which solves (7) in a tiled fashion. The global scaling factor $\alpha$ corresponds to the smallest of the local scaling factors $\alpha_{k\ell}$. The solution $Y$ is obtained by scaling the tiles consistently.

Algorithm 4 can be parallelized with tasks. Task parallelism promises improved scalability, locality and load balancing and has been successfully applied to tiled algorithms [2,3,11]. An algorithm is represented as Directed Acyclic Graph (DAG), whose nodes constitute tasks and whose edges represent

---

**Algorithm 3.** Robust Linear Tile Update

---

**Input:** $A \in \mathbb{R}^{m \times k}$ with $||A||_\infty \leq \Omega$ , $B \in \mathbb{R}^{k \times n}$ with $||B||_\infty \leq \Omega$ , $C \in \mathbb{R}^{m \times n}$
    with $||C||_\infty \leq \Omega$ and scalars $\alpha$, $\beta$, $\gamma \in (0, 1]$.
**Output:** $D \in \mathbb{R}^{m \times n}$ and $\delta \in (0, 1]$ such that
    $(\delta^{-1} D) = (\gamma^{-1} C) - (\alpha^{-1} A)(\beta^{-1} B).$
1  ROBUSTUPDATE($\langle \gamma, C \rangle$, $\langle \alpha, A \rangle$, $\langle \beta, B \rangle$)
2     $\eta \leftarrow \min\{\gamma, \alpha, \beta\}$;
3     $\zeta \leftarrow$ PROTECTUPDATE$((\eta/\gamma)||C||_\infty), (\eta/\alpha)||A||_\infty, (\eta/\beta)||B||_\infty)$;
4     $\delta \leftarrow \eta\zeta$;
5     $D \leftarrow (\delta/\gamma)C - [(\delta/\alpha)A][(\delta/\beta)B]$;
6     **return** $\langle \delta, D \rangle$;

---

dependences between tasks. The tasks are scheduled asynchronously by a runtime system such that the dependencies are satisfied, see e.g. [4,5] for efficient scheduling strategies. For Algorithm 4, each function call corresponds to a task. ROBUSTSYL on $Y_{k\ell}$ has outgoing dependences to ROBUSTUPDATE modifying $Y_{il}$, $i = 1, \ldots, k - 1$ and $Y_{kj}$, $j = \ell + 1, \ldots N$. The incoming dependences of a ROBUSTSYL task on $Y_{k\ell}$ are satisfied when all updates modifying $Y_{k\ell}$ have been completed. The updates to $Y_{k\ell}$ require exclusive write access, which we achieve with a lock.

ROBUSTUPDATE tasks rely on upper bounds $||\tilde{A}_{ik}||_\infty$ and $||\tilde{B}_{\ell j}||_\infty$. Since the computation of norms is expensive, the matrix norms are precomputed and recorded. This is realized through perfectly parallel BOUND tasks. To limit the amount of dependences to be handled by the runtime system, a synchronization point separates this preprocessing step and ROBUSTSYL/ROBUSTUPDATE tasks.

Another synchronization point precedes the consistency scaling. The local scaling factors are reduced sequentially to the global scaling factor $\alpha$. The consistency scaling of the tiles is executed with independent SCALE tasks.

## 3 Experiment Setup

This section describes the setup of the numerical experiments. We specify the hardware, the solvers and their configuration, and the matrices of the triangular Sylvester equation.

*Execution Environment.* The experiments were executed on an Intel Xeon E5-2690v4 ("Broadwell") node with 28 cores arranged in two NUMA islands with 14 cores each. The theoretical peak performance in double-precision arithmetic is 41.6 GFLOPS/s for one core and 1164.8 GFLOPS/s for a full node. In the STREAM triad benchmark the single core memory bandwidth was measured at 19 GB/s; the full node reached 123 GB/s.

We use the GNU compiler 6.4.0 and link against single-threaded OpenBLAS 0.2.20 and LAPACK 3.7.0. The compiler optimization flags are `-O2 -xHost`. We forbid migration of threads and fill one NUMA island with threads before assigning threads to the second NUMA island by setting `OMP_PROC_BIND` to `close`.

---

**Algorithm 4.** Tiled Robust Triangular Sylvester Equation Solver

**Input:** $\tilde{A}$, $\tilde{B}$, $\tilde{C}$ as in (8) where $||\tilde{A}_{ij}||_\infty \leq \Omega$, $||\tilde{B}_{k\ell}||_\infty \leq \Omega$, $||\tilde{C}_{i\ell}||_\infty \leq \Omega$.
**Output:** $\alpha \in (0,1]$, $Y \in \mathbb{R}^{m \times n}$ such that $\tilde{A}Y + Y\tilde{B} = \alpha\tilde{C}$.
**Ensure:** For $Y$ as in (8) where $||Y_{k\ell}||_\infty \leq \Omega$.

```
1  DRSYLV(Ã, B̃, C̃)
2     for k ← M, M − 1, . . . , 1 do
3        for ℓ ← 1, . . . , N do
4           ⟨α_{kℓ}, Y_{kℓ}⟩ ← ⟨1, C̃_{kℓ}⟩;
5     for k ← M, M − 1, . . . , 1 do
6        for ℓ ← 1, . . . , N do
7           α_{kl}, Y_{kℓ} ← ROBUSTSYL(Ã_{kk}, B̃_{ℓℓ}, Y_{kℓ});
8           for i ← k − 1, k − 2, . . . , 1 do
9              ⟨α_{iℓ}, Y_{iℓ}⟩ ← ROBUSTUPDATE(⟨α_{iℓ}, Y_{iℓ}⟩, ⟨1, Ã_{ik}⟩, ⟨α_{kℓ}, Y_{kℓ}⟩);
10             for j ← ℓ + 1, ℓ + 2, . . . , N do
11                ⟨α_{kj}, Y_{kj}⟩ ← ROBUSTUPDATE(⟨α_{kj}, Y_{kj}⟩, ⟨α_{kℓ}, Y_{kℓ}⟩, ⟨1, B̃_{ℓj}⟩);
12    α ←   min    {α_{kℓ}} ;              // Compute global scaling factor
          1≤k≤M,1≤ℓ≤N
13    for k ← M, M − 1, . . . , 1 do
14       for ℓ ← 1, . . . , N do
15          Y_{kℓ} ← (α/α_{kℓ}) Y_{kℓ} ;      // Consistency scaling
16    return α, Y ;
```

---

*Software.* This section describes the routines used in the numerical experiments. The first two routines are non-robust, i.e., the routines solve $\tilde{A}Y + Y\tilde{B} = \tilde{C}$.

- FLA_Sylv. The libflame version 5.1.0-58 [12,14] solver partitions the problem into tiles and executes the linear updates as matrix-matrix multiplications.
- FLASH_Sylv. This libflame routine is the supermatrix version of FLA_Sylv and introduces task parallelism.

The following three routines are robust and solve $\tilde{A}Y + Y\tilde{B} = \alpha\tilde{C}$.

- dtrsyl. The scalar LAPACK 3.7.0 routine realizes overflow protection with a global scaling factor. Any scaling events triggers scaling of the *entire* matrix $Y$.
- recsy. The Fortran library[1] release 2009-12-28 [8] offers recursive blocked solvers for a variety of Sylvester and Lyapunov equations. Recursion allows expressing as much of the computation as possible as level-3 BLAS operations, where the efficiency improves by going up in the recursion tree due to larger matrix sizes. If overflow protection is triggered at some recursion level, the required scaling is propagated.
- drsylv. Our solver can be viewed as a robust version of FLASH_Sylv. Due to the usage of local scaling factors, our solver exhibits the same degree of parallelism as FLASH_Sylv.

---

[1] https://people.cs.umu.se/isak/recsy/.

*Test Matrices.* We design the system matrices such that the growth during the solve is controlled. This allows us to (a) design systems that the non-robust solvers must be able to solve and (b) examine the cost of robustness by increasing the growth and, in turn, the amount of scaling necessary.

The matrix $\tilde{C}$ and the upper triangular part of $\tilde{A}$ and $\tilde{B}$ are filled with ones. The diagonal blocks are set to $\tilde{A}_{ii} = \mu T_{ii}$ and $\tilde{B}_{ii} = \nu T_{ii}$, where $T_{ii}$ is either the 1-by-1 block given by $t_{ii} = 1$ or the 2-by-2 block given by

$$T_{ii} = \begin{bmatrix} 1 & 1 \\ -1 & 1 \end{bmatrix}.$$

The magnitude of the diagonal entries of $\tilde{A}$ and $\tilde{B}$ is controlled by $\mu$ and $\nu$. This also holds for 2-by-2 blocks, which encode a complex conjugate pair of eigenvalues. The 2-by-2 blocks cannot be reduced to triangular form using a real similarity transformation. A unitary transformation, however, can transform the matrix into triangular shape. As an example, consider the transformation $Q^H \tilde{A} Q$ into triangular shape for the 5-by-5 matrix

$$\tilde{A} = \begin{bmatrix} \mu & 1 & 1 & 1 & 1 \\ & \mu & 1 & 1 & 1 \\ & & \mu & \mu & 1 \\ & & -\mu & \mu & 1 \\ & & & & \mu \end{bmatrix}, \quad Q^H \tilde{A} Q = \begin{bmatrix} \mu & 1 & \frac{-1+i}{\sqrt{2}} & \frac{1-i}{\sqrt{2}} & 1 \\ & \mu & \frac{-1+i}{\sqrt{2}} & \frac{1-i}{\sqrt{2}} & 1 \\ & & \mu(1+i) & 0 & \frac{-1-i}{\sqrt{2}} \\ & & & \mu(1-i) & \frac{1+i}{\sqrt{2}} \\ & & & & \mu(1+i) \end{bmatrix}.$$

Small values of $\mu$ yield large intermediate results for the divisions during the backward substitution. Hence, the choice of $\mu$ and $\nu$ control the growth during the solve.

*Tuning.* The tile sizes for FLASH_Sylv and drsylv were tuned using $\mu = m$ and $\nu = n$. For each core count, a sweep over $[100, 612]$ with a step size of 32 was evaluated three times and the tile size with the best median runtime was chosen.

*Reliability.* We extend the relative residual defined by Higham [7] with the scaling factor $\alpha$ and evaluate

$$\frac{\|R\|_F}{(\|\tilde{A}\|_F + \|\tilde{B}\|_F)\|Y\|_F + \|\alpha \tilde{C}\|_F}, \tag{9}$$

where $R \leftarrow \alpha \tilde{C} - (\tilde{A}Y + Y\tilde{B})$. We report the median runtime of three runs.

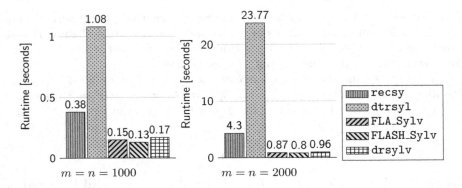

**Fig. 1.** Sequential runtime comparison on systems that do not require scaling.

# 4  Performance Results

This section presents three sets of performance results. First, the five solvers are executed in sequential mode. Second, the parallel scalability is analyzed. Third, the cost of robustness is investigated.

*Sequential Comparison without Numerical Scaling.* Figure 1 compares the solvers using $\mu = m$ and $\nu = n$. With this choice, dynamic downscaling of $Y$ is not necessary. Our solver `drsylv` is slightly slower than `FLA_Sylv` and `FLASH_Sylv`, possibly because the overhead from robustness cannot be amortized. Since the gap between `recsy` and `dtrsyl` on the one hand and `FLA_Sylv`, `FLASH_Sylv` and `drsylv` on the other hand grows with increasing matrix sizes, we restrict the parallel experiments to a comparison between `drsylv` and `FLASH_Sylv`.

*Strong Scalability without Numerical Scaling.* We examine the strong scalability of `FLASH_Syl` and `drsylv` on one shared memory node. Perfect strong scalability manifests itself in a constant efficiency when the number of cores is increased for a fixed problem. Figure 2 shows the results for $\mu = m$ and $\nu = n$. Robustness as implemented in `drsylv` does not hamper the scalability on systems that do not require scaling.

*Cost of Robustness.* Figure 3 (right) analyzes the cost of robustness. The amount of scaling necessary is controlled by fixing $\nu = 10^{-2}$ and varying $\mu$. The experiments with $\mu = 10^2, 10$ do not require scaling. The choice $\mu = 1$ triggers scaling for a few tiles. Frequent scaling is required for $\mu = 10^{-3}, 10^{-7}$. The cost of RobustUpdate increases when some of the three input tiles require scaling. In the worst case all three tiles have to be rescaled in every update. Figure 3 (left) shows that despite of robustness a decent fraction of the peak performance is reached.

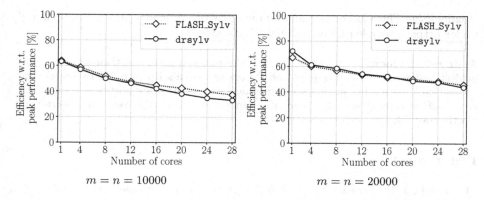

**Fig. 2.** Strong scalability on systems that do not require scaling.

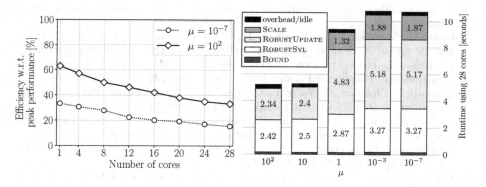

**Fig. 3.** Cost of robustness for $m = n = 10000$.

## 5    Conclusion

The solution of the triangular Sylvester equation is a central step in the Bartels-Stewart algorithm. During the backward substitution, the components of the solution can exhibit growth, possibly exceeding the representable floating-point range. This paper introduced a task-parallel solver with overflow protection. By adding overflow protection, task-parallel solvers can now solve the same set of problems that is solvable with LAPACK.

The numerical experiments revealed that the overhead of overflow protection is negligible when scaling is not needed. Hence, the non-robust solver offers no real advantage over our robust solver. When scaling is necessary, our robust algorithm automatically applies scaling to prevent overflow. While scaling increases the runtime, it guarantees a representable result. The computed solution can be evaluated in the context of the user's application. This certainty cannot be achieved with a non-robust algorithm.

*Availability.* The source code and an overview of available options can be found under https://people.cs.umu.se/angies/sylvester.

**Acknowledgements.** The authors thank the research group for their support. This project has received funding from the European Union's Horizon 2020 research and innovation programme under grant agreement No 671633. Support was received by eSSENCE, a collaborative e-Science programme funded by the Swedish Government via the Swedish Research Council (VR).

# References

1. Bartels, R.H., Stewart, G.W.: Solution of the matrix equation $AX + XB = C$. Commun. ACM **15**(9), 820–826 (1972)
2. Buttari, A., Langou, J., Kurzak, J., Dongarra, J.: A class of parallel tiled linear algebra algorithms for multicore architectures. Parallel Comput. **35**(1), 38–53 (2009)
3. Chan, E., Van Zee, F.G., Bientinesi, P., Quintana-Orti, E.S., Quintana-Orti, G., Van de Geijn, R.: Supermatrix: a multithreaded runtime scheduling system for algorithms-by-blocks. In: Proceedings of the 13th ACM SIGPLAN Symposium on Principles and Practice of Parallel programming, pp. 123–132. ACM (2008)
4. Cosnard, M., Jeannot, E., Yang, T.: Compact DAG representation and its symbolic scheduling. J. Parallel Distrib. Comput. **64**(8), 921–935 (2004)
5. Drebes, A., Pop, A., Heydemann, K., Cohen, A., Drach, N.: Scalable task parallelism for NUMA: a uniform abstraction for coordinated scheduling and memory management. In: International Conference on Parallel Architecture and Compilation Techniques, pp. 125–137. IEEE (2016)
6. Golub, G.H., Van Loan, C.F.: Matrix Computations, 3rd edn. John Hopkins University Press, Baltimore (1996)
7. Higham, N.J.: Accuracy and Stability of Numerical Algorithms, 2nd edn. SIAM, Philadelphia (2002)
8. Jonsson, I., Kågström, B.: Recursive blocked algorithms for solving triangular systems - Part I: one-sided and coupled Sylvester-type matrix equations. ACM TOMS **28**(4), 392–415 (2002)
9. Mikkelsen, C.C.K., Karlsson, L.: Robust solution of triangular linear systems. In: NLAFET Working Note 9 (2017)
10. Kjelgaard Mikkelsen, C.C., Karlsson, L.: Blocked algorithms for robust solution of triangular linear systems. In: Wyrzykowski, R., Dongarra, J., Deelman, E., Karczewski, K. (eds.) PPAM 2017. LNCS, vol. 10777, pp. 68–78. Springer, Cham (2018). https://doi.org/10.1007/978-3-319-78024-5_7
11. Perez, J.M., Badia, R.M., Labarta, J.: A dependency-aware task-based programming environment for multi-core architectures. In: 2008 IEEE International Conference on Cluster Computing, pp. 142–151. IEEE (2008)
12. Quintana-Ortí, E.S., Van De Geijn, R.A.: Formal derivation of algorithms: the triangular Sylvester equation. ACM TOMS **29**(2), 218–243 (2003)
13. Simoncini, V.: Computational methods for linear matrix equations. SIAM Rev. **58**(3), 377–441 (2016)
14. Van Zee, F.G.: libflame: The Complete Reference (version 5.1.0-56) (2009)

# Vectorized Parallel Solver for Tridiagonal Toeplitz Systems of Linear Equations

Beata Dmitruk[(✉)] and Przemysław Stpiczyński

Institute of Computer Science, Maria Curie–Skłodowska University,
ul. Akademicka 9, 20-033 Lublin, Poland
beata.dmitruk@umcs.pl, przem@hektor.umcs.lublin.pl

**Abstract.** The aim of this paper is to present two versions of a new *divide and conquer* parallel algorithm for solving tridiagonal Toeplitz systems of linear equations. Our new approach is based on a recently developed algorithm for solving linear recurrence systems. We discuss how to reduce the number of necessary synchronizations and show proper data layout that allows to use cache memory and SIMD extensions of modern processors. Numerical experiments show that our new implementations achieve very good seedup on multicore and manycore architectures. Moreover, they are more energy efficient than a simple sequential algorithm.

**Keywords:** Tridiagonal Toeplitz systems · Parallel algorithms · Vectorization · SIMD extensions · OpenMP · Energy efficiency

## 1 Introduction

Tridiagonal Toeplitz systems of linear equations play an important role in many theoretical and practical applications. They appear in numerical algorithms for solving boundary value problems for ordinary and partial differential equations [12,14]. For example, a numerical solution to the heat-diffusion equation of the following form

$$\frac{\partial u}{\partial t}(x,t) = \alpha^2 \frac{\partial^2 u}{\partial x^2}(x,t), \text{ for } 0 < x < l \text{ and } 0 < t,$$

with boundary conditions $u(0,t) = u(l,t) = 0$, for $0 < t$ and $u(x,0) = f(x)$, for $0 \leq x \leq l$ can be found using finite difference methods that reduce to the problem of solving tridiagonal Toeplitz systems. Such systems also arise in piecewise cubic interpolation and splines algorithms [2,11,13]. Moreover, such systems are useful when we solve the 2D Poisson equation by the variable separation method and the 3D Poisson equation by a combination of the alternating direction implicit and the variable separation methods [13]. Banded Toeplitz matrices also appear in signal and image processing [1].

There are several methods for solving such systems [3–5,7–9,13–15]. The basic idea comes from Rojo [9]. He proposed a method for solving symmetric

© Springer Nature Switzerland AG 2020
R. Wyrzykowski et al. (Eds.): PPAM 2019, LNCS 12043, pp. 93–103, 2020.
https://doi.org/10.1007/978-3-030-43229-4_9

tridiagonal Toeplitz systems using $LU$ decomposition of a system with almost Toeplitz structure together with Sherman-Morrison's formula [4]. This approach has been modified to obtain new solvers for a possible parallel execution [5,14]. A simple vectorized but non-parallel algorithm was proposed in [3]. A different approach was proposed by McNally et al. [8] who developed a scalable communication-less algorithm. It finds an *approximation* of the exact solution of a system with a given acceptable tolerance level. However, the algorithm does not utilize vectorization explicitly. Terekhov [13] proposed a highly scalable parallel algorithm for solving tridiagonal Toeplitz systems with multiple right hand sides. It should be noticed that well-known numerical libraries optimized for modern multicore architectures like Intel MKL, PLAPACK or NAG do not provide routines for solving tridiagonal Toeplitz systems. These libraries provide solvers for more general systems i.e.non-Toeplitz or dense Toeplitz systems. The case studied here is more detailed, but it allows to formulate more efficient solvers.

In this paper, we present two versions of a new *divide and conquer* parallel vectorized algorithm for finding the exact solution of tridiagonal Toeplitz systems of linear equations. As the starting point, we consider the splitting $T = LR + P$, where $L$, $R$ are bidiagonal and $P$ has only one non-zero entry [3]. Our new approach for solving bidiagonal Toeplitz systems is based on recently developed algorithms for solving linear recurrence systems [10,12]. We discuss possible OpenMP implementations and show how to reduce the number of necessary synchronizations to improve the performance. Further improvements come from the proper data layout that allows to use cache memory and SIMD extensions of modern processors. Numerical experiments performed on Intel Xeon CPUs and Intel Xeon Phi show that our new implementations achieve good performance on multicore and manycore architectures. Moreover, they are more energy efficient than a simple sequential algorithm.

## 2    Parallel Algorithm for Tridiagonal Toeplitz Systems

Let us consider a tridiagonal Toeplitz system of linear equations $T\mathbf{x} = \mathbf{b}$ of the following form

$$
\begin{bmatrix}
d & a & & & \\
1 & d & a & & \\
& \ddots & \ddots & \ddots & \\
& & 1 & d & a \\
& & & 1 & d
\end{bmatrix}
\begin{bmatrix}
x_0 \\
x_1 \\
\vdots \\
\vdots \\
x_{n-1}
\end{bmatrix}
=
\begin{bmatrix}
b_0 \\
b_1 \\
\vdots \\
\vdots \\
b_{n-1}
\end{bmatrix}.
\tag{1}
$$

For the sake of simplicity let us assume that $n = 2^m$, $m \in \mathbb{N}$, and $|d| > 1 + |a|$. Thus, $T$ is not singular and pivoting is not needed to assure numerical stability. To find the solution to (1) we can follow the approach presented in [3] and decompose $T$ as follows

$$
\begin{bmatrix} d & a & & & \\ 1 & d & a & & \\ & \ddots & \ddots & \ddots & \\ & & 1 & d & a \\ & & & 1 & d \end{bmatrix} = \underbrace{\begin{bmatrix} r_2 & & & \\ 1 & r_2 & & \\ & \ddots & \ddots & \\ & & 1 & r_2 \\ & & & 1 & r_2 \end{bmatrix}}_{L} \underbrace{\begin{bmatrix} 1 & r_1 & & \\ & 1 & r_1 & \\ & & \ddots & \ddots \\ & & & 1 & r_1 \\ & & & & 1 \end{bmatrix}}_{R} + \begin{bmatrix} r_1 & 0 & \cdots & 0 \\ 0 & 0 & & \vdots \\ \vdots & & \ddots & \vdots \\ 0 & \cdots & \cdots & 0 \end{bmatrix}, \quad (2)
$$

where $r_2 = (d \pm \sqrt{d^2 - 4a})/2$ and $r_1 = d - r_2$. Using this formula we can rewrite the Eq. (1) as follows

$$
\begin{bmatrix} x_0 \\ x_1 \\ \vdots \\ x_{n-1} \end{bmatrix} + r_1 x_0 \, (LR)^{-1} \underbrace{\begin{bmatrix} 1 \\ 0 \\ \vdots \\ 0 \end{bmatrix}}_{\mathbf{u}} = (LR)^{-1} \underbrace{\begin{bmatrix} b_0 \\ b_1 \\ \vdots \\ b_{n-1} \end{bmatrix}}_{\mathbf{v}} \qquad (3)
$$

or simply $\mathbf{x} + r_1 x_0 \mathbf{u} = \mathbf{v}$. Then the solution to (1) can be found using

$$
\begin{cases} x_0 = \frac{v_0}{1 + r_1 u_0} \\ x_i = v_i - r_1 x_0 u_i, & i = 1, \ldots, n-1. \end{cases} \qquad (4)
$$

Let $\mathbf{e}_k = \underbrace{(0, \ldots, 0, 1, 0, \ldots, 0)}_{k}{}^T \in \mathbb{R}^s$, $k = 0, \ldots, s - 1$. Note that to apply (4) we only need to have vectors $\mathbf{u}$ and $\mathbf{v}$ that are solutions to systems of linear equations $LR\mathbf{u} = \mathbf{e}_0$ and $LR\mathbf{v} = \mathbf{b}$, respectively. The solution to the system of linear equations $LR\mathbf{y} = \mathbf{f}$ can be found using a simple sequential algorithm based on the following recurrence relations

$$
\begin{cases} z_0 = f_0/r_2 \\ z_i = (f_i - z_{i-1})/r_2, & i = 1, \ldots, n-1, \end{cases} \qquad (5)
$$

and

$$
\begin{cases} y_{n-1} = z_{n-1} \\ y_i = z_i - r_1 y_{i+1}, & i = n-2, \ldots, 0. \end{cases} \qquad (6)
$$

Thus, to solve (1) using (4), such a simple sequential algorithm based on (5) and (6) should be applied twice for $LR\mathbf{u} = \mathbf{e}_0$ and $LR\mathbf{v} = \mathbf{b}$, respectively. This sequential algorithm requires $9n + O(1)$ flops. It should be pointed out that (5) and (6) contain obvious data dependencies, thus they cannot be parallelized and vectorized automatically by the compiler.

To develop a new parallel algorithm for solving $LR\mathbf{y} = \mathbf{f}$ that could utilize vector extensions of modern multiprocessors, we apply the *divide-and-conquer* algorithm for solving first-order linear recurrence systems with constant coefficients [10,12]. Let us assume that $n = rs$ and $r, s > 1$. First, we arrange entries of $L$ into blocks to obtain the following block matrix

$$L = \begin{bmatrix} L_s & & & \\ B & L_s & & \\ & \ddots & \ddots & \\ & & B & L_s \end{bmatrix}, \quad L_s = \begin{bmatrix} r_2 & & & \\ 1 & r_2 & & \\ & \ddots & \ddots & \\ & & 1 & r_2 \end{bmatrix}, \quad B = \begin{bmatrix} 0 & \dots & 0 & 1 \\ \vdots & & 0 & 0 \\ \vdots & \ddots & & \vdots \\ 0 & \dots & \dots & 0 \end{bmatrix}. \quad (7)$$

Let $\mathbf{z}_i = (z_{is}, \dots, z_{(i+1)s-1})^T$ and $\mathbf{f}_i = (f_{is}, \dots, f_{(i+1)s-1})^T$. Then the lower bidiagonal system of linear equations $L\mathbf{z} = \mathbf{f}$ can be rewritten in the following recursive form

$$\begin{cases} L_s\mathbf{z}_0 = \mathbf{f}_0 \\ L_s\mathbf{z}_i = \mathbf{f}_i - B\mathbf{z}_{i-1}, \quad i = 1, \dots, r-1. \end{cases} \quad (8)$$

Equation (8) reduces to the following form

$$\begin{cases} \mathbf{z}_0 = L_s^{-1}\mathbf{f}_0 \\ \mathbf{z}_i = L_s^{-1}\mathbf{f}_i - z_{is-1}L_s^{-1}\mathbf{e}_0, \quad i = 1, \dots, r-1. \end{cases} \quad (9)$$

Note that (9) has a lot of potential parallelism. Just after all vectors $L_s^{-1}\mathbf{f}_i$, $i = 0, \dots, r-1$, have been found, we can apply (9) to find $\mathbf{z}_i$, $i = 1, \dots, r-1$, "one-by-one" using the OpenMP "for simd" construct. It is clear that to find $\mathbf{z}_i$ we need the last entry of $\mathbf{z}_{i-1}$. Thus, before calculating the next vector, all threads should be synchronized. Alternatively, we can find last entries of all vectors $\mathbf{z}_i$ and then $s - 1$ first entries can be found in parallel without the need for the synchronization of threads.

Similarly, in case of the upper bidiagonal system $R\mathbf{y} = \mathbf{z}$, assuming the same as previously, we get

$$R = \begin{bmatrix} R_s & C & & \\ & R_s & \ddots & \\ & & \ddots & C \\ & & & R_s \end{bmatrix}, \quad R_s = \begin{bmatrix} 1 & r_1 & & \\ & 1 & \ddots & \\ & & \ddots & r_1 \\ & & & 1 \end{bmatrix}, \quad C = \begin{bmatrix} 0 & \dots & \dots & 0 \\ \vdots & & \ddots & \vdots \\ 0 & 0 & & \vdots \\ r_1 & 0 & \dots & 0 \end{bmatrix}. \quad (10)$$

The solution of the system (i.e. vectors $\mathbf{y}_i$, $i = 0, \dots, r-1$), satisfies

$$\begin{cases} R_s\mathbf{y}_{r-1} = \mathbf{z}_{r-1} \\ R_s\mathbf{y}_i = \mathbf{z}_i - C\mathbf{y}_{i+1}, \quad i = r-2, \dots, 0. \end{cases} \quad (11)$$

Finally, we get

$$\begin{cases} \mathbf{y}_{r-1} = R_s^{-1}\mathbf{z}_{r-1} \\ \mathbf{y}_i = R_s^{-1}\mathbf{z}_i - r_1 y_{(i+1)s}R_s^{-1}\mathbf{e}_{s-1}, \quad i = r-2, \dots, 0. \end{cases} \quad (12)$$

We have a similar situation as previously, but to find $\mathbf{y}_i$, $i = 0, \dots, r-2$, we need to know first entries of all vectors $\mathbf{y}_i$, $i = 1, \dots, r-1$. Then we can find other entries simultaneously. Such a parallel algorithm requires $16n - 3r - 6s + O(1)$ flops.

# 3   Implementation and Results of Experiments

Following (9), (12) and (4) we can formulate two OpenMP versions of our algorithm for solving (1). They are presented in Fig. 1. The overall structure of both versions is the same. We assume that the value of $s$ is a power of two, thus each column is properly aligned in memory (lines 14, 32). Moreover, each column occupies a contiguous block in memory. We have two kinds of loops. The first one (lines 12–19) does not utilize vector extensions, but can be executed in parallel. Note that the inner loop (lines 17–18) retrieves successive elements of columns, thus necessary entries can be found in cache memory. Lines 35–37 contain another kind of loop. It is a parallel loop that utilize vector extensions (using the OpenMP "for simd" construct). The difference between the versions is relatively small. In case of the first version, there is an implicit barrier after the inner loop 35–37, because we need the last entry of the previous column in the next iteration of the outer loop (lines 30–39). Alternatively, we find all last entries in a sequential loop (lines 26–28) and then the inner loop (lines 35–37) can be launched with the "nowait" clause. However, the explicit barrier must be issued after the outer loop (line 39) to ensure that all necessary computations have been completed.

All experiments have been carried out on a server with two Intel Xeon E5-2670 v3 processors (CPU) (totally 24 cores, 2.3 GHz, 256-bit AVX2), 128 GB RAM and a server with Intel Xeon Phi Coprocessor 7120P (KNC, 61 cores with multithreading, 1.238 GHz, 16 GB RAM, 512-bit vector extensions) which is an example of Intel MIC architecture, running under Linux with Intel Parallel Studio ver. 2017. Experiments on Xeon Phi have been carried out using its native mode. We have tested Sequential implementation based on (5), (6), (4), optimized automatically by the compiler, and two versions of our parallel algorithm (Version_1 and Version_2). On both architectures (CPU and MIC), in most cases, the best performance of the parallel algorithm is obtained for one thread per core. However, on MIC for larger problem sizes, Version_1 achieves better performance for two threads per core.

Examples of the results are presented in Figs. 2, 3 and Tables 1, 2. Figures 2, 3 show the execution time for various $n \in \{2^{26} \text{ or } 2^{27}, \ldots, 2^{29} \text{ or } 2^{30}\}$ (depending on the architecture) and $r$.

It should be observed that the right choice of $r$ and $s = n/r$ is very important for achieving good performance. When a bigger value of $r$ is used, the performance is better only to a certain value of r. A bigger $r$ implies smaller $s$. Thus we need to find a trade-off between getting the benefits form separating one part of loop and making this loop too long. Both parallel implementations achieve the best performance when the value of $r$ is a small multiple of the number of available cores.

```
1   void method(int n,int r,double a,double d,double *b,double *u){
2       const double r2=(d>0)?((d+sqrt(d*d-4*a))/2):((d-sqrt(d*d-4*a))/2);
3       const double r1=d-r2, tmp=-1/r2;
4       u[0]=1./r2;
5       int s=n/r;
6       double *es=_mm_malloc(s*sizeof(double), 64); es[s-1]=1;
7
8       #pragma omp parallel  // parallel region starts here
9       {
10        double tmp=u[0];
11        #pragma omp for nowait schedule(static)
12        for(int j=0;j<r;j++){
13          double *col;
14          __assume_aligned(col,64); // each column is properly aligned
15          col=&b[j*s];
16          col[0]/=r2;
17          for(int i=1;i<s;i++)
18            col[i]=(col[i]-col[i-1])/r2;
19        }
20        #pragma omp single
21        for(int i=1;i<s;i++)
22          u[i]=u[i-1]*tmp;
23        // implicit barrier
```

```
24  //version 1                           //version 2
25
26                                          #pragma omp single
27                                          for(int j=1;j<r;j++)
28                                            b[(j+1)*s-1]-=b[j*s-1]*u[s-1];
29
30  for(int j=1;j<r;j++){                   for(int j=1;j<r;j++){
31    double *col;                           double *col;
32    __assume_aligned(col,64);              __assume_aligned(col,64);
33    col=&b[j*s];                           col=&b[j*s];
34    double last=b[j*s-1];                  double last=b[j*s-1];
35    #pragma omp for simd \                 #pragma omp for simd nowait \
36           schedule(static)                       schedule(static)
37    for(int i=0;i<s;i++)                   for(int i=0;i<s-1;i++)
38      col[i]-=last*u[i];                     col[i]-=last*u[i];
39    // implicit barrier                  }
40  }                                    #pragma omp barrier
41  ...the rest of the implementation    ...the rest of the implementation
```

```
41        #pragma omp single
42        b[0]/=(1+r1*u[0]);
43
44        tmp=r1*b[0];
45        #pragma omp for simd schedule(static)
46        for(int i=1;i<n;i++)
47          b[i]-=tmp*u[i];
48
49    } // end of parallel region
50    _mm_free(es);
51  }
```

**Fig. 1.** Two OpenMP versions of the parallel algorithm (abbreviated)

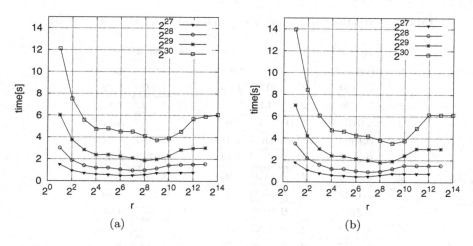

**Fig. 2.** Execution time for various $n$ and $r$ on CPU: `Version_1` (a), `Version_2` (b)

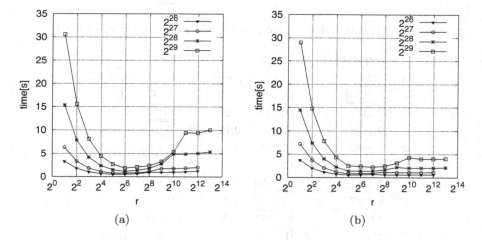

**Fig. 3.** Execution time for various $n$ and $r$ on MIC: `Version_1` (a), `Version_2` (b)

Tables 1, 2 show the execution time and speedup obtained for optimal values of $r$. On CPU, `Version_1` achieves better performance for smaller values of $n$, but on MIC the situation is reversed: `Version_1` achieves better performance for bigger values of $n$. Better speedup (up to 30) and efficiency can be observed on MIC, where the use of vector extensions is crucial for achieving good performance.

**Table 1.** Execution time [s] and speedup for optimal values of $r$ (CPU)

|  | Sequential | Version_1 | | | Version_2 | | |
|---|---|---|---|---|---|---|---|
| n | time | r | time | speedup | r | time | speedup |
| $2^{20}$ | 0.0160 | $2^4$ | 0.0095 | 1.69 | $2^5$ | 0.0090 | 1.79 |
| $2^{21}$ | 0.0336 | $2^4$ | 0.0141 | 2.37 | $2^6$ | 0.0119 | 2.83 |
| $2^{22}$ | 0.0704 | $2^6$ | 0.0233 | 3.02 | $2^6$ | 0.0219 | 3.22 |
| $2^{23}$ | 0.1393 | $2^4$ | 0.0409 | 3.41 | $2^8$ | 0.0435 | 3.21 |
| $2^{24}$ | 0.2786 | $2^4$ | 0.0746 | 3.74 | $2^4$ | 0.0783 | 3.56 |
| $2^{25}$ | 0.5572 | $2^4$ | 0.1410 | 3.95 | $2^4$ | 0.1433 | 3.89 |
| $2^{26}$ | 1.1143 | $2^6$ | 0.2609 | 4.27 | $2^6$ | 0.2621 | 4.25 |
| $2^{27}$ | 2.2263 | $2^6$ | 0.4873 | 4.57 | $2^6$ | 0.4755 | 4.68 |
| $2^{28}$ | 4.4509 | $2^7$ | 0.9606 | 4.63 | $2^7$ | 0.9183 | 4.85 |
| $2^{29}$ | 8.9077 | $2^8$ | 1.8586 | 4.79 | $2^8$ | 1.7777 | 5.01 |
| $2^{30}$ | 17.8155 | $2^9$ | 3.7072 | 4.81 | $2^9$ | 3.5225 | 5.06 |

**Table 2.** Execution time [s] and speedup for optimal values of $r$ (MIC)

|  | Sequential | Version_1 | | | Version_2 | | |
|---|---|---|---|---|---|---|---|
| n | time | r | time | speedup | r | time | speedup |
| $2^{20}$ | 0.1095 | $2^6$ | 0.1503 | 0.73 | $2^6$ | 0.1391 | 0.79 |
| $2^{21}$ | 0.2177 | $2^6$ | 0.1711 | 1.27 | $2^7$ | 0.1524 | 1.43 |
| $2^{22}$ | 0.4328 | $2^5$ | 0.1981 | 2.19 | $2^7$ | 0.1705 | 2.54 |
| $2^{23}$ | 0.8582 | $2^4$ | 0.2458 | 3.49 | $2^8$ | 0.2008 | 4.27 |
| $2^{24}$ | 1.7095 | $2^5$ | 0.2974 | 5.75 | $2^8$ | 0.2633 | 6.49 |
| $2^{25}$ | 3.4111 | $2^5$ | 0.3649 | 9.35 | $2^9$ | 0.3671 | 9.29 |
| $2^{26}$ | 6.8127 | $2^5$ | 0.5095 | 13.37 | $2^5$ | 0.5357 | 12.72 |
| $2^{27}$ | 13.6193 | $2^5$ | 0.7563 | 18.01 | $2^5$ | 0.8088 | 16.84 |
| $2^{28}$ | 27.2880 | $2^6$ | 1.1568 | 23.59 | $2^5$ | 1.3780 | 19.80 |
| $2^{29}$ | 54.5765 | $2^6$ | 1.8531 | 29.45 | $2^7$ | 2.2310 | 24.46 |

## 4   The Energy Efficiency

Figure 4 and Table 3 present the exemplary results of our experiments concerning the energy efficiency of our implementations. Data for this plot have been collected on the server with two Intel Xeon E5-2670 v3 processors using Intel's Running Average Power Limit (RAPL) [6]. This interface enables to measure the power consumption for CPUs and DRAMs. Figure 4 shows how the power consumption changes during the execution of the program, which comprises calls to Sequential, Version_1 and Version_2, respectively.

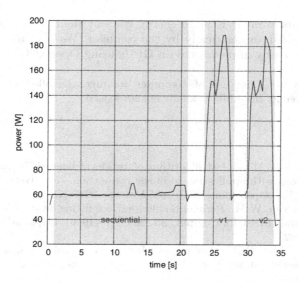

**Fig. 4.** Total power consumption required by all considered implementations.

We also present the total power consumption of CPUs and DRAMs for $n = 2^{30}$ and the optimal value of $r$ (in case of the parallel implementations). We start with the sequential method (it takes 18.2 s), next we perform Version_1 (4.2 s) and Version_2 (3.8 s). It is clear that current power draw during the execution of Version_1 and Version_2 is much higher, but it only lasts for a short time.

The power consumption [J] for various problem sizes is presented in Table 3. We can observe that both parallel versions need about 50% of the energy required by Sequential.

**Table 3.** Total power consumption [J] on CPU required by all considered implementations

| n | $2^{20}$ | $2^{21}$ | $2^{22}$ | $2^{23}$ | $2^{24}$ | $2^{25}$ | $2^{26}$ | $2^{27}$ | $2^{28}$ | $2^{29}$ | $2^{30}$ |
|---|---|---|---|---|---|---|---|---|---|---|---|
| Sequential | 1.06 | 2.10 | 4.10 | 8.57 | 17.57 | 35.88 | 70.25 | 139.68 | 277.47 | 544.99 | 1092.91 |
| Version_1 | 1.25 | 1.98 | 3.17 | 5.44 | 10.26 | 20.80 | 40.17 | 75.03 | 147.28 | 291.43 | 583.29 |
| Version_2 | 0.97 | 1.95 | 3.11 | 6.03 | 11.02 | 20.78 | 39.17 | 71.32 | 142.93 | 281.67 | 543.25 |

## 5   Conclusions and Future Work

We have presented the new vectorized parallel algorithm for solving tridiagonal Toeplitz systems of linear equations. Numerical experiments have shown that it achieves good performance and speedup on multicore and especially manycore architectures. Moreover, it is more energy efficient than the simple sequential algorithm optimized automatically by the compiler. We plan to show that our approach can be easily implemented on GPUs using OpenACC.

**Acknowledgements.** The use of computer resources installed at Maria Curie-Skłodowska University in Lublin is kindly acknowledged.

# References

1. Belhaj, S., Dridi, M.: A fast algorithm of two-level banded Toeplitz systems of linear equations with application to image restoration. New Trends Math. Sci. **2**, 277–283 (2017). https://doi.org/10.20852/ntmsci.2017.178
2. Chung, K.L., Yan, W.M.: Parallel B-spline surface fitting on mesh-connected computers. J. Parallel Distrib. Comput. **35**, 205–210 (1996). https://doi.org/10.1006/jpdc.1996.0082
3. Chung, K.L., Yan, W.M.: Vectorized algorithms for solving special tridiagonal systems. Comput. Math. Appl. **32**, 1–14 (1996). https://doi.org/10.1016/S0898-1221(96)00203-9
4. Du, L., Sogabe, T., Zhang, S.L.: A fast algorithm for solving tridiagonal quasi-Toeplitz linear systems. Appl. Math. Lett. **75**, 74–81 (2018). https://doi.org/10.1016/j.aml.2017.06.016
5. Garey, L., Shaw, R.: A parallel method for linear equations with tridiagonal Toeplitz coefficient matrices. Comput. Math. Appl. **42**(1), 1–11 (2001). https://doi.org/10.1016/S0898-1221(01)00125-0
6. Khan, K.N., Hirki, M., Niemi, T., Nurminen, J.K., Ou, Z.: RAPL in action: experiences in using RAPL for power measurements. ACM Trans. Model. Perform. Eval. Comput. Syst. **3**(2), 9:1–9:26 (2018). https://doi.org/10.1145/3177754
7. McNally, J.M., Garey, L.E., Shaw, R.E.: A split-correct parallel algorithm for solving tridiagonal symmetric Toeplitz systems. Int. J. Comput. Math. **75**(3), 303–313 (2000). https://doi.org/10.1080/00207160008804986
8. McNally, J.M., Garey, L., Shaw, R.: A communication-less parallel algorithm for tridiagonal Toeplitz systems. J. Comput. Appl. Math. **212**, 260–271 (2008). https://doi.org/10.1016/j.cam.2006.12.001
9. Rojo, O.: A new method for solving symmetric circulant tridiagonal systems of linear equations. Comput. Math. Appl. **20**, 61–67 (1990). https://doi.org/10.1016/0898-1221(90)90165-G
10. Stpiczyński, P.: Solving linear recurrence systems using level 2 and 3 BLAS routines. In: Wyrzykowski, R., Dongarra, J., Paprzycki, M., Waśniewski, J. (eds.) PPAM 2003. LNCS, vol. 3019, pp. 1059–1066. Springer, Heidelberg (2004). https://doi.org/10.1007/978-3-540-24669-5_137
11. Stpiczyński, P., Potiopa, J.: Piecewise cubic interpolation on distributed memory parallel computers and clusters of workstations. In: Fifth International Conference on Parallel Computing in Electrical Engineering (PARELEC 2006), Bialystok, Poland, 13–17 September 2006, pp. 284–289. IEEE Computer Society (2006). https://doi.org/10.1109/PARELEC.2006.68
12. Stpiczyński, P., Potiopa, J.: Solving a kind of boundary-value problem for ordinary differential equations using Fermi—the next generation CUDA computing architecture. J. Comput. Appl. Math. **236**, 384–393 (2011). https://doi.org/10.1016/j.cam.2011.07.028

13. Terekhov, A.V.: A highly scalable parallel algorithm for solving Toeplitz tridiagonal systems of linear equations. J. Parallel Distrib. Comput. **87**, 102–108 (2016). https://doi.org/10.1016/j.jpdc.2015.10.004
14. Vidal, A.M., Alonso, P.: Solving systems of symmetric Toeplitz tridiagonal equations: Rojo's algorithm revisited. Appl. Math. Comput. **219**, 1874–1889 (2012). https://doi.org/10.1016/j.amc.2012.08.030
15. Wang, H.H.: A parallel method for tridiagonal equations. ACM Trans. Math. Softw. **7**(2), 170–183 (1981). https://doi.org/10.1145/355945.355947

# Parallel Performance of an Iterative Solver Based on the Golub-Kahan Bidiagonalization

Carola Kruse[1]([✉]), Masha Sosonkina[2], Mario Arioli[3], Nicolas Tardieu[4], and Ulrich Rüde[1,5]

[1] Cerfacs, 42 Avenue Gaspard Coriolis, 31100 Toulouse, France
carola.kruse@cerfacs.fr
[2] Department of Computational Modeling and Simulation Engineering,
Old Dominion University, Norfolk, VA 23529, USA
[3] Libera Universita Mediterranea, Strada Statale, 73125 Casamassima, BA, Italy
[4] IMSIA, UMR 9219 EDF-CNRS-CEA-ENSTA, Université Paris Saclay,
828 Boulevard des Maréchaux, 91762 Palaiseau Cedex, France
[5] Friedrich-Alexander-Universität Erlangen-Nürnberg,
Cauerstr. 11, 91058 Erlangen, Germany

**Abstract.** We present an iterative method based on a generalization of the Golub-Kahan bidiagonalization for solving indefinite matrices with a $2 \times 2$ block structure. We focus in particular on our recent implementation of the algorithm using the parallel numerical library PETSc. Since the algorithm is a nested solver, we investigate different choices for parallel inner solvers and show its strong scalability for two Stokes test problems. The algorithm is found to be scalable for large sparse problems.

**Keywords:** Golub-Kahan bidiagonalization · Iterative solver · PETSc · Parallel performance

## 1 Introduction

As current and future high-performance computing (HPC) platforms scale, calculations will be able to increase in size and complexity and take advantage of the available processing power and memory. At this scale, HPC applications will increasingly rely on the parallel numerical libraries and environments, such as those offered by Trilinos [17] or PETSc [4] for the solution of large-scale linear systems using either direct or iterative methods, to abstract the low-level parallel programming details and enable application users and developers to focus on their domain problem at hand. A report from the U.S. Department of Energy [9] notes that "Numerical libraries will continue to play an important role at the

M. Sosonkina—The work of the second author was supported in part by the U.S. National Science Foundation under grant CNS-1828593.

exascale" and that they will allow to share the methods implemented therein among "applications with similar characteristics".

In this contribution, we focus on iterative solvers for indefinite saddle point systems of the type

$$\begin{pmatrix} \mathbf{W} & \mathbf{A} \\ \mathbf{A}^T & 0 \end{pmatrix} \begin{pmatrix} \mathbf{w} \\ \mathbf{p} \end{pmatrix} = \begin{pmatrix} \mathbf{g} \\ \mathbf{r} \end{pmatrix}, \tag{1}$$

with a symmetric positive semi-definite matrix $\mathbf{W} \in \mathbb{R}^{m \times m}$ and $\mathbf{A} \in \mathbb{R}^{m \times n}$. These systems arise in many applications and their efficient solution is an active research area. A comprehensive review of application fields, solvers and preconditioners can be found in [7]. In [2], Arioli proposed a new iterative algorithm by generalizing the standard Golub-Kahan bidiagonalization to matrices of type (1). In a recent project jointly with the French electric utility EDF, we further investigated this generalized Golub-Kahan bidiagonalization solver (GKB) [3]. We have shown on the industrial test case of the structural analysis of nuclear reactor containment buildings that the solver converges in only a few steps and that it is competitive with popular sparse direct solvers in sequential execution. We have, however, not yet provided results on the GKB parallel performance, which is the main focus of this contribution.

A major milestone in the project was the deployment of the developed GKB-based solver for industrial use at EDF. In several application fields, numerical studies are run in the company with the initially in-house and then later open source finite element software CODE_ASTER[1]. It is interfaced with PETSc, which motivated our selection to implement the GKB solver into this framework. A major advantage of this choice is that the algorithm leverages many PETSc features, such as high-degree of parallelism and efficient parallel implementation of basic sparse linear algebra operations [6].

This paper is organized as follows. Some theoretical aspects of the GKB algorithm are reviewed in Sect. 2. Then we will comment on its implementation and usage in PETSc in Sect. 3. In Sect. 4, we introduce two test problems and determine parameters and stopping criteria. Finally we show that the GKB solver is scalable when an efficient inner solver is used.

## 2    Generalized Golub-Kahan Bidiagonalization

We start by summarizing the main results of [2] which are needed in our further discussion. The generalized Golub-Kahan bidiagonalization algorithm requires a positive definite (1,1)-block and $\mathbf{g} = 0$ in the right-hand side. Depending on the application, $\mathbf{W}$ may, however, be only positive semi-definite. A common method to obtain a positive definite (1,1)-block is to apply an augmented Lagrangian approach as described in [2,7,16]. Let $\ker(\mathbf{W}) \cap \ker(\mathbf{A}^T) = \{0\}$ and $\mathbf{N} \in \mathbb{R}^{n \times n}$ be symmetric positive definite. We modify the upper left block to

$$\mathbf{M} := \mathbf{W} + \gamma \mathbf{A} \mathbf{N}^{-1} \mathbf{A}^T \tag{2}$$

---

[1] https://www.code-aster.org.

for some $0 \leq \gamma \leq 1$. With the additional transformation

$$\mathbf{u} = \mathbf{w} - \mathbf{M}^{-1}(\mathbf{g} + \gamma \mathbf{A} \mathbf{N}^{-1} \mathbf{r})$$
$$\mathbf{b} = \mathbf{r} - \mathbf{A}^T \mathbf{M}^{-1}(\mathbf{g} + \gamma \mathbf{A} \mathbf{N}^{-1} \mathbf{r}), \tag{3}$$

(1) is then equivalent to

$$\begin{bmatrix} \mathbf{M} & \mathbf{A} \\ \mathbf{A}^T & 0 \end{bmatrix} \begin{bmatrix} \mathbf{u} \\ \mathbf{p} \end{bmatrix} = \begin{bmatrix} 0 \\ \mathbf{b} \end{bmatrix}. \tag{4}$$

The non-singularity of $\mathbf{M}$ follows from $\ker(\mathbf{W}) \cap \ker(\mathbf{A}^T) = \{0\}$. This kind of regularization can also be applied when $\mathbf{W}$ is positive definite, with the goal that for a suitably chosen $\mathbf{N}$, we may find that (4) becomes easier to solve than the original system. In the following, we will use the notation $\mathbf{M}$ for a symmetric positive definite matrix. We will use $\gamma = 1$ whenever an augmented Lagrangian approach is used and $\gamma = 0$ to work with the original matrix $\mathbf{W}$ (thus avoiding the matrix transformation in (2)–(4)). We will not discuss any intermediate value of $\gamma$, as this factor can be included in $\mathbf{N}$. Furthermore, we use the Hilbert spaces

$$\mathcal{M} = \{\mathbf{v} \in \mathbb{R}^m : \|\mathbf{v}\|_{\mathbf{M}}^2 = \mathbf{v}^T \mathbf{M} \mathbf{v}\}, \ \mathcal{N} = \{\mathbf{q} \in \mathbb{R}^n : \|\mathbf{q}\|_{\mathbf{N}}^2 = \mathbf{q}^T \mathbf{N} \mathbf{q}\}.$$

## 2.1  Fundamentals of the Golub-Kahan Bidiagonalization Algorithm

The (standard) Golub-Kahan bidiagonalization procedure has been widely used in the computation of the singular value decomposition of rectangular matrices. Let $\tilde{\mathbf{A}} \in \mathbb{R}^{m \times n}$, then we search for two unitary matrices $\tilde{\mathbf{Q}}^{n \times n}$ and $\tilde{\mathbf{V}}^{m \times m}$, such that $\tilde{\mathbf{V}}^T \tilde{\mathbf{A}} \tilde{\mathbf{Q}} = \mathbf{B}$, where

$$\mathbf{B} = \begin{bmatrix} \alpha_1 & \beta_2 & 0 & \cdots & 0 \\ 0 & \alpha_2 & \beta_3 & \ddots & 0 \\ \vdots & \ddots & \ddots & \ddots & \vdots \\ 0 & \cdots & 0 & \alpha_{n-1} & \beta_n \\ 0 & \cdots & 0 & 0 & \alpha_n \end{bmatrix}.$$

In [15, 20, 21], several algorithms for the bidiagonalization are presented that can be applied to $\tilde{\mathbf{A}}$. Here, we will specifically analyze one of the variants known as the Craig-variant [20–22]. With the transformations $\mathbf{A} = \mathbf{M}^{1/2} \tilde{\mathbf{A}} \mathbf{N}^{1/2}$, $\mathbf{Q} = \mathbf{N}^{-1} \tilde{\mathbf{Q}}$ and $\mathbf{V} = \mathbf{M}^{-1} \tilde{\mathbf{V}}$, the Golub-Kahan bidiagonalization can be generalized into seeking the matrices $\mathbf{Q}, \mathbf{V}$ and $\mathbf{B}$, such that

$$\begin{cases} \mathbf{A}\mathbf{Q} = \mathbf{M}\mathbf{V} \begin{bmatrix} \mathbf{B} \\ 0 \end{bmatrix}, & \mathbf{V}^T \mathbf{M} \mathbf{V} = \mathbf{I}_m \\ \\ \mathbf{A}^T \mathbf{V} = \mathbf{N}\mathbf{Q} \begin{bmatrix} \mathbf{B}^T; 0 \end{bmatrix}, & \mathbf{Q}^T \mathbf{N} \mathbf{Q} = \mathbf{I}_n \end{cases} \tag{5}$$

For a more detailed derivation, we refer to [2, 19]. By the change of variables $\mathbf{u} := \mathbf{V}\hat{\mathbf{z}}$ and $\mathbf{p} := \mathbf{Q}\hat{\mathbf{y}}$ and by multiplying the system from the left by the block diagonal matrix blockdiag($\mathbf{V}^T, \mathbf{Q}^T$), the augmented system can be transformed with (5) into

$$\begin{bmatrix} \mathbf{I}_n & 0 & \mathbf{B} \\ 0 & \mathbf{I}_{m-n} & 0 \\ \mathbf{B}^T & 0 & 0 \end{bmatrix} \begin{bmatrix} \hat{\mathbf{z}}_1 \\ \hat{\mathbf{z}}_2 \\ \hat{\mathbf{y}} \end{bmatrix} = \begin{bmatrix} 0 \\ 0 \\ \mathbf{Q}^T\mathbf{b} \end{bmatrix}.$$

It follows immediately that $\hat{\mathbf{z}}^T = (\hat{\mathbf{z}}_1^T, \hat{\mathbf{z}}_2^T) = (\hat{\mathbf{z}}_1^T, 0)$. Consequently, $\mathbf{u}$ only depends on the first $n$ columns of $\mathbf{V}$ and, thus, the system reduces to

$$\begin{bmatrix} \mathbf{I}_n & \mathbf{B} \\ \mathbf{B}^T & 0 \end{bmatrix} \begin{bmatrix} \hat{\mathbf{z}}_1 \\ \hat{\mathbf{y}} \end{bmatrix} = \begin{bmatrix} 0 \\ \mathbf{Q}^T\mathbf{b} \end{bmatrix}.$$

The GKB algorithm can be set such that $\mathbf{Q}^T\mathbf{b} = \|\mathbf{b}\|_{\mathbf{N}}\mathbf{e}_1$ by choosing $\mathbf{q}_1 = \mathbf{N}^{-1}\mathbf{b}/\|\mathbf{b}\|_{\mathbf{N}^{-1}}$. We denote the iterates of $\hat{\mathbf{z}}_1$ by $\mathbf{z}_k$ and the entries of $\mathbf{z}_k$ by $\zeta_j$, $j = 1, .., k$, i.e. $\mathbf{z}_k^T = (\zeta_1, .., \zeta_k)$. In [2], it is proved that by taking advantage of the recursive properties of the standard Golub-Kahan algorithm [15], and using some of the results of [20], we can obtain the fully recursive Craig's variant algorithm (Algorithm 1). We highlight that in each iteration two linear systems, one for $\mathbf{M}$ and one for $\mathbf{N}$ must be solved. In the following, we use exclusively $\mathbf{N} = \frac{1}{\nu}\mathbf{I}$, so that the inversion of $\mathbf{N}$ has only negligible cost.

---

**Algorithm 1.** Golub-Kahan bidiagonalization algorithm

---

**Require: M, A, N, b**, maxit
  $k = 0$; $\beta_1 = \|\mathbf{b}\|_{\mathbf{N}^{-1}}$; $\mathbf{q}_1 = \mathbf{N}^{-1}\mathbf{b}/\beta_1$
  $\mathbf{w} = \mathbf{M}^{-1}\mathbf{A}\mathbf{q}_1$; $\alpha_1 = \|\mathbf{w}\|_{\mathbf{M}}$; $\mathbf{v}_1 = \mathbf{w}/\alpha_1$
  $\zeta_1 = \beta_1/\alpha_1$; $\mathbf{d}_1 = \mathbf{q}_1/\alpha_1$; $\mathbf{p}^{(1)} = -\zeta_1\mathbf{d}_1$; $\mathbf{u}_1 = \zeta_1\mathbf{v}_1$;
  **while** convergence = false and $k < $ maxit **do**
    $k = k + 1$
    $\mathbf{g} = \mathbf{N}^{-1}\left(\mathbf{A}^T\mathbf{v}_k - \alpha_k\mathbf{N}\mathbf{q}_k\right)$; $\beta_{k+1} = \|\mathbf{g}\|_{\mathbf{N}}$
    $\mathbf{q}_{k+1} = \mathbf{g}/\beta_{k+1}$
    $\mathbf{w} = \mathbf{M}^{-1}\left(\mathbf{A}\mathbf{q}_{k+1} - \beta_{k+1}\mathbf{M}\mathbf{v}_k\right)$; $\alpha_{k+1} = \|\mathbf{w}\|_{\mathbf{M}}$
    $\mathbf{v}_{k+1} = \mathbf{w}/\alpha_{k+1}$
    $\zeta_{k+1} = -\dfrac{\beta_{k+1}}{\alpha_{k+1}}\zeta_k$
    $\mathbf{d}_{k+1} = (\mathbf{q}_{k+1} - \beta_{k+1}\mathbf{d}_k)/\alpha_{k+1}$
    $\mathbf{u}_{k+1} = \mathbf{u}_k + \zeta_{k+1}\mathbf{v}_{k+1}$; $\mathbf{p}_{k+1} = \mathbf{p}_k - \zeta_{k+1}\mathbf{d}_{k+1}$
    [ convergence ] = check($\mathbf{z}_k, \dots$)
  **end while**
  **return** $\mathbf{u}_{k+1}, \mathbf{p}_{k+1}$

---

## 2.2  Stopping Criterion

The statement 'check($\mathbf{z}_k$)' in Algorithm 1 is yet undefined. In this section, we review a lower bound estimate of the error in energy norm as stopping criterion that was initially proposed in [2]. The error $\mathbf{e}_k = \mathbf{u} - \mathbf{u}_k$ can be expressed, using the relations in (5) and the $\mathbf{M}$-orthogonality property of $\mathbf{V}$, by

$$\|\mathbf{e}_k\|_{\mathbf{M}}^2 = \sum_{j=k+1}^{n} \zeta_j^2 = \left\|\hat{\mathbf{z}} - \begin{bmatrix} \mathbf{z}_k \\ 0 \end{bmatrix}\right\|_2^2.$$

To compute $\mathbf{e}_k$, we thus need $\zeta_{k+1}$ to $\zeta_n$, which are available only after the full $n$ iterations of the algorithm. Given a threshold $\tau < 1$ and an integer $d$, we define lower bounds of $\|\mathbf{e}_k\|_{\mathbf{M}}^2$ and $\|\mathbf{u}\|_{\mathbf{M}}$ by

$$\xi_{k,d}^2 = \sum_{j=k+1}^{k+d+1} \zeta_j^2 < \|\mathbf{e}_k\|_{\mathbf{M}}^2, \quad \sum_{j=1}^{k+d+1} \zeta_j^2 < \|\mathbf{u}\|_{\mathbf{M}}^2$$

and, then, by them a stopping criterion

$$\text{if} \quad \xi_{k,d}^2 \leq \tau \sum_{j=1}^{k+d+1} \zeta_j^2 \quad \text{stop.} \tag{6}$$

$\xi_{k,d}$ measures the error at step $k - d$, but as the following $\mathbf{u}_k$ minimize the error due to an important property of minimization of Craig's algorithm [21], we can safely use the last ones. For computational examples underlying the efficiency of this lower bound stopping criterion, we refer the reader to [2,3]. In terms of numerical cost, this lower bound estimate is very inexpensive to compute.

## 3   Implementation and Usage Details

We have implemented the GKB solver in the PETSc PCFIELDSPLIT [5, Chapter 4.5] environment and it is available in the 3.11 release. PCFIELDSPLIT provides preconditioners and solvers for block-matrices, as for example several variants of Schur complement preconditioners for $2 \times 2$ block matrices.

Similar to many solution methods in PETSc, the GKB method can be used as either a preconditioner or solver. To obtain GKB as solver, the standard PETSc options are to be set as *-ksp_type preonly -pc_type fieldsplit -pc_fieldsplit_type gkb*. The solver can only be used for symmetric block matrix systems with zero (2,2)-block as in Eq. (1). If the matrix is not symmetric, the code will stop with an error message. The (1,1)-block may be positive semi-definite or definite. An augmented Lagrangian approach must be used to ensure the non-singularity of the matrix in the first case and it can be used to obtain a potentially better convergence in the second case (see Sect. 2). The GKB PETSc options are

| -pc_fieldsplit_gkb_nu | $\nu > 0$: Eq. (2) is used with $\gamma = 1, \mathbf{N} = \frac{1}{\nu}\mathbf{I}$ |
|---|---|
|  | $\nu = 0$: Original system is used, i.e. $\gamma = 0$ and $\mathbf{N} = \mathbf{I}$ |
| -pc_fieldsplit_gkb_delay | The delay $d$ in the lower bound stopping criterion of Sect. 2.2 |
| -pc_fieldsplit_gkb_tol | Stopping tolerance $\tau$ of the solver. |
| -pc_fieldsplit_gkb_maxit | Maximal number of iterations. |
| -pc_fieldsplit_gkb_monitor | Displays the lower bound estimate at each iteration |

In general, $\mathbf{N}$ may be any kind of positive definite matrix. In our PETSc implementation, the matrix is however restricted to $\mathbf{N} = \frac{1}{\nu}\mathbf{I}$. The augmented Lagrangian approach is switched off with -pc_fieldsplit_gkb_nu 0. This is done for the convenience of not passing the parameter $\gamma$, but corresponds to $\gamma = 0$, $\mathbf{N} = \mathbf{I}$.

A considerable advantage of the integration of the GKB iterative method in PETSc is the availability of a large choice of solver-preconditioner combinations for the inner solution step of linear systems of type $\mathbf{Mx} = \mathbf{f}$ in Algorithm 1. Although the outer loop in the GKB method is sequential, each matrix or vector operation is fully parallel and scalability is achieved by the inner solvers (see Sect. 4).

## 4   Numerical Experiments

Iterative solvers for the Stokes system have been a field of intensive research. These include preconditioned Krylov space methods [10,11,23] and multigrid methods [8]. Parallel multigrid methods for the Stokes system have been studied in [13,14]. In this section, we will apply the proposed GKB iterative solver to two discretizations of the Stokes equations. A comparitive study between the GKB and the previously cited methods would however be out of the scope of this paper. Instead, we will focus on comparing its performance using different inner solvers to the sparse direct solver MUMPS [1]. In particular, we discuss choices for $\nu$ as well as the stopping tolerances $\tau$ of the GKB method and $\tau_{in}$ of its inner iterative solvers.

### 4.1   Poieseuille Flow

We consider a viscous, laminar flow in a 2D channel $\Omega = [0,2] \times [0,1]$ with parabolic velocity profile and linear pressure drop, i.e. $\mathbf{u}(x,y) = (4y(1-y),0)$ and $\mathbf{p} = 8(2-x)$. This Poiseuille flow problem is the exact solution of the 2D Stokes problem

$$-\Delta\mathbf{u} + \nabla\mathbf{p} = 0,$$
$$\mathrm{div}(\mathbf{u}) = 0 \tag{7}$$

with no-slip boundary conditions. We adapt `ex70.c`[2] given in the PETSc `SNES` section for the simulations. The domain $\Omega$ is discretized into a Cartesian grid, using $n_x, n_y$ elements in $x$- and $y$-direction, respectively. The equations are approximated by a cell-centered co-located finite volume method, where, after application of Gauss's divergence theorem, the gradient term is discretized by central differencing and linear interpolation is used for $\mathbf{p}$ and $\mathbf{u}$ in the momentum equation [12,18]. This discretization leads to a block system of the form (4) with a symmetric positive definite matrix $\mathbf{M}$. Let, in Matlab notation, $\mathbf{D} = \mathrm{diag}(\mathrm{diag}(\mathbf{M}))$ and $\mathbf{R} = \mathrm{diag}(\mathrm{diag}(\mathbf{A}^T\mathbf{D}^{-1}\mathbf{A}))$. To equilibrate the different blocks of (4), we scale the system from the left and the right by the block diagonal matrix $\mathrm{blockdiag}(\mathbf{D}^{-1/2}, \mathbf{R}^{-1/2})$.

**Discretization Error.** In a first experiment, we determine the discretization errors of the model for a sequence of mesh sizes. This will indicate the necessary stopping tolerance for the iterative algorithms, which thus ensures a fair comparison between the direct and iterative solvers. Once the accuracy of the iterative solution falls below the interpolation error on the nodes ($O(h^2)$ for FEM), the solver can stop, as no further significant improvement of the solution accuracy on the exact solution of the PDE can be achieved. To get an accurate estimate of the discretization error, we do not take advantage of the block structure of the matrix and solve the complete system with MUMPS [1]. In Table 1, we present the number of degrees of freedom for the (1,1)-block $\mathbf{M}$ ($\mathrm{dof_M}$), the number of constraints ($\mathrm{dof_A}$) and the discretization errors of $\mathbf{u}_h$ and $\mathbf{p}_h$ in the 2- and energy-norms

$$\mathrm{err}_2^{\mathbf{u}} = \frac{1}{n_x n_y}\|\mathbf{u}_h - \mathbf{u}\|_2, \quad \mathrm{err}_2^{\mathbf{P}} = \frac{1}{n_x n_y}\|\mathbf{p}_h - \mathbf{p}\|_2, \quad \mathrm{err}_{\mathbf{M}}^{u} = \frac{\|\mathbf{u}_h - \mathbf{u}\|_{\mathbf{M}}}{\|\mathbf{u}\|_{\mathbf{M}}}.$$

Furthermore, we give the computation time of the algorithm. The calculations are executed on the cluster Kraken of the Cerfacs computing resources. The Kraken cluster includes 121 compute nodes equipped with two Skylake processors at 2.3 Ghz, each of them has 18 cores and share 96 GB DDR4 memory. For algorithmic constraints we use a power of two number of cores. In the numerical computations, we first fill up one node up to $2^5$ cores, and for higher counts 32 out of the 36 cores are used per node. The MPI tasks are evenly distributed among the processors of a node. For the largest case `Prob 4` ($>25 \cdot 10^6$ unknowns) in the Poiseuille flow example, the computations with MUMPS are done on 32 of 36 cores on one "fat" compute node with 768 GB of memory. This is necessary as MUMPS exits with a memory error on the standard compute nodes. In particular, this underlines the necessity of iterative solvers for such large sparse systems. The computation times (in seconds) are obtained with the PETSc profiling option *log_view*, from which we present the time for *ksp_solve*.

---

[2] https://www.mcs.anl.gov/petsc/petsc-dev/src/snes/examples/tutorials/ex70.c.html.

**Table 1.** Discretization information for the model problem, solved with MUMPS.

| Name | $n_x$ | $n_y$ | dof | $\text{dof}_M$ | $\text{dof}_A$ | $\text{err}_2^u$ | $\text{err}_2^P$ | $\text{err}_M^u$ | Time, s |
|---|---|---|---|---|---|---|---|---|---|
| Prob 1 | 512 | 256 | 393 216 | 262 144 | 131 072 | 6.50e−06 | 1.56e−02 | 4.01e−05 | 5 |
| Prob 2 | 1024 | 512 | 1 572 864 | 1 048 576 | 524 288 | 1.63e−06 | 7.81e−03 | 1.10e−05 | 26 |
| Prob 3 | 2048 | 1024 | 6 291 456 | 4 194 304 | 2 097 152 | 4.06e−07 | 3.90e−03 | 2.57e−06 | 141 |
| Prob 4 | 4096 | 2048 | 25 165 824 | 16 777 216 | 8 388 608 | 1.02e−07 | 1.95e−03 | 6.45e−07 | 888 |

**GKB Algorithm – Direct Inner Solver.** We next discuss the GKB algorithm with the direct inner solver MUMPS. Although **M** is symmetric positive definite in our application, we apply the augmented Lagrangian approach (2)–(4) with $\mathbf{N} = \frac{1}{\nu}\mathbf{I}$ and show the influence of $\nu$ on the number of GKB iterations and the computation time. We also present results for the case without augmented Lagrangian approach (i.e. $\gamma = 0$ and $\mathbf{N} = \mathbf{I}$). Since the matrices are all symmetric positive definite, we use the Cholesky factorization in MUMPS, which switches off the pivoting and leads to a better performance. The results presented are for Prob 3. As in the section above, there is not enough memory available to solve Prob 4 on the standard compute nodes of Kraken.

**Table 2.** Choice of $\mathbf{N} = \frac{1}{\nu}$ for Prob 3. $\tau = 10^{-6}$, $d = 5$ and 32 cores

| $\nu$ | $\text{err}_2^u$ | $\text{err}_2^P$ | $\text{err}_M^u$ | l.b. estimate | GKB iter | Time, s |
|---|---|---|---|---|---|---|
| 0 | 4.20e−07 | 5.42e−04 | 2.59e−06 | 9.87e−07 | 106 | 127 |
| 1 | 4.10e−07 | 5.56e−04 | 2.59e−06 | 9.84e−07 | 59 | 115 |
| 10 | 4.06e−07 | 7.58e−04 | 2.57e−06 | 9.92e−07 | 42 | 106 |
| 100 | 4.06e−07 | 1.60e−03 | 2.57e−06 | 9.83e−07 | 53 | 114 |

We use 32 cores on one single node. The stopping tolerance of the GKB method $\tau$ is chosen as $\tau = 1/n_y^2 \approx 10^{-6}$ such that we can have *superconergence* at the mesh nodes and, thus, a smooth reconstruction of the solutions. Indeed, $\tau$ is of the same order of magnitude as the energy discretization error $\text{err}_M^u$ in Table 1. The delay in the lower bound stopping criterion is chosen as $d = 5$. Results for the choice of $\nu$ are presented in Table 2. Of the $\nu$-values tested, the fastest simulation is obtained for $\nu = 10$. This will thus be our choice in the following experiments. For a discussion about the influence of the parameter $\nu$ on the convergence of the algorithm on different problem settings, we refer to [2,3].

**GKB Algorithm – Iterative Inner Solver.** We will look at an inner iterative solver for the solution of the linear systems involving **M** in Algorithm 1. This is necessary when **M** is too large to be solved with a direct method or advantageous if it contains a structure favorable for highly scalable iterative solvers (e.g., multigrid). Since in our example **M** is the stiffness matrix for the Laplacian, we decided to use CG and flexible gmres (denoted *fgmres*, which allows any iterative solver as a preconditioner) preconditioned by BoomerAMG of the library `Hypre` in PETSc. After numerical tests, we report that both variants and BoomerAMG do not converge when applied to the augmented Lagrangian matrix (4). We, thus, use $\gamma = 0$ and $\mathbf{N} = \mathbf{I}$. We found also that the tolerance of the inner iterative solver $\tau_{in}$ has to be no less than one order of magnitude smaller than the outer one to obtain a solution of required accuracy. The precise size also depends on the kind of employed stopping criteria in the implementation of the inner iterative solver and its compatibility with (6), which results here into different tolerances for CG and fgmres. A more precise analysis is beyond the scope of this paper and is left as a future work. Results are presented in Table 3 where the errors given are those of the CG method. The errors for fgmres are equivalent and of sufficient accuracy compared to Table 1. Note that the fgmres method has a smaller computation time than CG, although the algorithm is more complex.

**Table 3.** Inner-outer GKB algorithm with $d = 5$ and CG/fgmres-BoomerAMG.

| Name | Tolerances | | | $err_2^u$ | $err_2^p$ | $err_M^u$ | Iter GKB | Time, s: GKB with | |
|------|------|------|------|------|------|------|------|------|------|
| | $\tau$ | CG | fgmres | | | | | CG | fgmres |
| Prob 1 | 1e−05 | 1e−06 | 1e−07 | 6.53e−06 | 2.57e−03 | 4.04e−05 | 68 | 3 | 2 |
| Prob 2 | 1e−06 | 1e−07 | 1e−08 | 1.62e−06 | 1.15e−03 | 1.02e−05 | 43 | 44 | 36 |
| Prob 3 | 1e−06 | 1e−07 | 1e−08 | 4.16e−07 | 5.41e−04 | 2.59e−06 | 106 | 127 | 107 |
| Prob 4 | 1e−07 | 1e−08 | 1e−09 | 1.03e−07 | 2.75e−04 | 6.47e−07 | 220 | 1826 | 1436 |

**Parallel Performance.** Strong scaling results for `Prob` 3 and the four previously discussed methods are presented in Fig. 1a and Table 4. The stopping tolerance for the outer GKB iteration is $10^{-6}$ and the stopping tolerances for the inner iterative solvers are given in Table 3. Furthermore, we choose $\nu = 10$ for GKB-MUMPS and the delay $d = 5$ in (6). We observe that once the computations are done on more than 8 cores, all the three variants of the GKB method are faster than MUMPS applied to the overall system. Among the three GKB-methods, the two inner-outer iterative methods with either CG or fgmres clearly outperform GKB-MUMPS. While the latter one starts to level off at about 64 cores, the iterative variants have not yet reached their plateau, even for 512 cores. Both of them have a speed-up of about 40 from 2 to 512 cores. We emphasize here that CG/fgmres take advantage of the multigrid preconditioner, which is known to scale for discretizations of elliptic equations and in particular the Laplace operator. As already noted before, the fgmres method performs better than CG does, but their scaling behavior is similar and their relative performance also stays similar with the increase in the number of cores.

**Table 4.** Solver time and strong scaling for `Prob 3`.

| Cores | MUMPS | | GKB-MUMPS ($\nu = 10$) | | GKB-CG ($\nu = 0$, $\tau = 10^{-6}$, $\tau_{in} = 10^{-7}$) | | GKB-fgmres ($\nu = 0$, $\tau = 10^{-6}$, $\tau_{in} = 10^{-8}$) | |
|---|---|---|---|---|---|---|---|---|
| | Time (s) | Scale | Time (s) | Scale | Time (s) | Scale | Time (s) | Scale |
| 2 | 641 | | 491 | | 919 | | 830 | |
| 4 | 419 | 1.5 | 305 | 1.6 | 493 | 1.9 | 428 | 1.9 |
| 8 | 257 | 2.5 | 184 | 2.7 | 286 | 3.2 | 237 | 3.5 |
| 16 | 170 | 3.8 | 139 | 5.5 | 161 | 5.7 | 137 | 6.1 |
| 32 | 140 | 4.6 | 106 | 4.6 | 127 | 7.2 | 107 | 7.8 |
| 64 | 107 | 6.0 | 90 | 5.5 | 85 | 10.8 | 68 | 12.2 |
| 128 | 99 | 6.5 | 83 | 5.9 | 53 | 17.3 | 44 | 18.9 |
| 256 | 97 | 6.6 | 76 | 6.5 | 33 | 27.8 | 27 | 30.7 |
| 512 | 93 | 6.9 | 76 | 6.5 | 23 | 40.0 | 21 | 39.5 |

## 4.2   Isoviscous Stokes

Second, we adapt example `ex62.c`[3] given in the `SNES` tutorials section in PETSc to comply with (7). It simulates the 2d isoviscous variant of the Stokes problem (7), approximated by a Q2-P1 mixed finite element method with a discontinuous pressure field. The domain $\Omega$ is the unit square and is discretized by an unstructured mesh with quadrilateral elements. The exact solution is chosen as $\mathbf{u} = (x^3 + y^3, 2x^3 - 3x^2y)$, $\mathbf{p} = \frac{3}{2}x^2 + \frac{3}{2}y^2 - 1$. Since we deal with a linear problem, the `SNES` non-linear solver converges in one iteration. We focus thus on the solution of the Jacobian matrix for which we use the `PCFIELDSPLIT` environment with the GKB solver and compare the four solution techniques as in the previous example. We choose $n_x, n_y = 1024$ cells in the $x$- and $y$-direction, leading to $m \approx 8.3 \cdot 10^6$ and $n \approx 3.1 \cdot 10^6$ degrees of freedom. The finite element discretization errors (obtained by MUMPS) are $\|\mathbf{u}_h - \mathbf{u}\|_{L^2} \approx 6.62 \cdot 10^{-11}$ and $\|\mathbf{p}_h - \mathbf{p}\|_{L^2} \approx 1.51 \cdot 10^{-7}$. As an implementation of the energy norm is not available in this example, we experimentally choose the stopping tolerances such that the solution respects the $L^2$-errors. We obtain $\tau = 10^{-8}$ for the outer GKB iteration, and $\tau_{in} = 10^{-13}$ for the inner iterative solvers in GKB-CG/fgmres. For GKB-MUMPS, we use $\nu = 10^7$. This large value is justified by the different scaling of the matrix blocks. The matrix $\mathbf{A}\mathbf{A}^T$ scales with $h^2 \approx 10^{-6}$, and $\nu$ thus equilibrates the values in the blocks of (4). Lastly, we choose $d = 5$ in (6).

The results are presented in Fig. 1b and Table 5. As in the previous example, MUMPS and GKB-MUMPS are the better choice for a smaller number of cores. When more than 64 cores are available, GKB-CG and GKB-fgmres outperform the latter two. Both variants involving MUMPS reach a plateau at about 64 cores and have similar computation times. For CG- and fgmres-BoomerAMG as inner solvers, the performance plateau is however still not reached for 512 cores. We observe a speed-up of about 163 for GKB-fgmres and 148 for GKB-CG.

---

[3] https://www.mcs.anl.gov/petsc/petsc-dev/src/snes/examples/tutorials/ex62.c. html.

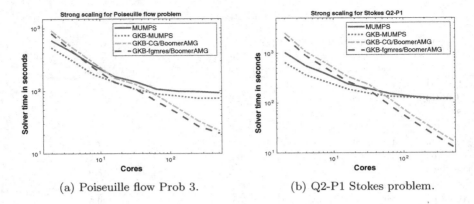

(a) Poiseuille flow Prob 3.    (b) Q2-P1 Stokes problem.

**Fig. 1.** Scaling behavior on Kraken (log-log scale).

**Table 5.** Solver time and strong scaling for Stokes Q2-P1 example.

| Cores | MUMPS | | GKB-MUMPS $(\nu = 10^7)$ | | GKB-CG $(\nu = 0,\ \tau = 10^{-8},\ \tau_{in} = 10^{-13})$ | | GKB-fgmres $(\nu = 0,\ \tau = 10^{-8},\ \tau_{in} = 10^{-13})$ | |
|---|---|---|---|---|---|---|---|---|
| | Time (s) | Scale | Time (s) | Scale | Time (s) | Scale | Time (s) | Scale |
| 2 | 1031 | | 649 | | 2518 | | 2115 | |
| 4 | 556 | 1.9 | 364 | 1.8 | 1067 | 2.4 | 886 | 2.4 |
| 8 | 379 | 2.7 | 263 | 2.5 | 615 | 4.1 | 488 | 4.3 |
| 16 | 241 | 4.3 | 189 | 3.4 | 339 | 7.4 | 268 | 7.9 |
| 32 | 192 | 5.4 | 155 | 4.2 | 233 | 11.8 | 180 | 11.8 |
| 64 | 153 | 6.7 | 135 | 4.8 | 118 | 21.3 | 88 | 24.0 |
| 128 | 134 | 7.7 | 128 | 5.1 | 58 | 43.4 | 45 | 47.0 |
| 256 | 124 | 8.3 | 121 | 5.4 | 31 | 81.2 | 24 | 88.1 |
| 512 | 122 | 8.5 | 119 | 5.5 | 17 | 148.1 | 13 | 162.7 |

## 5    Conclusions

We presented an iterative algorithm based on the Golub Kahan bidiagonalization for $2 \times 2$ block matrices. Furthermore, we outlined our PETSc implementation of the solver and applied it to the Poiseuille flow as well as an isoviscuous Stokes problem. The scalability tests showed that MUMPS and the GKB-MUMPS solvers start to level off at about 64 cores with speed-up factors of at most 8, while the GKB algorithm with BoomerAMG as preconditioner showed a speed-up between 40 to 160 from 2 to 512 cores. Regarding the gains in the absolute computation time, either MUMPS or GKB-MUMPS are the methods of choice for a smaller number of cores and unknowns on the order of $10^6$. When more than 64 cores are available, the GKB method with inner iterative solvers outperforms its direct counterparts, and hence, should be employed. When increasing the problem size to about $2 \cdot 10^7$ unknowns (**Prob 4**), MUMPS required more memory than that available on Cerfacs' cluster Kraken in either the direct or GKB-MUMPS case. Iterative methods, such as GKB-fgmres and GKB-CG, are usable for this problem, however, and present the only alternative to its solution.

# References

1. Amestoy, P., Duff, I., L'Excellent, J., Koster, J.: A fully asynchronous multifrontal solver using distributed dynamic scheduling. SIAM J. Matrix Anal. Appl. **23**(1), 15–41 (2001). https://doi.org/10.1137/S0895479899358194
2. Arioli, M.: Generalized Golub-Kahan bidiagonalization and stopping criteria. SIAM J. Matrix Anal. Appl. **34**(2), 571–592 (2013). https://doi.org/10.1137/120866543
3. Arioli, M., Kruse, C., Rüde, U., Tardieu, N.: An iterative generalized Golub-Kahan algorithm for problems in structural mechanics. CoRR (2018). http://arxiv.org/abs/1808.07677
4. Balay, S., Abhyankar, S., Adams, M.F., Brown, J., Brune, P., Buschelman, K., et al.: PETSc Web page (2019). http://www.mcs.anl.gov/petsc
5. Balay, S., Abhyankar, S., Adams, M.F., Brown, J., Brune, P., Buschelman, K., et al.: PETSc users manual. Technical report ANL-95/11 - Revision 3.11, Argonne National Laboratory (2019). http://www.mcs.anl.gov/petsc
6. Balay, S., Gropp, W.D., McInnes, L.C., Smith, B.F.: Efficient management of parallelism in object oriented numerical software libraries. In: Arge, E., Bruaset, A.M., Langtangen, H.P. (eds.) Modern Software Tools in Scientific Computing, pp. 163–202. Birkhäuser Press, Boston (1997). https://doi.org/10.1007/978-1-4612-1986-6_8
7. Benzi, M., Golub, G.H., Liesen, J.: Numerical solution of saddle point problems. Acta Numer. **14**, 1–137 (2005). https://doi.org/10.1017/S0962492904000212
8. Brandt, A., Livne, O.E.: Multigrid techniques: 1984 guide with applications to fluid dynamics. SIAM **67** (2011). https://doi.org/10.1137/1.9781611970753
9. Dongarra, J., et al.: Applied mathematics research for exascale computing. Technical report, Lawrence Livermore National Laboratory (LLNL), Livermore (2014). https://doi.org/10.2172/1149042
10. Elman, H.C., Silvester, D.J., Wathen, A.J.: Performance and analysis of saddle point preconditioners for the discrete steady-state navier-stokes equations. Numer. Math. **90**(4), 665–688 (2002). https://doi.org/10.1007/s002110100
11. Elman, H.C., Silvester, D.J., Wathen, A.J.: Finite Elements and Fast Iterative Solvers: with Applications in Incompressible Fluid Dynamics. Oxford University Press, USA (2014). https://doi.org/10.1093/acprof:oso/9780199678792.001.0001
12. Ferziger, J., Peric, M.: Computational Methods for Fluid Dynamics. Springer, Heidelberg (2001). https://doi.org/10.1007/978-3-642-56026-2
13. Gmeiner, B., Huber, M., John, L., Rüde, U., Wohlmuth, B.: A quantitative performance study for stokes solvers at the extreme scale. J. Comput. Sci. **17**, 509–521 (2016). https://doi.org/10.1016/j.jocs.2016.06.006
14. Gmeiner, B., Rüde, U., Stengel, H., Waluga, C., Wohlmuth, B.: Towards textbook efficiency for parallel multigrid. Numer. Math.: Theory Methods Appl. **8**(1), 22–46 (2015). https://doi.org/10.4208/nmtma.2015.w10si
15. Golub, G., Kahan, W.: Calculating the singular values and pseudo-inverse of a matrix. J. Soc. Indu. Appl. Math. Ser. B Numer. Anal. **2**(2), 205–224 (1965). https://doi.org/10.1137/0702016
16. Golub, G.H., Greif, C.: On solving block-structured indefinite linear systems. SIAM J. Sci. Comput. **24**(6), 2076–2092 (2003). https://doi.org/10.1137/S1064827500375096
17. Heroux, M.A., et al.: An overview of the trilinos project. ACM Trans. Math. Softw. **31**(3), 397–423 (2005). https://doi.org/10.1145/1089014.1089021

18. Klaij, C.: On the stabilization of finite volume methods with co-located variables for incompressible flow. J. Comput. Phys. **297**(C), 84–89 (2015). https://doi.org/ 10.1016/j.jcp.2015.05.012
19. Orban, D., Arioli, M.: Iterative solution of symmetric quasi-definite linear systems. SIAM Spotlights. Society for Industrial and Applied Mathematics (2017). https:// doi.org/10.1137/1.9781611974737
20. Paige, C.C., Saunders, M.A.: LSQR: an algorithm for sparse linear equations and sparse least squares. ACM Trans. Math. Softw. **8**(1), 43–71 (1982). https://doi. org/10.1145/355984.355989
21. Saunders, M.A.: Solution of sparse rectangular systems using LSQR and CRAIG. BIT Numer. Math. **35**(4), 588–604 (1995). https://doi.org/10.1007/BF01739829
22. Saunders, M.A.: Computing projections with LSQR. BIT Numer. Math. **37**(1), 96–104 (1997). https://doi.org/10.1007/BF02510175
23. Ur Rehman, M., Geenen, T., Vuik, C., Segal, G., MacLachlan, S.: On iterative methods for the incompressible stokes problem. Int. J. Numer. Meth. Fluids **65**(10), 1180–1200 (2011). https://doi.org/10.1002/fld.2235

# A High-Performance Implementation of a Robust Preconditioner for Heterogeneous Problems

Linus Seelinger[1]([✉])[iD], Anne Reinarz[2][iD], and Robert Scheichl[3][iD]

[1] Institute for Scientific Computing, Heidelberg University, Heidelberg, Germany
linus.seelinger@iwr.uni-heidelberg.de
[2] Department of Informatics, Technical University of Munich, Garching, Germany
reinarz@in.tum.de
[3] Institute for Applied Mathematics, Heidelberg University, Heidelberg, Germany
r.scheichl@uni-heidelberg.de

**Abstract.** We present an efficient implementation of the highly robust and scalable GenEO (Generalized Eigenproblems in the Overlap) preconditioner [16] in the high-performance PDE framework DUNE [6]. The GenEO coarse space is constructed by combining low energy solutions of a local generalised eigenproblem using a partition of unity. The main contribution of this paper is documenting the technical details that are crucial to the efficiency of a high-performance implementation of the GenEO preconditioner. We demonstrate both weak and strong scaling for the GenEO solver on over 15,000 cores by solving an industrially motivated problem in aerospace engineering. Further, we show that for highly complex parameter distributions arising in certain real-world applications, established methods become intractable while GenEO remains fully effective.

**Keywords:** Partial differential equations · Domain decomposition · Preconditioning · High performance computing

## 1 Introduction

Computer simulations have become a vital tool in science and engineering. The demand for solving PDEs on ever larger domains and increasing accuracy necessitates the use of high performance computers and the implementation of efficient parallel algorithms. When designing parallel algorithms two issues are crucial: Robustness and scalability.

(i) **Robustness:** The parameters involved in the PDE affect the performance of the algorithm to a large extent. A frequent issue is a distribution of parameters with large contrast jumps at different length scales, which may lead to slow or no solver convergence.

© Springer Nature Switzerland AG 2020
R. Wyrzykowski et al. (Eds.): PPAM 2019, LNCS 12043, pp. 117–128, 2020.
https://doi.org/10.1007/978-3-030-43229-4_11

(ii) **Scalability:** The immediate scalability of the finite element method is limited as each degree of freedom is coupled with all others.

One approach to achieve scalability in solving partial differential equations are domain decomposition methods (see e.g. [14, 17]), splitting the given domain into multiple subdomains. The solution of the original problem restricted to each subdomain is computed in parallel and the results are combined to form an approximate solution. This is repeated until convergence is reached. The number of these iterations, however, still depends strongly on the number of subdomains involved as well as coefficient variations. Introducing an additional coarse space that covers all subdomains can restore performance for large numbers of subdomains.

This global space can be tailored to specific problem, as in the generalized finite element method [3]. While these methods are applicable to an entire class of parameter distributions, each of these approaches is based on certain assumptions on the parameters, e.g. the parameters vary strongly only in one direction. Over recent years, more generic approaches have been developed, e.g. [8, 11]. The GenEO coarse space chosen in this work is a related approach, originally introduced in [16]. It does not require a-priori knowledge of the parameter distribution and is applicable to a wide range of problems making it suitable as a 'black-box' solver.

In this paper we focus on two different elliptic problems; the Darcy equation describing incompressible flow in a porous medium and the anisotropic linear elasticity equations. For both equations the case of heterogeneous coefficients is of great interest.

Composite materials, which make up over 50% of recent aircraft constructions, are manufactured from carbon fibres and soft resin layers. The large jump in material properties between the layers makes the simulation of these materials challenging. Commercial solvers such as ABAQUS often rely on direct solvers to deal with these jumps [12]. However, the scalability of direct solvers is limited. We demonstrate that the GenEO approach converges independently of the contrast in material properties and the number of subdomains.

After introducing our PDE models in Sect. 2, we sketch the construction of the GenEO preconditioner in Sect. 3. In Sect. 4, we then discuss how to efficiently implement the solver in the high-performance finite element framework DUNE [4, 5]. Finally, in Sect. 5, we provide several numerical experiments demonstrating both robustness and scalability of the solver, including a large-scale industrially motivated example.

## 2    Problem Formulation and Variational Setting

Let $V$ be a Hilbert space, $a : V \times V \to \mathbb{R}$ a symmetric and coercive bilinear form and $f \in V'$. We consider the following abstract variational problem. Find $v \in V$ such that

$$a(v, w) = \langle f, w \rangle, \quad \forall w \in V, \tag{1}$$

where $\langle \cdot, \cdot \rangle$ denotes the duality pairing.

This variational problem is associated with an elliptic boundary value problem on a domain $\Omega \subset \mathbb{R}^d$, $d = 2, 3$ with Dirichlet boundary $\partial \Omega_D$. In particular, we focus on the following two examples.

(i) **Darcy problem:** Given material properties $\kappa \in L^\infty(\Omega)$, find $v \in V = \{v \in H^1(\Omega) : v|_{\Omega_D} = 0\}$ such that

$$a(v, w) = \int_\Omega \kappa(x) \nabla v(x) \cdot \nabla w(x) \, dx = \int_\Omega f(x) w(x) \, dx, \qquad \forall w \in V.$$

(ii) **Linear Elasticity:** Given material properties $C$, find $v \in V = \{v \in H^1(\Omega)^d : v|_{\Omega_D} = 0\}$ such that

$$a(v, w) = \int_\Omega C(x) \varepsilon(v) : \varepsilon(w) dx = \int_\Omega f \cdot w \, dx + \int_{\partial \Omega} (\sigma \cdot n) \cdot v \, dx, \qquad \forall w \in V,$$

where $\varepsilon_{ij}(v) = \frac{1}{2}(\partial_i v_j + \partial_j v_i)$ is the strain, and $\sigma_{ij}(v) = \sum_{k,l=1}^d C_{ijkl} \varepsilon_{kl}$ is the stress.

Consider a discretization of the variational problem (1) using finite elements on a mesh $T_h$ of $\Omega$ such that $\overline{\Omega} = \cup_{\tau \in T_h} \tau$. Let $V_h \subset V$ be a conforming space of finite element functions. Then the discrete form of (1) is: Find $v_h \in V_h$ such that

$$a(v_h, w_h) = \langle f, w_h \rangle, \qquad \forall w_h \in V_h. \tag{2}$$

## 3   The GenEO Preconditioner

In order to construct a parallel and scalable preconditioner, we employ a two-level Additive Schwarz preconditioner as comprehensively analyzed in [17]. Our specific choice of coarse space is the GenEO (Generalized Eigenproblems in the Overlap) space as introduced in [16] and proven to be robust in [15]. For brevity, we only give a brief review of the main results and otherwise refer to [15] for full details and notation.

**Definition 1 (Two-level Additive Schwarz).** *We denote the finite element matrix originating from (2) by* $\mathbf{A}$. *Using appropriate restrictions, we denote the problem restricted to subdomains by* $\mathbf{A}_j := \mathbf{R}_j \mathbf{A} \mathbf{R}_j^T$ *and to the coarse space by* $\mathbf{A}_H := \mathbf{R}_H \mathbf{A} \mathbf{R}_H^T$. *Then the two-level Additive Schwarz preconditioner is given by*

$$M_{\mathrm{AS},2}^{-1} := \mathbf{R}_H^T \mathbf{A}_H^{-1} \mathbf{R}_H + \sum_{j=1}^N \mathbf{R}_j^T \mathbf{A}_j^{-1} \mathbf{R}_j.$$

**Definition 2 (GenEO eigenproblem).** *For each subdomain* $j = 1, \ldots, N$, *we define the generalized eigenproblem: Find* $p \in V_h(\Omega_j)$ *such that*

$$a_{\Omega_j}(p, v) = \lambda a_{\Omega_j^\circ}(\Xi_j(p), \Xi_j(v)), \quad \forall v \in V_h(\Omega_j).$$

*Note that the eigenproblems are local to their respective subdomain $\Omega_j$, i.e. they can be computed in parallel. To use them as a global basis they need to be extended to the entire domain using the partition of unity operators.*

**Definition 3 (GenEO coarse space).** *For each subdomain $j = 1, \ldots, N$, let $(p_k^j)_{k=1}^{m_j}$ be the eigenfunctions from the eigenproblem in Definition 2 corresponding to the $m_j$ smallest eigenvalues. Further, denote the partition of unity by $\Xi_j$. Then the GenEO coarse space is defined as*

$$V_H := \operatorname{span}\{R_j^T \Xi_j(p_k^j) : k = 1, \ldots, m_j; j = 1, \ldots, N\}.$$

In [15], the following condition bound proves robustness with respect to parameter contrast and number of subdomains.

**Theorem 1.** *For all $1 \leqslant j \leqslant N$, let the number of eigenvectors chosen in each subdomain be*

$$m_j := \min\left\{m : \lambda_{m+1}^j > \frac{\delta_j}{H_j}\right\},$$

*where $\delta_j$ is a measure of the width of the overlap $\Omega_j^o$ and $H_j = \operatorname{diam}(\Omega_j)$. Then,*

$$\kappa(M_{AS,2}^{-1}A) \leqslant (1 + k_0)\left[2 + k_0(2k_0 + 1)\max_{1 \leqslant j \leqslant N}\left(1 + \frac{H_j}{\delta_j}\right)\right].$$

# 4   HPC Implementation of GenEO in Modern PDE Frameworks

When implementing the GenEO preconditioner in a PDE framework, the primary goal is to preserve the beneficial properties offered by its theoretical construction, namely:

(i) **High parallel scalability:** Since the condition bound in Theorem 1 is independent of the number of subdomains we expect the implementation to yield high parallel scalability. The solution of the eigenproblems parallelizes trivially. However, care has to be taken when it comes to the communication necessary to set up and solve the coarse matrix.

(ii) **Robustness with respect to problem parameters:** While this is an inherent property of the preconditioner, some care is required in implementing the Dirichlet boundary conditions.

(iii) **Applicability to various types of PDEs:** The theoretical framework only requires a symmetric positive definite bilinear form as in (1). This flexibility can be preserved in any numerical framework that is based on abstract bilinear forms. This is the case for many modern PDE frameworks, e.g. FEniCS [1], DUNE [4], or deal.ii [2].

In this section, we present a new implementation of the GenEO coarse space and preconditioner within DUNE (Distributed and Unified Numerics Environment), which fulfills these properties. This section serves as a reference for the implementation, which is freely available as part of the `dune-pdelab` module [6] since version 2.6, as well as a general guideline for future implementations in other software packages. DUNE is a generic package that provides the user with key ingredients for solving any FEM problem. As an open source framework written using modern C++ programming techniques, it allows for modularity and reusability while providing HPC grade performance. Note that an alternative GenEO implementation is available in FreeFEM++ [9].

## 4.1 Prerequisites

Many of the components required to implement a two-level Schwarz method already exist within DUNE. In particular, we use the PDELab discretization module's functionality to assemble stiffness matrices based on bilinear forms and for efficient communication across overlapping subdomains. The GenEO basis functions have support not restricted to individual elements, which makes the existing high-level components of PDELab unsuited for storing the coarse space. As part of this project, components facilitating such coarse spaces were fully integrated within the framework. Further, an efficient sequential solver for generalized eigenproblems is needed. Here, we choose ARPACK [10].

## 4.2 General Structure

The implementation in PDELab closely follows the structure of the previous section. All mathematical objects are represented as individual classes (see Fig. 1). This separation of concerns leads to an easy to understand and well-structured code. Further, components are easily interchangeable when constructing related methods. In particular, the intricate process of constructing a global coarse space from per-subdomain basis functions is entirely contained in the class `SubdomainProjectedCoarseSpace`. Thus, the GenEO basis can easily be replaced by a different local basis.

The following subsections describe the steps of setting up the local eigenproblems, solving them, combining them to a coarse space and finally employing that space as a two-level Schwarz preconditioner.

## 4.3 Discrete Basis

To calculate GenEO basis functions we solve the discrete form of the eigenproblem in Definition 2, i.e.

$$\tilde{A}_j p_k^j = \lambda_k^j X_j \tilde{A}_j^o X_j p_k^j,$$

where $X_j$ is the discrete form of the partition of unity.

The matrix $\tilde{A}_j$ has to be assembled with Dirichlet constraints on the domain boundary as prescribed by the given PDE problem. However, in contrast to the

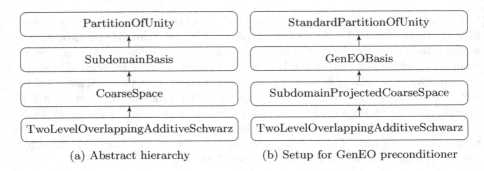

(a) Abstract hierarchy          (b) Setup for GenEO preconditioner

**Fig. 1.** Class hierarchy of GenEO implementation in DUNE PDELab

matrices $A_j$ needed for the one-level component of the two-level additive Schwarz method, no Dirichlet constraints are imposed on subdomain boundaries.

For assembling $\tilde{A}_j^o$, the same boundary conditions can be applied. However, additionally, the matrix should only be assembled on the overlap region. Internally overlap elements are identified by adding a vector of ones across subdomains and checking for results greater than one.

The matrices $X_j$ representing the partition of unity operator are diagonal and can be stored as vectors. Entries of $X_j$ corresponding to Dirichlet domain boundaries or processor boundaries should be zero, and in sum they should add up to one across subdomains. Such a partition of unity is generated by adding vectors of ones with a single communication between subdomains.

### 4.4   Solving the Eigenproblem

As the eigenproblems are defined per-subdomain, the eigensolver itself does not need to run in parallel. However, solving larger problems requires an efficient iterative solver. A suitable choice is ARPACK [10].

In order to stabilize the method, the *Shift and Invert Spectral Transformation Mode* supported by ARPACK is used. Instead of the generalized eigenproblem $Ax = Mx\lambda$, ARPACK solves the transformed problem $(A - \sigma M)^{-1}Mx = x\nu$. The eigenvalues of the transformed problem are related to those of the original problem by $\nu = \frac{1}{\lambda - \sigma}$ and the eigenvectors are identical. In the transformed problem, the eigenvalues of the original problem whose absolute values are closest to $\sigma$ are now the eigenvalues of largest magnitude, and can therefore be efficiently solved by the Krylov method. Choosing $\sigma$ near zero, the method delivers the eigenvalues of smallest magnitude at good performance. Finally, in order to form the actual basis vectors, the eigenvectors are multiplied by $X_j$ and then normalized in the $l^2$ norm, as ARPACK delivers vectors of strongly varying norms.

### 4.5   Scalable Coarse Setup

Assembling the coarse matrix $A_H$ requires particular care, as it is a non-localized, not trivially scalable operation. Due to domain decomposition, the global matrix $A$ is only available in distributed form as matrices $A_j$. Exploiting local support of basis functions, the coarse matrix $A_H$ breaks down into

$$(A_H)_{i,j} = (R_H A R_H^T)_{i,j} = \varphi_i A_i \varphi_j.$$

We note that $\varphi_i A_i \varphi_j$ is zero for $\Omega_i \cap \Omega_j = \varnothing$, leading to a sparse structure in $A_H$. Therefore, all rows $i$ of $A_H$ associated to basis functions $\varphi_i$ can be computed on the associated process locally while only requiring basis functions $\varphi_j$ from adjacent subdomains. In the implementation multiple basis functions are communicated in a single step.

The resulting blocks are combined into a matrix $A_H$ available on all processes, using direct MPI calls, while exploiting sparsity. Communication effort increases linearly with the dimension of $V_H$. This is a direct consequence of how two-level preconditioners are designed, and a good balance between coarse space size and preconditioner performance must be found.

The restriction and prolongation operators $R_H$ and $R_H^T$ are also only available locally. In case of the restriction $R_H v_h$ of a distributed vector $v_h \in V_{h,0}(\Omega)$, it holds

$$(R_H v_h)_i = \varphi_i \cdot v_h.$$

Each row $i$ can be computed by the process associated to $\varphi_i$, and the rows can be exchanged among all processes via *MPI_Allgatherv*. Again, the communication effort increases with the dimension of $V_H$.

Finally, the prolongation $R_H^T v_H$ of a global vector $v_H \in V_H$ fulfills

$$R_H^T v_H = \sum_i \varphi_i (v_H)_i.$$

Here, each part of the sum associated with a processor can be computed locally and combined by nearest-neighbor communication, scaling ideally. This completes the components needed for the two-level preconditioner according to Definition 1.

## 5   Numerical Experiments

In this section we demonstrate the solver's salient features, including its high parallel scalability up to $15,360$ cores, its robustness to heterogeneous material parameters and its applicability to different elliptic PDEs. With exception of the final large-scale experiment all numerical examples in this section have been computed using the Balena HPC cluster of the University of Bath. Balena consists of 192 nodes each with two 8-core Intel Xeon E5-2650v2 Ivybridge processors, each running at 2.6 GHz, giving a total of 3072 available cores.

**Fig. 2.** Coarse approximation error. From left to right: The parameter distribution and domain decomposition followed by the error $u - u_H$ with 1, 2 and 4 eigenvectors respectively.

### 5.1   GenEO Basis on Highly Structured Problems

With clearly structured problems, it can be visually seen that the GenEO coarse space systematically picks up inclusions or channels in the parameter distribution. In Fig. 2 the coarse approximation error is shown for a Darcy problem on a square domain. Dirichlet conditions are set to one at the top and zero at the bottom, Neumann conditions are set at the remaining boundary and a high-contrast parameter distribution with jumps and channels as shown on the left. We see that each inclusion has an effect on the approximation error. Adding additional eigenvectors from each subdomain to the coarse basis removes some of those error sources, the next eigenvectors pick up the skyscrapers and with only 4 eigenvectors per subdomain most channels are resolved. A total of 16 coarse basis functions is enough to almost entirely solve the given problem.

### 5.2   Demonstration of Robustness

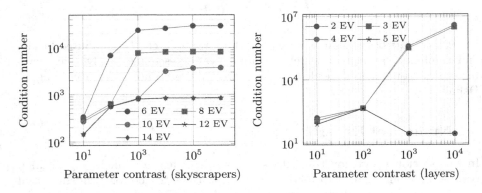

**Fig. 3.** Robustness of GenEO preconditioner

Robustness with respect to parameter contrast can be demonstrated solving the same Darcy problem as in Sect. 5.1. We choose a subdomain decomposition into 8 by 8 squares, a two-cell overlap region diameter and a total of 800 $Q_1$ elements in each direction. Figure 3 (left) shows the resulting condition number for increasing contrast when setting up a GenEO basis with various numbers of eigenvectors per subdomain. Clearly, the asymptotic robustness guaranteed by the analysis is achieved in practice.

When running the same setup with a parameter distribution of 40 horizontal equally thick layers, it becomes clear from Fig. 3 (right) that robustness is achieved exactly at four eigenvectors per subdomain. That stems from the fact that four coarse basis functions (together with the contribution of the one-level Schwarz method) are sufficient to represent the five layers contained in that subdomain. Similar relations can be observed with other strongly structured parameter distributions.

### 5.3   Comparison to Other Solvers

In this section we compare the performance of various preconditioned CG solvers. We compared with two different implementations of AMG, the implementation included in `dune-istl` and `boomerAMG` [18]. For this test we consider a flat composite plate made up of 12 composite layers stacked in a sequence of different angles, referred to as a stacking sequence. The composite layers are seperated by very thin layers of resin. There is a large jump in material strength between the composite and resin layer and due to the rotated layers the anisotropy cannot be grid aligned. We discretise with quadratic, 20-node serendipity elements to avoid shear locking and use full Gaussian integration.

**Table 1.** Demonstration of performance of different preconditioners for a problem of fixed size.

| | 1-level | | GenEO | | | BOOMERAMG |
| $N_{\text{core}}$ | iter. | $\kappa(A)$ | iter. | $\kappa(A)$ | dim($V_H$) | iter. |
|---|---|---|---|---|---|---|
| 4 | 89 | 79,735 | 16 | 10 | 78 | 258 |
| 8 | 97 | 84,023 | 15 | 9 | 126 | 258 |
| 16 | 107 | 98,579 | 16 | 10 | 182 | 257 |
| 32 | 158 | 226,871 | 16 | 9 | 526 | 263 |

In Table 1 we compare the convergence of three iterative solvers. We record the condition number, the dimension of the coarse space dim($V_H$) and the number of CG iterations required to achieve a residual reduction of $10^{-5}$. As expected the iteration counts increase steadily with the number of subdomains when no coarse space is used. In contrast, the iterations and the condition number estimates remain constant for the GENEO preconditioner as predicted by Theorem 1. The `boomerAMG` solver also retains a near constant number of iterations, although they are considerably higher. Due to a lower setup cost the `boomerAMG`

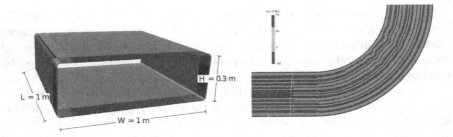

**Fig. 4.** Left: Geometry of the wingbox with dimensions; the colouring shows the number of eigenmodes used in GENEO in each of the subdomains. Right: the stacking sequence change around the corner containing a wrinkle. (Color figure online)

solver is faster in actual CPU time than the GenEO solver in this small test case. However, for more complex geometries, boomerAMG does not perform very well and in our tests it does not scale beyond about 100 cores in composite applications [7]. In its current form the *dune-istl* implementation does not seem to be robust for composite problems, especially in parallel [12]. In the test setup used here the dune-istl AMG converges very slowly or not at all, thus we do not include it in Table 1.

### 5.4    Industrially-Motivated Example: Wingbox

In this section we describe an industrially motivated example in which we asses the strength of an airplane wingbox with a small localised wrinkle defect in one corner. Wrinkle defects often form during the manufacturing process and lead to strong local stress concentrations, which may cause premature failure [12,13]. We perform a weak scaling and a strong scaling experiment. The experiments in this section were performed using the UK national HPC cluster ARCHER, which consists of 4,920 Cray XC30 nodes with two 2.7 GHz, 12-core E5-2697 v2 CPUs each.

We model a single bay of a wingbox as shown in Fig. 4 (left). As in a typical aerospace application, the stacking sequence differs across the wingbox as shown in Fig. 4 (right). In total the wingbox is made up of 77 layers. One of the corner radii contains a localised wrinkle with a parametrisation matching an observed defect in a CT-Scan of a real corner section. An internal pressure of 0.109 MPa, arising from the fuel, is applied to the internal surface. The influence of ribs that constrain the wingbox are approximated by clamping all degrees of freedom at one end and tying the degrees of freedom at the other end using a multipoint constraint. A thermal pre-stress induced by the manufacturing process is also imposed. More details on this test setup can be found in [7].

For the weak scaling experiment we refine the mesh, doubling the number of elements as we double the number of cores. Table 2 (left) contains the number of degrees of freedom, iteration numbers for the preconditioned CG, the dimension of the coarse space $\dim(V_H)$, as well as the total run time in seconds, time spent in setup of the preconditioner and in CG iterations for each test. As expected the weak scaling of the iterative CG solver with GENEO preconditioner is indeed almost optimal up to at least 15, 360 cores.

It should be noted that while the runtime consists mainly of CG iterations and the setup time for the GenEO preconditioner, which of these dominates depends on the eigenvalue threshold chosen and can be tuned to ensure a low total runtime. In this test case the same eigenvalue threshold is used for all problem sizes, in each the setup time dominates. The setup can be performed almost entirely in parallel and has very low MPI overhead as it is dominated by the solution of local eigenproblems. Conversely the CG iterations do show a slight increase in runtime for very large problems due to communication overhead.

**Table 2.** Parallel performance of the composites application on ARCHER. Left: Details of the weak scaling test. Right: details of the strong scaling test.

| $N_{core}$ | DOF | iter. | $\dim(V_H)$ | **time** | setup | CG | $N_{core}$ | $\dim(V_H)$ | it. | time | efficiency |
|---|---|---|---|---|---|---|---|---|---|---|---|
| 480 | $6 \cdot 10^6$ | 156 | 5025 | **734** | 478 | 276 | 2880 | 18843 | 167 | 2906 | 1.00 |
| 960 | $1 \cdot 10^7$ | 154 | 7840 | **806** | 528 | 278 | 3840 | 26333 | 153 | 1766 | 1.23 |
| 1920 | $2 \cdot 10^7$ | 152 | 18752 | **800** | 513 | 287 | 7680 | 52622 | 132 | 1057 | 0.83 |
| 3840 | $5 \cdot 10^7$ | 144 | 29444 | **772** | 490 | 282 | 11320 | 78233 | 162 | 706 | 1.01 |
| 7680 | $1 \cdot 10^8$ | 132 | 50930 | **764** | 489 | 275 | | | | | |
| 15360 | $2 \cdot 10^8$ | 102 | 94527 | **845** | 510 | 335 | | | | | |

Table 2 (right) shows a strong-scaling experiment. The iterative CG solver with GenEO preconditioner scales almost optimally to at least 11, 320 cores. Memory constraints prevented tests with fewer than 2880 cores. Correspondingly we take 2880 as a baseline for these tests, leading in some cases to a parallel efficiency larger than 1. Table 2 shows that the number of iterations indeed remains almost constant. The last column gives the parallel efficiency, it remains high up to 11, 320 cores. Fluctuations are due mainly to the effects of the domain decomposition and eigenvalue threshold.

**Acknowledgements.** This work was supported by an EPSRC Maths for Manufacturing grant (EP/K031368/1). This research made use of the Balena High Performance Computing Service at the University of Bath. This work used the ARCHER UK National Supercomputing Service (http://www.archer.ac.uk).

# References

1. Alnæs, M.S., et al.: The FEniCS project version 1.5. Arch. Numer. Softw. **3**(100), 9–23 (2015). https://doi.org/10.11588/ans.2015.100.20553
2. Alzetta, G., et al.: The deal.II library version 9.0. J. Numer. Math. **26**(4), 173–183 (2018). https://doi.org/10.1515/jnma-2018-0054

3. Babuška, I., Caloz, G., Osborn, J.E.: Special finite element methods for a class of second order elliptic problems with rough coefficients. SIAM J. Numer. Anal. **31**(4), 945–981 (1994)
4. Bastian, P., Blatt, M.: On the generic parallelisation of iterative solvers for the finite element method. Int. J. Comput. Sci. Eng. **4**(1), 56–69 (2008)
5. Bastian, P., et al.: A generic grid interface for parallel and adaptive scientific computing. Part ii. Implementation and tests in dune. Computing **82**(2–3), 121–138 (2008)
6. Bastian, P., Heimann, F., Marnach, S.: Generic implementation of finite element methods in the distributed and unified numerics environment (DUNE). Kybernetika **46**(2), 294–315 (2010)
7. Butler, R., Dodwell, T., Reinarz, A., Sandhu, A., Scheichl, R., Seelinger, L.: Dune-composites - an open source, high performance package for solving large-scale anisotropic elasticity problems. arXiv e-prints arXiv:1901.05188 (January 2019)
8. Chung, E., Efendiev, Y., Tat Leung, W., Ye, S.: Generalized multiscale finite element methods for space-time heterogeneous parabolic equations. Comput. Math. Appl. **76**(2), 419–437 (2016). https://doi.org/10.1016/j.camwa.2018.04.028
9. Jolivet, P., Hecht, F., Nataf, F., Prud'homme, C.: Scalable domain decomposition preconditioners for heterogeneous elliptic problems. In: Proceedings of the International Conference on High Performance Computing, Networking, Storage and Analysis, pp. 80:1–80:11. SC 2013. ACM, New York (2013). https://doi.org/10. 1145/2503210.2503212
10. Lehoucq, R.B., Sorensen, D.C., Yang, C.: ARPACK users guide: solution of large scale eigenvalue problems by implicitly restarted Arnoldi methods (1997)
11. Pechstein, C., Dohrmann, C.R.: A unified framework for adaptive BDDC. Electron. Trans. Numer. Anal. **46**, 273–336 (2017)
12. Reinarz, A., Dodwell, T., Fletcher, T., Seelinger, L., Butler, R., Scheichl, R.: Dune-composites - a new framework for high-performance finite element modelling of laminates. Compos. Struct. **184**, 269–278 (2018)
13. Sandhu, A., Reinarz, A., Dodwell, T.: A bayesian framework for assessing the strength distribution of composite structures with random defects. Compos. Struct. **205**, 58–68 (2018). https://doi.org/10.1016/j.compstruct.2018.08.074
14. Smith, B.F., Bjørstad, P.E., Gropp, W.: Domain Decomposition. Cambridge University Press, Cambridge (1996). includes bibliographical references
15. Spillane, N., Dolean, V., Hauret, P., Nataf, F., Pechstein, C., Scheichl, R.: Abstract robust coarse spaces for systems of PDEs via generalized eigenproblems in the overlaps. Numer. Math. **126**(4), 741–770 (2014). https://doi.org/10.1007/s00211-013-0576-y
16. Spillane, N., Dolean, V., Hauret, P., Nataf, F., Pechstein, C., Scheichl, R.: A robust two-level domain decomposition preconditioner for systems of PDEs. C. R. Math. **349**(23–24), 1255–1259 (2011)
17. Toselli, A., Widlund, O.: Domain Decomposition Methods - Algorithms and Theory. Springer Series in Computational Mathematics. Springer, Heidelberg (2005). https://doi.org/10.1007/b137868
18. Yang, U.M., Henson, V.E.: BoomerAMG: a parallel algebraic multigrid solver and preconditioner. Appl. Numer. Math. **41**(1), 155–177 (2002)

# Hybrid Solver for Quasi Block Diagonal Linear Systems

Viviana Arrigoni$^{(\boxtimes)}$ ⓘ and Annalisa Massini$^{(\boxtimes)}$ ⓘ

Department of Computer Science, Sapienza University of Rome, Rome, Italy
{arrigoni,massini}@di.uniroma1.it

**Abstract.** We present a solver for a class of sparse linear systems that we call *quasi block diagonal*. The solver combines multi-processors and multi-threaded parallelisms using MPI and OpenMP to implement preconditioned Jacobi. Specific formats for sparse matrices are exploited in order to reduce memory storage requirements. Our experiments show that communication costs are negligible, so as that speed-up and efficiency with respect to the sequential implementation are very high. Our hybrid implementation is tested on a cluster and compared to Intel MKL PARDISO linear solver.

**Keywords:** Sparse matrices · Linear systems · Preconditioned Jacobi · MPI · OpenMP

## 1 Introduction

Sparse matrices arise from problems in several fields, including circuit simulations, power network analysis and graph theory, as well as from the discretization of partial differential equations (PDEs) when modelling phenomena spanning over the widest scientific range, including meteorology, fluid dynamics and complex biological processes. In engineering, Finite Element Model (FEM), see e.g., [7,10,16], is widely used to model structural components of vehicles, big infrastructures, prostheses, etc., and reproduces these objects through matrices depicting their physical and mechanical structures. Such matrices are usually very large and sparse, and denser blocks are distributed along their diagonal, without any other specific assumption on their sparsity pattern, hence, we refer to them as *quasi block diagonal matrices*. They represent the coefficient matrices of the PDEs that model the system under study, and that are discretized and solved numerically integrating multiple large, sparse linear systems. The growing necessity of fast simulations has motivated many authors to concentrate their efforts in devising fast linear solvers that would adapt to the most recent parallel architectures, and that would take advantage of specific structural and mathematical features of the input problem.

In this work, we present a hybrid implementation of preconditioned Jacobi to solve quasi block diagonal linear systems. The preconditioning phase is inspired

© Springer Nature Switzerland AG 2020
R. Wyrzykowski et al. (Eds.): PPAM 2019, LNCS 12043, pp. 129–140, 2020.
https://doi.org/10.1007/978-3-030-43229-4_12

by the factorization computed in the Spike algorithm, that is instead applied on banded matrices (see Sect. 2). This preliminary operation is particularly convenient for quasi block diagonal matrices, that can then be stored very efficiently, reducing memory requirements. The solver here presented is implemented using a parallel hybrid MPI/OpenMP approach and run on a cluster. The multiprocessor system is a particularly suitable environment for our application, since most operations can be computed in perfect parallelism and communication is limited to a small amount of data. Comparisons with the parallel solver provided by the Intel high performance library for Linear ALgebra, MKL, are reported to validate our software.

## 2    Related Work

Linear solvers for sparse matrices have been under investigation for long years. Iterative methods are particularly convenient for sparse linear systems, since they do not alter their sparsity patterns: Jacobi and Gauss-Seidel algorithms, together with Krylov subspaces-based methods (e.g., CG, MINRES, GMRES, etc. see e.g., [16]) have been used, modified and adapted to specific problems in numerous occasions, see e.g., [2,17,19]. Direct solvers may be adapted in order to make them more suitable for sparse matrices, as for instance ILU and IChol factorizations (see e.g., [16]). A direct-method approach is used also in Spike algorithm, that somehow inspired the preconditioning phase of the solver here proposed. Spike was introduced in [14] and [15] and it is a hybrid solver for sparse and dense banded linear system. Although several promising preconditioners for sparse, non symmetric and indefinite linear systems have been developed, see e.g., [8], the one introduced in Spike algorithm properly fits our problem.

In [13] a multi-threaded version of Spike was implemented for shared-memory architectures using OpenMP and offering a competitive alternative to BLAS LU-based solver. A different multi-threaded implementation is described in [3] and compared to PARDISO routines for direct solvers. In [12] the Spike family of solvers is combined with the general sparse solver PARDISO to produce a fast, robust hybrid software, PSpike, and compared with some direct and iterative solver packages. A hybrid approach is instead introduced in [11]. An extensive analysis of different approaches to perform the factorization phase of Spike is given in [6]. Our solver uses the matrix splitting shown in [3] for preconditioning, and then solves the preconditioned linear system using the Jacobi algorithm, that is suitably implemented in hybrid multi-processors/threading architectures, see also [18]. Another class of sparse matrices are almost block diagonal matrices, that were described in [1] and arise in discretizations of boundary value problems for ordinary differential equations (BVODEs) and of related partial differential equations (PDEs). They can be interpreted as block banded matrices, and therefore they have a less general sparsity pattern than quasi block diagonal matrices, defined in Sect. 3. Our solver combines together MPI and the OpenMP technologies, but another well known approach consists in using distributed GPUs and is often applied for developing linear solvers in contexts similar to the one that is analysed here, see e.g., [21].

The fastest linear solvers are implemented in routines provided by high-performance Linear Algebra libraries. Some of them also supply implementations for parallel environments. In particular we compare our hybrid solver with Intel MKL PARDISO routines for cluster interfaces (see [5]).

## 3 Preconditioned Jacobi for Quasi Block Diagonal Linear Systems

Before describing the idea of our parallel implementation, we give the definition of quasi block diagonal matrices and we introduce the notion of *D-sparse matrices*.

**Definition 1 (D-sparse matrix).** *Let $D_{n \times n} = diag(D_0, \ldots, D_{k-1})$ be a block diagonal matrix. We say that a matrix $R_{n \times n}$ is D-sparse if it is sparse and if its sparsity pattern does not overlap the one of D, that is $r_{i,j} \neq 0$ implies $d_{i,j} = 0$ $\forall i, j \in \{1, \ldots, n\}$.*

**Definition 2 (Quasi Block Diagonal matrix).** *Let $R_{n \times n}$ and $D_{n \times n}$ be a D-sparse and a block diagonal matrix, respectively. A quasi block diagonal matrix $A_{n \times n}$ is a matrix defined as $A = D + R$.*

A simple example of how quasi block diagonal matrices are split in a block diagonal matrix and a D-sparse matrix, is given in Figs. 1, 2 and 3.

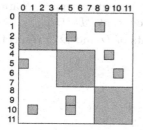

**Fig. 1.** Sparsity pattern of matrix $A = D + R$.

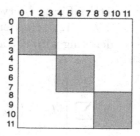

**Fig. 2.** Sparsity pattern of matrix $D$.

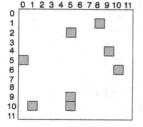

**Fig. 3.** Sparsity pattern of matrix $R$.

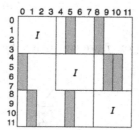

**Fig. 4.** Sparsity pattern of matrix $S = D^{-1}A = I + D^{-1}R = I + G$.

The parallel implementation that we propose uses preconditioned Jacobi to solve linear systems $A\boldsymbol{x} = \boldsymbol{b}$, where matrix $A$ is quasi block diagonal. Similarly to how described in [3], we precondition the linear system $A\boldsymbol{x} = \boldsymbol{b}$, where $A = D+R$, by left-multiplying both sides of the system by $D^{-1}$. The resulting system is $(I+D^{-1}R)\boldsymbol{x} = D^{-1}\boldsymbol{b}$, that we can write in compact form as $S\boldsymbol{x} = \boldsymbol{f}$, denoting $S = D^{-1}A = I + D^{-1}R$, and $\boldsymbol{f} = D^{-1}\boldsymbol{b}$. The sparsity pattern of $S$ for the example in Figs. 1, 2 and 3 is shown in Fig. 4. We denote as $G$ the matrix $D^{-1}R = S - I$, and we observe that matrix $G$ is $D$-sparse and its nonzero entries are distributed in small sub-columns having the same columns indexes as the nonzero entries of $R$. Usually $S$ is sparser than $A$, in particular when $R$ is very sparse and its elements have common column indexes. The Jacobi algorithm is implemented to solve the transformed system $S\boldsymbol{x} = \boldsymbol{f}$.

## 3.1   Algorithm Convergence

In the algorithm we propose, we conveniently apply the Jacobi method to the preconditioned system $S\boldsymbol{x} = \boldsymbol{f}$. To guarantee the convergence of the method, matrix $S$ has to be strictly diagonally dominant, as we prove in Proposition 1. A more detailed proof of this result is in [4]. In the following, given a matrix $M$, we denote with $m_{i,j}$ the element of row $i$ and column $j$.

**Proposition 1.** *If $A = D+R$ is strictly diagonally dominant, also $S = D^{-1}A = D^{-1}(D + R)$ is strictly diagonally dominant.*

*Proof* (Sketch). First, observe that all diagonal elements of matrix $S$ are equal to 1: $s_{i,i} = 1$, $\forall i = 1, \ldots, n$. Hence, we have to show that $||D^{-1}R||_\infty < 1$. In order to do so, denoting with $d_{i,j}$ and $\tilde{d}_{i,j}$ the elements of $D$ and $D^{-1}$ respectively, we can write:

$$\sum_{k=1}^{n} \left[ |\tilde{d}_{i,k}| \cdot \left( |d_{k,k}| - \sum_{\substack{j=1 \\ j \neq k}}^{n} |d_{k,j}| \right) \right] \leq 1 \qquad \forall i = 1, \ldots, n. \tag{1}$$

Considering matrix $G(= S - I)$, for all $i, j \in \{1, \ldots, n\}$, $i \neq j$, it holds that:

$$|s_{i,j}| = |g_{i,j}| = \left| \sum_{k=1}^{n} \tilde{d}_{i,k} r_{k,j} \right|.$$

Summing on $j$, we can write the following inequality:

$$\sum_{\substack{j=1 \\ j \neq i}}^{n} |s_{i,j}| \leq \sum_{k=1}^{n} \left[ |\tilde{d}_{i,j}| \cdot \sum_{\substack{j=1 \\ j \neq i}}^{n} |r_{i,j}| \right]. \tag{2}$$

Since $A$ is by hypothesis strictly diagonally dominant, so is $D$, and therefore

$$\sum_{j=1}^{n} |r_{k,j}| < |d_{k,k}| - \sum_{\substack{j=1 \\ j\neq i}}^{n} |d_{k,j}|. \tag{3}$$

Combining inequalities (1), (2) and (3), the thesis follows.

## 4   Hybrid Preconditioned Jacobi Implementation

Our solver implements preconditioned Jacobi in a hybrid distributed and shared memory fashion. We shall refer to our hybrid solver as HPJ (standing for Hybrid Preconditioned Jacobi). The distributed memory parallelism is managed by the MPI library that allows distributed and independent parallel processes to cooperate and to synchronize through communication. The execution of each process is accelerated by deploying multi-threading, enabled by the OpenMP technology. Let $A\boldsymbol{x} = \boldsymbol{b}$ be the quasi block diagonal linear system to solve. HPJ works when the input problem is assigned to distributed processes as follows: the coefficient matrix $A$ is partitioned horizontally in $P$ rectangular sub-matrices, where $P$ is the number of distributed processes. We shall call $A_p$ (and, consequently, $D_p$ and $R_p$) the $p$-th sub-matrix of $A$, $p \in \{0, 1, \dots, P-1\}$. Similarly, let us call $\boldsymbol{b}_p$ the portion of the vector of known terms $\boldsymbol{b}$ having the same row indexes as $A_p$, and let us define $\boldsymbol{f}_p$ analogously. The cuts are made along the diagonal blocks of $A$, and hence of $D$. In the simple example of Fig. 5 and in the discussion that follows in this section, the number of partitions is the same as the number of blocks, hence each partitions has exactly one block. More general configurations will be taken into account in Sect. 5, where we shall discuss about experimental results. Three phases may be distinguished in HPJ: *I. Preconditioning, II. Communication* and *III. Solving* phases. We shall now describe them in detail.

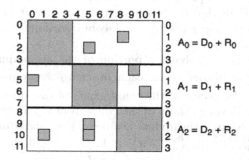

**Fig. 5.** Matrix $A$ partitioned horizontally into $A_0$, $A_1$ and $A_2$, where $A_i = D_i + R_i$, for $i = 0, 1, 2$.

*I. Preconditioning Phase.* As explained in Sect. 3, the first step to execute is the preconditioning $D^{-1}Ax = D^{-1}b$. The particular shape of the precondition matrix allows to compute $D^{-1}$ explicitly in a convenient way. We use the fact that the inverse of a block diagonal matrix $D = diag(D_0, D_1, \ldots, D_{P-1})$ is again a block diagonal matrix, composed of the inverse of each block, that is $D^{-1} = diag(D_0^{-1}, D_1^{-1}, \ldots, D_{P-1}^{-1})$. This property allows to perform matrix inversion explicitly and in perfect parallelism without the need for distributed processes to communicate. This is also true for the matrix multiplication $G = D^{-1}R$, as thanks to the sparsity patterns of $D$ and $R$, it holds that $G = [G_0; G_1; \ldots; G_{P-1}]$ where $G_p = D_p^{-1}R_p$. The last operation of this phase is the matrix-vector product, $f_p = D_p^{-1}b_p$.

Hence the preconditioning phase is carried out by distributed processes in a perfectly parallel and independent way, without the need for communication.

*II. Communication Phase.* After the preconditioning phase, distributed processes are ready to solve the preconditioned linear system. Each process $p$ computes $x_p$, that is the portion of the approximated solution $x$ having the same rows as $G_p$.

Since $R$ and $G$ are $D$-sparse, the column indexes of $G_p$ do not lay in the same row indexes as $x_p$, meaning that the approximated solution $x_p$ depends on components of $x$ that are not in $x_p$ and that are going to be computed by other processes. For this reason, before applying the Jacobi algorithm, communication among process must be established: depending on the nonzero column indexes of $G_p$, each process $p$ sends an integer to all other processes as follows.

If the indexes of some nonzero columns of $G_p$ are in the same range as the rows of block $D_q$, process $p$ sends to process $q$ an integer $ncols_p^q$ saying how many columns of $G_p$ lay on the $q$-th block, and an array of length $ncols_p^q$ containing the indexes of such columns. In this way, processor $q$ is aware of what entries of $x_q$ it must send to processor $p$ at each Jacobi iteration. If, on the contrary, none of the column indexes of $G_p$ is in the same range as the row indexes of block $D_r$, process $p$ sends $-1$ to process $r$; this means that none of the entries of $x_r$ appears in the equations that process $p$ has to solve, and process $r$ will not have to send anything to process $p$ during the Jacobi iterations.

*III. Solving Phase.* When solving a portion of the approximated solution by implementing the Jacobi algorithm, each process $p$ computes $x_p^{(k+1)} = f_p - G_p x|_{G_p}^{(k)}$, being $x|_{G_p}$ the vector of the components of the approximated solution $x$ having the same indexes as those of the columns of $G_p$. During each iteration, processes properly exchange the updated values of the approximated solution and the partial errors, $\epsilon_p^{(k+1)} = ||x_p^{(k+1)} - x_p^{(k)}||$. In this way all distributed processes can compute the global error $\epsilon^{(k+1)} = \sqrt{\epsilon_0^{(k+1)^2} + \epsilon_1^{(k+1)^2} + \ldots + \epsilon_{P-1}^{(k+1)^2}}$ so that they meet the stopping criteria at the same iteration: the algorithm stops either when $\epsilon^{(k+1)}$ is below a threshold or when the maximum number of iterations is reached.

## 4.1  Technical Details

In our solver we apply the Intel MKL LAPACK and CBLAS OpenMP-threaded routines dgesv and dgemv in order to compute $D_p^{-1}$ and $\boldsymbol{f}_p = D_p^{-1}\boldsymbol{b}_p$, respectively. This can be done because matrices $D_p$ are stored in a full format. Matrices $R_p$ are instead stored in coordinates format, while matrices $G_p$ use a modified Ellpack format, where the column indexes matrix is an array that only has as many rows as the number of diagonal blocks that process $p$ stores. Because of the different and specific sparse formats that we adopted, we implemented the matrix multiplication $G_p = D_p^{-1}R_p$ with the OpenMP directives (#pragma omp).

# 5  Performance Evaluation

We have tested our solver on randomly generated strictly diagonally dominant, quasi block diagonal matrices. We compared the execution time required to compute a solution having order of $10^{-16}$ precision with the sequential version of the algorithm, and with Intel MKL PARDISO routines for cluster interfaces.

## 5.1  Experimental Setup

Performances have been tested on Galileo cluster [20], installed in CINECA (Bologna, Italy - IBM NeXtScale, Linux Infiniband Cluster consisting of 360 nodes $2 \times 18$-cores Intel Xeon E5-2697 v4 (Broadwell) processors (2.30 GHz), and 15 nodes $2 \times 8$-cores Intel Haswell (2.40 Ghz) processors (endowed with 2 NVIDIA K80 GPUs, that we do not use in our implementation). In our tests, we generate $P = k^2$ distributed processes for different values of $k$. We involve $k$ cluster compute nodes, and $k$ tasks per node, each using 4 cores for multi-threading.

## 5.2  HPJ vs Sequential

We assess the performances of HPJ in terms of speed-up and efficiency, with respect to the sequential version of HPJ, that is, HPJ run with only one MPI process, and with the LAPACK dgesv function for solving general linear systems provided by Intel MKL. These two benchmarks are run using 4 cores. We compute the speed-up as the ratio $T_s/T_p$, where $T_s$ and $T_p$ are the execution times required by the sequential and the parallel implementations, respectively. Efficiency is computed as $S/P$, where $S$ is the speed-up and $P$ is the number of distributed processes, since we are running sequential preconditioned Jacobi and dgesv with 4 cores. For both batches of experiments, the average execution time is computed on randomly generated quasi block diagonal linear systems of size $n = 10^4$. The number of diagonal blocks is randomly chosen in the range $\{25, \ldots, 100\}$; the size of each block ranges in the set $\{100, 200, 300, 400\}$. We generate random row and column coordinates for the non zero entries of matrix $R$.

The outcomes of the first set of experiments, comparing HPJ and the sequential preconditioned Jacobi, are shown in Fig. 6(a) and (b). Speed-up is super-linear when the number of MPI processes $P$ is 4 and 9, and goes slightly below the linear trend for $P = 16$, where the efficiency is 0.83.

In Fig. 7(a) and (b), we show the speed-up and the efficiency of HPJ with respect to the Intel MKL implementation of the LAPACK dgesv routine.

The phenomenal performances of HPJ in this case are due to the fact that LAPACK does not provide sparse formats for storing sparse matrices; as a consequence, its performances are poor on quasi block diagonal linear systems. In both experiments, super-linearity is also a consequence of the greater memory availability of our hybrid solver, since it is run on more than one compute node, as explained in Sect. 5.1.

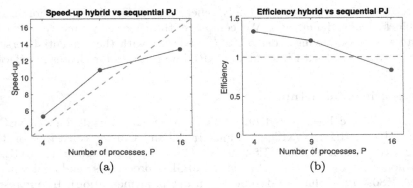

**Fig. 6.** Speed-up (a) and efficiency (b) of hybrid vs sequential preconditioned Jacobi.

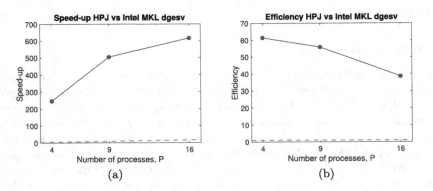

**Fig. 7.** Speed-up (a) and efficiency (b) of HPJ versus Intel MKL dgesv.

## 5.3  HPJ vs Intel MKL PARDISO

We have compared our solver HPJ with Intel MKL PARDISO, that is a software package of the Intel MKL library for solving large sparse linear systems $Ax = b$.

The Intel MKL library provides routines for cluster interfaces that integrate PARDISO functions and that follow a hybrid MPI/OpenMP approach. In particular, we used the cluster_sparse_solver routine for symbolic factorization, numerical factorization, solution retrieval and termination. We used the CSR (Compressed Sparse Row) format provided by PARDISO for storing the coefficient matrix $A$. The cluster_sparse_solver routine supports both distributed (d) and non distributed (nd) inputs. In the first case, matrix $A$ is partitioned among distributed processes. In the non distributed fashion instead, the MPI master process (the one having rank $= 0$) is the only one to read the whole matrix and to manage workload distribution among processes. Both configurations were compared with HPJ. We used the MPI timer to track the execution time for computation steps. In cluster_sparse_solver, this corresponds to the time for numerical factorization, backward substitution and iterative refinement. In our experiments, we set two steps of iterative refinement. Initially, the nested dissection algorithm from the METIS package is used for reordering the input matrix in order to reduce the amount of fill-in during the numerical factorization phase.[1] Figs. 8(a-c) show the speed-up of HPJ with respect to Intel MKL PARDISO in both the distributed and non-distributed fashion. We show how the speed-up changes varying the matrix size, $n$, for different numbers of MPI processes, $(P = 36, 49, 81)$. In this set of experiments, when quasi block diagonal random matrices of size $n$ are generated, the number of diagonal blocks ranges in the set $\{n/1000, \ldots, n/500\}$, and the size of each block in the set $\{500, 600, 700, 800, 900, 1000\}$. We generate random row and column coordinates for the non zero entries of matrix $R$. For every interval of row indexes defined by the row indexes of each diagonal block, we generate a random number of non zero entries of $R$ in the set $\{25, 40, 55, 70, 85, 100\}$.

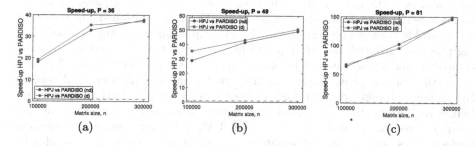

**Fig. 8.** Speed-up of HPJ vs distributed (d) and non-distributed (nd) Intel MKL PARDISO for $P = 36$ (a), $P = 49$ (b) and $P = 81$ (c) MPI processes on matrix sizes $10^5$, $2 \cdot 10^5$ and $3 \cdot 10^5$.

---

[1] More details on Intel MKL cluster_sparse_solver available options can be found on page: https://software.intel.com/en-us/mkl-developer-reference-c-cluster-sparse-solver-iparm-parameter.

The average execution times of HPJ and Intel MKL PARDISO are barely comparable, and the latter suffers from poor scalability. This fact holds for both distributed and non distributed configurations and is supported by the technical description of Intel PARDISO for clusters package, given in [9], where it is shown how PARDISO scaling factor is low when the number of threads dedicated to each MPI process is small. The authors tested the Intel PARDISO for clusters solver on two smaller matrices, with $15.7 \cdot 10^6$ and $12 \cdot 10^6$ non zero entries, and on a larger one, having $5 \cdot 10^8$ non zero entries, and they analyse the behaviour of the solver for different numbers of threads per MPI process. Only one MPI process per compute node is generated. A configuration with 4 threads per process (that is the same one adopted in our experiments) is considered too, and results in a low speed-up (approximately, the value of the speed-up is 9 and 12 on 16 MPI processes on the two smaller matrices, and less than 4 on 16 MPI processes for the larger linear system). The number of non zero elements of our largest matrices ranges between $9 \cdot 10^7$ and $1.5 \cdot 10^8$, hence our experiments represent an intermediate situation between the two scenarios considered in [9]. Performance improves when all the available threads in each compute nodes are activated, suggesting that the best configuration for Intel MKL PARDISO to set up is to reserve an entire compute node per MPI process. HPJ, instead, does not suffer from this drawback, as it does not require such a wide hardware availability to achieve high performances.

## 6   Conclusions

We have proposed a hybrid MPI/OpenMP parallel solver for clusters of multi-threaded processors, that uses preconditioned Jacobi to solve sparse linear systems and is particularly suitable for quasi block diagonal matrices, that commonly emerge from finite element analysis. The preconditioned linear system has provable convergence properties. In our implementation, the preconditioning phase is executed in perfect parallelism among processes, and independent computations are accelerated by using OpenMP for multi-threading. Furthermore, the solving phase implies an irrelevant communication cost. To reduce the memory storage consumption, we used specific sparse matrix formats. Our hybrid implementation was tested against the analogous sequential version and the dgesv routine implemented in the Intel MKL library, revealing impressive performances and exceptional scalability. Comparisons with solvers provided by the high performance library Intel MKL PARDISO show that our solver is the fastest in all considered experimental configurations.

Thanks to its very good scalability properties, our solver can support massive parallelism using a large number of MPI processes whose execution can be accelerated by multi-threading. As future work, we plan to design a hybrid CPU/GPU implementation of our solver.

**Acknowledgements.** This work has been partially supported by MIUR grant Excellence Departments 2018–2022, assigned to the Computer Science Department of

Sapienza University of Rome. The experimental part has been run on the Galileo cluster, located at Cineca, Bologna, Italy, thanks to Class C ISCRA Project n. HP10CCM8RG.

# References

1. Amodio, P., et al.: Almost block diagonal linear systems: sequential and parallel solution techniques, and applications. Numerical Linear Algebra Appl. **7**(5), 275–317 (2000). https://doi.org/10.1002/1099-1506(200007/08)7:5<275:: AID-NLA198>3.0.CO;2-G
2. Bertsekas, D., Tsitsiklis, J.: Some aspects of parallel and distributed iterative algorithms - a survey. Automatica **27**(1), 3–21 (1991). https://doi.org/10.1016/0005-1098(91)90003-K
3. Bolukbasi, E.S., Manguoglu, M.: A multithreaded recursive and nonrecursive parallel sparse direct solver. In: Bazilevs, Y., Takizawa, K. (eds.) Advances in Computational Fluid-Structure Interaction and Flow Simulation. MSSET, pp. 283–292. Springer, Cham (2016). https://doi.org/10.1007/978-3-319-40827-9_22
4. D'Alessandro, N.: Comparison of direct and iterative methods applied to almost block diagonal matrices. Master thesis, Sapienza University of Rome, Italy, Computer Science Department (2019)
5. Developer Reference for Intel® Math Kernel Library 2019 - C (2019). https:// software.intel.com/en-us/download/developer-reference-for-intel-math-kernel-library-c. Accessed 30 Oct 2019
6. Eijkhout, V., van de Geijn, R.: The spike factorization as domain decomposition method; equivalent and variant approaches. In: Berry, M., et al. (eds.) High-Performance Scientific Computing, pp. 157–169. Springer, London (2012). https:// doi.org/10.1007/978-1-4471-2437-5_7
7. Elman, H., Silvester, D., Wathen, A.: Finite Elements and Fast Iterative Solvers: With Applications in Incompressible Fluid Dynamics. Oxford University Press, Oxford (2014). https://doi.org/10.1093/acprof:oso/9780199678792.001.0001
8. Ferronato, M., Janna, C., Pini, G.: A generalized block FSAI preconditioner for nonsymmetric linear systems. J. Comput. Appl. Math. **256**, 230–241 (2014). https://doi.org/10.1016/j.cam.2013.07.049
9. Kalinkin, A., Arturov, K.: Asynchronous approach to memory management in sparse multifrontal methods on multiprocessors. Appl. Math. **4**(12), 33 (2013). https://doi.org/10.4236/am.2013.412A004
10. Larson, M., Bengzon, F.: The Finite Element Method: Theory, Implementation, and Applications, vol. 10. Springer, Heidelberg (2013). https://doi.org/10.1007/978-3-642-33287-6
11. Manguoglu, M.: A domain-decomposing parallel sparse linear system solver. J. Comput. Appl. Math. **236**(3), 319–325 (2011). https://doi.org/10.1016/j.cam.2011.07.017
12. Manguoglu, M., Sameh, A.H., Schenk, O.: PSPIKE: a parallel hybrid sparse linear system solver. In: Sips, H., Epema, D., Lin, H.-X. (eds.) Euro-Par 2009. LNCS, vol. 5704, pp. 797–808. Springer, Heidelberg (2009). https://doi.org/10.1007/978-3-642-03869-3_74
13. Mendiratta, K., Polizzi, E.: A threaded SPIKE algorithm for solving general banded systems. Parallel Comput. **37**(12), 733–741 (2011). https://doi.org/10.1016/j.parco.2011.09.003

14. Polizzi, E., Sameh, A.: A parallel hybrid banded system solver: the SPIKE algorithm. Parallel Comput. **32**(2), 177–194 (2006). https://doi.org/10.1016/j.parco.2005.07.005
15. Polizzi, E., Sameh, A.: SPIKE: a parallel environment for solving banded linear systems. Comput. Fluids **36**(1), 113–120 (2007). https://doi.org/10.1016/j.compfluid.2005.07.005
16. Saad, Y.: Iterative Methods for Sparse Linear Systems, 2nd edn. Society for Industrial and Applied Mathematics, Philadelphia (2003). https://doi.org/10.1137/1.9780898718003
17. Saad, Y., Van Der Vorst, H.: Iterative solution of linear systems in the 20th century. J. Comput. Appl. Math. **123**(1–2), 1–33 (2000). https://doi.org/10.1016/S0377-0427(00)00412-X
18. Shi, A., Shen, W., Li, Y., He, L., Zhao, D.: Implementation and analysis of Jacobi iteration based on hybrid programming. In: International Conference on Computer Design and Applications (2010). https://doi.org/10.1109/ICCDA.2010.5541479
19. Simoncini, V., Szyld, D.: Recent computational developments in Krylov subspace methods for linear systems. Numerical Linear Algebra Appl. **14**(1), 1–59 (2007). https://doi.org/10.1002/nla.499
20. UG3.3: GALILEO UserGuide (2018). https://wiki.u-gov.it/confluence/display/SCAIUS/UG3.3%3A+GALILEO+UserGuide. Accessed 30 Oct 2019
21. Yang, W., Li, K., Li, K.: A parallel solving method for block-tridiagonal equations on CPU-GPU heterogeneous computing systems. J. Supercomput. **73**(5), 1760–1781 (2017). https://doi.org/10.1007/s11227-016-1881-x

# Parallel Adaptive Cross Approximation for the Multi-trace Formulation of Scattering Problems

Michal Kravčenko[1,2(✉)] , Jan Zapletal[1,2] , Xavier Claeys[3] ,
and Michal Merta[1,2]

[1] IT4Innovations, VŠB – Technical University of Ostrava,
17. listopadu 2172/15, 708 00 Ostrava-Poruba, Czech Republic
[2] Department of Applied Mathematics, VŠB – Technical University of Ostrava,
17. listopadu 2172/15, 708 00 Ostrava-Poruba, Czech Republic
{michal.kravcenko,jan.zapletal,michal.merta}@vsb.cz
[3] Jacques-Louis Lions Laboratory, Sorbonne Université,
Boite courrier 187, 75252 Paris Cedex 05, France
claeys@ann.jussieu.fr

**Abstract.** We present a highly parallel version of the boundary element method accelerated by the adaptive cross approximation for the efficient solution of scattering problems with composite scatterers. Individual boundary integral operators are treated independently, i.e. the boundary of every homogeneous subdomain is decomposed into clusters of elements defining a block structure of the local matrix. The blocks are distributed across computational nodes by a graph algorithm providing a load balancing strategy. The intra-node implementation further utilizes threading in shared memory and in-core SIMD vectorization to make use of all features of modern processors. The suggested approach is validated on a series of numerical experiments presented in the paper.

**Keywords:** Boundary element method · Adaptive cross approximation · Multi-trace formulation · Distributed parallelization

## 1 Introduction

Boundary integral equations present a valuable tool for the description of natural phenomena including wave scattering problems. Their numerical counterpart, the boundary element method (BEM), has become an alternative to volume based approaches such as finite element or finite volume methods. Except for a natural transition of the problem to the skeleton of the computational domain, BEM has found its place in HPC implementations of PDE solvers. One of the reasons is high computational intensity of system matrix assemblers, which fits well with today's design of HPC clusters, where memory accesses are much more costly than arithmetic operations. To overcome the quadratic complexity of BEM, several methods have been proposed including the fast multipole [4,15],

R. Wyrzykowski et al. (Eds.): PPAM 2019, LNCS 12043, pp. 141–150, 2020.
https://doi.org/10.1007/978-3-030-43229-4_13

or adaptive cross approximation (ACA) [2,16,17] methods, with the latter one utilized in this paper.

To describe wave scattering in a composite scatterer and its complement we utilize the local version of the multi-trace formulation (MTF) as introduced in [5–7] and presented here briefly in Sect. 2. The formulation leads to a block matrix with individual boundary integral operators for every homogeneous subdomain and coupling matrices on the off-diagonals. Although not described in detail in this paper, MTF allows for a natural operator preconditioning. In Sect. 3 we propose a strategy to parallelize the assembly of the MTF matrix blocks and their application in an iterative solver based on the approach presented in [11–13] for single domain problems. Except for the distributed parallelism, the method takes full advantage of the BEM4I library [14,20,21] and its assemblers parallelized in shared memory and vectorized by OpenMP. We provide the results of numerical experiments in Sect. 4.

## 2   Multi-trace Formulation

In the following we consider a partitioning of the space into $n + 1$ Lipschitz domains

$$\mathbb{R}^3 = \bigcup_{i=0}^{n} \overline{\Omega_i}, \quad \Gamma_i := \partial \Omega_i, \quad \Sigma := \bigcup_{i=0}^{n} \Gamma_i$$

with $\Omega_0$ denoting the background unbounded domain, $\bigcup_{i=1}^{n} \Omega_i$ representing a composite scatterer, and the skeleton $\Sigma$. For a given incident field $u_{\text{inc}}$ satisfying $-\Delta u_{\text{inc}} - \kappa_0^2 u_{\text{inc}} = 0$ in $\mathbb{R}^3$, we aim to solve the equation for $u = u_{\text{sc}} + u_{\text{inc}}$,

$$- \Delta u - \kappa_i^2 u = 0 \text{ in } \Omega_i, \quad u - u_{\text{inc}} \text{ satisfies the Sommerfeld condition} \quad (2.1)$$

with the transmission conditions

$$u_i - u_j = 0, \quad \mu_i^{-1} t_i + \mu_j^{-1} t_j = 0 \quad \text{on } \Gamma_{i,j} := \Gamma_i \cap \Gamma_j \quad (2.2)$$

with $u_i$, $t_i$ denoting the Dirichlet and Neumann traces of $u|_{\Omega_i}$. Note that the normal vector $\boldsymbol{n}_i$ is always directed outside of $\Omega_i$, see Fig. 1.

Due to (2.1) and $u_{\text{inc}}$ satisfying the Helmholtz equation with $\kappa_0$ in $\mathbb{R}^3 \backslash \overline{\Omega_0}$ we have $u = \widetilde{V}_{\kappa_0} t_0 - W_{\kappa_0} u_0 + u_{\text{inc}}$ in $\Omega_0$ and $u = \widetilde{V}_{\kappa_i} t_i - W_{\kappa_i} u_i$ in $\Omega_i$, $i > 0$ [16,18] with the single- and double-layer potentials

$$\widetilde{V}_{\kappa_i} : H^{-1/2}(\Gamma_i) \to H_{\text{loc}}^1(\Omega_i), \quad \widetilde{V}_{\kappa_i} t_i(\boldsymbol{x}) := \int_{\Gamma_i} v_{\kappa_i}(\boldsymbol{x}, \boldsymbol{y}) t_i(\boldsymbol{y}) \, \mathrm{d}s_{\boldsymbol{y}},$$

$$W_{\kappa_i} : H^{1/2}(\Gamma_i) \to H_{\text{loc}}^1(\Omega_i), \quad W_{\kappa_i} u_i(\boldsymbol{x}) := \int_{\Gamma_i} \frac{\partial v_{\kappa_i}}{\partial \boldsymbol{n}_{i,\boldsymbol{y}}}(\boldsymbol{x}, \boldsymbol{y}) u_i(\boldsymbol{y}) \, \mathrm{d}s_{\boldsymbol{y}},$$

respectively, and the fundamental solution $v_\kappa(\boldsymbol{x}, \boldsymbol{y}) = \exp(i\kappa|\boldsymbol{x} - \boldsymbol{y}|)/(4\pi|\boldsymbol{x} - \boldsymbol{y}|)$. Observing that

$$\begin{bmatrix} -K_{\kappa_0} & V_{\kappa_0} \\ D_{\kappa_0} & K_{\kappa_0}^* \end{bmatrix} \begin{bmatrix} u_{\text{inc}} \\ t_{\text{inc}} \end{bmatrix} = -\frac{1}{2} \begin{bmatrix} u_{\text{inc}} \\ t_{\text{inc}} \end{bmatrix} \quad \text{on } \Gamma_0$$

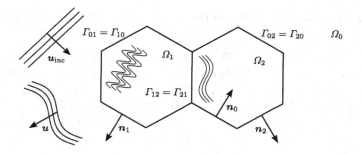

**Fig. 1.** Notation and setting for the scattering problem.

results in the relation between the traces of $u|_{\Omega_i}$

$$\begin{bmatrix} -K_{\kappa_i} & V_{\kappa_i} \\ D_{\kappa_i} & K_{\kappa_i}^* \end{bmatrix} \begin{bmatrix} u_i \\ t_i \end{bmatrix} - \frac{1}{2} \begin{bmatrix} u_i \\ t_i \end{bmatrix} = \begin{bmatrix} f_i \\ g_i \end{bmatrix} \text{ on } \Gamma_i, \quad \begin{bmatrix} f_0 \\ g_0 \end{bmatrix} := - \begin{bmatrix} u_{inc} \\ t_{inc} \end{bmatrix}, \quad \begin{bmatrix} f_i \\ g_i \end{bmatrix} := 0 \text{ for } i > 0,$$
(2.3)

with the boundary integral operators defined as the composition of the trace operators $\gamma_i^D \colon H^1_{loc}(\Omega_i) \to H^{1/2}(\Gamma_i)$, $\gamma_i^N \colon H^1_{loc}(\Omega_i) \to H^{-1/2}(\Gamma_i)$ and the potentials, i.e.

$$V_{\kappa_i} := \gamma_i^D \widetilde{V}_{\kappa_i}, \quad -\frac{1}{2}I + K_{\kappa_i} := \gamma_i^D W_{\kappa_i}, \quad \frac{1}{2}I + K_{\kappa_i}^* := \gamma_i^N V_{\kappa_i}, \quad D_{\kappa_i} := -\gamma_i^N W_{\kappa_i}.$$

The idea of MTF is to replace the identity part of (2.3) by the transmission conditions (2.2). To this end we define the operators

$$X_{i,j} u_i := \begin{cases} \frac{1}{2} u_i|_{\Gamma_{i,j}} & \text{on } \Gamma_{i,j}, \\ 0 & \text{otherwise}, \end{cases} \qquad Y_{i,j} t_i := \begin{cases} \frac{1}{2} \frac{\mu_j}{\mu_i} t_i|_{\Gamma_{i,j}} & \text{on } \Gamma_{i,j}, \\ 0 & \text{otherwise}. \end{cases}$$

The variational formulation for the situation in Fig. 1 thus reads

$$\left\langle \begin{bmatrix} -K_{\kappa_0} & V_{\kappa_0} & -X_{1,0} & 0 & -X_{2,0} & 0 \\ D_{\kappa_0} & K_{\kappa_0}^* & 0 & Y_{1,0} & 0 & Y_{2,0} \\ -X_{0,1} & 0 & -K_{\kappa_1} & V_{\kappa_1} & -X_{2,1} & 0 \\ 0 & Y_{0,1} & D_{\kappa_1} & K_{\kappa_1}^* & 0 & Y_{2,1} \\ -X_{0,2} & 0 & -X_{1,2} & 0 & -K_{\kappa_2} & V_{\kappa_2} \\ 0 & Y_{0,2} & 0 & Y_{1,2} & D_{\kappa_2} & K_{\kappa_2}^* \end{bmatrix} \begin{bmatrix} u_0 \\ t_0 \\ u_1 \\ t_1 \\ u_2 \\ t_2 \end{bmatrix}, \begin{bmatrix} s_0 \\ v_0 \\ s_1 \\ v_1 \\ s_2 \\ v_2 \end{bmatrix} \right\rangle_\Sigma = -\left\langle \begin{bmatrix} u_{inc} \\ t_{inc} \\ 0 \\ 0 \\ 0 \\ 0 \end{bmatrix}, \begin{bmatrix} s_0 \\ v_0 \\ s_1 \\ v_1 \\ s_2 \\ v_2 \end{bmatrix} \right\rangle_\Sigma$$

with

$$(u_i, t_i) \in H^{1/2}(\Gamma_i) \times H^{-1/2}(\Gamma_i), \qquad (v_i, s_i) \in \widetilde{H}_{pw}^{1/2}(\Gamma_i) \times \widetilde{H}_{pw}^{-1/2}(\Gamma_i).$$

For a definition of the above Sobolev spaces on manifolds we refer the interested reader to [1,7,18,19] and references therein.

To discretize the system we decompose the skeleton $\Sigma$ into plane triangles $\tau_k$ and define the discrete function space span($\varphi_\ell$) of globally continuous piecewise linear functions to approximate all the above spaces. The discrete system thus reads

$$
\begin{bmatrix}
-\mathsf{K}_{\kappa_0,h} & \mathsf{V}_{\kappa_0,h} & -\mathsf{M}_{1,0,h} & \mathsf{O} & -\mathsf{M}_{2,0,h} & \mathsf{O} \\
\mathsf{D}_{\kappa_0,h} & \mathsf{K}_{\kappa_0,h}^\top & \mathsf{O} & \frac{\mu_0}{\mu_1}\mathsf{M}_{1,0,h} & \mathsf{O} & \frac{\mu_0}{\mu_2}\mathsf{M}_{2,0,h} \\
-\mathsf{M}_{0,1,h} & \mathsf{O} & -\mathsf{K}_{\kappa_1,h} & \mathsf{V}_{\kappa_1,h} & -\mathsf{M}_{2,1,h} & \mathsf{O} \\
\mathsf{O} & \frac{\mu_1}{\mu_0}\mathsf{M}_{0,1,h} & \mathsf{D}_{\kappa_1,h} & \mathsf{K}_{\kappa_1,h}^\top & \mathsf{O} & \frac{\mu_1}{\mu_2}\mathsf{M}_{2,1,h} \\
-\mathsf{M}_{0,2,h} & \mathsf{O} & -\mathsf{M}_{1,2,h} & \mathsf{O} & -\mathsf{K}_{\kappa_2,h} & \mathsf{V}_{\kappa_2,h} \\
\mathsf{O} & \frac{\mu_2}{\mu_0}\mathsf{M}_{0,2,h} & \mathsf{O} & \frac{\mu_2}{\mu_1}\mathsf{M}_{1,2,h} & \mathsf{D}_{\kappa_2,h} & \mathsf{K}_{\kappa_2,h}^\top
\end{bmatrix}
\begin{bmatrix}
\boldsymbol{u}_0 \\ \boldsymbol{t}_0 \\ \boldsymbol{u}_1 \\ \boldsymbol{t}_1 \\ \boldsymbol{u}_2 \\ \boldsymbol{t}_2
\end{bmatrix}
= -
\begin{bmatrix}
\mathsf{M}_{0,h}\boldsymbol{u}_{\mathrm{inc}} \\ \mathsf{M}_{0,h}\boldsymbol{t}_{\mathrm{inc}} \\ 0 \\ 0 \\ 0 \\ 0
\end{bmatrix}
$$

with the matrices

$$
\begin{aligned}
\mathsf{V}_{\kappa_i,h}[k,\ell] &:= \langle V_{\kappa_i}\varphi_\ell, \varphi_k \rangle_{\Gamma_i}, \quad & \mathsf{K}_{\kappa_i,h}[k,\ell] &:= \langle K_{\kappa_i}\varphi_\ell, \varphi_k \rangle_{\Gamma_i}, \\
\mathsf{D}_{\kappa_i,h}[k,\ell] &:= \langle D_{\kappa_i}\varphi_\ell, \varphi_k \rangle_{\Gamma_i}, \\
\mathsf{M}_{i,j,h}[k,\ell] &:= \frac{1}{2}\langle \varphi_\ell, \varphi_k \rangle_{\Gamma_{i,j}}, \quad & \mathsf{M}_{0,h}[k,\ell] &:= \langle \varphi_\ell, \varphi_k \rangle_{\Gamma_0}
\end{aligned}
\tag{2.4}
$$

and the duality pairings

$$
\langle u_i, t_i \rangle_{\Gamma_i} = \int_{\Gamma_i} u_i(\boldsymbol{x}) t_i(\boldsymbol{x}) \, \mathrm{d}\boldsymbol{s}_{\boldsymbol{x}}, \qquad
\langle u_i, t_j \rangle_{\Gamma_{i,j}} = \int_{\Gamma_{i,j}} u_i(\boldsymbol{x}) t_j(\boldsymbol{x}) \, \mathrm{d}\boldsymbol{s}_{\boldsymbol{x}}.
$$

The local indices $k, \ell$ in (2.4) span over all globally defined basis functions supported on the respective interfaces $\Gamma_i$ or $\Gamma_{i,j}$.

## 3   Parallel ACA

The matrices produced by a classical BEM are usually dense. Although the number of degrees of freedom is smaller compared to the volume-based methods (e.g., finite element method), some of the so-called fast BEM have to be applied in order to solve large-scale problems. These methods are usually based on a hierarchical clustering of the underlying mesh and approximations of matrix blocks corresponding to pairs of sufficiently separated clusters. Here we use the adaptive cross approximation (ACA) method since it is a purely algebraic approach and once properly implemented, it can be used for various kind of problems. Other approaches include the fast multipole method [9,15] or the wavelet compression [8].

After hierarchical clustering, ACA proceeds by approximating sub-matrices corresponding to pairs of well-separated (admissible) clusters by a product of two low-rank matrices. Non-admissible clusters are assembled as dense matrices. Due to a limited scope of this paper, we refer the reader to [3,16] for more details. The parallelization of the method based on the cyclic graph decomposition was presented in [13] where only certain special numbers of processors were discussed. In [12] we further extended the approach to support general number of processors. In the following section we recollect its basic principle and extend it to support the solution of MTF systems.

Let us first briefly describe the parallelization of a single-domain problem. To distribute BEM system matrices among $P$ processors we decompose the input surface mesh $\Gamma$ into $P$ disjoint sub-meshes $\Gamma^1, \Gamma^2, \ldots, \Gamma^P$ of approximately the same number of elements using, e.g. the METIS library [10]. In this manner we obtain a $P \times P$ block structure of matrices such that the block in row $i$ and column $j$ is induced by the integration over $\Gamma^i$ and $\Gamma^j$. Afterwards, the $P^2$ blocks are assembled in parallel via $P$ MPI processes. The workload distribution follows the so-called cyclic graph decomposition introduced in [13] for special values of $P$ and later generalized in [12]. This way, each MPI process requires $\mathcal{O}(n/\sqrt{P})$ mesh data for the assembly of the blocks and actively works with $\mathcal{O}(n/\sqrt{P})$ degrees of freedom during matrix-vector multiplication, which has a positive effect on the efficiency of the required MPI synchronization phase. See Fig. 2 for an example of an $8 \times 8$ block distribution together with the respective generator graph.

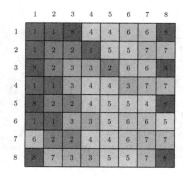 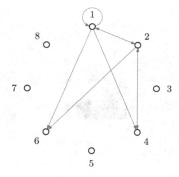

**Fig. 2.** Workload block distribution for 8 MPI processes (left) and the respective generator graph (right). The vertex indices connected by an edge correspond to a block of the matrix assigned to the first process. Other processes are assigned blocks determined by a clockwise rotation of the generator graph.

The extension of the proposed parallelization scheme to local multi-trace operators can be implemented in a straight-forward manner. Let $\Omega_0, \Omega_1, \cdots, \Omega_m$ denote the subdomains with their respective boundaries $\Gamma_0, \Gamma_1, \cdots, \Gamma_m$. We apply our parallel ACA scheme to each subdomain individually, i.e. each $\Gamma_j$ is split into $P$ submeshes. The BEM operators $\mathsf{K}_{\kappa_j,h}, \mathsf{V}_{\kappa_j,h}, \mathsf{D}_{\kappa_j,h}$ are treated as $P \times P$ block operators, each assembled in parallel according to the corresponding cyclic graph decomposition. This approach works reasonably well, however, distributing each boundary among the same number of processes may lead to a poor parallel efficiency in the case of $\Gamma_j$ with small number of elements. Thus, the goal of our future research is to design an advanced parallel scheme for the assembly of local operators such that $\mathsf{K}_{\kappa_j,h}, \mathsf{V}_{\kappa_j,h}, \mathsf{D}_{\kappa_j,h}$ are assembled by various numbers of MPI processes.

## 4  Numerical Experiments

Our main interest lies in the study of the parallel matrix assembly and efficiency of the matrix-vector multiplication. The results of strong and weak scalability experiments are presented in Tables 1 and 2, respectively.

The numerical experiments were carried out on the Salomon supercomputer at IT4Innovations National Supercomputing Center, Czech Republic. The cluster is composed of 1 008 compute nodes, each of them equipped with two 12-core Intel Xeon E5-2680v3 processors and 128 GB of RAM. The nodes are interconnected by the 7D enhanced hypercube InfiniBand network. The theoretical peak performance totals 2 PFlop/s. We used METIS 5.1.0 for the decomposition of the mesh [10] and Intel Parallel Studio 2017.1.132 with Intel Math Kernel Library (MKL) as a BLAS implementation. We use two MPI processes per compute node, 12 OMP threads per one MPI process and set KMP_AFFINITY="granularity=fine,scatter".

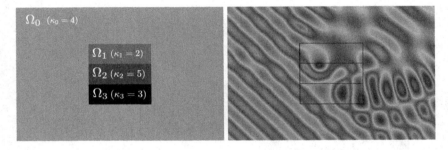

**Fig. 3.** A cut through the test domain geometry (left) and plot of the resulting total field (right). The test domain is a cube split into three parts.

All the tests have been performed on a cubical geometry split into three parts, see Fig. 3 for a central cut of the domain and also the depiction of the resulting total field. The wave numbers are $\kappa_0 = 4$, $\kappa_1 = 2$, $\kappa_2 = 5$, $\kappa_3 = 3$ and $\mu_0 = \mu_1 = \mu_2 = \mu_3 = 1$. We used globally continuous piecewise linear trial and test functions for all operators included in the formulation. The parameters controlling the complexity of ACA were set to $\varepsilon_{ACA} = 10^{-6}$, $\mu_{ACA} = 1.2$. The relative precision for the GMRES solver was set to $\varepsilon_{GMRES} = 10^{-6}$.

The measured quantities are the time in seconds to assemble the matrices, the time to perform one matrix-vector multiplication without MPI synchronization, and also the overhead required to synchronize the results via MPI. We also present the efficiency of the parallel solver. We performed a series of five experiments and the presented values are the averages of the results.

In the case of strong scaling experiments (see Table 1), $\Gamma_0$ contains 258 048 elements and $\Gamma_1, \Gamma_2, \Gamma_3$ contain 110 592 elements each. Let the real time for parallel matrix assembly and matrix-vector multiplication on $P$ processes be $t_P$, then the efficiency of the strong scaling on $P$ processes is calculated as $(2t_2)/(Pt_P)$. Taking into account the relatively small size of the problem, we obtain a reasonable parallel efficiency of the system matrix assembly up to 64 compute nodes (128 processes). The matrix-vector multiplication scales relatively well up to 16 nodes (32 processes). The reason is twofold. The first major factor is increasing time required for vector synchronization after each multiplication. The second factor is the increasing density of the ACA matrices on higher number of processes. In the current implementation, our synchronization scheme is the following:

- we split the outer vector into 4 disjoint intervals (one for each subdomain),
- then we perform all necessary computations for each interval followed by a non-blocking allreduce across all processes,
- to facilitate the progress of the non-blocking reductions, we perform periodic calls to MPI_Test on the master OMP thread.

**Table 1.** Strong scalability of operator assembly, matrix-vector multiplication and total runtime of the solver on $1, 2, \ldots, 64$ compute nodes (MPI synchronization included).

| $P$ | Matrix assembly | | Matrix-vector multiply | | | Total time | |
|---|---|---|---|---|---|---|---|
| | [s] | eff. [%] | apply [s] | sync. [s] | eff. [%] | [s] | eff. [%] |
| 2 | 627.1 | 100.0 | 2.55 | 0.010 | 100.0 | 1614.6 | 100.0 |
| 4 | 316.2 | 99.2 | 1.27 | 0.028 | 98.8 | 884.1 | 91.3 |
| 8 | 177.2 | 88.5 | 0.66 | 0.028 | 92.8 | 542.4 | 74.4 |
| 16 | 112.5 | 69.7 | 0.34 | 0.039 | 83.5 | 377.6 | 53.4 |
| 32 | 54.7 | 71.7 | 0.19 | 0.052 | 66.7 | 273.7 | 36.9 |
| 64 | 26.8 | 73.0 | 0.10 | 0.076 | 45.1 | 226.9 | 22.2 |
| 128 | 14.3 | 68.5 | 0.06 | 0.091 | 26.0 | 214.1 | 11.8 |

The weak scaling experiments were performed on 1, 4 and 16 compute nodes each running two MPI processes (see Table 2). On a single node, $\Gamma_0$ contains $k_0 := 81\,648$ elements, and $\Gamma_1, \Gamma_2, \Gamma_3$ contain $k_1 := 34\,992$ elements each. On $P/2$ nodes, the number of mesh elements is proportionally increased, i.e. $Pk_0/2$ for $\Gamma_0$ and $Pk_1/2$ for $\Gamma_1, \Gamma_2$ and $\Gamma_3$. Let the expected complexity of parallel matrix assembly and matrix-vector multiplication on $P$ processes be $e_P := (k_0 \log(\frac{P}{2}k_0)) + 3k_1 \log(\frac{P}{2}k_1))/2$ and let the real time be $t_P$, respectively. The efficiency of the weak scaling on $P$ processes is then calculated as $(t_2 e_P)/(t_P e_2)$.

**Table 2.** Efficiency of the weak scaling of operator assembly and matrix-vector multiplication on $1, 4, 16$ compute nodes (MPI synchronization included).

| | Matrix assembly | | Matrix-vector multiply | | |
|---|---|---|---|---|---|
| $P$ | $t$ [s] | eff. [%] | apply $t$ [s] | sync. $t$ [s] | eff. [%] |
| 2 | 206.7 | 100.0 | 0.79 | 0.0069 | 100.0 |
| 8 | 247.9 | 94.0 | 1.11 | 0.0476 | 80.2 |
| 32 | 322.2 | 80.6 | 1.99 | 0.2080 | 49.6 |

## 5    Conclusion

We briefly presented a local multi-trace formulation for scattering problems with heterogeneous scatterers and applied the so-called cyclic graph decomposition to define a block-based workload distribution for distributed parallel matrix assembly and matrix-vector multiplication. We performed experiments on up to 64 compute nodes and presented strong and weak scaling properties of our approach. In our following work, we plan to refine and generalize the proposed methods to directly decompose the skeleton of the scatterer instead of the individual subdomains. We believe this could lead to a better scalability and ability to deal with problems with differently sized subdomains.

**Acknowledgement.** The authors acknowledge the support provided by The Ministry of Education, Youth and Sports from the National Programme of Sustainability (NPS II) project 'IT4Innovations excellence in science – LQ1602' and from the Large Infrastructures for Research, Experimental Development and Innovations project 'IT4Innovations National Supercomputing Center – LM2015070'. MK was further supported by the ESF in 'Science without borders' project, reg. nr. CZ.02.2.69/0.0./0.0./16_027/0008463 within the Operational Programme Research, Development and Education.

## References

1. Amann, D.: Boundary element methods for Helmholtz transmission problems. Master's thesis, TU Graz (2014)
2. Bebendorf, M.: Hierarchical Matrices: A Means to Efficiently Solve Elliptic Boundary Value Problems. Lecture Notes in Computational Science and Engineering. Springer, Heidelberg (2008). https://doi.org/10.1007/978-3-540-77147-0
3. Bebendorf, M., Rjasanow, S.: Adaptive low-rank approximation of collocation matrices. Computing **70**(1), 1–24 (2003). https://doi.org/10.1007/s00607-002-1469-6
4. Chaillat, S., Collino, F.: A wideband fast multipole method for the Helmholtz kernel: theoretical developments. Comput. Math. Appl. **70**(4), 660–678 (2015)
5. Claeys, X., Dolean, V., Gander, M.: An introduction to multi-trace formulations and associated domain decomposition solvers. Appl. Numer. Math. **135**, 69–86 (2019)

6. Claeys, X., Hiptmair, R.: Multi-trace boundary integral formulation for acoustic scattering by composite structures. Commun. Pure Appl. Math. **66**(8), 1163–1201 (2013)
7. Claeys, X., Hiptmair, R., Jerez-Hancknes, C., Pintarelli, S.: Novel multi-trace boundary integral equations for transmission boundary value problems. In: Unified Transform for Boundary Value Problems: Applications and Advances, pp. 227–258. SIAM (2015)
8. Dahlke, S., Harbrecht, H., Utzinger, M., Weimar, M.: Adaptive wavelet BEM for boundary integral equations: theory and numerical experiments. Numer. Functional Anal. Optim. **39**(2), 208–232 (2018)
9. Greengard, L., Rokhlin, V.: A fast algorithm for particle simulations. J. Comput. Phys. **135**(2), 280–292 (1997)
10. Karypis, G., Kumar, V.: A fast and high quality multilevel scheme for partitioning irregular graphs. SIAM J. Sci. Comput. **20**(1), 359–392 (1999)
11. Kravcenko, M., Maly, L., Merta, M., Zapletal, J.: Parallel assembly of ACA BEM matrices on Xeon Phi clusters. In: Wyrzykowski, R., Dongarra, J., Deelman, E., Karczewski, K. (eds.) PPAM 2017. LNCS, vol. 10777, pp. 101–110. Springer, Cham (2018). https://doi.org/10.1007/978-3-319-78024-5_10
12. Kravčenko, M., Merta, M., Zapletal, J.: Distributed fast boundary element methods for Helmholtz problems. Appl. Math. Comput. **362**, 124503 (2019)
13. Lukáš, D., Kovář, P., Kovářová, T., Merta, M.: A parallel fast boundary element method using cyclic graph decompositions. Numer. Algorithms **70**(4), 807–824 (2015)
14. Merta, M., Zapletal, J.: BEM4I. IT4Innovations, VŠB - Technical University of Ostrava, 17. listopadu 2172/15, 708 00 Ostrava-Poruba, Czech Republic (2013). http://bem4i.it4i.cz/
15. Of, G.: Fast multipole methods and applications. In: Schanz, M., Steinbach, O. (eds.) Boundary Element Analysis. Lecture Notes in Applied and Computational Mechanics, vol. 29, pp. 135–160. Springer, Berlin Heidelberg (2007). https://doi.org/10.1007/978-3-540-47533-0_6
16. Rjasanow, S., Steinbach, O.: The Fast Solution of Boundary Integral Equations. Mathematical and Analytical Techniques with Applications to Engineering. Springer, Heidelberg (2007). https://doi.org/10.1007/0-387-34042-4
17. Rjasanow, S., Weggler, L.: Matrix valued adaptive cross approximation. Math. Methods Appl. Sci. **40**(7), 2522–2531 (2017)
18. Sauter, S.A., Schwab, C.: Boundary Element Methods. Springer Series in Computational Mathematics. Springer, Heidelberg (2010). https://doi.org/10.1007/978-3-540-68093-2
19. Steinbach, O.: Numerical Approximation Methods for Elliptic Boundary Value Problems: Finite and Boundary Elements. Texts in Applied Mathematics. Springer, New York (2008). https://doi.org/10.1007/978-0-387-68805-3
20. Zapletal, J., Merta, M., Malý, L.: Boundary element quadrature schemes for multi- and many-core architectures. Comput. Math. Appl. **74**(1), 157–173 (2017). 5th European Seminar on Computing ESCO 2016
21. Zapletal, J., Of, G., Merta, M.: Parallel and vectorized implementation of analytic evaluation of boundary integral operators. Engineering Analysis with Boundary Elements **96**, 194–208 (2018)

# Implementation of Parallel 3-D Real FFT with 2-D Decomposition on Intel Xeon Phi Clusters

Daisuke Takahashi[✉]

Center for Computational Sciences, University of Tsukuba,
1-1-1 Tennodai, Tsukuba, Ibaraki 305-8577, Japan
daisuke@cs.tsukuba.ac.jp

**Abstract.** In this paper, we propose an implementation of a parallel 3-D real fast Fourier transform (FFT) with 2-D decomposition on Intel Xeon Phi clusters. The proposed implementation of the parallel 3-D real FFT is based on the conjugate symmetry property of the discrete Fourier transform (DFT) and the row-column FFT algorithm. We vectorized FFT kernels using the Intel Advanced Vector Extensions 512 (Intel AVX-512) instructions. Performance results of parallel 3-D real FFTs on Intel Xeon Phi clusters are reported. We successfully achieved a level of performance over 10 TFlops on 2048 nodes of Fujitsu PRIMERGY CX1640 M1 cluster for an $8192^3$-point FFT.

**Keywords:** Fast Fourier transform · 2-D decomposition · Intel Xeon Phi clusters

## 1 Introduction

The fast Fourier transform (FFT) [6] is an algorithm which is currently widely used in science and engineering. There is a substantial literature investigating parallel 3-D FFT algorithms on distributed-memory parallel computers [1,3,4,7–11].

As a typical data distribution method in parallel 3-D FFTs, only one dimension (for example, $z$-axis) between three dimensions (the $x$-, $y$-, and $z$-axes) is divided. In this case, the number of data points on the $z$-axis must be greater than or equal to the number of MPI processes. Parallel 3-D FFTs of the FFTW [7] and the Intel Math Kernel Library (Intel MKL) use this 1-D decomposition.

In contemporary supercomputers, the number of cores and the number of processors tend to increase to improve performance. For example, Sunway TaihuLight was ranked third in the TOP500 list in November 2018 [2], and its number of cores exceeds 10 million. Even reducing the number of MPI processes by hybrid MPI and OpenMP parallel programming, the maximum number of MPI processes is more than 10000 for such a system. Therefore, when 1-D decomposition on the $z$-axis is used, the number of data points on the $z$-axis must

© Springer Nature Switzerland AG 2020
R. Wyrzykowski et al. (Eds.): PPAM 2019, LNCS 12043, pp. 151–161, 2020.
https://doi.org/10.1007/978-3-030-43229-4_14

exceed 10000, and the problem size for the 3-D FFT is restricted. As a method of addressing this issue, parallel 3-D FFTs with 2-D decomposition have been proposed [1,3,9–11]. However, to the best of our knowledge, an implementation of a parallel 3-D real FFT with 2-D decomposition on Intel Xeon Phi clusters has not yet been reported. Therefore, we pursue this implementation and report the resulting performance. The remainder of the paper is organized as follows. Section 2 describes the 3-D real FFT algorithm. Section 3 describes the parallel 3-D real FFT algorithm with 2-D decomposition. Section 4 reports on a comparison between 1-D and 2-D decomposition. Section 5 presents the performance results. Finally, in Sect. 6, concluding remarks are offered.

## 2   3-D Real FFT Algorithm

The discrete Fourier transform (DFT) is given by

$$y(k) = \sum_{j=0}^{n-1} x(j)\omega_n^{jk}, \quad 0 \le k \le n-1, \tag{1}$$

where $\omega_n = e^{-2\pi i/n}$ and $i = \sqrt{-1}$.

When the input data $x(j)$ of the DFT are real, two $n$-point real DFTs can be computed efficiently using a single $n$-point complex DFT [5]. Perform complex DFT by putting two real data points in each real and imaginary part of complex input data.

Let

$$x(j) = x_1(j) + ix_2(j), \quad 0 \le j \le n-1, \tag{2}$$

where $x_1(j)$ and $x_2(j)$ for $0 \le j \le n-1$ are two $n$-point real input data.

With respect to the output data $y(k)$ of the complex DFT, the following equations hold from the complex conjugate property of DFT:

$$y(k) = y_1(k) + iy_2(k), \quad 0 \le k \le n-1, \tag{3}$$

$$\overline{y}(n-k) = y_1(k) - iy_2(k), \quad 0 \le k \le n-1, \tag{4}$$

where $y_1(k)$ and $y_2(k)$ for $0 \le k \le n-1$ are two $n$-point real DFTs of $x_1(j)$ and $x_2(j)$ for $0 \le j \le n-1$, respectively.

The output data $y_1(k)$ and $y_2(k)$ of two $n$-point real DFTs are obtained from $y(k)$ and $\overline{y}(n-k)$ in Eqs. (3) and (4).

$$y_1(k) = \frac{1}{2}\{y(k) + \overline{y}(n-k)\}, \quad 0 \le k \le n-1, \tag{5}$$

$$y_2(k) = -\frac{i}{2}\{y(k) - \overline{y}(n-k)\}, \quad 0 \le k \le n-1. \tag{6}$$

The 3-D DFT is given by

$$y(k_1, k_2, k_3) = \sum_{j_1=0}^{n_1-1} \sum_{j_2=0}^{n_2-1} \sum_{j_3=0}^{n_3-1} x(j_1, j_2, j_3)\omega_{n_1}^{j_1 k_1} \omega_{n_2}^{j_2 k_2} \omega_{n_3}^{j_3 k_3},$$

$$0 \le k_1 \le n_1 - 1, \quad 0 \le k_2 \le n_2 - 1, \quad 0 \le k_3 \le n_3 - 1, \tag{7}$$

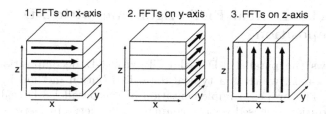

**Fig. 1.** 1-D decomposition for parallel 3-D real FFT

where $\omega_{n_r} = e^{-2\pi i/n_r}$ $(1 \le r \le 3)$ and $i = \sqrt{-1}$.

The 3-D real FFT based on the conjugate symmetry property of the DFT and the row-column FFT algorithm is as follows:

Step 1:   $n_2 n_3$ individual $n_1$-point multicolumn real FFTs

$$x_1(k_1, j_2, j_3) = \sum_{j_1=0}^{n_1-1} x(j_1, j_2, j_3)\omega_{n_1}^{j_1 k_1}.$$

Step 2:   Transposition

$$x_2(j_2, j_3, k_1) = x_1(k_1, j_2, j_3).$$

Step 3:   $n_3 \cdot (n_1/2 + 1)$ individual $n_2$-point multicolumn complex FFTs

$$x_3(k_2, j_3, k_1) = \sum_{j_2=0}^{n_2-1} x_2(j_2, j_3, k_1)\omega_{n_2}^{j_2 k_2}.$$

Step 4:   Transposition

$$x_4(j_3, k_1, k_2) = x_3(k_2, j_3, k_1).$$

Step 5:   $(n_1/2 + 1) \cdot n_2$ individual $n_3$-point multicolumn complex FFTs

$$x_5(k_3, k_1, k_2) = \sum_{j_3=0}^{n_3-1} x_4(j_3, k_1, k_2)\omega_{n_3}^{j_3 k_3}.$$

Step 6:   Transposition

$$y(k_1, k_2, k_3) = x_5(k_3, k_1, k_2).$$

In Step 1, $n_2 n_3$ individual $n_1$-point multicolumn real FFTs can be performed by using $n_2 n_3/2$ individual $n_1$-point multicolumn complex FFTs and Eqs. (5) and (6). If $n_2 n_3$ is not divisible by two, the remaining $n_1$-point real FFT is performed by using an $n_1$-point complex FFT with the imaginary part of the input data being zero. Since each output datum of $n_1$-point real FFTs in Step 1 obeys conjugate-even symmetry, it is sufficient to store only $(n_1/2 + 1)$-point complex output data.

# 3    Parallel 3-D Real FFT Algorithm with 2-D Decomposition

Figure 1 shows the parallel 3-D real FFT with 1-D decomposition of initial data on the $z$-axis, which is a typical data distribution method in parallel 3-D real FFT. The proposed implementation of the parallel 3-D real FFT with 2-D decomposition is based on the conjugate symmetry property of the DFT and the parallel 3-D complex FFT with 2-D decomposition [11].

Assume that the number of data points $N$ is $N = N_1 \times N_2 \times N_3$, and the MPI process is mapped in two dimensions $P \times Q$. In a distributed-memory parallel computer with $P \times Q$ MPI processes, the 3-D array $x(N_1, N_2, N_3)$ is distributed along the second $N_2$ and third $N_3$ dimension. If $N_2$ is divisible by $P$ and $N_3$ is also divisible by $Q$, $N_1 \times (N_2/P) \times (N_3/Q)$ pieces of data are distributed to each MPI process. Although somewhat complex, we next introduce the notation $\hat{N}_r = N_r/P$ and $\hat{\hat{N}}_r = N_r/Q$. Moreover, $\hat{J}_r$ denotes that data along the index $J_r$ are distributed to $P$ MPI processes and $\hat{\hat{J}}_r$ denotes that data along the index $J_r$ are distributed to $Q$ MPI processes. We note that $r$ means that the index belongs to dimension $r$. Therefore, the distributed 3-D array can be expressed as $\hat{x}(N_1, \hat{N}_2, \hat{\hat{N}}_3)$. According to the block distribution, the local index $\hat{J}_r(l)$ in the $l$-th MPI process on the $y$-axis corresponds to the global index $J_r$ as follows:

$$J_r = l \times \hat{N}_r + \hat{J}_r(l), \quad 0 \le l \le P - 1, \quad 1 \le r \le 3. \tag{8}$$

Moreover, the local index $\hat{\hat{J}}_r(m)$ in the $m$-th MPI process on the $z$-axis corresponds to the global index $J_r$ as follows:

$$J_r = m \times \hat{\hat{N}}_r + \hat{\hat{J}}_r(m), \quad 0 \le m \le Q - 1, \quad 1 \le r \le 3. \tag{9}$$

To show all-to-all communication, the notation $\tilde{N}_i \equiv N_i/P_i$ and $\tilde{\tilde{N}}_i \equiv N_i/Q_i$ is introduced. Here, $N_i$ is decomposed into 2-D representations of $\tilde{N}_i$ and $P_i$, and $\tilde{\tilde{N}}_i$ and $Q_i$. We note that $P_i$ and $Q_i$ are the same as $P$ and $Q$, respectively, but this indicates that these indices belong to dimension $i$.

Letting the initial data be $\hat{x}(N_1, \hat{N}_2, \hat{\hat{N}}_3)$, the parallel 3-D real FFT with 2-D decomposition is as follows:

Step 1:    $(N_2/P) \cdot (N_3/Q)$ individual $N_1$-point multicolumn real FFTs

$$\hat{x_1}(K_1, \hat{J}_2, \hat{\hat{J}}_3) = \sum_{J_1=0}^{N_1-1} \hat{x}(J_1, \hat{J}_2, \hat{\hat{J}}_3)\omega_{N_1}^{J_1 K_1}.$$

Step 2:    Transposition

$$\hat{x_2}(\hat{J}_2, \hat{\hat{J}}_3, \tilde{K}_1, P_1) \equiv \hat{x_2}(\hat{J}_2, \hat{\hat{J}}_3, K_1) = \hat{x_1}(K_1, \hat{J}_2, \hat{\hat{J}}_3).$$

Step 3:    $Q$ individual all-to-allv communications across $P$ MPI processes on the $y$-axis

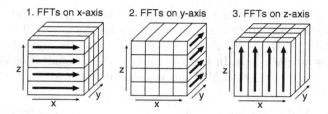

**Fig. 2.** 2-D decomposition for parallel 3-D real FFT

$$\hat{x}_3(\tilde{J}_2, \hat{\tilde{J}}_3, \hat{K}_1, P_2) = \hat{x}_2(\hat{J}_2, \hat{\tilde{J}}_3, \tilde{K}_1, P_1).$$

Step 4:　Rearrangement

$$\hat{x}_4(J_2, \hat{\tilde{J}}_3, \hat{K}_1) \equiv \hat{x}_4(\tilde{J}_2, P_2, \hat{\tilde{J}}_3, \hat{K}_1) = \hat{x}_3(\tilde{J}_2, \hat{\tilde{J}}_3, \hat{K}_1, P_2).$$

Step 5:　$(N_3/Q) \cdot (N_1/(2P))$ individual $N_2$-point multicolumn complex FFTs

$$\hat{x}_5(K_2, \hat{\tilde{J}}_3, \hat{K}_1) = \sum_{J_2=0}^{N_2-1} \hat{x}_4(J_2, \hat{\tilde{J}}_3, \hat{K}_1)\omega_{N_2}^{J_2 K_2}.$$

Step 6:　Transposition

$$\hat{x}_6(\hat{\tilde{J}}_3, \hat{K}_1, \tilde{K}_2, Q_2) \equiv \hat{x}_6(\hat{\tilde{J}}_3, \hat{K}_1, K_2) = \hat{x}_5(K_2, \hat{\tilde{J}}_3, \hat{K}_1).$$

Step 7:　$P$ individual all-to-all communications across $Q$ MPI processes on
　　　　the $z$-axis

$$\hat{x}_7(\hat{\tilde{J}}_3, \hat{K}_1, \hat{K}_2, Q_3) = \hat{x}_6(\hat{\tilde{J}}_3, \hat{K}_1, \tilde{K}_2, Q_2).$$

Step 8:　Rearrangement

$$\hat{x}_8(J_3, \hat{K}_1, \hat{K}_2) \equiv \hat{x}_8(\hat{\tilde{J}}_3, Q_3, \hat{K}_1, \hat{K}_2) = \hat{x}_7(\hat{\tilde{J}}_3, \hat{K}_1, \hat{K}_2, Q_3).$$

Step 9:　$(N_1/(2P)) \cdot (N_2/Q)$ individual $N_3$-point multicolumn complex FFTs

$$\hat{x}_9(K_3, \hat{K}_1, \hat{K}_2) = \sum_{J_3=0}^{N_3-1} \hat{x}_8(J_3, \hat{K}_1, \hat{K}_2)\omega_{N_3}^{J_3 K_3}.$$

Step 10:　Transposition

$$\hat{y}(\hat{K}_1, \hat{K}_2, K_3) = \hat{x}_9(K_3, \hat{K}_1, \hat{K}_2).$$

Figure 2 shows the parallel 3-D real FFT with 2-D decomposition of initial data on the $y$- and $z$-axes. Since each output datum of $N_1$-point real FFTs in Step 1 obeys conjugate-even symmetry, it is sufficient to store only $(N_1/2 + 1)$-point complex output data. If $N_1/2$ is divisible by $P$ (where $P$ is greater than 1), $N_1/2+1$ is not divisible by $P$. In Step 3, the 0-th MPI process on the $y$-axis sends $(N_2/P) \cdot (N_3/Q) \cdot ((N_1/2)/P+1)$ pieces of complex data to $P-1$ MPI processes on the $y$-axis. The other MPI processes send $(N_2/P) \cdot (N_3/Q) \cdot ((N_1/2)/P)$ pieces of complex data to $P-1$ MPI processes on the $y$-axis. Therefore, it is necessary to use all-to-allv communication instead of all-to-all communication in Step 3. In

Step 7, the 0-th MPI process on the $y$-axis sends $(N_3/Q)\cdot((N_1/2)/P+1)\cdot(N_2/Q)$ pieces of complex data to $Q-1$ MPI processes on the $z$-axis. The other MPI processes send $(N_3/Q)\cdot((N_1/2)/P)\cdot(N_2/Q)$ pieces of complex data to $Q-1$ MPI processes on the $z$-axis. In this case, all-to-all communications of different sizes are performed in the 0-th MPI process and the other MPI processes on the $y$-axis.

We note that while the input data $\hat{x}(J_1, \hat{J}_2, \hat{J}_3)$ are two-dimensionally decomposed on the $y$- and $z$-axes, the Fourier transformed output data $\hat{y}(\hat{K}_1, \hat{K}_2, K_3)$ are two-dimensionally decomposed on the $x$- and $y$-axes. In this manner, by using different data distributions for input and output, it is sufficient to perform all-to-allv communication on the $y$-axis in Step 3 and all-to-all communication on the $z$-axis in Step 7. If the same data distribution is used for input and output, then additional all-to-allv communication on the $y$-axis and all-to-all communication on the $z$-axis are needed.

We vectorized FFT kernels using the Intel Advanced Vector Extensions 512 (Intel AVX-512) instructions. When we parallelize the multicolumn FFTs by using OpenMP, the outermost loop of each FFT step is distributed across the threads. The outermost loop length may not have sufficient parallelism for the Intel Xeon Phi processor. A loop collapsing makes the length of a loop long by collapsing nested loops into a single-nested loop. By using the OpenMP collapse clause, the parallelism of the outermost loop can be expanded [12]. Transpositions in Steps 2, 6, and 10 are performed using cache blocking to reduce the number of cache misses.

## 4    Communication Time in 1-D and 2-D Decomposition

Let $N$ be the total number of data points and let $P \times Q$ be the number of MPI processes. Moreover, let $W$ be the communication bandwidth (byte/s), and let $L$ be the communication latency (s). Hereinafter, the communication time in the cases of 1-D and 2-D decomposition will be examined. For expository purposes, we assume that the message size for all MPI processes is the same and there is no communication contention in all-to-all(v) communications.

### 4.1    Communication Time in 1-D Decomposition

In the case of 1-D decomposition, since all-to-allv communication is performed among $PQ$ MPI processes, each MPI process sends $N/(2(PQ)^2)$ pieces of double-precision complex data to $PQ-1$ MPI processes.

Therefore, the communication time $T_{1\text{dim}}$ in the 1-D decomposition is expressed as follows:

$$T_{1\text{dim}} = (PQ-1)\left(L + \frac{16N}{2(PQ)^2 \cdot W}\right)$$

$$\approx PQ \cdot L + \frac{8N}{PQ \cdot W}. \tag{10}$$

## 4.2   Communication Time in 2-D Decomposition

In the case of 2-D decomposition, since $Q$ pairs of all-to-allv communications are performed among $P$ MPI processes on the $y$-axis, each MPI process on the $y$-axis sends $N/(2P^2Q)$ pieces of double-precision complex data to $P-1$ MPI processes on the $y$-axis. Moreover, since $P$ pairs of all-to-all communications are performed among $Q$ MPI processes on the $z$-axis, each MPI process on the $z$-axis sends $N/(2PQ^2)$ pieces of double-precision complex data to $Q-1$ MPI processes on the $z$-axis.

Therefore, the communication time $T_{\text{2dim}}$ in the 2-D decomposition is expressed as follows:

$$T_{\text{2dim}} = (P-1)\left(L + \frac{16N}{2P^2Q \cdot W}\right) + (Q-1)\left(L + \frac{16N}{2PQ^2 \cdot W}\right)$$
$$\approx (P+Q) \cdot L + \frac{16N}{PQ \cdot W}. \tag{11}$$

## 4.3   Comparison of 1-D and 2-D Decomposition

The communication times in the cases of 1-D and 2-D decomposition are expressed by Eqs. (10) and (11), respectively. In comparing these two equations, the 1-D decomposition of Eq. (10) becomes approximately half of 2-D decomposition of Eq. (11) as the total communication amount. However, when the total number of MPI processes $P \times Q$ is large and the communication latency $L$ is large, the communication time is shorter in the 2-D decomposition of Eq. (11). Here, a condition is obtained whereby the communication time $T_{\text{2dim}}$ represented by Eq. (11) is smaller than the communication time $T_{\text{1dim}}$ represented by Eq. (10).

From Eqs. (10) and (11),

$$(P+Q) \cdot L + \frac{16N}{PQ \cdot W} < PQ \cdot L + \frac{8N}{PQ \cdot W}. \tag{12}$$

We have

$$N < \frac{(LW \cdot PQ)(PQ - P - Q)}{8}. \tag{13}$$

For example, substituting $L = 10^{-5}$ (s), $W = 10^9$ (byte/s), and $P = Q = 64$ into Eq. (13), in the range of $N < 2 \times 10^{10}$, the communication time of 2-D decomposition is smaller than that of 1-D decomposition.

## 5   Performance Results

To evaluate the implemented parallel 3-D real FFT with 2-D decomposition, referred to as FFTE (version 7.0, 2-D decomposition), we compared its performance with that of the FFTE (version 7.0, 1-D decomposition), the FFTW (version 3.3.8) [7] and the P3DFFT (version 2.7.7) [9].

**Table 1.** Specification of the Fujitsu PRIMERGY CX1640 M1 cluster.

| | |
|---|---|
| Number of nodes | 8208 |
| CPU | Intel Xeon Phi 7250 (68-core, 1.4 GHz) |
| Main memory | MCDRAM 16 GB + DDR4-2400 96 GB |
| Theoretical peak performance | 25.004 PFlops |
| Total main memory size | 897.75 TB |
| Interconnect | Intel Omni-Path Architecture |
| Network topology | Fat-tree |
| Fortran compiler | Intel Fortran Compiler Version 18.0.1.163 |
| C compiler | Intel C Compiler Version 18.0.1.163 |
| MPI library | Intel MPI 2018.1.163 |

The elapsed times obtained from 10 executions of real FFTs were averaged. The same data distribution is used for input and output. The input of the parallel 3-D real FFTs is double-precision real data. The table for twiddle factors was prepared in advance. In the FFTW, the "measure" planner was used.

Performance was measured on the Fujitsu PRIMERGY CX1640 M1 cluster at the Joint Center for Advanced High Performance Computing (JCAHPC), which is a collaborative endeavor between the University of Tokyo and University of Tsukuba. The specification of the Fujitsu PRIMERGY CX1640 M1 cluster is shown in Table 1. We note that Hyper-Threading was enabled on the Intel Xeon Phi 7250. The experiments used from 1 to 2048 nodes.

If only one core of the Intel Xeon Phi processor is used for communication, the bandwidth of the Intel Omni-Path interconnect cannot be fully utilized. Therefore, it is necessary to communicate using multiple MPI processes per processor. On the other hand, as the number of MPI processes increases, the message size for all-to-all(v) communication decreases. This causes a decrease in communication bandwidth. In addition, the number of threads per core also affects performance. From a result of preliminary experiments, with the Intel Xeon Phi cluster, each node has 4 MPI processes, and 64 threads per MPI process were used.

The compiler options used were specified as `mpiifort -O3 -xMIC-AVX512 -qopenmp` for the FFTE and the P3DFFT. The compiler options used were specified as `mpiicc -O3 -xMIC-AVX512 -qopenmp` for the FFTW and the P3DFFT. The executions were performed in "flat mode" and "quadrant mode". Although `compact`, `scatter`, and `balanced` can be specified for `KMP_AFFINITY` environment variable, `balanced` that showed the best performance was specified. All programs were run in MCDRAM.

Figure 3 shows the weak scaling performance of parallel 3-D real FFTs ($N = 256 \times 512 \times 512 \times$ the number of MPI processes) in terms of the average execution performance in GFlops. Each GFlops value is based on $2.5N \log_2 N$ for a transform of size $N = 2^m$. As shown in Fig. 3, the FFTE with 1-D decom-

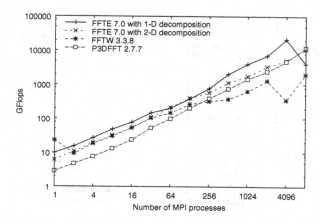

**Fig. 3.** Weak scaling performance of parallel 3-D real FFTs ($N = 256 \times 512 \times 512 \times$ the number of MPI processes)

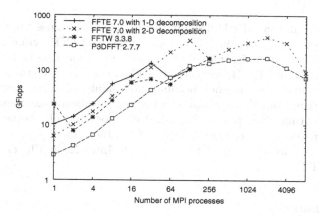

**Fig. 4.** Strong scaling performance of parallel 3-D real FFTs ($N = 256 \times 512 \times 512$)

position is faster than the FFTE with 2-D decomposition except for 64, 128, and 8192 MPI processes. This is because the total communication amount of the 1-D decomposition is approximately half that of the 2-D decomposition. Also, the FFTE with 2-D decomposition is faster than the FFTW except for 1 and 16 MPI processes, and the P3DFFT except for 8192 MPI processes.

Figure 4 shows the strong scaling performance of parallel 3-D real FFTs ($N = 256 \times 512 \times 512$). As shown in Fig. 4, the FFTE with 2-D decomposition is faster than the FFTE with 1-D decomposition for 64 and 128 MPI processes, due to communication latency. Also, the FFTE with 2-D decomposition is faster than the FFTW except for 1 and 256 MPI processes, and the P3DFFT.

Figure 5 shows the breakdown of execution time in FFTE ($N = 1024^3$, 512 MPI processes). As shown in Fig. 5, the communication time of 2-D decomposi-

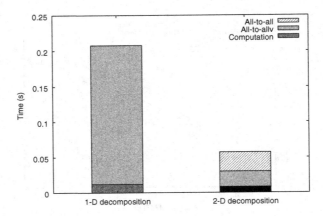

**Fig. 5.** Breakdown of execution time in FFTE (version 7.0) ($N = 1024^3$, 512 MPI processes)

tion is smaller than that of 1-D decomposition. This is because the message size of all-to-allv communication in 1-D decomposition is 32 KB, while the message sizes of all-to-allv communication and all-to-all communication in 2-D decomposition are 512 KB and 1 MB, respectively. When the message size is small such as 32 KB, communication latency becomes dominant in communication time.

As can be seen from Fig. 5, all-to-all(v) communication time dominates most of the execution time, even when using 2-D decomposition. This is the reason why over 10 TFlops (only approximately 0.2% of the theoretical peak performance) was obtained despite the use of the 2048-node Intel Xeon Phi cluster for an $8192^3$-point FFT.

## 6    Conclusion

We proposed an implementation of the parallel 3-D real FFT with 2-D decomposition on Intel Xeon Phi clusters. The proposed implementation of the parallel 3-D real FFT is based on the conjugate symmetry property of the DFT and the row-column FFT algorithm. We vectorized FFT kernels using the Intel AVX-512 instructions. The performance of the implemented parallel 3-D real FFT remains at a high level even for the strong scaling, owing to the 2-D decomposition. We succeeded in obtaining a performance over 10 TFlops on 2048 nodes of Fujitsu PRIMERGY CX1640 M1 cluster for an $8192^3$-point FFT. The performance results demonstrate that the proposed implementation of parallel 3-D real FFT with 2-D decomposition effectively improves performance by reducing the communication time for larger numbers of MPI processes on Intel Xeon Phi clusters.

**Acknowledgments.** This research used computational resources of the Oakforest-PACS provided by the Multidisciplinary Cooperative Research Program in Center for Computational Sciences, University of Tsukuba. This research was partially supported by JSPS KAKENHI Grant Number JP19K11989.

# References

1. 2DECOMP&FFT - Library for 2D Pencil Decomposition and Distributed FFTs. http://www.2decomp.org/
2. TOP500 Supercomputer Sites. https://www.top500.org/
3. Ayala, O., Wang, L.P.: Parallel implementation and scalability analysis of 3D Fast Fourier Transform using 2D domain decomposition. Parallel Comput. **39**, 58–77 (2013)
4. Brass, A., Pawley, G.S.: Two and three dimensional FFTs on highly parallel computers. Parallel Comput. **3**, 167–184 (1986)
5. Brigham, E.O.: The Fast Fourier Transform and Its Applications. Prentice-Hall, Upper Saddle River (1988)
6. Cooley, J.W., Tukey, J.W.: An algorithm for the machine calculation of complex Fourier series. Math. Comput. **19**, 297–301 (1965)
7. Frigo, M., Johnson, S.G.: The design and implementation of FFTW3. Proc. IEEE **93**, 216–231 (2005)
8. Liu, Y.Q., Li, Y., Zhang, Y.Q., Zhang, X.Y.: Memory efficient two-pass 3D FFT algorithm for Intel® Xeon Phi$^{TM}$ coprocessor. J. Comput. Sci. Technol. **29**, 989–1002 (2014)
9. Pekurovsky, D.: P3DFFT: a framework for parallel computations of Fourier transforms in three dimensions. SIAM J. Sci. Comput. **34**, C192–C209 (2012)
10. Pippig, M.: PFFT: an extension of FFTW to massively parallel architectures. SIAM J. Sci. Comput. **35**, C213–C236 (2013)
11. Takahashi, D.: An implementation of parallel 3-D FFT with 2-D decomposition on a massively parallel cluster of multi-core processors. In: Wyrzykowski, R., et al. (eds.) PPAM 2009, Part I. LNCS, vol. 6067, pp. 606–614. Springer, Heidelberg (2010)
12. Takahashi, D.: An implementation of parallel 1-D real FFT on Intel Xeon Phi processors. In: Gervasi, O., et al. (eds.) ICCSA 2017, Part I. LNCS, vol. 10404, pp. 401–410. Springer, Cham (2017)

# Exploiting Symmetries of Small Prime-Sized DFTs

Doru Thom Popovici[1]([⊠]) [iD], Devangi N. Parikh[2] [iD], Daniele G. Spampinato[3] [iD],
and Tze Meng Low[3]

[1] Lawrence Berkeley National Lab, Berkeley, CA, USA
dtpopovici@lbl.gov
[2] University of Texas at Austin, Austin, TX, USA
dnp@cs.utexas.edu
[3] Carnegie Mellon University, Pittsburgh, PA, USA
{spampinato,lowt}@cmu.edu

**Abstract.** Small prime-sized discrete Fourier transforms appear in various applications from quantum mechanics, material sciences and machine learning. The typical implementation of the discrete Fourier transform for such problem sizes is done as a cyclic convolution using algorithms like Rader or Bluestein. However, these approaches exhibit extra computation and expensive data movement. In this work, we present an alternative method by casting the Fourier transform as a direct symmetric matrix-vector multiplication. Exploiting the symmetries of the Fourier matrix and using knowledge from dense linear algebra, we present an implementation that reduces the amount of computation and requires less memory usage. We show that this approach achieves up to 2x performance gains on Intel and AMD architectures, compared to implementations offered by Intel MKL and FFTW that use Rader and Bluestein.

**Keywords:** Prime-sized DFTs · Rader algorithm · Bluestein algorithm · Symmetric matrix-vector multiplication

## 1 Introduction

The discrete Fourier transform (DFT) is a widely used mathematical tool in material sciences [5], quantum mechanics [9], and machine learning [10]. While, most applications use power of two Fourier transforms, for which algorithms like the Cooley-Tukey algorithm [2] offer high performance implementations, there are situations where odd-sized DFTs are needed. For example, applications [7,9] from quantum mechanics require the calculation of a central frequency, that can only be computed if the DFT is applied on problem sizes that are not divisible by two. Similarly, in machine learning, depending on the number of channels, convolution neural networks [10] require batches of small prime-sized DFTs like 7, 11 or 13. The DFT implementations for such problem sizes use either the Rader [8] or the Bluestein [1] algorithms, where the DFT computation is expressed as a

© Springer Nature Switzerland AG 2020
R. Wyrzykowski et al. (Eds.): PPAM 2019, LNCS 12043, pp. 162–173, 2020.
https://doi.org/10.1007/978-3-030-43229-4_15

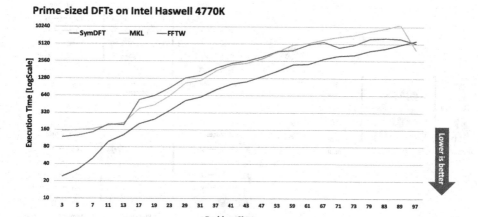

**Fig. 1.** The execution time in log scale for three DFT implementations for prime sizes on the Intel Haswell 4770k. The red line represents the approach in this paper, while the yellow and blue lines represent MKL and FFTW, respectively. (Color figure online)

cyclic convolution. Expressing the DFT as a cyclic convolution comes at the cost of incurring extra computation, data re-organization and/or memory footprint due to pre/post processing steps as seen in Fig. 2.

In this work, we present an alternative method for computing the DFT for prime sizes by casting the computation as a symmetric matrix-vector multiplication. Casting the computation as a symmetric matrix-vector multiplication allows us to leverage lessons learnt from dense linear algebra, and produce a high performance implementation. We show that our approach (SymDFT) achieves improvements by reducing the amount of floating point operations, avoiding costly memory accesses and having a smaller memory footprint compared to the versions implemented using either Rader's or Bluestein's algorithms. For problem sizes no greater than 100, i.e. batched operations in applications from machine learning or signal processing, we show in Fig. 1 the performance gap between our approach (red line) and library implementations, MKL [4] (yellow line) and FFTW [3] (blue line). It can be seen that, for small prime numbers, our approach outperforms both library codes by almost 2x.

**Contributions.** The paper makes the following contributions:

- An alternative approach for computing the DFT for prime numbers, where the computation is cast as a symmetric matrix-vector multiplication.
- A systematic way of exposing the symmetries of the Fourier computation, that enables complex number multiplication to be simplified.
- Experimental evidence of the benefits of using our approach, showing an improvement of up to 2x against MKL- and FFTW-based implementations for prime-sized DFTs up to 100.

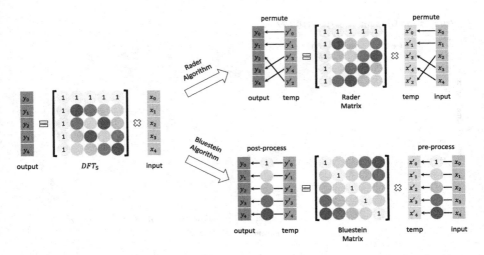

**Fig. 2.** Rader and Bluestein algorithms cast the DFT computation as a cyclic convolution. Rader requires a reorganizing of the input and output data, while Bluestein imposes extra computation by increasing the problem size. Since both algorithms compute a cyclic convolution, the convolution itself is done using the DFT-based convolution.

## 2    The Discrete Fourier Transform

In this section, we briefly present the Fourier transform, focusing on the two algorithms used for the DFT on prime sizes. Basically, the DFT of size $n$ is a matrix-vector multiplication, that computes the output $y$ given the input $x$ as

$$y = DFT_n \cdot x, \tag{1}$$

where the $DFT_n$ represents the Fourier matrix of size $n \times n$, defined as

$$DFT_n = \left[\omega_n^{kl}\right]_{0 \leq k,l < n}. \tag{2}$$

The $\omega_n = e^{-j\frac{2\pi}{n}}$ represents the $n$th primitive root of unity. For composite problem sizes, the Cooley-Tukey algorithm decomposes the dense matrix and reduces the computational complexity to $O(n \log(n))$. However, for prime sizes the matrix cannot be decomposed and algorithms like Rader or Bluestein are used.

Figure 2 outlines the steps for computing the DFT for prime sizes using using either Rader's or Bluestein's algorithms. Rader casts the computation into a cyclic convolution by re-organizing the input and output data, which in turn modifies the original Fourier matrix. As seen in Fig. 2, the Rader matrix keeps the row and column of 1s of the original Fourier matrix, and modifies the inner $(n-1) \times (n-1)$ matrix. The inner matrix is a circulant matrix, where each row is a cyclic shift of the previous row. Bluestein decomposes the Fourier matrix

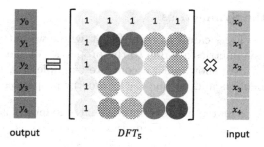

**Fig. 3.** The Fourier matrix has the first row and first column all 1s. The matrix is symmetric. The inner $4 \times 4$ matrix is itself symmetric. It is persymmetric and exhibits conjugate symmetries across the columns and rows. The patterned circles are complex conjugate of the corresponding full circles.

into pre-/post-processing scaling operations and a symmetric complex matrix. The Bluestein matrix is completely different from the original Fourier matrix and it is part of a larger circulant matrix of size $(2n - 1) \times (2n - 1)$. Typically, the cyclic convolution is implemented using the Fourier convolution property. For more details, we recommend the reader to follow [1,8].

Both algorithms re-organize the data within the Fourier matrix in order to transform the computation into a cyclic convolution, which is then computed using the Fourier convolution property. However, *none of the algorithms exploit the structure of the Fourier matrix as is, without re-ordering its elements*.

## 3    Exploiting the Structure of the Fourier Matrix

In this section, we emphasize the symmetries of the Fourier matrix and its submatrix, and then outline the algorithmic choices used in our approach. While the classical methods expose symmetries at the cost of modifying the original Fourier matrix, *our approach preserves the original matrix*. The Fourier matrix is a symmetric matrix, that has 1s in the first row and column such that

$$DFT_n = \begin{bmatrix} 1 & \mathbf{1}_{n-1}^T \\ \mathbf{1}_{n-1} & E_{n-1,} \end{bmatrix}, \tag{3}$$

where $\mathbf{1}_{n-1}$ and $\mathbf{1}_{n-1}^T$ represent the column and row of 1s of size $n-1$. Note that the $(n-1) \times (n-1)$ submatrix $E_{n-1} = \left[ \omega_n^{kl} \right]_{1 \le k, l < n}$ contains all the roots of unity that are different from 1 for prime values of $n$. The $E_{n-1}$ matrix inherits the symmetry of the Fourier matrix. In addition, the matrix also exhibits persymmetry, which suggests the matrix to be symmetric across the anti-diagonal, and conjugate symmetries across the middle lines in row and column direction. The four types of symmetries can be seen in Fig. 3.

## 3.1   Exposing the Symmetries

The structure of the $E_{n-1}$ can be seen in Fig. 3, where we outline the four lines of symmetry. The symmetries are due to the properties of the complex roots of unity $w_n^{kl} = e^{-j\frac{2\pi}{n}kl}$. We show, with a concrete example, $E_4$, the properties of the complex exponentials and outline the symmetry properties.

*Symmetry.* The $E_{n-1}$ matrix is a symmetric matrix, since it is a sub-matrix of the Fourier matrix. For $n = 5$, the $E_4$ matrix is expressed as

$$E_4 = \begin{bmatrix} w_5 & w_5^2 & w_5^3 & w_5^4 \\ w_5^2 & w_5^4 & w_5^6 & w_5^8 \\ w_5^3 & w_5^6 & w_5^9 & w_5^{12} \\ w_5^4 & w_5^8 & w_5^{12} & w_5^{16} \end{bmatrix}. \tag{4}$$

Note that $E_4^T = E_4$, where $(\cdot)^T$ represents the transpose operator defined as $M^T[i][j] = M[j][i]$ for a matrix $M$ of $n \times n$ elements.

*Persymmetry.* The $E_{n-1}$ matrix is persymmetric. The roots of unity follow the property that $w_n^k = w_n^{k \bmod n}$, which simplifies $E_4$ as

$$E_4 = \begin{bmatrix} w_5 & w_5^2 & w_5^3 & w_5^4 \\ w_5^2 & w_5^4 & w_5 & w_5^3 \\ w_5^3 & w_5 & w_5^4 & w_5^2 \\ w_5^4 & w_5^3 & w_5^2 & w_5 \end{bmatrix}. \tag{5}$$

Note that $E_4^A = E_4$. The $(\cdot)^A$ represents the persymmetry operator defined as $M^A[i][j] = M[n - j - 1][n - i - 1]$ for a matrix $M$ of $n \times n$ elements.

*Conjugate Symmetries.* The complex roots of unity follow the property that $w_n^k = w_n^{k-n}$ for all values of $k > n/2$. This simplifies $E_4$ such that

$$E_4 = \left[ \begin{array}{cc|cc} w_5 & w_5^2 & w_5^{-2} & w_5^{-1} \\ w_5^2 & w_5^{-1} & w_5 & w_5^{-2} \\ \hline w_5^{-2} & w_5 & w_5^{-1} & w_5^2 \\ w_5^{-1} & w_5^{-2} & w_5^2 & w_5 \end{array} \right]. \tag{6}$$

Note that $E_4^C = E_4^R = E_4$. The $(\cdot)^C$ and $(\cdot)^R$ represent the complex conjugate symmetry operators across the columns and rows, defined as $M^C[i][j] = M^*[i][n-j-1]$ and $M^R[i][j] = M^*[n-i-1][j]$ for a matrix $M$ of $n \times n$ elements.

The $E_{n-1}$ matrix can be reduced from an $(n - 1) \times (n - 1)$ matrix to an upper triangular matrix that contains $(n - 1)(n + 1)/8$ elements. For example, the $E_4$ matrix can be reduced to $F_4$ matrix defined as

$$F_4 = \begin{bmatrix} w_5 & w_5^2 \\ & w_5^{-1} \end{bmatrix}, \tag{7}$$

which is sufficient to compute the entire Fourier transform for $n = 5$. *Both the Rader and Bluestein matrices exhibit transposed symmetry, however due to the data re-arrangement the persymmetry and conjugate symmetries are lost.*

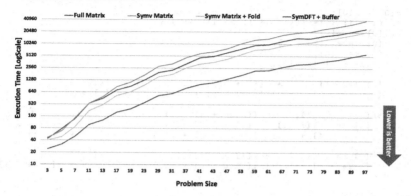

**Fig. 4.** The execution time as CPU cycles for different variants of the same algorithm. The baseline is the full matrix-vector multiplication, and each line represents a modification brought to this baseline. The red-line represents the implementation shown in Fig. 9, that has all the optimization. (Color figure online)

## 3.2 Exploiting the Symmetries

In this subsection, we present our direct approach of computing the DFT for prime sizes. We assume the input and output to use the complex interleaved data layout, where the real and imaginary parts of each complex number are stored in consecutive memory locations. First, we start with the matrix-vector implementation. This variant has $4n^2$ fused-multiply add (`fma`) instructions and requires storage for the entire Fourier matrix. The dark blue line in Fig. 4 shows the execution of this approach in CPU cycles. We gradually apply transformations to this baseline as follows:

- **Exploit the Symmetry of the Fourier Matrix.** The Fourier matrix is symmetric. Hence, the matrix-vector implementation is transformed into a symmetric matrix-vector multiplication. The number of floating operations does not change, it remains $4n^2$ `fmas`. However, the required memory for the Fourier matrix drops from $n^2$ to $n(n + 1)/2$ complex elements. For small problem sizes, the execution of the symmetric matrix-vector multiply (light blue line in Fig. 4) follows the baseline case, however, as the problem size increases the execution degrades.

- **Exploit the Rows and Columns of 1 and the Persymmetry in $E_{n-1}$.** The first row and first column of the Fourier matrix are all 1s. We perform loop peeling of one iteration on the outer loop of the symmetric matrix-vector multiplication to exploit this aspect. The remaining loops compute the symmetric matrix-vector multiplication with the matrix $E_{n-1}$. However, the matrix $E_{n-1}$ also exhibits persymmetry, which allows the loops to further be manipulated as shown in Fig. 5. Line 2, $4-7$ are due to the row and column of 1. Lines 9–10, 13–14, 16–17 correspond to the DFT computation when the

```
1     void SymDFT(int size, complex double *x, complex double *DFT, complex
          double *y) {
2     y[0] = x[0];
3     for(i = 1; i != ((size + 1) >> 1); ++i) {
4       y[0] += x[i];
5       y[0] += x[size - i];
6       y[i] += x[0];
7       y[size - i] += x[0];
8
9       y[i] += DFT[i][i] * x[i] + DFT[i][size - i] * x[size - i];
10      y[size - i] += DFT[i][size - i] * x[i] + DFT[i][i] * x[size - i];
11
12      for(j = (i + 1); j != ((size + 1) >> 1); ++j) {
13        y[i] += DFT[i][j] * x[j] + DFT[i][size - j] * x[size - j];
14        y[size - 1] += DFT[i][size - j] * x[j] + DFT[i][j] * x[size - j];
15
16        y[j] += DFT[i][j] * x[i] + DFT[i][size - j] * x[size - i];
17        y[size - j] += DFT[i][size - j] * x[i] + DFT[i][j] * x[size - i];
18      }
19    }
20  }
```

**Fig. 5.** C Code implementation for computing the DFT for small prime sizes, when the rows and columns of 1s and the symmetry and persymmetry are exploited. x and y represent the input, and output arrays. The computation requires only 1/4 of the original DFT matrix.

symmetry ($E_{n-1}^T = E_{n-1}$) and persymmetry ($E_{n-1}^A = E_{n-1}$) are used. These optimizations reduce the number of floating point operations to $4n - 4$ **add** and $4(n-1)^2$ **fma** instructions and reduce the memory footprint for the partial DFT matrix by almost 4 times, compared to the original DFT matrix. The execution time of this variant is slightly better than that of the matrix-vector multiplication variant, as shown by the yellow line in Fig. 4.

- **Exploit the Column Conjugate Symmetry in $E_{n-1}$.** A consequence of the column symmetry ($E_{n-1}^C = E_{n-1}$) is the recognition that each pair of $x_k$ and $x_{n-k}$ input values are scaled by complex conjugate elements of the $E_{n-1}$ matrix, such as

$$y_k = a \cdot x_k + a^* \cdot x_{n-k};$$
$$y_{n-k} = a^* \cdot x_k + a \cdot x_{n-k};$$

$$(8)$$

where $a = a_{re} + j \cdot a_{im}$. $a_{re}$ and $a_{im}$ are the real and imaginary components of $a$, and $j = \sqrt{-1}$. This exposes the following computation

$$t_0 = x_k + x_{n-k};$$
$$t_1 = j \cdot (x_k - x_{n-k});$$
$$a_0 = a_{re} \cdot t_0;$$
$$a_1 = a_{im} \cdot t_1;$$
$$y_k = a_0 + a_1;$$
$$y_{n-k} = a_0 - a_1;$$

$$(9)$$

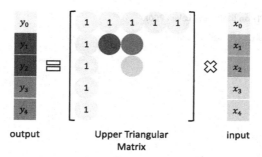

**Fig. 6.** Exploiting the symmetry, persymmetry and conjugate symmetries, the original Fourier matrix is reduced to the row and column of 1s and the upper triangular matrix $F_4$. This is sufficient to compute any prime-sized DFT.

Exploiting the conjugate symmetry and simplifying the computation in lines 9−10, 13−14, and 16−17 in Fig. 5, helps reduce the amount of floating operations to $2(n-1)^2 + 4n - 4$ adds and $(n-1)^2$ fmas. In addition, the memory requirement drops to $(n-1)(n+1)/8$. As shown in Fig. 6 the approach requires only the upper triangular matrix of the original Fourier matrix to fully compute the DFT for prime sizes.

- **Reduce the Amount of Redundant Computation.** The pair of operations, $t_0 = x_k + x_{n-k}$, $t_1 = x_k - x_{n-k}$ and $y_k = a_0 + a_1$, $y_{n-k} = a_0 - a_1$, are being redundantly computed. As shown in Fig. 5, the outer loop reads all the input values and fully updates the output values. For each iteration of the outer loop, the inner loop re-reads the x[j] and x[size-j] input pairs and updates the y[j] and y[size-j] output pairs for $i < j \le (size-1)/2$. Reading the inputs each time requires the computation of $t_0$ and $t_1$, and updating the outputs each time requires the additions of $a_0$ and $a_1$. The redundant additions can be reduced by performing loop peeling and partial computations, as shown in Fig. 9. Lines 4–28 represent the peeled loop that goes once through the input data, computes $t_0$ and $t_1$ once, and stores the temporary values in a favorable layout as shown in [6]. Lines 20–21 compute partial results that are then assembled at the end of the loop at lines 30–31. These optimization reduce the number of floating point operations to $5(n-1)$ adds and $(n-1)^2$ fmas. The memory requirement increases by $(n-1)$ complex samples, for storing the temporary values. The red line in Fig. 4 shows the execution time of this variant. It can be seen that this variant is almost four times faster than the original matrix-vector multiplication.

## 4    Experimental Results

In this section, we present the experimental setup and show the results obtained using our approach (SymDFT) on both Intel and AMD architectures. We focus only on computations on double precision floating point numbers. For the Intel

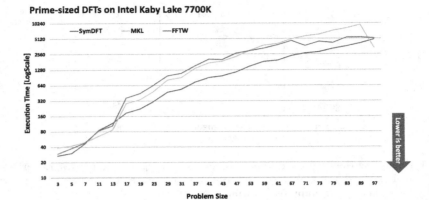

**Fig. 7.** The execution time on a log scale for three DFT implementations for prime sizes on the Intel Kaby Lake 7700k. Execution is reported as CPU cycles.

architectures, we use Intel's Haswell 4770k and Kaby Lake 7700k, and compare our approach against both Intel MKL 2019.3.199 and FFTW 3.3.8. For the AMD architectures, we use AMD's Piledriver FX8350, and only compare our implementation against FFTW version 3.3.8 which replaces AMD's proprietary DFT library. While the Intel's library is compiled using their off-the-shelf installation script, the FFTW library is compiled with the SSE, AVX and FMA flags enabled. SymDFT is implemented using SSE SIMD intrinsics and follows the code presented in Fig. 9. On both architectures the code is compiled using the GCC compiler version 5.4.0 and the -O2 and -msse4.2 flags enabled.

For all experiments, we run the code using the warm cache approach. We execute each DFT call for 100 dry runs, where the execution time is not recorded. We then run each DFT function call for 50000 runs and take the overall average. For each implementation, we report CPU cycles using the `rdtsc` performance counter. Figure 1 shows the results obtained on the Intel Haswell 4770k for all three implementations. Note that for most prime numbers smaller than 100, with the exception of 97, the DFT implementation using SymDFT outperforms MKL and FFTW by almost 2x. For 97, FFTW increases the initial problem size to 200, which is a composite number. On such composite numbers, FFTW then applies the Cooley-Tukey algorithm to get the $log(n)$ reduction in computation. Even though the problem size is doubled, the Bluestein approach gives shorter time to solution for prime sizes greater or equal to 97. For prime sizes smaller than 100, the SymDFT approach requires $5(n-1)$ floating point add and $(n-1)^2$ fused-multiply add scalar instructions. Both Rader and Bluestein have a higher instruction count especially if the circular convolution is done as a DFT-based convolution, for small numbers no greater than 100.

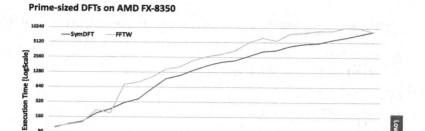

**Fig. 8.** The execution time in log scale for three DFT implementations for prime sizes on the AMD Piledriver FX8350. Execution is reported as CPU cycles.

In Fig. 7 we report results on the Intel Kaby Lake. For sizes smaller than 17, MKL and FFTW perform similar or better than SymDFT. MKL uses assembly codelets that are tuned for the architecture, while FFTW offers multiple codelets for which the instructions are statically scheduled. We implement SymDFT using SSE vector intrinsics, and we use the GCC compiler with the -O2 flag, which may prove problematic on Intel Kaby Lake. We leave as future work, the scheduling of the instructions using inline assembly as shown in [11] to further improve the execution time. Finally, in Fig. 8, we report the execution time of SymDFT compared to FFTW's on the AMD Piledriver. Once again, SymDFT performs similar to FFTW for problem sizes smaller than 17 and outperforms all the other sizes up to 97. It is well known that FFTW provides multiple codelets that are specifically optimized for AMD. In contrast, SymDFT, as shown in Fig. 9 provides a single implementation for all the prime sizes.

```
1    void SymDFT(int size, complex double *x, complex double *t, complex double *
         DFT, complex double *y) {
2        double complex yt, yb, t0, t1, t00, t11;
3        double *DFT_ri = (double*) DFT;
4        t0 = x[1] + x[size - 1];
5        t1 = (x[1] - x[size - 1]) * I;
6
7        y[0] = x[0] + t0;
8
9        yt = x[0] + DFT_ri[0] * t0;
10       yb = DFT_ri[1] * t1;
11       DFT_ri += 2;
12
13       for(i = 2; i != ((size + 1) >> 1); ++i) {
14           int off = 2 * (i - 1);
15           t00 = x[i] + x[size - i];
16           t11 = (x[i] - x[size - i]) * I;
17
18           y[0] += t00;
19
20           yt += DFT_ri[0] * t00;
21           yb += DFT_ri[1] * t11;
22           y[i] = x[0] + DFT_ri[0] * t0;
23           y[size - i] = DFT_ri[1] * t1;
24           DFT_ri += 2;
25
26           t[off + 0] = t00;
27           t[off + 1] = t11;
28       }
29
30       y[1] = yt + yb;
31       y[size - 1] = yt - yb;
32
33       for(i = 2; i != ((size + 1) >> 1); ++i) {
34           int offi = 2 * (i - 1);
35           t0 = t[offi + 0];
36           t1 = t[offi + 1];
37
38           yt = y[i];
39           yb = y[size - i];
40
41           yt += DFT_ri[0] * t0;
42           yb += DFT_ri[1] * t1;
43           DFT_ri += 2;
44
45           for(j = (i + 1); j != ((size + 1) >> 1); ++j) {
46               int offj = 2 * (j - 1);
47               t00 = t[offj + 0];
48               t11 = t[offj + 1];
49
50               yt += DFT_ri[0] * t00;
51               yb += DFT_ri[1] * t11;
52               y[j] += DFT_ri[0] * t0;
53               y[size - j] += DFT_ri[1] * t1;
54               DFT_ri += 2;
55           }
56
57           y[i] = yt + yb;
58           y[size - i] = yt - yb;
59       }
60   }
```

**Fig. 9.** C Code implementation of the symmetric matrix-vector approach for computing the DFT for small prime sizes. The x, t and y arrays represent the input, temporary and output arrays, respectively. The DFT array is used to store the upper diagonal quadrant of the Fourier matrix.

# 5   Conclusion

In this work, we showed an alternative approach to implementing the DFT computation for small prime sizes. Exploiting the symmetries of the Fourier matrix, without modifying the original layout of the input and output data, we expressed the DFT computation for prime sizes as a symmetric matrix-vector multiplication. We showed that the implementation of this approach matches or outperforms by almost 2x on Intel and AMD architectures, the code offered by state-of-the-art libraries such as MKL and FFTW. Both libraries make use of algorithms like Rader and Bluestein for prime-sized DFT computations.

# References

1. Bluestein, L.: A linear filtering approach to the computation of discrete Fourier transform. IEEE Trans. Audio Electroacoust. **18**, 451–455 (1970)
2. Cooley, J.W., Tukey, J.W.: An algorithm for the machine calculation of complex Fourier series. Math. Comput. **19**, 297–301 (1965)
3. Frigo, M., Johnson, S.G.: The design and implementation of FFTW3. Proc. IEEE **93**(2), 216–231 (2005). Special issue on "Program Generation, Optimization, and Adaptation"
4. Intel: Math Kernel Library (2018). http://developer.intel.com/software/products/mkl/
5. Lebensohn, R.A., Kanjarla, A.K., Eisenlohr, P.: An elasto-viscoplastic formulation based on fast Fourier transforms for the prediction of micromechanical fields in polycrystalline materials. Int. J. Plast. **32**, 59–69 (2012)
6. Popovici, D., Franchetti, F., Low, T.M.: Mixed data layout kernels for vectorized complex arithmetic. In: 2017 IEEE High Performance Extreme Computing Conference, HPEC 2017 (2017)
7. Popovici, D.T., Russell, F.P., Wilkinson, K., Skylaris, C.K., Kelly, P.H., Franchetti, F.: Generating optimized Fourier interpolation routines for density functional theory using SPIRAL. In: 2015 IEEE International Parallel and Distributed Processing Symposium, pp. 743–752. IEEE (2015)
8. Rader, C.M.: Discrete Fourier transforms when the number of data samples is prime. Proc. IEEE **56**, 1107–1108 (1968)
9. Skylaris, C.K., Haynes, P.D., Mostofi, A.A., Payne, M.C.: Introducing ONETEP: linear-scaling density functional simulations on parallel computers. J. Chem. Phys. **122**, 084119 (2005)
10. Vasilache, N., Johnson, J., Mathieu, M., Chintala, S., Piantino, S., LeCun, Y.: Fast convolutional nets with fbfft: A GPU performance evaluation. arXiv preprint arXiv:1412.7580 (2014)
11. Veras, R., Popovici, D.T., Low, T.M., Franchetti, F.: Compilers, hands-off my hands-on optimizations. In: Proceedings of the 3rd Workshop on Programming Models for SIMD/Vector Processing, WPMVP 2016, pp. 4:1–4:8 (2016). https://doi.org/10.1145/2870650.2870654

# Parallel Computations for Various Scalarization Schemes in Multicriteria Optimization Problems

Victor Gergel$^{(\boxtimes)}$ ⓘ and Evgeny Kozinov ⓘ

Lobachevsky State University of Nizhni Novgorod, Nizhni Novgorod, Russia
gergel@unn.ru, evgeny.kozinov@itmm.unn.ru

**Abstract.** In the present paper, a novel approach to parallel computations for solving time-consuming multicriteria global optimization problems is presented. This approach includes various methods for the scalarization of vector criteria, dimensionality reduction with the use of the Peano space-filling curves, and efficient global search algorithms. The applied criteria scalarization methods can be altered in the course of computations in order to better meet the stated optimality requirements. To reduce the computational complexity of multicriteria problems, the methods developed feature an extensive use of all the computed optimization information and are well parallelized for effective performance on high-performance computing systems. Numerical experiments confirmed the efficiency of the developed approach.

**Keywords:** Multicriteria optimization · Criteria scalarization · Global optimization · Parallel computations

## 1 Introduction

Multicriteria optimization (MCO) problems arise in various applications of science and technology. The wide scope of application of MCO problems determines the intensity of research in the field [1–5].

Usually, the criteria of the MCO problems are contradictory, and obtaining the decisions, which provide the best values with respect to all criteria simultaneously, is impossible. In such situations, a set of efficient (non-dominated) decisions is considered for such MCO problems, where the improvement of the values of some criteria cannot be achieved without sacrificing the efficiency values with respect to other criteria. When solving MCO problems, it may become necessary to obtain the whole set of the efficient decisions (the Pareto set) that may require performing a large amount of computations. Besides, the analysis of a large set of the efficient decisions may appear to be difficult for a decision

This research was supported by the Russian Science Foundation, project No. 16-11-10150 "Novel efficient methods and software tools for time-consuming decision making problems using supercomputers of superior performance.".

maker. Therefore, in practice the finding of only a relatively small set of efficient decisions may be justified. Constructing such a limited set of efficient decisions is performed usually according to the optimality requirements defined by the decision-maker.

The restricted set of the computed efficient decisions leads to a notable reduction in the amount of required computations. However, the criteria of efficiency may have a complex multiextremal form, and the computation of the criteria values may appear to be time-consuming. One way to overcome the considerable computational complexity when solving MCO problems is to use high-performance supercomputing systems.

In the present paper, the research results on the development of highly efficient parallel methods of the multicriteria optimization using the search information obtained in the course of computations [6–9,16] are presented. A novel contribution consists in the development of an approach whereby it is possible to use different MCO problem criteria scalarization schemes taking into account different optimality requirements to desirable decisions. For the scalarization schemes considered in this study, a general method of parallel computations was proposed and the computational evaluation of the developed approach's efficiency was performed.

The further structure of the paper is as follows. In Sect. 2, the statement of the multicriteria optimization problems is given. In Sect. 3, a general scalarization scheme for the MCO problem criteria involving various kinds of criteria convolution is proposed. In Sect. 4, the search information obtained in the course of computations, which can be reused after the altering of the applied scalarization schemes, is considered. Also, this section presents a general scheme for the parallel execution of the global search algorithms, in the framework of which an efficient parallel method for solving time-consuming global optimization problems is considered. Section 5 contains the results of the numerical experiments confirming that the approach developed by the authors is quite promising. In the Conclusions section, the obtained results are discussed and possible main areas for further studies are outlined.

## 2    Multicriteria Optimization Problem Statement

In the most general form, the multicriteria optimization problem is defined in the following way

$$f(y) \to min, y \in D : g(y) \leq 0, \tag{1}$$

where $y = (y_1, y_2, \ldots, y_N)$ is a vector of varied parameters,

$f(y) = (f_1(y), f_2(y), \ldots, f_s(y))$ is the vector of the efficiency criteria, $D \subset R^N$ is the search domain

$$D = \{y \in R^N : a_i \leq y_i \leq b_i, 1 \leq i \leq N\}, \tag{2}$$

for given vectors $a$ and $b$, and $g(y) = (g_1(y), g_2(y), \ldots, g_m(y))$ are the constraints of the MCO problem (the conditions of feasibility of the chosen decisions $y \in D$).

In the most complex cases, the criteria $f_i(y)$, $1 \leq i \leq s$, can be multiextremal and computing the values of these ones can appear to be time-consuming. Usually, the criteria $f_i(y)$, $1 \leq i \leq s$, satisfy the Lipschitz condition

$$|f_i(y_1) - f_i(y_2)| \leq L_i \|y_1 - y_2\|, 1 \leq i \leq s, \tag{3}$$

where $L_i$ are the Lipschitz constants for the criteria $f_j(y)$, $1 \leq j \leq s$ and $\| * \|$ denotes the Euclidean norm in $R^N$.

Without any loss in generality, in further consideration we will assume the criteria $f_i(y)$, $1 \leq i \leq s$ to be non-negative and their decrease corresponds to the increase in the efficiency of the chosen decisions.

## 3   Reducing Multicriteria Optimization to Multiple Global Search Problems

As mentioned above, for solving MCO problems it is required to find a set of the efficient (Pareto optimal) decisions, for which the values cannot be improved with respect to all criteria simultaneously. However, a large amount of computations may be required to find a complete set of efficient decisions. A commonly used approach to reducing the computational costs consists in finding of only a relatively small set of efficient decisions, defined according to the optimality requirements of the decision maker. In many cases, the required decisions are found by means of reducing the vector criteria to some general scalar efficiency function that allows applying a variety of already existing global optimization methods for solving MCO problems. Possible scalarization methods include, for example, the weighted sum method, the compromise programming method, the weighted min-max method, as well as many other methods – see, for example, [2,4]. Within the developed approach, it was proposed to employ the following methods of scalarization of the vector criterion of MCO problems.

1. One of the widely applied scalarization methods consists in the use of the min-max convolution of the criteria [4,6]:

$$F_1(\lambda, y) = \max (\lambda_i f_i(y), 1 \leq i \leq s),$$

$$\lambda = (\lambda_1, \lambda_2, \ldots, \lambda_s) \in \Lambda \subset R^s : \sum_{i=1}^{s} \lambda_i = 1, \lambda_i \geq 0, 1 \leq i \leq s. \tag{4}$$

2. In the case when the criteria can be arranged in the order of importance, the method of successive concessions (MSC) is usually applied [5,6]. In this method, the multistage computations can be reduced to solving the following scalar optimization problem

$$F_2(\delta, y) = f_s(y), f_i(y) \leq f_i^{min} + \delta_i(f_i^{max} - f_i^{min}), 1 \leq i < s, g(y) \leq 0, y \in D, \tag{5}$$

where $f_i^{min}$, $f_i^{max}$, $1 \leq i < s$, are the minimum and the maximum values of the criteria in the domain $D$ from (2), and $0 \leq \delta_i \leq 1$, $1 \leq i < s$, are the concessions with respect to each criterion.

3. If an estimate of the criteria values for the required decision exists a *priori* (for example, based on some ideal decision), the solution of a MCO problem can consist in finding an efficient decision that matches the predefined optimality values [2,4]:

$$F_3(\theta, y) = 1/s \sum_{i=1}^{s} \theta_i (f_i(y) - f_i^*)^2, g(y) \le 0, y \in D, \tag{6}$$

where $F_3(\theta, y)$ is the mean square deviation of the decision $y \in D$ from the sought ideal decision $y^*$, and the values $0 \le \theta_i \le 1$, $1 \le i < s$, are the importance parameters of the approximation accuracy with respect to each variable $y_i$, $1 \le i \le N$.

In the general case, the statements of the global optimization problems generated by the criteria scalarization schemes (4)–(6) can be represented in the form:

$$\min \varphi(y) = \min F(\alpha, y), g(y) \le 0, y \in D, \tag{7}$$

where $F$ is the objective function generated as a result of the criteria scalarization, $\alpha$ is the vector of parameters of the criteria convolution, $g(y)$ are the constraints of the MCO problem, and $D$ is the search domain from (2). It should be noted that due to the possibility of altering the optimality requirements in the course of computations, the form of the function $\varphi(y)$ from (7) can vary. Thus, it may be necessary to alter the scalarization method employed (4)–(6) and/or to change the convolution parameters $\lambda$, $\delta$, and $\theta$. Such variations form a set of scalar global optimization problems

$$\mathbb{F}_T = \{F_k(\alpha_i, y) : 1 \le i \le T, k = 1, 2, 3\}. \tag{8}$$

This set of problems can be formed sequentially in the course of computations. The problems in this set can be solved either sequentially or simultaneously – in the time sharing mode or in parallel with the use of high performance computational systems. The opportunity of forming the set $\mathbb{F}_T$ allows formulating *a new class of multicriteria global optimization problems*.

Within the framework of the proposed approach, one more step of transformation of the problems $F(\alpha, y)$ from the set $\mathbb{F}_T$ is applied – namely, the dimensionality reduction with the use of the Peano space-filling curves (evolvents) $y(x)$ providing an unambiguous mapping of the interval $[0, 1]$ onto an $N$-dimensional hypercube $D$ [10,11]. As a result of such reduction, a multidimensional problem (7) is reduced to a one-dimensional global optimization problem:

$$F(\alpha, y(x^*)) = \min \{F(\alpha, y(x)) : g(y(x)) \le 0, x \in [0, 1]\}. \tag{9}$$

The dimensionality reduction allows applying many highly efficient one-dimensional global search algorithms – see, for example, [10–12,14] – for solving the problems of the set $\mathbb{F}_T$ from (8). It should be also noted that according to (3) the function $F(\alpha, y(x))$ satisfies the uniform Hölder condition with a constant $H$ [10,11] i.e.

$$|F(\alpha, y(x')) - F(\alpha, y(x''))| \le H|x' - x''|^{1/N}. \tag{10}$$

# 4    Parallel Computations for Solving Multiple Global Optimization Problems

The numerical solving of the MCO problems consists usually in sequential computing of the values of the criteria $f(y)$ and the constraints $g(y)$ from (1) at the points $y^i$, $0 \le i \le k$, of the search domain $D$ [10,12–14]. After the scalarization of the vector criterion (7) and the application of the dimensionality reduction (9), the obtained search information takes the form of the set:

$$A_k = \{(x_i, z_i, f_i, g_i)^T : 1 \le i \le k\}, \tag{11}$$

where $x_i$, $1 \le i \le k$ are the reduced points of the executed global search iterations arranged in ascending order of the coordinates and $z_i$, $f_i$, $g_i$, $1 \le i \le k$, are the values of the scalar criterion, the objective function criteria and the constraints respectively at these points for the current optimization problem $F(\alpha, y(x))$.

The availability of the set $A_k$ from (11) allows the transformation of the values of the criteria and constraints calculated earlier into the values of the current optimization problem $F(\alpha, y(x))$ from (7) without repeating the time-consuming computations of the values of criteria and constraints i.e.

$$(x_i, f_i, g_i) \to z_i = F(\alpha, y(x_i)), 1 \le i \le k. \tag{12}$$

In this way, the whole search information $A_k$ from (11) recalculated according to (12) can be reused to continue the solution of the MCO problem. This can provide a considerable reduction of the amount of computations performed for solving every next MCO problem of the set $\mathbb{F}_T$ from (8) up to a limited number of global search iterations.

A further increase in computational efficiency when solving MCO problems can be provided by means of parallel computations. Within the framework of the developed approach, a general parallel computation scheme for solving global optimization problems is applied – namely, parallel computations are provided by the simultaneous computing of the values of the minimized function $F(\alpha, y(x))$ from (7) at several different points of the search domain $D$ [6,10]. Such an approach provides the parallelization of the most time-consuming part of the global search computations and can be applied to any global optimization method for a variety of global optimization problems.

The state-of-the-art in the field of global optimization is reviewed, for example, in [10,12,13]. In the present paper, the proposed approach is based on the information-statistical theory of global search [10]. This theory has been applied as a basis for the development of a large number of efficient methods of multi-extremal optimization – see, for example, [6–11,15–18]. The main ideas of the algorithms developed consist in the dimensionality reduction with the use of the Peano space-filling curves and in parallel computations by simultaneous computing the values of the minimized functions at several points of the search domain. From the general point of view, the developed *multidimensional parallel global search algorithms* (MPGSA) can be presented by a unified computation scheme as follows [6,10,15].

Let $p > 1$ be the number of the applied processors (cores) with shared memory. At the initial iteration of the MGGSA, the computing of the minimized function values at $p$ arbitrary points of the interval $(0, 1)$ is performed (the obtaining of the function value will be called hereafter a *trial*). Then, let us assume $\tau$, $\tau > 1$ global search iterations to be completed. The choice of the trial points for the next $(\tau + 1)^{th}$ iteration is determined by the following rules.

*Rule 1.* Renumber the points of the trial points by the lower indices in the order of increasing values of coordinate[1]

$$0 = x_0 \leq x_1 \leq \cdots \leq x_i \leq \cdots \leq x_k \leq x_{k+1} = 1 \tag{13}$$

The points $x_0$, $x_{k+1}$ were introduced additionally for the convenience of further presentation, and $k = \tau p$ is the total number of trials performed earlier.

*Rule 2.* For each interval $(x_{i-1}, x_i)$, $1 \leq i \leq k+1$, compute the quantity $R(i)$ called hereafter the *characteristic* of the interval.

*Rule 3.* Arrange the characteristics of the intervals in the decreasing order

$$R(t_1) \geq R(t_2) \geq \cdots \geq R(t_{k-1}) \geq R(t_k) \tag{14}$$

Select $p$ intervals with the indices $t_j$, $1 \leq j \leq p$, having the highest characteristics.

*Rule 4.* Perform new trials at the points of the interval $(0, 1)$

$$x^{k+1}, x^{k+2}, \ldots, x^{k+p}, \tag{15}$$

placed in the intervals with the highest characteristics from (14).

The stopping condition, according to which the trials are terminated, is defined as

$$(x_t - x_{t-1})^{1/N} \leq \varepsilon, \tag{16}$$

fulfilled for at least one of the indices $t_j$, $1 \leq j \leq p$, from (14), $N$ is the dimensionality, and $\varepsilon > 0$ is the predefined accuracy of solving the optimization problem. If the stopping condition is not met, the iteration index $\tau$ is incremented by 1, and a new global search iteration is performed.

According to this general scheme, particular global search algorithms are distinguished by expressions applied for computing the characteristics $R(i)$, $1 \leq i \leq k + 1$, from (14) and the points of the next trials $x^{k+j}$, $1 \leq j \leq p$, from (15) in the intervals with the highest characteristics. Thus, for example, for the multidimensional generalized global search algorithm (MGGSA) [10], the interval characteristic takes the form

$$R(i) = m\varrho_i + \frac{(z_i - z_{i-1})^2}{m\varrho_i} - 2(z_i + z_{i-1}), 1 \leq i \leq k + 1, \tag{17}$$

where $m$ is the numerical estimate of the Hölder constant from (10), $z_i$, $1 \leq i \leq k + 1$, are the computed values of the minimized function $F(\alpha, y(x))$ from (9)

---

[1] It should be noted that the condition (13) determines the need to order the search information stored in the set $A_k$ from (11).

at the points of the performed global search iterations, and $\varrho_i = (x_i - x_{i-1})^{1/N}$, $1 \leq i \leq k+1$.

The MGGSA rule for computing the next trial point $x^k \in (0,1)$ is defined as

$$x^{k+1} = \frac{x_t + x_{t-1}}{2} - sign(z_t - z_{t-1})\frac{1}{2r}\Big[\frac{|z_t - z_{t-1}|}{m}\Big]^N, \qquad (18)$$

where $r$, $r > 1$ is a parameter of the MGGSA algorithm.

To clarify the computational scheme considered, the following can be noted. The computed characteristics $R(i)$, $1 \leq i \leq k+1$, from (14) can be considered as some measures of importance of the intervals from the viewpoint of their containing the global minimum point. Thus, the scheme of selecting the interval for executing the new trial becomes clear – the point of every next trial $x^{k+1}$ from (18) is selected in the interval with the highest value of the interval characteristic (i.e. in the interval where the global minimum is most likely to be found).

The conditions of convergence for the algorithms developed in the framework of the information-statistical theory of global search and the conditions of non-redundancy of parallel computations were considered in [10].

## 5  Results of Numerical Experiments

Numerical experiments were carried out on the Lobachevsky supercomputer at the State University of Nizhny Novgorod (the operating system – CentOS 6.4, the management system – SLURM). Each supercomputer node had 2 Intel Sandy Bridge E5-2660 2.2 GHz, 64 Gb RAM processors. Each processor had 8 cores (i.e. a total of 16 CPU cores were available on each node).

In these experiments, bi-criteria two-dimensional MCO problems (i.e. $N = 2$, $s = 2$) were solved. Multiextremal functions defined by the relations [10]:

$$f(y_1, y_2) = -\{AB + CD\}^{1/2}$$

$$AB = \Big(\sum_{i=1}^{7}\sum_{j=1}^{7}[A_{ij}a_{ij}(y_1, y_2) + B_{ij}b_{ij}(y_1, y_2)]\Big)^2 \qquad (19)$$

$$CD = \Big(\sum_{i=1}^{7}\sum_{j=1}^{7}[C_{ij}a_{ij}(y_1, y_2) - D_{ij}b_{ij}(y_1, y_2)]\Big)^2$$

were used as the problem criteria; the expressions

$$a_{ij}(y_1, y_2) = \sin(\pi i y_1)\sin(\pi j y_2), b_{ij}(y_1, y_2) = \cos(\pi i y_1)\cos(\pi j y_2) \qquad (20)$$

were defined in the range $0 \leq y_1, y_2 \leq 1$, the parameters $1 \leq A_{ij}, B_{ij}, C_{ij}, D_{ij} \leq 1$ were the independent random numbers. The minimization of such functions arises, for example, in the problem of evaluation of the maximum strain (determining the strength) in a thin plate under the normal load.

In Fig. 1, the contour plots of three functions of this family are shown. As one can see, the functions of such kind are essentially multiextremal.

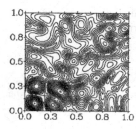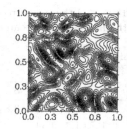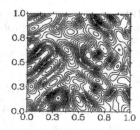

**Fig. 1.** Contour plots of three multiextremal functions from the test optimization problem family

In [9], the proposed approach using the min-max scalarization scheme was compared with the well-known multicriteria optimization methods:

- The Monte-Carlo (MC) method, where the trial points are selected within the search domain $D$ randomly and uniformly,
- The genetic algorithm SEMO from the PISA library,
- The non-uniform coverage (NUC) method,
- The bi-objective Lipschitz optimization (BLO) method.

In this paper, the efficiency of the proposed approach is compared with other scalarization schemes: the method of successive concessions and the reference point method.

In order to draw more justified conclusions on the efficiency of the developed approach, we solved 100 multicriteria problems formed using the multiextremal functions of the family (19). To solve each problem, a total of 50 different coefficients $\alpha$ of the convolutions (4)–(6) were used. The values of the coefficient $\alpha$ are uniformly distributed within the interval $[0,1]$. The parameter values in our experiments were taken as follows: the reliability parameter $r = 2,3$, the required accuracy $\varepsilon = 0,01$. All the results presented below were averaged over the family of the problems solved.

In the first experiment, multicriteria problems were solved using all the considered scalarization schemes (4)–(6), for which a set of different scalarization coefficients was taken. The calculations were performed without using the accumulated search information $A_k$ from (11). The numerical results are shown in Table 1, where the columns labeled "K" contain the average number of iterations; in the columns labeled as "S", the achieved speedup of the parallel computations is shown. As follows from the presented results of our experiments, the parallel algorithm MPGSA based on the MGGSA method [10, 16] showed a high efficiency. When employing 16 cores, the speedup was more than 12.5 times for any type of convolutions (4)–(6).

The results presented in Table 2 show that by reusing the search information the amount of performed computations can be reduced at least 10-fold.

The efficiency of the developed approach becomes more evident when the obtained reduction of executed optimization iterations is shown in comparison

**Table 1.** Average number of iterations and the speedup of parallel computations when solving a single MCO problem without reusing the search information

| Convolution | Number of cores | | | | | | | | |
|---|---|---|---|---|---|---|---|---|---|
| | 1 | | 2 | | 4 | | 8 | | 16 | |
| | K | S | K | S | K | S | K | S | K | S |
| $F_1(\lambda, y)$ from (4) | 13 221 | 1 | 7 187 | 1,8 | 3 456 | 3,8 | 1 897 | 7,0 | 1 058 | 12,5 |
| $F_2(\delta, y)$ from (5) | 16 002 | 1 | 9 458 | 1,7 | 4 113 | 3,9 | 2 356 | 6,8 | 1 204 | 13,3 |
| $F_3(\theta, y)$ from (6) | 14 041 | 1 | 7 866 | 1,8 | 3 558 | 3,9 | 1 982 | 7,1 | 1 068 | 13,2 |

**Table 2.** Average number of iterations and speedup of parallel computations when solving a single MCO problem with the reuse of the search information

| Convolution | Number of cores | | | | | | | | |
|---|---|---|---|---|---|---|---|---|---|
| | 1 | | 2 | | 4 | | 8 | | 16 | |
| | K | S | K | S | K | S | K | S | K | S |
| $F_1(\lambda, y)$ from (4) | 1 193 | 1 | 595 | 2,0 | 300 | 4,0 | 173 | 6,9 | 103 | 11,5 |
| $F_2(\delta, y)$ from (5) | 2 021 | 1 | 934 | 2,2 | 475 | 4,3 | 272 | 7,4 | 147 | 13,7 |
| $F_3(\theta, y)$ from (6) | 990 | 1 | 491 | 2,0 | 249 | 4,0 | 145 | 6,8 | 90 | 10,9 |

**Table 3.** Overall reduction of executed iterations provided by using the developed approach for solving a MCO problem

| Convolution | Number of cores | | | | |
|---|---|---|---|---|---|
| | 1 | 2 | 4 | 8 | 16 |
| $F_1(\lambda, y)$ from (4) | 11,1 | 22,2 | 44,1 | 76,2 | 127,5 |
| $F_2(\delta, y)$ from (5) | 7,9 | 17,1 | 33,7 | 58,7 | 108,2 |
| $F_3(\theta, y)$ from (6) | 14,2 | 28,6 | 56,3 | 96,3 | 154,9 |

with the initial sequential algorithm, which does not use the search information (Table 3). As follows from the results of performed experiments, the overall reduction when using 16 computation cores always exceeded 100.

## 6   Conclusion

In the present paper, a new approach to parallel computations for solving time-consuming multicriteria global optimization problems is presented. This approach is based on various methods for scalarization of vector criteria, on dimensionality reduction with the use of the Peano space-filling curves, and on efficient global search algorithms.

A novel contribution consists in the development of an approach permitting to use different criteria scalarization schemes with the account of different optimality requirements to desirable decisions. For the scalarization schemes considered, a general method of parallel computations was proposed and the efficiency of the developed approach was evaluated computationally. Our numerical experiments confirmed the efficiency of the proposed approach – for example, the speedup of parallel computations when using 16 computational cores always exceeded 100.

In further investigations, we intend to perform numerical experiments on parallel solving the multicriteria optimization problems for a larger number of efficiency criteria and for larger dimensionality. Parallel computations on computational nodes with distributed memory can be considered as well.

# References

1. Marler, R.T., Arora, J.S.: Multi-Objective Optimization: Concepts and Methods for Engineering. VDM Verlag, Saarbrücken (2009)
2. Ehrgott, M.: Multicriteria Optimization, 2nd edn. Springer, Heidelberg (2010). https://doi.org/10.1007/3-540-27659-9
3. Collette, Y., Siarry, P.: Multiobjective Optimization: Principles and Case Studies. Decision Engineering. Springer, Heidelberg (2011). https://doi.org/10.1007/978-3-662-08883-8
4. Pardalos, P.M., Žilinskas, A., Žilinskas, J.: Non-Convex Multi-Objective Optimization. Springer Optimization and Its Applications, vol. 123. Springer, Cham (2017)
5. Modorskii, V.Y., Gaynutdinova, D.F., Gergel, V.P., Barkalov, K.A.: Optimization in design of scientific products for purposes of cavitation problems. In: AIP Conference Proceedings, vol. 1738, p. 400013 (2016). https://doi.org/10.1063/1.4952201
6. Gergel, V.P., Strongin, R.G.: Parallel computing for globally optimal decision making. In: Malyshkin, V.E. (ed.) PaCT 2003. LNCS, vol. 2763, pp. 76–88. Springer, Heidelberg (2003). https://doi.org/10.1007/978-3-540-45145-7_7
7. Gergel, V., Kozinov, E.: Efficient methods of multicriterial optimization based on the intensive use of search information. In: Kalyagin, V.A., Nikolaev, A.I., Pardalos, P.M., Prokopyev, O.A. (eds.) NET 2016. SPMS, vol. 197, pp. 27–45. Springer, Cham (2017). https://doi.org/10.1007/978-3-319-56829-4_3
8. Barkalov, K., Gergel, V., Lebedev, I.: Solving global optimization problems on GPU cluster. In: AIP Conference Proceedings, vol. 1738, p. 400006 (2016)
9. Gergel, V., Kozinov, E.: Efficient multicriterial optimization based on intensive reuse of search information. J. Global Optim. **71**(1), 73–90 (2018). https://doi.org/10.1007/s10898-018-0624-3
10. Strongin, R., Sergeyev, Ya.: Global Optimization with Non-Convex Constraints. Sequential and Parallel Algorithms. Kluwer Academic Publishers, Dordrecht (2nd edn. 2013, 3rd edn. 2014)
11. Sergeyev, Y.D., Strongin, R.G., Lera, D.: Introduction to Global Optimization Exploiting Space-Filling Curves. SpringerBriefs in Optimization. Springer, New York (2013). https://doi.org/10.1007/978-1-4614-8042-6
12. Zhigljavsky, A., Žilinskas, A.: Stochastic Global Optimization. Springer Optimization and Its Applications. Springer, New York (2008). https://doi.org/10.1007/978-0-387-74740-8
13. Locatelli, M., Schoen, F.: Global Optimization: Theory, Algorithms, and Applications. SIAM, Philadelphia (2013)
14. Floudas, C.A., Pardalos, M.P.: Recent Advances in Global Optimization. Princeton University Press, Princeton (2016)

15. Gergel, V., Sidorov, S.: A two-level parallel global search algorithm for solution of computationally intensive multiextremal optimization problems. In: Malyshkin, V. (ed.) PaCT 2015. LNCS, vol. 9251, pp. 505–515. Springer, Cham (2015). https://doi.org/10.1007/978-3-319-21909-7_49

16. Gergel, V., Kozinov, E.: Parallel computing for time-consuming multicriterial optimization problems. In: Malyshkin, V. (ed.) PaCT 2017. LNCS, vol. 10421, pp. 446–458. Springer, Cham (2017). https://doi.org/10.1007/978-3-319-62932-2_43

17. Sergeyev, Y.D., Kvasov, D.E.: A deterministic global optimization using smooth diagonal auxiliary functions. Commun. Nonlinear Sci. Numer. Simul. **21**(1–3), 99–111 (2015)

18. Gergel, V., Grishagin, V., Gergel, A.: Adaptive nested optimization scheme for multidimensional global search. J. Global Optim. **66**(1), 35–51 (2015). https://doi.org/10.1007/s10898-015-0355-7

# Emerging HPC Architectures

# Early Performance Assessment of the ThunderX2 Processor for Lattice Based Simulations

Enrico Calore[2], Alessandro Gabbana[1,2], Fabio Rinaldi[1],
Sebastiano Fabio Schifano[1,2(✉)], and Raffaele Tripiccione[1,2]

[1] Università degli Studi di Ferrara, Ferrara, Italy
[2] INFN Sezione di Ferrara, Ferrara, Italy
schifano@fe.infn.it

**Abstract.** This paper presents an early performance assessment of the *ThunderX2*, the most recent Arm-based multi-core processor designed for HPC applications. We use as benchmarks well known stencil-based LBM and LQCD algorithms, widely used to study respectively fluid flows, and interaction properties of elementary particles. We run benchmark kernels derived from OpenMP production codes, we measure performance as a function of the number of threads, and evaluate the impact of different choices for data layout. We then analyze our results in the framework of the roofline model, and compare with the performances measured on mainstream Intel Skylake processors. We find that these Arm based processors reach levels of performance competitive with those of other state-of-the-art options.

**Keywords:** ThunderX2 · Lattice-Boltzmann · Lattice-QCD

## 1 Introduction and Related Works

The number of CPU cores integrated onto a single chip has steadily increased over the years. This trend has been further pushed forward by the introduction of many-core architectures, such as the Intel Xeon Phi processors, which integrate up to hundreds of full CPUs, providing a large amount of computational power on a single chip. Recently, Arm architecture following this approach have appeared on the market. One such example is the *ThunderX2* processor, a very recent evolution of the *ThunderX*, already considered with promising results for HPC applications [1,2].

The Arm processor architecture started to attract the attention of the HPC community several years ago [3], mainly for reasons related to energy-efficiency of this class of processors specifically designed and optimized for mobile devices. Since then, several research works and projects [4] have investigated the use of Arm processors as building blocks for energy-efficient HPC systems [5].

© Springer Nature Switzerland AG 2020
R. Wyrzykowski et al. (Eds.): PPAM 2019, LNCS 12043, pp. 187–198, 2020.
https://doi.org/10.1007/978-3-030-43229-4_17

These has lead to the development of a comprehensive software ecosystem, allowing to easily deploy HPC scientific applications on cluster based on Arm processors [6, 7].

Recently, the *ThunderX2* in particular has attracted lot of attention, and several new high-end HPC machines have adopted it. Among them, the *Astra* system at Sandia National Labs in the US is the first Arm-based machine listed in the Top500 rank [8]; in Europe the *Dibona* machine, built by the Mont Blanc3 project [9], the *Isambard* system of the GW4 alliance [10], and the *Carmen* cluster installed at CINECA in Italy, are based on the *ThunderX2*. The performance of this processor for HPC applications have been studied for a small range of applications and mini-apps relevant for the UK national HPC service (ARCHER) [10], and the Sandia National Laboratories [8].

In this context, this paper present an early performance assessment of the *ThunderX2* processor, using as test cases two production ready HPC codes based on stencils, and additionally analyzing the impact on performance of several memory data layouts. Results are compared to that achieved on recent commodity multi-core processors of the same class, not including accelerators as they are based on a different class of architecture design named many-core.

Stencil cods are a wide class of applications of general interest, largely deployed on massively parallel HPC systems. The computing performance and impact on energy usage of these applications has been investigated in several works [11], and has also lead to the development of specific programming and tuning frameworks [12, 13], to deploy such codes both on homogeneous and heterogeneous systems based on GPUs or other accelerators.

In particular, here we consider implementations of *Lattice Boltzmann Methods* (LBM), widely used in *Computational Fluid Dynamics* (CFD) and one specific *Lattice QCD* (LQCD) code used to simulate the interactions between hadronic elementary particles of matter. Both benchmarks execute stencil algorithms on a lattice. In this case, task parallelism is easily obtained by assigning tiles of the physical lattice to different cores, and results in an efficient road to performances [14, 15]. Recent works have shown however that performance on multi- and many-core processors strongly depends on the choices of the memory data layout [16–18]. For this reason, we experiment with different data structures to store the lattice domain, trying to find the best options from the performance point of view. We evaluate the *ThunderX2* measuring its computing performances for our benchmarks and their scalability as a function of the number of cores, analyzing our results in the framework of the Roofline Model [19] and comparing them with those measured on a recent Intel Skylake processor.

The remainder of this paper is organized as follows: Sect. 2 highlights the main features of the *ThunderX2* processor, and Sect. 3 briefly describes the applications that we have used. Section 4 gives some details on the implementation of our codes, and Sect. 5 presents and analyzes our results. Finally, Sect. 6 wraps up the work done and draws some conclusions.

## 2    The ThunderX2 Processor

The *ThunderX2* is a very recent multi-core HPC-oriented processor; it relies on massive thread-parallelism, and operates in standalone mode running the standard GNU/Linux operating system. It was announced in 2016 by *Cavium Inc.* as an improvement over the previous *ThunderX* [20], and released later in 2018 by *Marvell* with an architecture based on the Broadcom *Vulcan* design. The *ThunderX2* has up to 32 64-bit Armv8 cores interconnected on a bidirectional ring bus. Each core integrates two 128-bit floating-point (FP) units, and two levels of caches, 32 KB of L1 and 256 KB of L2; furthermore, 1 MB L3 cache slice is shared by each core through the ring, assembling a large L3 distributed memory level. The FP units execute two double-precision SIMD NEON *Fused-Multiply-Accumulate* vector instructions per clock cycle, corresponding to a theoretical peak performance of 8 double-precision FLOP/cycle, 4× better than the *ThunderX*. The processor has two ECC memory controllers connected to the ring bus, and 8 DDR4-2666 channels providing a theoretical peak bandwidth of ≈160 GB/s; the effective value achieved by the STREAM [21] benchmark on a 32-core 2.0 GHz CN9980 *ThunderX2* model is ≈110 GB/s [9]. The Thermal Design Power (TDP) is ≈180 W, giving an interesting theoretical energy-performance ratio of 2.8 GFLOP/Watt. For more details on *ThunderX2* architecture see [22].

## 3    Applications

In this section we give a brief introduction on the benchmark applications used in this work. They are based on stencil algorithms, and are widely used by several scientific communities to perform extensive CFD and LQCD simulations.

### 3.1    Lattice Boltzmann Methods

To start, we take into consideration Lattice Boltzmann Methods (LBM) codes, an efficient class of CFD solvers capable of capturing the physics of complex fluid flows, through a mesoscopic approach. LBM are stencil-based algorithms, discrete in space, time and momenta, which typically operate on regular Cartesian grids. Each grid point can be processed in parallel, making this class of solvers ideal targets for efficient implementations on multi- and many-core processors. These codes consider the time evolution of pseudo-particles (*populations*) sitting at the edges of a two or three dimensions lattice. A model labeled as $DxQy$ describes a fluids in $x$ dimensions using $y$ populations per lattice site. At each time step, populations *propagate* to neighboring lattice sites, and then *collide* mixing and changing their values appropriately. These two computing steps are usually the most time consuming sections of production LBM simulations. In both routines, there are no data dependencies between different lattice points, so they can execute in parallel on as many lattice sites as possible and according to the most convenient schedule. On the other hand, *propagate* and *collide* have different computing requirements; *propagate* is memory-intensive and does not involve any floating-point computation, while *collide*, used to compute the collisional operator, is compute intensive. In this work we take into account the D3Q19 and the D3Q27 models, two popular choices for the simulation of iso-thermal flows in 3D, and a recently developed bi-dimensional thermal model, the D2Q37 [23,24].

## 3.2 Lattice Quantum ChromoDynamics

As a second class of benchmark applications, we consider a Lattice Quantum ChromoDynamics (LQCD) code. Quantum ChromoDynamics (QCD) is the the-ory of the strong nuclear force, describing the interaction between quarks and gluons, the fundamental particles that make up composite hadrons such as pro-tons and neutrons. LQCD is a non-perturbative discretization of QCD to tackle aspects of QCD physics that would be impossible to investigate systematically by using standard perturbation theory. In practice, the continuum four dimen-sional space-time of QCD is discretized on a four dimensional lattice, in such a way that continuum physics is recovered as the lattice spacing get closer to zero. LQCD codes come in a very large number of different versions, each devel-oped to study specific areas of the underlying theory [25]. In the following, we restrict ourselves to the staggered formulation, commonly adopted for investigat-ing QCD thermodynamics, for which we have recently developed the *OpenSta-PLE* (OpenACC Staggered Parallel LatticeQCD Everywhere) code, based on the OpenACC framework [26,27]. For benchmark purposes we further restrict our analysis to the *Dirac* operator, a linear-algebra routine specifically optimized for this application that typically account for ≈80% of the computational load of a production run.

# 4    Implementation Details

For both benchmark codes we adopt OpenMP to manage thread parallelism and take into consideration three different data-layouts: the *Array of Structure* (*AoS*) which in general optimizes cache accesses and data re-use; the *Structure of Arrays* (*SoA*) typically fitting better data-parallelism strategies appropriate for recent many-core processors; and finally the *Array of Structure of Arrays* (*AoSoA*) which tries to combine the advantages of the previous two layouts.

## 4.1    Lattice Boltzmann Methods

As highlighted in Sect. 3 the most time consuming kernels for LBM simulations are *propagate* and *collide*.

The *propagate* kernel moves populations from each site towards neighboring sites according to the stencil scheme defined by the specific LBM model adopted. In our cases, *propagate* moves 19, 27 or 37 data populations at each lattice site, accessing memory at sparse addresses, with non-optimal patterns and limited options for cache re-use.

The *collide* kernel computes the so called collisional operator, processing data associated to the populations moved at each lattice site by the previous *propagate* step. This step is completely local, as it uses the populations associate to each given site. In our codes, there are approximately 1220 floating points operations per lattice site for the D3Q19 model, and respectively 1840 and 6600 for the D3Q27 and the D2Q37 models. These values have been obtained using run-time profilers, and are in agreement with the number of floating-point operations counted in the corresponding kernel algorithm.

For all LBM models our implementation supports the three different memory layouts mentioned above. In the *AoS*, populations associated to each lattice site are stored one after the other at contiguous memory addresses; this improves data locality site-wise fitting better the computing requirements of the *collide* kernel. Conversely, the *SoA* stores at contiguous memory locations all populations having same index but belonging to different lattice sites, fitting better the requirements of the *propagate* kernel. Finally, the *AoSoA* stores at contiguous memory locations a block of BL populations of same index of different sites, and then all blocks of population for that BL sites are stored in memory. The rationale behind this approach is that each block can be moved in parallel using the SIMD instructions of the processors, while blocks corresponding to different populations indexes are relatively close in memory. This choice potentially retains the benefits of the *AoS* and *SoA* layouts, and therefore satisfies the requirements of both the *propagate* and *collide* kernels.

## 4.2   Lattice Quantum ChromoDynamics

For LQCD code, we have ported the OpenACC implementation of the *Dirac* operator of *OpenStaPLE* to OpenMP, mapping directly OpenACC directives to the corresponding OpenMP ones.

Recent versions of the *GCC* compiler support OpenACC annotated codes; however, as this feature is still at an early stage of development, we have opted for a handmade supervised translation. The *Dirac* operator performs multiplications between $3 \times 3$ complex-valued $SU(3)$ matrices and 3-component vectors, for a total of $\approx 1560$ floating-point operations per lattice site. This kernel has limited arithmetic intensity, so its performances depend largely on the behavior of the memory sub-system. The *Dirac* kernel uses a 4-dimensional stencil, and adopts 4-nested loops over the 4D-lattice. Our implementation supports also cache blocking on 2, 3 and 4 dimensions, and vector and matrix data elements associated to each lattice site can be stored using three different memory layouts. Similarly to LBM, in the *AoS* data elements are store site-wise, while in the *SoA* elements of different sites are allocated at contiguous addresses for both matrices and vectors. In the *AoSoA* layout a block of BL vectors (or matrix elements) of adjacent lattice sites along – in our implementation – the $x$-dimension are stored at contiguous memory locations, and then consecutive blocks are stored in memory. This layout allows to vectorize over adjacent site along the $x$-dimension, and adjacent memory accesses to collect all the elements for the same site are also possible, providing a compromise between the two previous layouts.

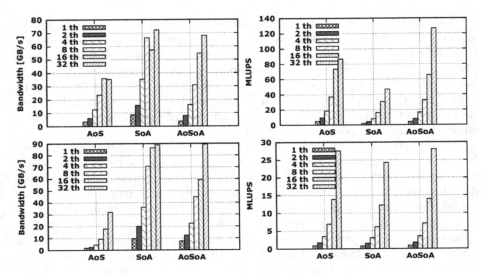

**Fig. 1.** Performance results of the D3Q27 (top) and D2Q37 (bottom) model for *propagate* (left) and *collide* (right) for several different data layouts. For each layout, we show results running with 1, 2, 4, 8, 16 and 32 threads. Performance for *propagate* is measured in GB/s, while for the *collide* we report the MLUPS (Million Lattice Update per Second).

## 5   Results

In this section we present our benchmark results on the *Carmen* machine, a prototype cluster installed at CINECA (Italy) with eight nodes, each with two 32-core CN9980 *ThunderX2* processors, 32 MB of L3 cache and 256 GB of memory. Then, we analyze our results in the framework of the Roofline Model, and compare with performances measured on Intel Skylake processors.

### 5.1   Lattice Boltzmann Methods

In many production codes, the *propagate* and *collide* kernels are merged in one single kernel as this reduces data traffic to/from memory. In our case, for benchmark purposes, we have run the two kernels separately, in order to more accurately profile the performance of the *ThunderX2* processor in terms of memory bandwidth and computing throughput. For the D3Q19 and D3Q27 models we have performed a simulation of the *Couette* flow, a well known example of steady-state flow. The setup consists of a stationary bottom plate, a top plate moving with a constant velocity, and periodic boundary conditions on the remaining sides, considering also a force acting in the same direction of the moving plate.

In Fig. 1 top, we show the results for the D3Q27 model, that we have measured on a single *ThunderX2* processor; results for the D3Q19 model are similar. The plots show performance results for the *propagate* and *collide* kernels as a

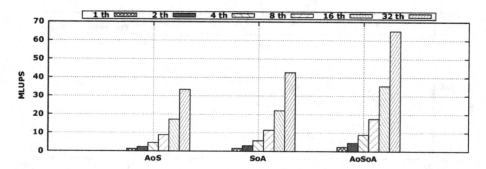

**Fig. 2.** Performance results of the Dirac Operator run on a single *ThunderX2* of the Carmen cluster, as a function of the data layout and number of OpenMP threads. The lattice has $32^4$ sites and performance are measured in MLUPS.

function of the number of threads used and for each data-layout. For *propagate* we plot the effective bandwidth, while for *collide* we measure performance in *Million Update Lattice Update per Second* (MLUPS). In general, performances of both kernels increase with the number of threads used, and strongly depend on data layouts. As expected, *propagate* performs better using the *SoA* data layout; D3Q19 and D3Q27 with 32 threads have a bandwidth of ≈70 GB/s, that is ≈60% of the bandwidth measured by the STREAM benchmark. With the *AoS* layout, performances are lower because SIMD load and store instructions cannot be used. On the other hand, for the *propagate* the *AoSoA* layout gives almost the same performance of the *SoA*, but is clearly the best choice for the *collide*; with 32 threads we measure ≈160 MLUPS for the D3Q19 model, and ≈120 MLUPS for the D3Q27 model, improving respectively by 1.3× and 1.5× the corresponding values with *AoS*. These results are obtained with a block length of $BL = 4$.

Figure 1 bottom shows the same results for the D2Q37 model. In this case a more complex stencil strongly stresses the memory controller and the floating point unit of the processor. As in the previous case, performance increases with the number of threads. For *propagate* we measure a bandwidth of ≈90 GB/s using both the *SoA* and *AoSoA* layouts and 32 threads. This corresponds to ≈80% of the bandwidth measured by the STREAM benchmark, and improves by a factor 3× the figure measured with the *AoS* layout. For *collide*, we obtain ≈28 MLUPS using the *AoS* and the *AoSoA* layouts and 32 threads. This is about 15% better than the *SoA* layout. Interestingly, in this case the best performance is achieved with a BL value of 8.

## 5.2    Lattice QCD

For LQCD we measure the performance of the *Dirac Operator* kernel, which shapes the performance of the whole *OpenStaPLE* code. We have experimented with several configuration parameters, using an automated script; we have run with different number of threads to profile the strong scaling across cores,

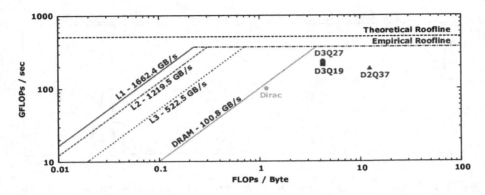

**Fig. 3.** Roofline chart for the *ThunderX2* processor. The blue and red lines corresponds to the performance and bandwidth measured running the ERT benchmark. Labeled points show the performance results of our benchmarks. (Color figure online)

enabling cache blocking on 2, 3 and 4 dimensions, and with all memory layouts; for the *AoSoA* layout, we have tested several BL values (*i.e.* 2, 4, 8 and 16). In Fig. 2 we show the performance in MLUPS units on a lattice of $32^4$ sites on one *ThunderX2* processor. For each data layout, we show the results corresponding to the best cache blocking configuration.

The best performance is obtained using the *AoSoA* layout with BL = 8, and adopting the cache blocking technique on the inner 2 dimensions. This is an improvement of $\approx 2\times$ w.r.t. the *AoS* data layout and $\approx 1.3\times$ over the *SoA* data layout, and translates to a measured performance of $\approx 100$ GFLOP/s and a sustained bandwidth of $\approx 85$ GB/s, using 32 threads.

## 5.3    Results Analysis

We now analyze our results with respect to the performance measured by the *Empirical Roofline Tool* [28] (ERT).

This tool runs several synthetic benchmarking kernels with different arithmetic intensity (AI) to estimate the *machine balance*, and the effective computing throughput and the bandwidth of the different memory levels. Each kernel executes a sequence of *fused multiply-add to accumulator* vector instructions trying to fill the floating point pipelines of all the FP units of the processor where it is being run.

The ERT tool originally did not support Arm architecture, and at a first naive compilation was not able to reach more than the 50% of the theoretical FP peak of the *ThunderX2*. Thus we provide a modified version of the original code, exploiting NEON *intrinsics*, allowing to reach an higher fraction of the theoretical peak. The modified ERT code is now freely available for download[1].

---

[1] https://bitbucket.org/ecalore/cs-roofline-toolkit/branch/feature/thunderx2.

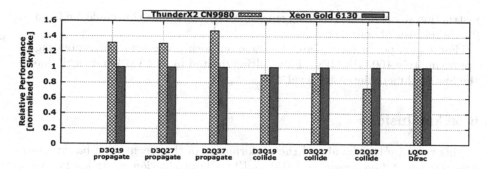

**Fig. 4.** Performance comparison between the *ThunderX2* model CN9980 and Intel Xeon Gold 6130 processors. For LBM we compare performances for *propagate* and *collide* kernels. 3D models are run on a lattice of $256^3$ and the 2D model on a lattice of $2048 \times 8192$ points. For LQCD we compare the performance of the *Dirac* operator run on a lattice of $32^4$ points. All kernels use the *AoSoA* data structure; results are normalized to the Intel Xeon figures.

On the *ThunderX2* used in this work, the ERT benchmark modified by us, measures an empirical computing throughput of ≈370 GFLOPs, approximately 70% of the maximum theoretical value.

Figure 3 shows the roofline chart of the ERT benchmark together with the performance values of the *collide* kernels for all our LBM models and of the *Dirac* operator. In our cases, the arithmetic intensity is approximately 4.26 FLOP/Byte for the D3Q19 and D3Q27 codes, 12.43 FLOP/Byte for the D2Q37 model, and 1.17 FLOP/Byte for the *Dirac* operator. We do not consider *propagate* because this kernel does not perform floating point operations.

Our results show that the roofline model very neatly describes our results; with reference to the empirical maximum value measured by ERT, we find out that the performance of the *collide* kernels reach approximately 50–60%, while the *Dirac* operator achieves about 27% and results limited by the DRAM bandwidth, due to the lower AI level.

Finally, in Fig. 4 we compare our results with those of a recent 2.1 GHz 16 core Intel-Xeon 6130 processor based on the *Skylake* architecture.

We start first with LBM codes. For *propagate* kernels, memory performance is the relevant figure-of-merit. On the *ThunderX2* processor, the theoretical peak value is 160 GB/s; we measure 68 GB/s for the LBM 3D models, and 89 GB/s for the 2D model, that is ≈1.3 − 1.4× better than the Xeon, whose peak bandwidth is 120 GB/s.

For *collide* kernels, the computing performances on the *ThunderX2* are approximately 217 GFLOPs for the D3Q19 and 233 GFLOPs for the D3Q27 models, and 185 GFLOPs for the D2Q37 model; this is approximately 10% lower than the Xeon for the D3Q19 and D3Q27 models, and 30% lower for D2Q37. Inspecting assembly codes of *collide*, generated both with GCC v8.2 and Arm v19.0 compilers, we see that this is due to inefficient use of FP units, caused by instruction dependencies to load operands; we think that this could be fixed in future releases of compilers.

For *Dirac*, performances are the same on both processors, corresponding to approximately 100 GFLOPs, and resulting limited by DRAM bandwidth due to the low arithmetic intensity level.

# 6  Conclusions

In this work we have analyzed the performance of the recent Arm-based multicore *ThunderX2* processor for LBM and LQCD production ready applications, assessing the impact on performance of using different layouts to store data in memory.

The fraction of effective memory bandwidth and computing throughput achieved by our application kernels is quite large: for *propagate* and *Dirac* is approximately 70–80% of memory peak bandwidth measured by the STREAM benchmark, and for *collide* is about 50–60% of the empirical computing throughout peak measured by the ERT tool. For *Dirac* the computing throughput is about 27% of the peak, but in this case this is due to the low AI level, making it limited by the memory bandwidth. Compared to the Intel Xeon 6130, the performance levels of our applications are similar; in particular, are higher for *propagate*, the same for *Dirac*, and slightly lower for *collide*. This is in agreement with the hardware features of the two processors. In fact, kernels with low AI achieve better performance on the *ThunderX2* benefiting from the 8 memory channels available which provide higher memory bandwidth; while kernels with higher AI achieve better performance on the Intel Xeon 6130 because of the longer hardware vectors providing higher computational throughput.

As for most recent processors, the use of appropriately tailored data layouts, such as the *AoSoA*, is demonstrated to be relevant for performances, allowing to better fit the computing requirements of different parts of applications.

From the software support point of view (compilers, libraries, etc.), we underline that for the *ThunderX2*, and more in general for Arm architectures, a wide ecosystem of tools is already available and has reached a good level of maturity. In fact, our codes have run almost out-of-the-box.

To conclude, our results show that these new processors are now competitive with state-of-the-art Intel high-end CPUs, opening new options for future systems, with potential advantages also from the energy-efficiency point of view.

As a future steps we plan to investigate the performance of multi-node systems analyzing also the scalability of our applications. Also, we are interested in analyzing the energy-efficiency of the *ThunderX2*; at this stage we only highlight that the TDP/core ratio is ≈5.6 Watt/core, about 1.4× less than Intel-Xeon 6130, leading to potential benefits also in terms of FLOPS/Watt performance.

**Acknowledgments.** This work has been done in the framework of the COKA, and COSA projects funded by INFN. We would like to thank CINECA (Italy) and Università di Ferrara for access to their HPC systems. All runs on the *ThunderX2* have been performed on computational resources provided and supported by E4 Computer Engineering and installed at CINECA.

# References

1. Pruitt, D.D., Freudenthal, E.A.: Preliminary investigation of mobile system features potentially relevant to HPC. In: 2016 4th International Workshop on Energy Efficient Supercomputing (E2SC), pp. 54–60, November 2016. https://doi.org/10.1109/E2SC.2016.013
2. Calore, E., Mantovani, F., Ruiz, D.: Advanced performance analysis of HPC workloads on Cavium ThunderX. In: 2018 International Conference on High Performance Computing Simulation (HPCS), pp. 375–382, July 2018. https://doi.org/10.1109/HPCS.2018.00068
3. Fürlinger, K., Klausecker, C., Kranzlmüller, D.: Towards energy efficient parallel computing on consumer electronic devices. In: Kranzlmüller, D., Toja, A.M. (eds.) ICT-GLOW 2011. LNCS, vol. 6868, pp. 1–9. Springer, Heidelberg (2011). https://doi.org/10.1007/978-3-642-23447-7_1
4. Rajovic, N., et al.: The Mont-Blanc prototype: an alternative approach for HPC systems. In: SC 2016: Proceedings of the International Conference for High Performance Computing, Networking, Storage and Analysis, pp. 444–455, November 2016
5. Yokoyama, D., Schulze, B., Borges, F., Mc Evoy, G.: The survey on arm processors for HPC. J. Supercomput. **75**(10), 7003–7036 (2019). https://doi.org/10.1007/s11227-019-02911-9
6. Oyarzun, G., Borrell, R., Gorobets, A., Mantovani, F., Oliva, A.: Efficient CFD code implementation for the ARM-based Mont-Blanc architecture. Future Gener. Comput. Syst. **79**, 786–796 (2018). https://doi.org/10.1016/j.future.2017.09.029
7. Stegailov, V., Smirnov, G., Vecher, V.: Vasp hits the memory wall: processors efficiency comparison. Concurr. Comput. Pract. Exp. **31**(19), e5136 (2019). https://doi.org/10.1002/cpe.5136
8. Hammond, S., et al.: Evaluating the Marvell ThunderX2 Server Processor for HPC Workloads (2019). https://cfwebprod.sandia.gov/cfdocs/CompResearch/docs/bench2019.pdf
9. Banchelli, F., et al.: MB3 D6.9 - performance analysis of applications and mini-applications and benchmarking on the project test platforms. Technical report, Mont-Blanc Project, Version 1.0 (2019)
10. McIntosh-Smith, S., Price, J., Deakin, T., Poenaru, A.: A performance analysis of the first generation of HPC-optimized arm processors. Concurr. Comput. Pract. Exp., e5110 (2018). https://doi.org/10.1002/cpe.5110
11. Ciznicki, M., Kurowski, K., Weglarz, J.: Energy aware scheduling model and online heuristics for stencil codes on heterogeneous computing architectures. Cluster Comput. **20**(3), 2535–2549 (2017). https://doi.org/10.1007/s10586-016-0686-2
12. Yount, C., Tobin, J., Breuer, A., Duran, A.: Yask—yet another stencil kernel: a framework for HPC stencil code-generation and tuning. In: Sixth International Workshop on Domain-Specific Languages and High-Level Frameworks for HPC (WOLFHPC), pp. 30–39 (2016). https://doi.org/10.1109/WOLFHPC.2016.08
13. Pereira, A.D., Ramos, L., Góes, L.F.W.: PSkel: a stencil programming framework for CPU-GPU systems. Concurr. Comput. Pract. Exp. **27**(17), 4938–4953 (2015). https://doi.org/10.1002/cpe.3479

14. Calore, E., Gabbana, A., Schifano, S.F., Tripiccione, R.: Optimization of lattice Boltzmann simulations on heterogeneous computers. Int. J. High Perform. Comput. Appl., 1–16 (2017). https://doi.org/10.1177/1094342017703771
15. Bonati, C., et al.: Portable multi-node LQCD Monte Carlo simulations using OpenACC. Int. J. Mod. Phys. C **29**(1) (2018). https://doi.org/10.1142/S0129183118500109
16. Shet, A.G., et al.: On vectorization for lattice based simulations. Int. J. Mod. Phys. C **24** (2013). https://doi.org/10.1142/S0129183113400111
17. Joó, B., Kalamkar, D.D., Kurth, T., Vaidyanathan, K., Walden, A.: Optimizing Wilson-Dirac operator and linear solvers for Intel® KNL. In: Taufer, M., Mohr, B., Kunkel, J.M. (eds.) ISC High Performance 2016. LNCS, vol. 9945, pp. 415–427. Springer, Cham (2016). https://doi.org/10.1007/978-3-319-46079-6_30
18. Calore, E., Gabbana, A., Schifano, S.F., Tripiccione, R.: Early experience on using Knights Landing processors for Lattice Boltzmann applications. In: Wyrzykowski, R., Dongarra, J., Deelman, E., Karczewski, K. (eds.) PPAM 2017. LNCS, vol. 10777, pp. 519–530. Springer, Cham (2018). https://doi.org/10.1007/978-3-319-78024-5_45
19. Williams, S., Waterman, A., Patterson, D.: Roofline: an insightful visual performance model for multicore architectures. Commun. ACM **52**(4), 65–76 (2009). https://doi.org/10.1145/1498765.1498785
20. Gwennap, L.: ThunderX rattles server market. Microproc. Rep. **29**(6), 1–4 (2014)
21. McCalpin, J.D.: Stream: sustainable memory bandwidth in high performance computers (2019). https://www.cs.virginia.edu/stream/. Accessed 14 Apr 2019
22. Marvell: ThunderX2 arm-based processors (2019). https://www.marvell.com/products/server-processors/thunderx2-arm-processors.html. Accessed 18 Apr 2019
23. Biferale, L., Mantovani, F., Sbragaglia, M., Scagliarini, A., Toschi, F., Tripiccione, R.: Second-order closure in stratified turbulence: simulations and modeling of bulk and entrainment regions. Phys. Rev. E **84**(1), 016305 (2011). https://doi.org/10.1103/PhysRevE.84.016305
24. Calore, E., Gabbana, A., Kraus, J., Schifano, S.F., Tripiccione, R.: Performance and portability of accelerated lattice Boltzmann applications with OpenACC. Concurr. Comput. Pract. Exp. **28**(12), 3485–3502 (2016). https://doi.org/10.1002/cpe.3862
25. DeGrand, T., DeTar, C.: Lattice Methods for Quantum ChromoDynamics. World Scientific (2006). https://doi.org/10.1142/6065
26. Bonati, C., et al.: Design and optimization of a portable LQCD Monte Carlo code using OpenACC. Int. J. Mod. Phys. C **28**(5) (2017). https://doi.org/10.1142/S0129183117500632
27. Bonati, C., et al.: Early experience on running OpenStaPLE on DAVIDE. In: Yokota, R., Weiland, M., Shalf, J., Alam, S. (eds.) ISC High Performance 2018. LNCS, vol. 11203, pp. 387–401. Springer, Cham (2018). https://doi.org/10.1007/978-3-030-02465-9_26
28. Lo, Y.J., et al.: Roofline model toolkit: a practical tool for architectural and program analysis. In: Jarvis, S.A., Wright, S.A., Hammond, S.D. (eds.) PMBS 2014. LNCS, vol. 8966, pp. 129–148. Springer, Cham (2015). https://doi.org/10.1007/978-3-319-17248-4_7

# An Area Efficient and Reusable HEVC 1D-DCT Hardware Accelerator

Mate Cobrnic$^{(\boxtimes)}$ , Alen Duspara , Leon Dragic , Igor Piljic ,
Hrvoje Mlinaric , and Mario Kovac

HPC Architecture and Application Research Center, Faculty of Electrical
Engineering and Computing, University of Zagreb, 10000 Zagreb, Croatia
`mate.cobrnic@fer.hr`

**Abstract.** In this paper is presented an area efficient reusable architecture for integer one dimensional Discrete Cosine Transform (1D DCT) with adjustable transform sizes in High Efficiency Video Coding (HEVC). Optimization is based on exploiting of symmetry and subset properties of the transform matrix. The proposed multiply-accumulate architecture is fully pipelined and applicable for all transform sizes. It provides the interface over which the processing system can control the datapath of the transform process and the synchronization channel that enables the system to receive the feedback information about utilization from the device. An intuitive line approach for calculating transform coefficients for all transform sizes was used instead of the commonly applied recursive decomposition approach. This approach simplifies disabling of lines that are not employed for a particular transform size. The proposed architecture is implemented on the FPGA platform, can operate at 407,5 MHz, achieves throughput of 815 Msps and can support encoding of a 4K UHD@30 fps video sequence in real time.

**Keywords:** Integer Discrete Cosine Transform (DCT) · High Efficiency Video Coding (HEVC) · Field-Programmable Gate Array (FPGA) · Pipelined architecture

## 1 Introduction and Related Work

Due to its properties which are useful for both compression efficiency and efficient implementation, discrete cosine transform (DCT) is widely used in video and image compression. The H.265/High Efficiency Video Coding standard [1], with significant improvements in compression performance compared to its predecessors, uses finite precision approximation of transformation, which is called integer DCT in the remainder of this paper. In overall encoding time, transform and quantization take about 25% for the all-intra and 11% for the random access configurations [2]. To ensure higher compression of ultra-high definition (UHD) video resolutions the standard introduced additional transform sizes at the cost

© Springer Nature Switzerland AG 2020
R. Wyrzykowski et al. (Eds.): PPAM 2019, LNCS 12043, pp. 199–208, 2020.
https://doi.org/10.1007/978-3-030-43229-4_18

of processing and implementation complexity. That is why efficient implementation, which was set as design goal during HEVC transform development [3–5], significantly impacts the overall performance of the codec.

Reusability of the $N/2$-point 1D integer DCT architecture for implementation of the N-point 1D integer DCT is identified in [6]. Architecture with a partial butterfly unit, Multiple-Constant-Multiplication (MCM) units, adder units and a $N/2$-point integer DCT unit was proposed as an alternative to the resource consuming, direct implementation of matrix multiplication on hardware. Transform matrix decomposition and hardware reusability were exploited in [7] as well. Additionally, a constant throughput for every transform size was achieved by adding the additional $N/2$-point integer DCT unit. Constant throughput with better resource utilization was achieved in [8]. Optimization of MCM units was investigated in [9] and [10] to avoid high latency of the hardware multipliers. To decrease design and verification time and improve design reusability, the usage of High Level Synthesis (HLS) is proposed in [11]. Since management and control of residual quadtree partitioning is usually not in scope of transform architectures, [12] proposes flexible input architecture and supporting a configuration encoding scheme. In [13] two unified pipelined architectures for forward/inverse integer DCT are proposed. There are also proposals where the transform accuracy is reduced to save hardware resources. In [7] resource savings are made through pruning the least significant bits in MCM and adder units. This affects rate-distortion performance [14] insignificantly but brings a reduction in the area and power. In [15] hardware-oriented implementation of Arai-based DCT is proposed and sinusoidal multiplication constants are approximated with a configurable number of representation bits.

All these architectures consider that all the elements of a residual vector are transferred to the input at the same cycle. In case of a video with an 8-bit color depth, 9 bits are required for residual value representation. 288 bits are needed if one residual vector for the biggest transform block size is transmitted at once, or 512 bits if residuals are byte-aligned. Moreover, during the design of hardware architectures for transform process high parallelization was set as a design objective. In modern System on Chips (SoCs), especially in their budget versions, hardware resources and communication interface bit widths are limited.

In this paper, an area efficient and reusable integer DCT architecture for systems with limited resources is proposed which can receive two byte-aligned residuals (total of a 32-bit bus width) and output two scaled intermediate 16 bit transform coefficients in one cycle. Improving area efficiency is stressed as the main design goal.

The rest of the paper is organized as follows. In the next section integer DCT properties and algorithms, exploited in the design of proposed architecture, will be presented. In Sect. 3 the proposed architecture and the lane concept are described. Implementation results are discussed and compared to other architectures in Sect. 4. Section 5 concludes the paper.

## 2 HEVC DCT Overview and Design Considerations

HEVC 2D integer DCT transform of size $N \times N(N = 4, 8, 16, 32)$ is given by

$$Y = D \times X \times D' \tag{1}$$

where $D$ is the HEVC transform matrix with constant values, $X$ the residual matrix of size $N \times N$ and $D'$ is a transpose of the transform matrix. When properties of the transpose operator are considered (1) can be rewritten as

$$Y = (D \times (D \times X)')' \tag{2}$$

In this equation it can be noticed that the matrix $D$ is used in multiplication twice. Compared to (1) basis vectors with constant values in both multiplications, are multiplicands that can be mapped in the hardware to a single multiply-accumulate block. Its input will be a series of residual vectors when implementing inner multiplication and series of intermediate vectors, which can be brought to input of the same block as feedback. This property known as transform separability can be used to design hardware architecture for 2D integer DCT by utilizing the same 1D integer DCT core and applying transpose to result matrices.

Since elements of the even basis vectors of the transform matrix are symmetric and the odd basis vectors antisymmetric, the decomposition of the 1D integer DCT matrix multiplication can be made

$$
\begin{bmatrix} Y_0^N \\ Y_2^N \\ \vdots \\ Y_{N-2}^N \end{bmatrix} = D^{\frac{N}{2}} \times \begin{bmatrix} a_0^N \\ a_1^N \\ \vdots \\ a_{\frac{N}{2}-1}^N \end{bmatrix} \tag{3a}
$$

$$
\begin{bmatrix} Y_1^N \\ Y_3^N \\ \vdots \\ Y_{N-1}^N \end{bmatrix} = B^{\frac{N}{2}} \times \begin{bmatrix} b_0^N \\ b_1^N \\ \vdots \\ b_{\frac{N}{2}-1}^N \end{bmatrix} \tag{3b}
$$

where

$$
\begin{aligned}
a_i^N &= X_i^N + X_{N-1-i}^N \\
b_i^N &= X_i^N - X_{N-1-i}^N
\end{aligned} \tag{4}
$$

for $i = 0, 1, 2, ..., N/2 - 1$. $X$ is the input residual column vector, $Y$ is the $N$-point 1D integer DCT of $X$, $D^{N/2}$ is the $N/2$-point HEVC transform matrix and $B^{N/2}$ is the matrix of size $N/2 \times N/2$ defined as

$$B_{i,j}^{N/2} = D_{2i+1,j}^N \tag{5}$$

for $i, j = 0, 1, 2, ..., N/2 - 1$. Matrix $D^{N/2}$ from Eq. (3a) can be further decomposed to $D^{N/4}$ and $B^{N/4}$ in the same way. Decomposition is applicable for all transform sizes supported in the HEVC standard. By observing the elements of matrix $B^N$ it can be seen that every row contains the same unique values with a different distribution. Reusability as a consequence of decompositions and a small number of unique elements will decrease implementation costs.

## 3  Proposed Architecture

To minimize the logic circuit for the input adder unit, elements of the residual vector are transferred sequentially into the transform engine. It can be observed in (3) and (4) that the optimal grouping of elements would be such that pairs of residuals are transferred in one cycle because they can be immediately processed. For area minimization purpose the input bus-width of 32 bits is selected. Two byte-aligned residuals will be either added or subtracted depending on the position in the result vector which is being calculated. Results of the operation are multiplied with all constants from matrix $D^N$ column and products are added to their respective accumulated value. Accumulation of products is done during $N/2$ cycles when the resulting vector element is available. Based on given steps in obtaining the result, the architecture will include a pre-operation unit (PREOP), providing sum and difference of the residuals, MCM units (MCMs) and accumulation units (ACCUs). Units that build a single lane are shown in Fig. 1. Parallel MCMs are the preferred solution for systems where the complete residual column vector can be retrieved instantaneously. In the proposed architecture with a 32-bit bus such a circuit wouldn't be fully utilized and would result in additional area. Therefore, multiplexed MCMs are used. They are implemented using the algorithm from [16] where a block diagram can be retrieved from a multiplexed multiplier block generator presented in [17]. Area advantage with those MCMs is obtained at the cost of increased latency and decreased throughput.

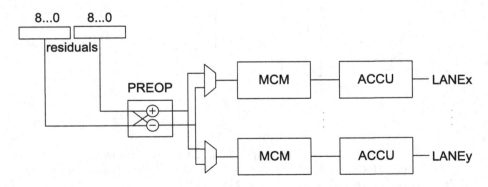

**Fig. 1.** General block diagram of the integer DCT lanes

To enable reusability of the same hardware for different transform sizes, design is structured into lanes. The lane used for the transform size $N$ is also used for all the larger transform sizes. The difference in the amount of multiply and accumulate cycles depends on transform size as described above. Compared to competing hierarchical architectures, which have multiplexers at the input of every $N$-point integer DCT unit when $N = 4$, 8 or 16, the proposed lane-based architecture requires multiplexing only once.

Constant multipliers used in a lane for one transform size are repeated for all transform sizes in which that lane is involved. What row of the result vector a lane outputs is determined with its MCM configuration register value. Every register contains sixteen codes for sixteen multipliers. Code length depends on the number of unique elements in the same row of the transform matrix. 1-bit code is sufficient to represent two unique multipliers in odd rows for a 4-point transform. In case of a 32-point transform the 4-bit code is needed to switch between sixteen different multiplication datapaths. The signs of multipliers are determined with the 16-bit ACCU configuration register value. Depending on each bit value the MCM output will be added or subtracted from current accumulated value. Instances of a lane differ from each other only in the content of these two registers. For example, LANE8 is used for 8, 16 and 32-point integer DCT. It has four constant multipliers, which are used in every one of these three applicable transform sizes. LANE2 implements multiplication with a single multiplier sixty-four and therefore doesn't contain the MCM configuration register. A lane can be interpreted as a set of functional blocks which implements the multiplication of a basis vector with a residual vector and outputs the single transform coefficient. Schematic for one lane is shown in Fig. 2.

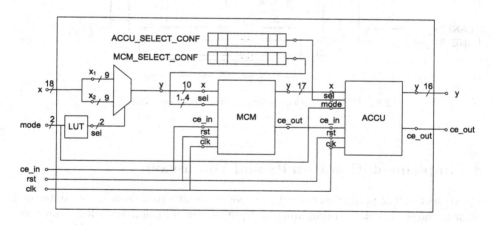

**Fig. 2.** Schematic of a lane

The serialization block is placed at the output because this architecture receives two residuals per cycle to keep the area usage low. It is controlled by the processing system. The architecture will output two 16-bit coefficients and signal their availability over the dedicated interface. There are two synchronization channels between the device and the processing system. Over one the system can request serialization hold and when it becomes effective it is signalized at the output. Over the second channel, the hold mode, as shown on Fig. 3, is activated when scaled transform coefficients are ready for serialization but the previously started serialization is not yet finished. In that case the result vector will be moved to the delay buffer and, using additional output interface, a signal will be sent that the transferring of new residuals has to be stopped and will remain so until the signal is deactivated.

The proposed 1D integer DCT architecture, which is designed for systems with limited hardware resources, supports all transform sizes. The proposed lane structure which exploits the subset property of the transform matrix simplifies power management. The interface responsible for bidirectional communication with the processing system enables the transform to hold and continue without any data loss.

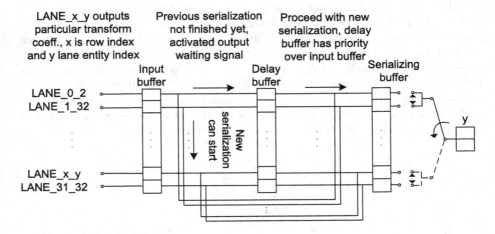

**Fig. 3.** Serialization of scaled transform coefficients

## 4    Implementation Results and Discussion

In this section the performance of the proposed architecture is evaluated in terms of area, latency, delay, throughput and power consumption with focus on the area. Implementation and result presentation were done for comparable circuits of leading FPGA manufacturers (Xilinx, Intel Altera). To have an equitable evaluation and comparison with competing parallel integer DCT architectures, two architectures will be implemented, the proposed architecture with 32 bits wide data input and output and a separate configuration where the serialization unit is omitted.

Design is coded in VHDL and synthesized targeting a Virtex 7 device, Xilinx XC7VX330T FFG1157 FPGA with speed grade 3. It is implemented using

Vivado 2016.4. The wordlength of the input residual is set to 16 bits. The area, the number of used registers, the maximum delay and the total power are obtained for the baselined design from the implementation reports, and the maximum operating frequency and throughput are calculated. The power analysis which is based on vectorless (probabilistic) power analysis methodology is run with default device and environment settings and a 100 MHz clock frequency. The results of integer DCT for two device configurations and different transform sizes are presented in Tables 1 and 2.

**Table 1.** Performance results of the proposed integer DCT architecture with output serialization

| Transform size | LUTs | Registers | Latency | Frequency (MHz) | Throughput (Gsps) | Total power (mW) |
|---|---|---|---|---|---|---|
| 4 | 313 | 452 | 8 | 741.84 | 1.48 | 189 |
| 8 | 1004 | 1113 | 10 | 590.32 | 1.18 | 208 |
| 16 | 2862 | 2493 | 14 | 480.77 | 0.96 | 251 |
| 32 | 6776 | 4948 | 22 | 407.50 | 0.81 | 355 |

**Table 2.** Performance results of the proposed integer DCT architecture with parallel output

| Transform size | LUTs | Registers | Latency | Frequency (MHz) | Throughput (Gsps) | Total power (mW) |
|---|---|---|---|---|---|---|
| 4 | 237 | 363 | 7 | 744.05 | 1.49 | 187 |
| 8 | 835 | 963 | 9 | 613.50 | 1.23 | 215 |
| 16 | 2367 | 2198 | 13 | 527.15 | 1.05 | 247 |
| 32 | 6085 | 4402 | 21 | 500.25 | 1.00 | 348 |

The results show that look-up tables' (LUT) utilization for $N$-point transform is about 2.8 times and registers' utilization about 2.2 times the utilization for $N/2$-point transform for the complete architecture. Power consumption increases by factor 1.2 on average. When the serialization unit is omitted there are fewer LUTs and registers employed and the maximum operating frequency is higher for all transform sizes. By eliminating one functional block in the architecture the wiring complexity is reduced, which affects the maximum path delay and operating frequency. The latency is contributed by two addends, five or six pipeline stages, depending on the architecture variant, and $N/2$ cycles needed to process one residual vector.

This architecture supports the processing of 4K ultra high definition (UHD) videos in HEVC standard at 30 frames/s and 4:2:0 YUV format. Minimum operating frequency for this video format is calculated as follows

$$F_{min} = 2 \times W \times H \times format \times fps \left( \frac{L}{32 \times 32} + \frac{1}{ppc} \right) \tag{6}$$

where $W \times H$ is the resolution of video sequence, $format$ equals 1.5 and $fps$ is the video frame rate. It is considered that a pair of residuals is transferred every cycle to the input of 1D integer DCT core and that two transform coefficients can be output per cycle. So $ppc$ is 2. Factor two in (6) is characteristic of the folded structure for 2D transform. $L$ is the pipeline latency. To process a 32 × 32 transform block using the 2D folded structure, built with the proposed 1D architecture, $2L + 1024$ cycles are needed. Transpose buffer is not in the scope of this work, but is required for 2D transform according to (2). It does not affect this performance evaluation if its throughput is not less than two pixel per cycle. The minimum operating frequency for the given format and processing rate is 393 MHz. This condition is satisfied by the proposed architecture. This is a satisfactory result considering the design objectives and compromise on throughput.

Performance of the proposed architecture is compared to related architectures which target the FPGA platform and results are shown in Table 3. Since about 90% of transform blocks are of size 4 × 4 or 8 × 8 [18], 4-point and 8-point integer DCT accelerators are considered. To be able to have a fair comparison with a pipelined design, like the one from [9], synthesis is made on the same device, Stratix V FPGA 5SGXMABN3F45I4 using Quartus Prime Standard Edition. Logic cells from Stratix V and Virtex 7 device families have a different architecture and therefore the number of used adaptive logic modules (ALMs) in Stratix V can't be directly compared with LUTs in Virtex 7 to assess area efficiency.

**Table 3.** Comparison with related works

| Architecture | Technology | Max Freq (MHz) | ALMs/LUTs | Registers | Pixels/CC | Throughput (Gsps) |
|---|---|---|---|---|---|---|
| Bolaños-Jojoa [9] | Stratix V | 630.12 | 108 | 288 | 4 | 2.52 |
| Proposed | Stratix V | 696.38 | 92 | 244 | 2 | 1.39 |
| Chatterjee [8] | Virtex 7 | 183.3 | 924 | – | 8 | 1.46 |
| Proposed | Virtex 7 | 613.50 | 835 | 963 | 2 | 1.23 |

All related designs are made for parallel processing of all elements in a residual vector. The maximum operating frequency for the proposed design is higher than [8] and [9]. The usage of parallel MCMs rather than multiplexed MCMs comes at the cost of the area. That is apparent in the number of utilized ALMs and LUTs. Moreover, these two competing designs are not developed as accelerators but as single DCT modules and therefore don't include synchronization and control signals which are necessary for communication with the processing system. The amount of utilized logic cells is 14.81% lower than in [9] while throughput is decreased 1.81 times. Compared to [8] hardware cost is 9.6% lower while throughput is lower 1.21 times. Though [8] is not implemented as a pipelined circuit this will implicate the use of flip-flops and the number of used LUTs is not under impact.

## 5 Conclusion

This paper presents the area efficient reusable integer DCT architecture for systems with limited resources. The properties of the transform matrix and through-put reduction are exploited to design the optimal architecture in the term of area resources. The design is based on regular lane structure where lanes are reused and shared among all supported transform sizes. Computation of one scaled transform coefficient is made within one lane, which processes added/subtracted symmetric residual values and involves multiplexed MCM and accumulation with scaling. Result values from 32 lanes are serialized to have the appropriate data format at the output. The proposed architecture and its variant with a parallel output are implemented on the FPGA platform. Concerning hardware cost, it outperforms related architectures. Maximum operating frequency for 1D DCT architecture is 407.5 MHz. The proposed architecture can encode 4K UHD@30 fps in real time.

**Acknowledgements.** The work presented in this paper has been partially funded by the European Processor Initiative project that has received funding from the European Union's Horizon 2020 research and innovation programme under grant agreement No. 826647.

## References

1. Telecommunication standardization sector of ITU: Recommendation ITU-T H.265 — International Standard ISO/IEC 23008-2. International Telecommunication Union, Geneva (2015)
2. Bossen, F., Flynn, D., Bross, B., Suhring, K.: HEVC complexity and implementation analysis. IEEE Trans. Circuits Syst. Video Technol. **22**(12), 1685–1696 (2012). https://doi.org/10.1109/TCSVT.2012.2221255
3. Budagavi, M., Fuldseth, A., Bjøntegaard, G., Sze, V., Sadafale, M.: Core transform design in the high efficiency video coding (HEVC) standard. IEEE J. Sel. Top. Sign. Proces. **7**(6), 1029–1041 (2013). https://doi.org/10.1109/JSTSP.2013.2270429
4. Tikekar, M., Huang, C.-T., Juvekar, C., Chandrakasan, A.: Core transform property for practical throughput hardware design. Paper Presented at the 7th Meeting of the Joint Collaborative Team on Video Coding (JCT-VC), Geneva, 21–30 November 2011
5. Fuldseth, A., Bjøntegaard, G., Sze, V., Budagavi, M.: Core transform design for HEVC. Paper Presented at the 7th Meeting of the Joint Collaborative Team on Video Coding (JCT-VC), Geneva, 21–30 November 2011
6. Zhao, W., Onoye, T., Song, T.: High-performance multiplierless transform architecture for HEVC. In: IEEE International Symposium on Circuits and Systems, Beijing, pp. 1668–1671. IEEE (2013). https://doi.org/10.1109/ISCAS.2013.6572184

7. Meher, P.K., Park, S.Y., Mohanty, B.K., Lim, K.S., Yeo, S.: Efficient integer DCT architectures for HEVC. IEEE Trans. Circuits Syst. Video Technol. **24**(1), 168–178 (2014). https://doi.org/10.1109/TCSVT.2013.2276862
8. Chatterjee, S., Sarawadekar, K.P.: A low cost, constant throughput and reusable 8X8 DCT architecture for HEVC. In: 59th International Midwest Symposium on Circuits and Systems, Abu Dhabi, pp. 1–4. IEEE (2016). https://doi.org/10.1109/MWSCAS.2016.7869994
9. Bolaños-Jojoa, J.D., Velasco-Medina, J.: Efficient hardware design of N-point 1D-DCT for HEVC. In: 20th Symposium on Signal Processing, Images and Computer Vision, Bogota, pp. 1–6. IEEE (2015). https://doi.org/10.1109/STSIVA.2015.7330449
10. Abdelrasoul, M., Sayed, M.S., Goulart, V.: Scalable integer DCT architecture for HEVC encoder. In: IEEE Computer Society Annual Symposium on VLSI, Pittsburgh, pp. 314–318. IEEE (2016). https://doi.org/10.1109/ISVLSI.2016.98
11. Sjövall, P., Viitamäki, V., Vanne, J., Hämäläinen, T.D.: High-level synthesis implementation of HEVC 2-D DCT/DST on FPGA. In: IEEE International Conference on Acoustics, Speech and Signal Processing, New Orleans, pp. 1547–1551. IEEE (2017). https://doi.org/10.1109/ICASSP.2017.7952416
12. Arayacheeppreecha, P., Pumrin, S., Supmonchai, B.: Flexible input transform architecture for HEVC encoder on FPGA. In: 12th International Conference on Electrical Engineering/Electronics, Computer, Telecommunications and Information Technology, Hua Hin, pp. 1–6. IEEE (2015). https://doi.org/10.1109/ECTICon.2015.7206947
13. Abdelrasoul, M., Sayed, M.S., Goulart, V.: Real-time unified architecture for forward/inverse discrete cosine transform in high efficiency video coding. IET Circuits Devices Syst. **11**(4), 381–387 (2017). https://doi.org/10.1049/iet-cds.2016.0423
14. Bjøntegaard, G.: Calculation of average PSNR differences between RD-curves. Paper Presented at the 16th Meeting of the Video Coding Experts Group (VCEG), Austin, 2–4 April 2001
15. Renda, G., Masera, M., Martina, M., Masera, G.: Approximate Arai DCT architecture for HEVC. In: New Generation of CAS, Genova, pp. 133–136. IEEE (2017). https://doi.org/10.1109/NGCAS.2017.38
16. Tummeltshammer, P., Hoe, J.C., Puschel, M.: Time-multiplexed multiple-constant multiplication. IEEE Trans. Comput. Aided Des. Integr. Circuits Syst. **26**(9), 1551–1563 (2007). https://doi.org/10.1109/TCAD.2007.893549
17. Spiral Project: Multiplexed Multiplier Block Generator. http://spiral.net/hardware/mmcm.html. Accessed 10 Dec 2018
18. Hong, L., Weifeng, H., Zhu, H., Mao, Z.: A cost effective 2-D adaptive block size IDCT architecture for HEVC standard. In: 56th International Midwest Symposium on Circuits and Systems, Columbus, pp. 1290–1293. IEEE (2013). https://doi.org/10.1109/MWSCAS.2013.6674891

# Performance Analysis and Scheduling in HPC Systems

# Improving Locality-Aware Scheduling with Acyclic Directed Graph Partitioning

M. Yusuf Özkaya[1(✉)], Anne Benoit[1,2], and Ümit V. Çatalyürek[1]

[1] CSE, Georgia Institute of Technology, Atlanta, GA, USA
{myozka,umit}@gatech.edu
[2] LIP, ENS Lyon, Cedex 07, France
anne.benoit@ens-lyon.fr

**Abstract.** We investigate efficient execution of computations, modeled as Directed Acyclic Graphs (DAGs), on a single processor with a two-level memory hierarchy, where there is a limited fast memory and a larger slower memory. Our goal is to minimize execution time by minimizing redundant data movement between fast and slow memory. We utilize a DAG partitioner that finds localized, acyclic parts of the whole computation that can fit into fast memory, and minimizes the edge cut among the parts. We propose a new scheduler that executes each part one-by-one, obeying the dependency among parts, aiming at reducing redundant data movement needed by cut-edges. Extensive experimental evaluation shows that the proposed DAG-based scheduler significantly reduces redundant data movement.

**Keywords:** Scheduling · Partitioning · Acyclic directed graph · Locality

## 1 Introduction

In today's computers, the cost of data movement through the memory hierarchy is dominant relative to the cost of arithmetic operations, and it is expected that the gap will continue to increase [10,25]. Hence, a significant portion of research efforts focuses on optimizing the data locality [1,4,5,8,11,13,15,17,18,20,23], for instance by considering the reuse distance.

Sparse computations are used by many scientific applications but they are notorious for their poor performance due to their irregular memory accesses and hence poor data locality in the cache. One of widely used kernels is Sparse Matrix-Vector Multiplication (SpMV), which is the main computation kernel for various applications (e.g., spectral clustering [21], dimensionality reduction [26], PageRank algorithm [22], etc.). Similarly, Sparse Matrix-Matrix Multiplication (SpMM), which is one of the main operations in many Linear Algebraic problems, suffers from similar cache under-utilization caused by unoptimized locality decisions. Several recent studies focus on the cache locality optimization for these problems [2,9,14,27]. In addition, many other irregular applications, such

© Springer Nature Switzerland AG 2020
R. Wyrzykowski et al. (Eds.): PPAM 2019, LNCS 12043, pp. 211–223, 2020.
https://doi.org/10.1007/978-3-030-43229-4_19

as machine/deep learning framework with optimization problems [19], generate
an internal computational (acyclic) task graph, that one can use to further optimize the execution to minimize data movement. In this work, we aim to optimize
the repeated execution of such sparse and irregular kernels, whose dependency
structure can be expressed as a DAG.

We investigate efficient execution of sparse and irregular computations, modeled as Directed Acyclic Graphs (DAGs) on a single processor, with a two-level
memory hierarchy. There are a limited number of fast memory locations, and
an unlimited number of slow memory locations. In order to compute a task, the
processor must load all of the input data that the task needs, and it will also
need some scratch memory for the task and its output. Because of the limited
fast memory, some computed values may need to be temporarily stored in slow
memory and reloaded later. The loads correspond to *cache misses*.

Utilizing graph and hypergraph partitioning for effective use of cache has
been studied by others before (e.g., [1,28]). However, to the best of our knowledge, our work is the first one that utilizes a multilevel acyclic DAG partitioning.
Acyclic partitioning allows us to develop a new scheduler, which will load each
part only once, and execute parts non-preemptively. Since the parts are computed to fit into fast memory, and they are acyclic, once all "incoming" edges of
the part are loaded, all the tasks in that part can be executed non-preemptively,
and without any need to bring additional data. Earlier studies that use undi-
rected partitioning, cannot guarantee such execution model.

The rest of the paper is organized as follows. In Sect. 2, we present the
computational model. Section 3 describes the proposed DAG-partitioning based
scheduling algorithms. Experimental evaluation of the proposed methods is displayed in Sect. 4. We end with a brief conclusion and future work in Sect. 5.

## 2    Model

We consider a directed acyclic graph (DAG) $G = (V, E)$, where the vertices
in set $V = \{v_1, \ldots, v_n\}$ represent computational tasks, and the dependency
among them, hence the communication of data between tasks, is captured by
graph edges in set $E$. Given $v_i \in V$, $pred_i = \{v_j \mid (v_j, v_i) \in E\}$ is the set of
predecessors of task $v_i$ in the graph, and $succ_i = \{v_j \mid (v_i, v_j) \in E\}$ is the set of
successors of task $v_i$.

Vertices and edges have weights representing the size of the (scratch) memory
required for task $v_i \in V$, $w_i$, and the size of the data that is communicated to
its successors. Here, we assume that $v_i$ produces a data of size $out_i$ that will be
communicated to all of its successors, hence the weight of each edge $(v_i, v_j) \in E$
will be $out_i$. An *entry* task $v_i \in V$ is a task with no predecessors (i.e., $pred_i = \emptyset$),
and such a task is a virtual task generating the initial data, hence $w_i = 0$. All
other tasks (with $pred_i \neq \emptyset$) cannot start until all predecessors have completed,
and the data from each predecessor has been generated. Here, we will use $in_i$
to represent the total input data size that task $v_i$ requires. For simplicity in the
presentation, we will use $w_i = 0$ and $out_i = 1$. Hence, the total input size of
task $v_i$ is $in_i = |pred_i|$, and $in_i = 0$ if task $v_i$ is an entry task.

We assume that the size of fast memory is $C$, and that the slow memory is large enough to hold the whole data. In order to compute task $v_i \in V$, the processor must access $in_i + w_i + out_i$ fast memory locations. Because of the limited fast memory, some computed values may need to be temporarily stored in slow memory and reloaded later.

Consider the example in Fig. 1. Task $v_1$ is the entry task and therefore just requires one cache location to generate its output data ($out_1 = 1$). The data corresponding to $out_1$ remains in cache while there is no need to evict it, while $in_1$ and $w_1$ will no longer be needed and can be evicted (but for an entry task, these are equal to 0). Consider that $v_2$ is executed next: It will need two cache locations ($in_2 = out_2 = 1$). Then, $v_3$ is executed. We want to keep $out_1$ in cache, if possible, since it will be reused later by $v_5$ and $v_7$. Hence, 3 memory locations are required. Similarly, when executing $v_4$, 4 memory locations are required ($out_2$ should also remain in cache). However, if the cache is too small, say $C = 3$, one will need to evict a data from the cache, hence resulting in a *cache miss*.

Given a traversal of the DAG, the *livesize* (live set size) is defined as the minimum cache size required for the execution so that there are no cache misses. In the example, with the traversal $v_1 \rightarrow v_2 \rightarrow v_3 \rightarrow v_4 \rightarrow v_5 \rightarrow v_6 \rightarrow v_7$, it would be 4. For another traversal however, the livesize may be smaller, and hence the order of traversal may greatly influence the number of cache misses. Consider the execution $v_1 \rightarrow v_7 \rightarrow v_2 \rightarrow v_5 \rightarrow v_6 \rightarrow v_3 \rightarrow v_4$: the livesize never exceeds 3 in this case, and this is the minimum cache size to execute this DAG, since task $v_6$ requires 3 cache locations to be executed.

*Parts, Cuts, Traversals, and Livesize.* We generalize and formalize the definition of the livesize introduced above. Consider an acyclic $k$-way partition $P = \{V_1, \ldots, V_k\}$ of the DAG $G = (V, E)$: the set of vertices $V$ is divided into $k$ disjoint subsets, or *parts*. There is a path between $V_i$ and $V_j$ ($V_i \rightsquigarrow V_j$) if and only if there is a path in $G$ between a vertex $v_i \in V_i$ and a vertex $v_j \in V_j$. The acyclic condition means that given any two parts $V_i$ and $V_j$, we cannot have $V_i \rightsquigarrow V_j$ and $V_j \rightsquigarrow V_i$. In the example of Fig. 1, an example of acyclic partition is $V_1 = \{v_1, v_2, v_3, v_4\}$ and $V_2 = \{v_5, v_6, v_7\}$.

Given a partition $P$ of the DAG, an edge is called a *cut edge* if its endpoints are in different parts. Let $E_{cut}(P)$ be the set of cut edges for this partition. The *edge cut* of a partition is defined as the sum of the costs of the cut edges, and can

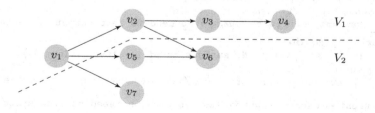

**Fig. 1.** A sample DAG and its acyclic partition.

be formalized as follows: $\mathtt{EdgeCut}(P) = \sum_{(v_i,v_j)\in E_{cut}(P)} out_i$, hence it is equal to $|E_{cut}(P)|$ with unit weights.

Let $V_i \subseteq V$ be a *part* of the DAG $(1 \leq i \leq k)$. A *traversal* of the part $V_i$, denoted by $\tau(V_i)$, is an ordered list of the vertices that respects precedence constraints within the part: if there is an edge $(v, v') \in E$, then $v$ must appear before $v'$ in the traversal. In the example, there is only a single possible traversal for $V_1$: $\tau(V_1) = [v_1, v_2, v_3, v_4]$, while there are three possibilities for $V_2$: $[v_5, v_6, v_7]$, $[v_5, v_7, v_6]$, and $[v_7, v_5, v_6]$.

Given a part $V_i$ and a traversal of this part $\tau(V_i)$, we can now define the livesize of the traversal as the maximum memory usage required to execute the whole part. We define $L(\tau(V_i))$ as the livesize computed such that inputs and outputs (of part $V_i$) are evicted from the cache if they are no longer required inside the part. Algorithm 1 details how the livesets are computed.

*Cache Eviction Algorithm.* If the livesize is greater than the cache size $C$, during execution, some data must be transferred from the cache back into slow memory. The data that will be evicted may affect the number of cache misses. Given a traversal, the optimal strategy consists in evicting the data whose next use will occur farthest in the future during execution. This strategy is called the optimal replacement algorithm OPT (clairvoyant replacement algorithm, also known as Belady's optimal page replacement algorithm) [3].

Hence, if a data needs to be evicted, the OPT algorithm looks at the upcoming schedule and orders the data in the cache by their next use times in ascending order. If a data is not used by any other task, it is assigned infinite value. The last data in this ordered list (the data not used or with farthest upcoming use) is then evicted from the cache. After scheduling a task, the next upcoming use of data generated by its immediate predecessors are updated.

---

**Algorithm 1.** Computing the livesize of a part

**Data:** Directed graph $G = (V, E)$, part $V_\ell \subseteq V$, traversal $\tau(V_\ell)$
**Result:** Livesize $L(\tau(V_\ell))$

1   $L(\tau(V_\ell)) \leftarrow 0$; $\mathtt{current} \leftarrow 0$; $V'_\ell \leftarrow V_\ell$;
2   **foreach** $v_i$ **in** $\tau(V_\ell)$ **order do**
3      $V'_\ell \leftarrow V'_\ell \setminus \{v_i\}$;
4      $\mathtt{current} \leftarrow \mathtt{current} + w_i + out_i$; /* Add current task to liveset     */
5      $L(\tau(V_\ell)) \leftarrow \max\{L(\tau(V_\ell)), \mathtt{current}\}$; /* Update $L$ if needed     */
6      **if** $succ_i = \emptyset$ **or** $\forall v_j \in succ_i, v_j \notin V_\ell$ **then**
7         $\lfloor$ $\mathtt{current} \leftarrow \mathtt{current} - out_i$; /* No need to keep output in cache   */
8      **for** $v_j \in pred_i$ **do**
9         **if** $V'_\ell \cap succ_j = \emptyset$ /* No successors of $v_j$ after $v_i$ in $\tau(V_\ell)$     */
10        **then**
11           $\lfloor$ $\mathtt{current} \leftarrow \mathtt{current} - out_j$; /* No need to keep $v_j$ in cache   */
12     $\mathtt{current} \leftarrow \mathtt{current} - w_i$; /* Task is done, no need scratch space    */

*Optimization Problems.* The goal is to minimize the number of loads and stores among all possible valid schedules. Formally, the MINCACHEMISS problem is the following: Given a DAG $G$, a cache of size $C$, find a topological order of $G$ that minimizes the number of cache misses when using the OPT strategy.

However, finding the optimal traversal to minimize the livesize is already an NP-complete problem [24], even though it is polynomial on trees [16]. Therefore, MINCACHEMISS is NP-complete (consider a problem instance where the whole DAG could be executed without any cache miss if the livesize was minimum).

Instead of looking for a global traversal of the whole graph, we propose to partition the DAG in an acyclic way. The key is, then, to have all the parts executable without cache misses, hence the only cache misses can be incurred by data on the cut between parts. Therefore, we aim at minimizing the edge cut of the partition.

# 3 DAG-Partitioning Assisted Locality-Aware Scheduling

We propose novel DAG-partitioning-based cache optimization heuristics that can be applied on top of classical ordering approaches to improve cache locality, using a modified version of a recent directed acyclic graph partitioner [12].

Three classical approaches are considered for the traversal of the whole DAG, and they return a total order on the tasks. The traversal must respect precedence constraints, and hence a task $v_i \in V$ is said to be *ready* when all its predecessors have already been executed: all predecessors must appear in the traversal before $v_i$.

- *Natural Ordering (Nat)* treats the node id's as the priority of the node, where the lower id has a higher priority, hence the traversal is $v_1 \rightarrow v_2 \rightarrow \cdots \rightarrow v_n$, except if node id's do not follow precedence constraints (schedule ready task of highest priority first).
- *DFS Traversal Ordering (DFS)* follows a depth-first traversal strategy among the ready tasks.
- *BFS Traversal Ordering (BFS)* follows a breadth-first traversal strategy among the ready tasks.

When applied to the whole DAG, these three traversal algorithms are baseline algorithms, and they will serve as a basis for comparison in terms of cache miss.

Also, these traversals can be applied to a part of the DAG in the DAG-partitioned approach, and hence they can be extended to parts themselves: part $V_i$ is ready if, for all $v_i \in V_i$, if $(v_j, v_i) \in E$ and $v_j \notin V_i$, then $v_j$ has already been scheduled. Furthermore, the part id is the minimum of ids of the nodes in the part. Then, considering parts as macro tasks, the same traversals can be used.

The modified acyclic DAG partitioner takes a DAG with vertex and edge weights (in this work, unit weights are considered), and a maximum livesize $L_m$ for each part as input. Its output is a partition of the vertices of $G$ into $K$ nonempty pairwise disjoint and collectively exhaustive parts satisfying three conditions: (i) the livesize of the parts are smaller than $L_m$; (ii) the edge cut is minimized; (iii) the partition is acyclic; in other words, the inter-part edges between the vertices from different parts should preserve an acyclic dependency structure among the parts. The partitioner does not take the number of parts as input, but takes a maximum livesize. That is, the partitioner continues recursive bisection until the livesize of the part at hand is less than $L_m$, meaning that it can be loaded and completely run in the cache of size $L_m$ without any extra load operations.

Hence, the DAGP approach (with DAG partitioner) uses multilevel recursive bisection approach, matching (coarsening) methods, and refinement methods that are specialized to create acyclic partitions. The refinement methods are modified versions of Fiduccia-Mattheyses' move-based refinement algorithm.

The partitioning algorithm starts by computing the livesize of the graph. If the computed livesize is greater than $L_m$, the graph is partitioned into two subgraphs with the edge cut minimization goal. The same procedure is repeated for the two subgraphs until each part has a livesize smaller than $L_m$. When the partitioning procedure is finished, i.e., the livesize is less than $L_m$ for each partition, there is a possible optimization opportunity to decrease the number of partitions. Some parts in these two subgraphs can be merged together and still have a small enough livesize. We try merging the last part of the first subgraph with the first part of the second subgraph. If the combined livesize of these two parts are still small enough, we update the part assignments for the nodes in the first part of the second subgraph. Then, we proceed to try merging the second part of second subgraph, and continue until all the parts in the second subgraph are merged, or the combined livesize exceeds $L_m$. This procedure is repeated for each recursive call to the function *recursive bisection*. We only try merging the last part in the first subgraph with the immediate next part (first part in the second partition), because selecting any other part might create a cycle in the part graph, and would require a costly check for the acyclicity.

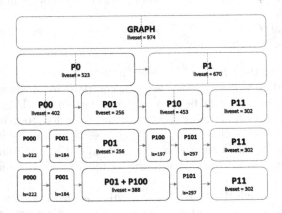

**Fig. 2.** A toy example showing a possible DAG partitioning procedure, with $L_m = 400$.

Figure 2 shows an example of partitioning with $L_m = 400$. The graph is partitioned into two and the livesize is computed for both subgraphs. Since both livesizes exceed $L_m$, both will be bisected. When P0 is completed, the recursive function is called for P1. In the fourth row, both P0 and P1 are done partitioning, and each part fits within $L_m$. During the recursive calls on P0 (resp. P1), we try to merge P001 with P01 (resp. P101 with P11), but these result in parts with a livesize greater than $L_m$, and hence these merges are not done. However, the last part of P0 (P01) can be merged with the first part of P1 (P100), resulting in a new part with a livesize still smaller than $L_m$. Finally, the merge of P01 + P100 with P101 exceeds $L_m$, hence the partitioning stops.

The partitioner also duplicates the boundary nodes of predecessor parts to successor parts, so as to account for the input data (caching) requirement of the tasks from predecessor parts.

Given $K$ parts $V_1, \ldots, V_K$ forming a partition of the DAG, we consider three variants of DAGP (DAGP-NAT, DAGP-DFS, and DAGP-BFS), building upon the traversals described above. Due to page limits, we always use the same algorithm for parts and for tasks within parts, but any combination would be possible (e.g., order parts with *Nat*, and then tasks within parts with *DFS*).

# 4    Experimental Evaluation

Experiments were conducted on a computer equipped with dual 2.1 GHz Xeon E5-2683 processors and 512 GB memory. We have performed an extensive evaluation of the proposed heuristics on DAG instances obtained from the matrices available in the SuiteSparse Matrix Collection (formerly known as the

**Table 1.** Graph instances from [6] and their livesizes.

| Graph | $|V|$ | $|E|$ | $\max_{in.deg}$ | $\max_{out.deg}$ | $L_{Nat}$ | $L_{DFS}$ | $L_{BFS}$ |
|---|---|---|---|---|---|---|---|
| 144 | 144,649 | 1,074,393 | 21 | 22 | 74,689 | 31,293 | 29,333 |
| 598a | 110,971 | 741,934 | 18 | 22 | 81,801 | 41,304 | 26,250 |
| caidaRouterLev | 192,244 | 609,066 | 321 | 1040 | 56,197 | 34,007 | 35,935 |
| coAuthorsCites | 227,320 | 814,134 | 95 | 1367 | 34,587 | 26,308 | 27,415 |
| delaunay-n17 | 131,072 | 393,176 | 12 | 14 | 32,752 | 39,839 | 52,882 |
| email-EuAll | 265,214 | 305,539 | 7,630 | 478 | 196,072 | 177,720 | 205,826 |
| fe-ocean | 143,437 | 409,593 | 4 | 4 | 8,322 | 7,099 | 3,716 |
| ford2 | 100,196 | 222,246 | 29 | 27 | 26,153 | 4,468 | 25,001 |
| halfb | 224,617 | 6,081,602 | 89 | 119 | 66,973 | 25,371 | 38,743 |
| luxembourg-osm | 114,599 | 119,666 | 4 | 5 | 4,686 | 2,768 | 6,544 |
| rgg-n-2-17-s0 | 131,072 | 728,753 | 18 | 19 | 759 | 1,484 | 1,544 |
| usroads | 129,164 | 165,435 | 4 | 5 | 297 | 8,024 | 9,789 |
| vsp-finan512 | 139,752 | 552,020 | 119 | 666 | 25,830 | 24,714 | 38,647 |
| vsp-mod2-pgp2 | 101,364 | 389,368 | 949 | 1726 | 41,191 | 36,902 | 36,672 |
| wave | 156,317 | 1,059,331 | 41 | 38 | 13,988 | 22,546 | 19,875 |

University of Florida Sparse Matrix Collection) [6]. From this collection, we picked 15 matrices satisfying the following properties: listed as binary, square, and has at least 100000 rows and at most $2^{26}$ nonzeros. All edges have unit costs, and all vertices have unit weights. For each such matrix, we took the strict upper triangular part as the associated DAG instance whenever this part has more nonzeros than the lower triangular part; otherwise, we took the lower triangular part. The graphs and their characteristics are listed in Table 1. We execute each graph with the *Nat*, *DFS*, and *BFS* traversals, and report the corresponding livesizes (computed with Algorithm 1).

Note that when reporting the cache miss counts, we do not include *compulsory (cold, first reference) misses*, the misses that occur at the first reference to a memory block, as these misses cannot be avoided.

We start by comparing the average performance of the three baseline traversal algorithms on varying cache sizes. Figure 3 shows the geometric mean of cache miss counts, normalized by the number of nodes, with cache size ranging from 512 to 10240 words. In smaller cache sizes, the natural ordering of the nodes provides the best results on average. As the cache size increases, *DFS* traversal surpasses the others and becomes the best option starting at 3072.

**Effect of Cache Size on Reported Relative Cache Miss.** Figure 4 shows the improvement of our algorithms over their respective baselines, averaged over 50 runs. The left figure shows the relative cache miss on a cache $C = 512$ and the right one on $C = 10240$, with $L_m = C$ (i.e., the partitioner stops when the livesize of each part fits in cache). A relative cache miss of 1 means that we get the same number of cache misses as without partitioning; the proposed solution is better than the baseline for a value lower than 1 (0 means that we reduced the cache misses to 0), and it is worsened in a few cases (i.e., values greater than 1).

One important takeaway from this figure is that as the cache increases, the input graph's livesize may become less than the cache size (e.g., fe_ocean, luxembourg-osm, rgg-n-2-17-s0, and usroads graphs have smaller livesize than the cache size), meaning that there is no cache miss even without partitioning. Thus, the partitioning phase does not need to divide the graph at all and just returns the initial graph.

Finally, the variance from one run to another is relatively low, demonstrating the stability of the algorithm, hence we perform the average over only 10 runs in the remaining experiments.

**Effect of $L_m$ and $C$ on Cache Miss Improvement and Edge Cut.** Figure 5 shows the normalized cache misses when the execution strategy of the graph is *Nat*, *DFS*, and *BFS* traversal respectively. We compare the number of cache misses when traversing the input graph with the three partitioning based-heuristics. We use five different values for the $L_m$ parameter of the algorithms: $L_m = \{2C, C, 0.5C, 0.25C, 0.125C\}$ (i.e., from twice the size of cache to down to one eighth of the cache size), and compare the results. Each bar in the chart shows the respective relative cache miss of partitioning-assisted ordering

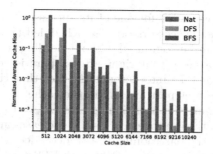

**Fig. 3.** Geometric mean of cache misses using *Nat*, *DFS*, and *BFS* traversals.

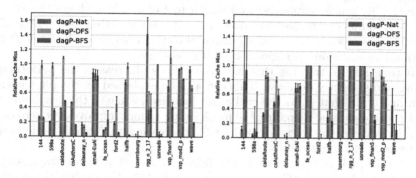

**Fig. 4.** Relative cache misses (geomean of average of 50 runs) for each graph separately (left cache size 512; right cache size 10240).

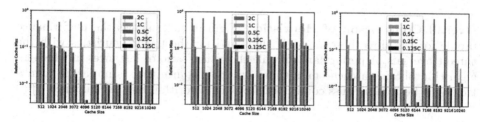

**Fig. 5.** Relative cache misses of DAGP-* with the given partition livesize for *Nat* (left), *DFS* (middle), and *BFS* (right) traversals.

compared to the baseline. Throughout all the bars in all the charts, we can see $L_m \leq C$ gives better results. This is expected since the partitioning phase is trying to decrease the livesize, which in turn, decreases the upper bound of cache misses, since the whole part can be executed without cache misses if $L_m \leq C$. Otherwise, for each part, we might still have cache misses.

There are two reasons why the partitioning result is not an exact cache miss count but an upper bound. First, the edge cut counts the edges more than once when a node $u$ from part $V_i$ is predecessor for multiple nodes in $V_j$. In cache, however, after node $u$ is loaded for one of its successors in $V_j$, ideally, it is not removed before its other successors are also scheduled ($L_m \leq C$ guarantees all the nodes in part $V_j$ can be computed without needing to remove node $u$ from cache). The second reason is that right after a part is completely scheduled, the partitioner does not take into account the fact that the last tasks (nodes) scheduled from the previous part are still in the cache.

On average, continuing the partitioning after $L_m = C$ usually improves the performance. However, the improvement as $L_m$ decreases and the increase in the complexity/runtime of the partitioning is a tradeoff decision.

Comparing the plots of the *Nat*, *DFS*, and *BFS* traversals, one can argue that the improvement over *DFS* ordering is not as high as for the *Nat* or *BFS* ordering. This can be explained by relating to Fig. 3, which shows the average cache miss counts for the baseline algorithms. It is obvious that, on average, *DFS* traversal gives much better results compared to the other two. Thus, DAGP-DFS variation has less room to improve over this baseline.

Another general trend is that as the cache size increases, the performance improvements slightly decrease. This can also be explained by the phenomenon shown in Fig. 4. That is, as the cache size increases, some of the graphs do not have any cache misses, therefore, the partitioning cannot improve over zero cache misses (relative cache miss is equal to 1), decreasing the overall performance improvement of the heuristics.

When we look at the values in the figure, we see that the average relative cache miss, for $L_m = 0.5 \times C$, is in the range 10x to 100x better than the baseline. In addition, relative cache miss of DAGP-BFS goes lower than $3 \cdot 10^{-3} = 0.003$ for $C = 6144$ and $L_m = 0.25C$, which is nearly 400x better than the baseline.

**Overall Comparison of Heuristics.** Figure 6 (left) shows the performance profile for the three baseline algorithms and for the DAGP heuristics with the three traversals. A performance profile shows the ratio of instances in which an algorithm obtains a cache miss count on the instance that is no larger than $\theta$ times the best cache miss count found by any algorithm for that instance [7].

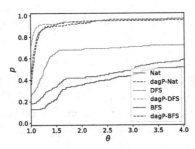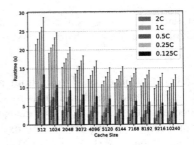

**Fig. 6.** (Left) Performance profile comparing baselines and heuristics with $L_m = 0.5 \times C$. (Right) Average runtime of all graphs for DAGP-DFS partitioning.

Here, we compare the heuristics using $L_m = 0.5 \times C$, since it constantly provides better results than $L_m = C$, and the increase in the partitioning overhead is minimal. We can see that DAGP-DFS gives the best ordering approximately 75% of the time; and 90% of the time, it gives cache misses no worse than 1.5 times the best heuristic. We can also see that all three proposed DAGP heuristics perform better over their respective baselines. Also, all three heuristics perform better than any of the baselines.

Finally, the runtime averages, including all graphs in the dataset for the given parameter configurations, are depicted in Fig. 6 (right). The black error bars show the minimum and maximum run times for that bar. It is clear that for smaller $L_m$ values, the partitioner needs to do more work, but partitioner is fast and does not take more than 10 s in most common cases (i.e., $L_m = 0.5 \times C$).

## 5  Conclusion and Future Work

As the cost of data movement through the memory hierarchy dominates the cost of computational operations in today's computer systems, data locality is a significant research focus. Although there have been many approaches to improve data locality of applications, to the best of our knowledge, there is no work that employs a DAG-partitioning assisted approach. Building upon such a partitioner, we design locality-aware scheduling strategies, and evaluate the proposed algorithms extensively on a graph dataset from various areas and applications, demonstrating that we can significantly reduce the number of cache misses. As the next step, it would be interesting to study the effect of a customized DAG-partitioner specifically for cache optimization purposes, and also to design traversal algorithms to optimize cache misses. It would also be interesting to use a better fitting hypergraph representation for the model.

## References

1. Akbudak, K., Kayaaslan, E., Aykanat, C.: Hypergraph partitioning based models and methods for exploiting cache locality in sparse matrix-vector multiplication. SIAM J. Sci. Comput. **35**(3), C237–C262 (2013)
2. Aktulga, H.M., Buluç, A., Williams, S., Yang, C.: Optimizing sparse matrix-multiple vectors multiplication for nuclear configuration interaction calculations. In 2014 IEEE 28th International Parallel and Distributed Processing Symposium, pp. 1213–1222, May 2014
3. Belady, L.A.: A study of replacement algorithms for a virtual-storage computer. IBM Syst. J. **5**(2), 78–101 (1966)
4. Bramas, B.: Impact study of data locality on task-based applications through the Heteroprio scheduler. PeerJ. Comput. Sci. **5**, e190 (2019)
5. Chen, D., Liu, F., Ding, C., Pai, S.: Locality analysis through static parallel sampling. In: Proceedings of the 39th ACM SIGPLAN Conference on PLDI, pp. 557–570. ACM, New York (2018)
6. Davis, T.A., Hu, Y.: The University of Florida sparse matrix collection. ACM Trans. Math. Softw. **38**(1), 1:1–1:25 (2011)

7. Dolan, E.D., Moré, J.J.: Benchmarking optimization software with performance profiles. Math. Program. **91**(2), 201–213 (2002)
8. Fauzia, N., et al.: Beyond reuse distance analysis: dynamic analysis for characterization of data locality potential. ACM Trans. Archit. Code Optim. **10**(4), 53:1–53:29 (2013)
9. Filippone, S., Cardellini, V., Barbieri, D., Fanfarillo, A.: Sparse matrix-vector multiplication on GPGPUs. ACM Trans. Math. Softw. **43**(4), 30:1–30:49 (2017)
10. Fuller, S.H., Millett, L.I.: The Future of Computing Performance: Game Over or Next Level? National Academy Press, Washington DC (2011)
11. Gustedt, J., Jeannot, E., Mansouri, F.: Optimizing locality by topology-aware placement for a task based programming model. In: 2016 IEEE International Conference on Cluster Computing (CLUSTER), pp. 164–165, September 2016
12. Herrmann, J., Özkaya, M.Y., Uçar, B., Kaya, K., Çatalyürek, Ü.V.: Multilevel algorithms for acyclic partitioning of directed acyclic graphs. SIAM J. Sci. Comput. (SISC) **41**(4), A2117–A2145 (2019)
13. Hollman, D.S., Bennett, J.C., Kolla, H., Lifflander, J., Slattengren, N., Wilke, J.: Metaprogramming-enabled parallel execution of apparently sequential C++ code. In: 2016 Second International Workshop on Extreme Scale Programming Models and Middleware (ESPM2), pp. 24–31, November 2016
14. Hong, C., Sukumaran-Rajam, A., Nisa, I., Singh, K., Sadayappan, P.: Adaptive sparse tiling for sparse matrix multiplication. In: Proceedings of the 24th Symposium on Principles and Practice of Parallel Programming, PPoPP 2019, pp. 300–314. ACM, New York (2019)
15. Hu, X., Wang, X., Zhou, L., Luo, Y., Ding, C., Wang, Z.: Kinetic modeling of data eviction in cache. In: 2016 USENIX Annual Technical Conference (USENIX ATC 2016), Denver, CO, pp. 351–364. USENIX Association (2016)
16. Jacquelin, M., Marchal, L., Robert, Y., Uçar, B.: On optimal tree traversals for sparse matrix factorization. In: IPDPS 2011, pp. 556–567 (2011)
17. Jin, J., et al.: A data-locality-aware task scheduler for distributed social graph queries. Futur. Gener. Comput. Syst. **93**, 1010–1022 (2019)
18. Kiani, M., Rajabzadeh, A.: Analyzing data locality in GPU kernels using memory footprint analysis. Simul. Model. Pract. Theory **91**, 102–122 (2019)
19. Mayer, R., Mayer, C., Laich, L.: The tensorflow partitioning and scheduling problem: it's the critical path! In: Proceedings of the 1st Workshop on Distributed Infrastructures for Deep Learning, DIDL 2017, pp. 1–6. ACM, New York (2017)
20. Naik, N.S., Negi, A., Tapas Bapu, B.R., Anitha, R.: A data locality based scheduler to enhance MapReduce performance in heterogeneous environments. Future Gener. Comput. Syst. **90**, 423–434 (2019)
21. Ng, A.Y., Jordan, M.I., Weiss, Y.: On spectral clustering: analysis and an algorithm. In: Advances in Neural Information Processing System, pp. 849–856 (2002)
22. Page, L., Brin, S., Motwani, R., Winograd, T.: The PageRank citation ranking: Bringing order to the web. Technical report, Stanford InfoLab (1999)
23. Rogers, S., Tabkhi, H.: Locality aware memory assignment and tiling. In: 2018 55th ACM/ESDA/IEEE Design Automation Conference (DAC), June 2018
24. Sethi, R.: Complete register allocation problems. In: Proceedings of the 5th Annual ACM Symposium on Theory of Computing (STOC 1973), pp. 182–195 (1973)
25. Shalf, J., Dosanjh, S., Morrison, J.: Exascale computing technology challenges. In: Palma, J.M.L.M., Daydé, M., Marques, O., Lopes, J.C. (eds.) VECPAR 2010. LNCS, vol. 6449, pp. 1–25. Springer, Heidelberg (2011). https://doi.org/10.1007/978-3-642-19328-6_1

26. Wold, S., Esbensen, K., Geladi, P.: Principal component analysis. Chemometr. Intell. Lab. Syst. **2**(1–3), 37–52 (1987)
27. Xie, B., et al.: CVR: Efficient vectorization of SpMV on x86 processors. In: Proceedings of the 2018 International Symposium on Code Generation and Optimization, CGO 2018, pp. 149–162. ACM (2018)
28. Yzelman, A.N., Bisseling, R.H.: Two-dimensional cache-oblivious sparse matrix-vector multiplication. Parallel Comput. **37**(12), 806–819 (2011)

# Isoefficiency Maps for Divisible Computations in Hierarchical Memory Systems

Maciej Drozdowski[1]([✉])(iD), Gaurav Singh[2], and Jędrzej M. Marszałkowski[1]

[1] Institute of Computing Science, Poznań University of Technology, Poznań, Poland
{Maciej.Drozdowski,Jedrzej.Marszalkowski}@cs.put.poznan.pl
[2] Technology Strategy and Innovation, BHP, Perth 6000, Australia
Gaurav.Singh@bhp.com

**Abstract.** In this paper we analyze impact of memory hierarchy on divisible load processing. Current computer systems have hierarchical memory systems. The core memory is fast but small, the external memory is large but slow. It is possible to avoid using external memory by parallelizing computations or by processing smaller work chunks sequentially. We will analyze how a combination of these two options improves efficiency of the computations. For this purpose divisible load theory representing data-parallel applications is used. A mathematical model for scheduling divisible computations is formulated as a mixed integer linear program. The model allows for executing multiple load installments sequentially or in parallel. The efficiency of the schedule is analyzed with respect to the impact of load size and machines number. The results are visualized as isoefficiency maps.

**Keywords:** Performance evaluation and prediction · Hierarchical memory · Divisible load theory · Isoefficiency maps

## 1 Introduction

Current computer systems have hierarchical memory. At the top of the hierarchy, CPU registers are the fastest form of computer memory, but also available in the smallest amount. Further memory hierarchy levels include CPU caches, core memory commonly referred to as RAM, networked caches, SSDs, HDDs, tapes, etc. When going down the hierarchy, memory size increases but speed of data transfers decreases. The part of the hierarchy from CPU registers to RAM is managed by hardware. The lower levels are controlled by software stacks [16]. Consequently, computations exploiting external storage, e.g. in the form of virtual memory, can be by two orders of magnitude slower than using only the upper part of memory hierarchy [9,14]. Therefore, for the purposes of this study, we will be distinguishing only two memory levels: CPU registers, caches and RAM by convention referred to as core (or main) memory, and out-of-core memory including all kinds of external storage.

© Springer Nature Switzerland AG 2020
R. Wyrzykowski et al. (Eds.): PPAM 2019, LNCS 12043, pp. 224–234, 2020.
https://doi.org/10.1007/978-3-030-43229-4_20

One of the key applications of parallel systems is processing big volumes of data. Core memory size is a limitation in processing big volumes of data. Memory footprint can be reduced in at least two ways: (i) by parallelizing computations, and hence, reducing processor individual load shares, (ii) by multi-installment processing, i.e. dividing the load into multiple chunks and processing them sequentially. There is a need for understanding how the number of machines, number of installments, memory size, and problem size interplay in determining performance of processing big volumes of data. Thus, our goal in this paper is to analyze the impact of various system parameters on performance of processing big volumes of data in parallel. Most of the earlier approaches to performance analysis focus on scalability with machine number (speedup) but at fixed problem size, while the problem size vs machine number interrelationship is ignored. Our goal is to allow easily grasping the system parameter interrelationships, and in particular, between machine number and problem size. For this purposes divisible load theory (DLT) will provide mathematical model of data parallel application and isoefficiency maps will be used as a visual front-end.

Divisible load theory is a scheduling and performance model of data-parallel applications typical for big volumes of data. In the DLT it is assumed that the computation consists in processing large amount of similar data units, conventionally referred to as load. The load can be partitioned between remote computers and processed in parallel. The individual data units are small enough in relation to the total load size so it can be assumed, without great loss of accuracy, that the load is arbitrarily divisible. Accuracy of DLT in predicting performance and scheduling data-parallel applications such as text, image and video processing, file compression, linear algebra, sorting has been reported in [1,11,14]. Further introduction to the DLT can be found in surveys [4,5,10,15]. The first DLT paper assuming flat memory model was [13]. A heuristic load distribution method has been proposed. In [9] hierarchical memory system has been analyzed and single-installment load distribution has been found by use of a linear program. In [3] data gathering networks with limited memory are studied. Time-energy trade-offs in divisible load processing have been studied in [6,14]. A single-installment load distribution has be found by linear programming formulation. It was assumed that all available machines were always activate (powered on). In [6] the single-installment schedules built by the linear program have been compared with heuristic load distribution methods derived from loop scheduling (see e.g. [5] for loop scheduling). Isoefficiency maps borrow the performance visualization concept from other types of iso-lines (isotherms, isobars, isogones). Such visualizations proved very effective in building understanding of sensitivities and relationships of complex phenomena in other areas of science and technology.

The contributions of this paper can be summarized as follows: (i) optimum divisible load scheduling formulation is proposed for systems with both hierarchical memory system, and multi-installment load distribution, while the individual machines are powered on only if needed; (ii) isoefficiency maps are proposed to analyze the impact of memory hierarchy on divisible load processing; (iii) computationally hard mixed integer linear program model is used as performance

oracle. Further organization of this paper is the following. In the next section
the model of parallel computation is introduced, and the problem of planning
optimum schedule for the computations is formulated as a mixed integer linear program. In Sect. 3 the method of drawing isoefficiency maps is explained.
Section 4 is dedicated to a study of isoefficiency maps for divisible loads processing. We conclude in the last section.

## 2     Mathematical Model of Parallel Application

We will be assuming that execution of the data-parallel application is initiated
by a root processor (a.k.a. master, originator) which starts worker machines,
schedules communications and distributes the load. Computing environment
is homogeneous and comprises $m$ identical machines $M_1, \ldots, M_m$. The system
interconnect is equivalent to a star (single level tree) and the originator com-
municates directly with worker processors. The machine starting process lasts
for $S$ time units and may include, e.g., cold-start, start from certain suspension
mode, loading the application code and initializing data structures. A machine
starting message is short and the delay it induces in the communication system
may be neglected. Load of total size $V$ is distributed to the worker processors in
installments (messages, load chunks). Sending a load chunk of size $\alpha$ takes time
$O + C\alpha$, where $O$ is a fixed delay required to start the communication and $C$ is
communication rate (in seconds per byte). Only after receiving the whole load
chunk can the worker machine start processing the load. A machine may receive
more than one load chunk, but only after finishing computations on the previous
one. Let $n$ be the total number of load chunks distributed by the originator.

It has been experimentally demonstrated [9,14] that the time of processing
load of size $\alpha$ in a system with two memory levels can be represented by function

$$\tau(\alpha) = \max\{a_1\alpha, a_2\alpha + b_2\} \tag{1}$$

where $a_1$ is rate (seconds per byte) of processing in-core, $a_2, b_2$ are parameters of
linear time function for computing out of core. Function $\tau(\alpha)$ has properties: (i)
$\tau(0) = 0$, (ii) $0 < a_1 < a_2$, $b_2 < 0$, (iii) $\tau(\rho) = a_1\rho = a_2\rho + b_2$. The third property
means that the two linear segments of $\tau$ in Eq. (1) intersect at load size $\rho$ which is
the core memory size available to application. The process of collecting results is
not explicitly scheduled because, e.g., the size of results is small and their transfer
time is very short, or the results are stored on the worker machines for further
processing. The optimum schedule of the computations requires determining: (i)
where to send the load (i.e. the sequence of load distributions to the processors),
(ii) when to send the load, (iii) sizes of the sent load chunks.

Let $x_{ij}$ be a binary variable equal to 1 if load chunk $j$ is sent to machine
$M_i$ and equal to 0 otherwise. We will denote by $\alpha_{ij}$ the size of load chunk $j$
sent to processor $i$. If the chunk is sent to some other processor, then $\alpha_{ij} = 0$.
The moment when sending chunk $j$ begins will by denoted by $t_j$. Let $T$ be the
length of the schedule (makespan). We will use auxiliary variables $q_{ij} = t_j x_{ij}$

and $\tau_{ij} = \max\{a_1\alpha_{ij}, a_2\alpha_{ij} + b_2\}$. The problem of constructing an optimum computation schedule can be formulated as mixed integer linear program (MIP):

$$\text{minimize } T \tag{2}$$

subject to:

$$t_j + C\sum_{i=1}^{m}\alpha_{ij} + O \le t_{j+1} \quad j = 1, \ldots, n \tag{3}$$

$$q_{ij} + C\alpha_{ij} + Ox_{ij} + \tau_{ij} \le T$$
$$j = 1, \ldots, n \quad i = 1, \ldots, m \tag{4}$$

$$q_{ij} + C\alpha_{ij} + Ox_{ij} + \tau_{ij} \le q_{il} + (1 - x_{il})Z$$
$$i = 1, \ldots, m \quad j = 1, \ldots, n - 1 \quad l = j + 1, \ldots, n \tag{5}$$

$$S\sum_{i=1}^{m} x_{ij} \le t_j \quad j = 1, \ldots, n \tag{6}$$

$$\sum_{i=1}^{m}\sum_{j=1}^{n}\alpha_{ij} \ge V \tag{7}$$

$$\alpha_{ij} \le Vx_{ij} \quad i = 1, \ldots, m \quad j = 1, \ldots, n \tag{8}$$

$$\sum_{i=1}^{m} x_{ij} = 1 \quad j = 1, \ldots, n \tag{9}$$

$$Zx_{ij} \ge q_{ij} \ge 0$$
$$t_j \ge q_{ij} \ge t_j - Z(1 - x_{ij}) \tag{10}$$
$$i = 1, \ldots, m \quad j = 1, \ldots, n$$

$$a_1\alpha_{ij} + Zu_{ij} \ge \tau_{ij} \ge a_1\alpha_{ij}$$
$$a_2\alpha_{ij} + b_2 + Z(1 - u_{ij}) \ge \tau_{ij} \ge a_2\alpha_{ij} + b_2 \tag{11}$$
$$i = 1, \ldots, m \quad j = 1, \ldots, n$$

In the above formulation $x_{ij}, \alpha_{ij}, q_{ij}, t_j, T, \tau_{ij}, u_{ij}$ are decision variables. $C, O, S,$ $V, a_1, a_2, b_2, m, n$ are constants defined in the parallel application, while $Z$ is a large number defined in the above MIP. Decision variables $x_{ij}$ determine the sequence of communications and any $n$-message sequence to the $m$ processors can be constructed. The purpose of constraint (3) is to guarantee that the $j$th message fits in interval $[t_j, t_{j+1}]$ and messages do not overlap in the communication channel. Inequalities (4) ensure that computations finish before the end of the schedule. Constraints (5) establish that if load chunks $j, l$ are sent to processor $i$, then there is enough time to receive the $j$th chunk and process it before receiving the $l$th chunk starts. By (6) the processor which is receiving the $j$th load chunk is already started when sending the chunk begins. Inequality (7) guarantees that the whole load is processed. Constraint (8) ensures that a processor that is not receiving the $j$th load chunks also receives no load in the

$j$th communication. By (9) only one machine can receive the $j$th load chunk. Inequalities (10) ensure that the auxiliary variable $q_{ij}$ is equal to $t_j x_{ij}$. Using product $t_j x_{ij}$ directly is not allowed because the formulation would become a quadratic mathematical program. Yet, it is possible to obtain the same value by use of linearizing constraints (10) and an additional variable $q_{ij}$. Inequalities (11) guarantee that $\tau_{ij} = \max\{a_1 \alpha_{ij}, a_2 \alpha_{ij} + b_2\}$. The trigger binary variable $u_{ij} = 0$ determines whether the first $(a_1 \alpha_{ij})$ or the second component $(a_2 \alpha_{ij} + b_2)$ in the max is active.

## 3   Isoefficiency Map Construction

In this section we introduce the concept of isoefficiency map, and then explain how such maps can be constructed for processing divisible loads in hierarchical memory systems. Performance of parallel computations is measured by two classic metrics: speedup $\mathcal{S}$ and efficiency $\mathcal{E}$:

$$S(m) = \frac{T(1)}{T(m)} \qquad \mathcal{E}(m) = \frac{S}{m} = \frac{T(1)}{mT(m)}, \qquad (12)$$

where $T(i)$ is execution time on $i$ machines. Speedup and efficiency measure scalability of the parallel application. $\mathcal{E}$ is often interpreted as the fraction of the processor set which really computes. In a well-designed application $\mathcal{S}$ should grow (preferably linearly) with $m$ and $\mathcal{E}$ should be as close to 1 as possible. However, in most cases speedup saturates at certain number of machines and efficiency decreases with $m$. The location of the maximum speedup depends on the size of the solved problem. Bigger problems allow to exploit more processors while preserving certain efficiency level. In order to grasp this relationship a concept of isoefficiency function has been introduced [12]. Isoefficiency function $I(e, m)$ is size of the problem required to maintain efficiency $\mathcal{E}(m) = e$. Consider an example of finding a minimum spanning tree in a graph with $n$ vertices. A straightforward parallel version of Prim's algorithm for this problem, has complexity $T(m) = c_1 n^2/m + c_2 n \log m$, where $c_1, c_2$ are constants (see e.g. [2], Section 10.6). Efficiency of this algorithm is $\mathcal{E}(m) = c_1 n^2/(c_1 n^2 + c_2 nm \log m)$. Hence, isoefficiency function for $m$ machines and efficiency level $e < 1$ is $I(e, m) = c_2 em \log m/(c_1(1 - e))$. For a fixed value of $e$, function $I(e, m)$ can be viewed as a line of equal efficiency in the $m \times n$ space. Such a line of equal efficiency will be called an *isoefficiency line*. Thus, performance of parallel computations can be visualized as a set of isoefficiency lines in $m \times problem\,size$ space. Such a visualization will be called an *isoefficiency map* of a parallel computation. The idea of isoefficiency maps has been extended to other pairs of parallel computation efficiency parameters, such as speed of communication, speed of computation, etc. [7,8]. Such visualizations are useful in comprehending the phenomena limiting performance of parallel processing.

Schedule length $T$ calculated by solving (2)–(11) can be used in performance evaluation of data-parallel applications. Let $T(m, n, V)$ denote the value of $T$

obtained for a particular number of machines $m$, communications $n$, and problem size $V$. The time of processing the same amount of load on a single machine is $T(1,1,V) = S + O + CV + \max\{a_1V, a_2V + b_2\}$. Thus, efficiency of the computation is $\mathcal{E}(m,n,V) = T(1,1,V)/(mT(m,n,V))$. The isoefficiency function for a given value of efficiency $e$ can be defined as $I(e,m,n) = \{V : \mathcal{E}(m,n,V) = e\}$. Function $I(e,m,n)$ allows to draw one *isoefficiency line*, i.e. a line of efficiency $e$ in the $m \times V$ space. The isoefficiency line depicts how problem size $V$ should grow in order to maintain equal efficiency $e$ with changing number of machines $m$.

Due to the complex nature of the formulation (2)–(11) it is not possible to derive a closed-form formula of $I(e,m,n)$. Therefore, $I(e,m,n)$ has been found numerically, using the following approach: It has been established that for fixed $m, n$, efficiency function $\mathcal{E}(m,n,V)$ has a single maximum $\mathcal{E}_{\max}(m,n)$ at load size $V_{\max}(m,n)$ and is monotonous on both sides of $V_{\max}(m,n)$. This will be further discussed in Sect. 4. A bisection search method has been used to find load sizes $V < V_{\max}(m,n)$ for which certain efficiency level $e < \mathcal{E}_{\max}(m,n)$ is achieved. Precisely, for a probe value $V$ times $T(1,1,V)$ and $T(m,n,V)$ were calculated and if the resulting efficiency satisfied $T(1,1,V)/(mT(m,n,V)) < e$ then the probe load size was increased, respectively decreased in the opposite case. Analogous method has been applied to calculate $I(e,m,n)$ for load sizes greater than $V_{\max}(m,n)$. The values of $V_{\max}(m,n)$ and $\mathcal{E}_{\max}(m,n)$ have been found by a modification of the bisection method: Efficiency has been calculated for two probe values $V_1, V_2$ in some tested interval. Then the load size interval has been narrowed to $V_1$ or $V_2$, whichever resulted in the smaller efficiency. Both in the bisection search and in the search for the maximum efficiency the procedures have been stopped if the width of the searched $V$ intervals dropped below 1 MB.

As the MIP solver Gurobi 7.5.2 has been used. Observe that MIP is an **NP**-hard problem, and in the worst-case MIP solvers run in exponential time in the number of variables. In order to obtain solutions in acceptable time, the MIP solver run times have been limited to 300 s on 6 CPU threads. Consequently, the obtained solutions mostly were not guaranteed optimum. Still, the solutions are always feasible and can be considered as approximations of the optimum solutions of (2)–(11).

## 4   Performance Modeling

In this section we present isoefficiency maps and discuss the performance phenomena they show. Unless stated to be otherwise the reference instance parameters were: for the computing time function $\tau(\alpha)$ : $a_1 = 0.109\,\text{s/MB}$, $a_2 = 4.132\,\text{s/MB}$, $b_2 = -27109\,\text{s}$, for the communication delays $C = 5\,\text{ms/MB}$, $O = 75\,\text{ms}$, machine startup time $S = 25.4\,\text{s}$, and a limit of $n = 20$ load chunks. The $a_1, a_2, b_2$ parameters correspond with usable RAM size $\rho = 6739\,\text{MB}$. Since these parameters are machine- and application-dependent and can vary widely (cf. [7,14]), we will concentrate on the frequent qualitative phenomena rather than on particular performance numbers. We will also attempt analytically explaining the relationships behind the observed shapes of the isoefficiency lines.

**Table 1.** Maximum efficiency and corresponding load sizes vs $m$ at $n = 20$ installments

| $m$ | | 2 | 3 | 4 | 5 | 6 | 7 | 8 | 9 | 10 |
|---|---|---|---|---|---|---|---|---|---|---|
| $\mathcal{E}_{\max}(m, n)$ | | 34.2 | 34.0 | 33.7 | 33.4 | 33.2 | 32.8 | 32.4 | 31.5 | 30.9 |
| $V_{\max}(m, n)$ [MB] | | 134485 | 120827 | 123329 | 105097 | 119663 | 95748 | 106921 | 115514 | 75048 |
| $m$ | 11 | 12 | 13 | 14 | 15 | 16 | 17 | 18 | 19 | 20 |
| $\mathcal{E}_{\max}(m, n)$ | 30.4 | 29.6 | 28.7 | 27.7 | 26.7 | 25.7 | 24.8 | 23.8 | 22.9 | 22.1 |
| $V_{\max}(m, n)$ [MB] | 81394 | 86545 | 90457 | 94532 | 98591 | 99460 | 100506 | 100354 | 99708 | 102240 |

In Fig. 1 isoefficiency map for the load sizes smaller than $V_{\max}(m, n)$ is shown, and in Fig. 2 for the loads above $V_{\max}(m, n)$. For better clarity, maximum values of efficiency $\mathcal{E}_{\max}(m, n)$, and the corresponding load sizes $V_{\max}(m, n)$ are shown in Table 1. The line of maximum efficiency $\mathcal{E}_{\max}(m, n)$ is denoted as MAX in Figs. 1, 2 and the isolines are labeled with their efficiency levels. The efficiency for $m = 1$ is always 1, and no isolines for $m = 1$ are shown. Note that $m$, shown along the horizontal axis, is a discrete variable and consequently the isolines are step functions. It can be observed in both figures that efficiencies greater than 1 (consequently also super-linear speedups) are possible. Though such situation is rare in typical parallel applications, it is not unusual in the context of memory hierarchies. If only one machine is used (as in the calculation of $T(1, 1, V)$) then for $V > \rho$ the processing rate tends asymptotically to $a_2$. Conversely, if the load is distributed between many processors then it can be processed in-core with rate $a_1$. In our case $a_2/a_1 \approx 37.9$ and efficiency levels close to 37 can be expected. The values in Table 1 are slightly smaller than $a_2/a_1$ which is a result of communication delays and machine startup times. The $\mathcal{E}_{\max}(m, n)$ line shows problem sizes $V$ which achieve the best balance between the advantage of processing load in-core over single machine out-of-core processing, and the costs of starting the machines, communicating and avoiding idle time. MIP (2)–(11) is a discrete optimization problem and, e.g., there are fixed overheads $S, O$ which can be switched on and off by the choice of the communication sequence. Furthermore, the best communication sequences are not always repetitive patterns. Consequently $\mathcal{E}_{\max}(m, n)$ is neither smooth nor does it show an obvious trend.

Let us consider the part of the isoefficiency map for problem sizes smaller than $V_{\max}(m, n)$ shown in Fig. 1. For such load sizes machines in set $M_1, \ldots, M_m$ compute in-core, but if the same load were processed on just one machine then the load may spill to out-of-core. As it is not possible to derive a closed-form formula of the (2)–(11) solution, we will analyze range of $\mathcal{E}(m, n, V)$. The efficiency in this part of the isoefficiency map can be bounded in the ensuing range:

$$\frac{S + O + CV + a'V}{mS + nmO + mCV + a_1V} \leq \mathcal{E}(m, n, V) \leq \frac{S + O + CV + a'V}{mS + mO + a_1V}. \tag{13}$$

In the numerator of (13) $a_1 \leq a' \leq a_2$ is an equivalent rate of processing on one machine. Product $mT(m, n, V)$ in denominator of (12) can be interpreted as area in $time \times m$ space which is easier to assess than the schedule length $T(m, n, V)$. The area of $mT(m, n, V)$ in (13) is bounded from below by $mS + mO + a_1V$ which

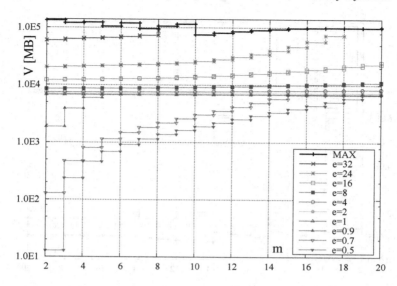

**Fig. 1.** Isoefficiency map for the load sizes $V$ below maximum efficiency. Logarithmic vertical axis.

is total machine startup time $mS$, minimum fixed overhead of communications $mO$ and total work of the computations in core $a_1V$. For the upper bound of the area, $mnO + mCV$ is an upper bound of machine waiting during the communications. It can be verified that both bounds of $\mathcal{E}(m, n, V)$ are increasing with $V$, and value of $V$ derived from the bound formulas increases with $m$ for fixed efficiency $e$. Indeed, it can be seen in Fig. 1 that problem sizes $V$ must grow with the number of machines $m$ to maintain some fixed level of efficiency. The isolines grow slightly faster than linearly with $m$ because the total processor waiting time in the actual solutions increases faster than linearly with $m$. One more peculiarity can be seen in Fig. 1: around $V = \rho = 6739$ MB a bunch of isolines coalesce. This is a result of using out-of-core memory in $T(1, 1, V)$ used in the efficiency formula. The single reference machine starts to use out-of-core memory at $V \approx \rho$, which extremely expands $T(1, 1, V)$ and problem size expressed by $I(e, m, n)$ has to increase only marginally to attain the required efficiency level. Consider, e.g., the upper bound of (13). The size of the load required to attain efficiency $e$ is $V = (S + O)(em - 1)/(C + a' - a_1)$. When $m$ grows also $V$ grows, but the single machine must use out-of-core memory and $a'$ tends to $a_2 \gg a_1$. As a result, the increase in the numerator $(S + O)(em - 1)$ is intensively suppressed by $a'$ growing in the denominator $(C + a' - a_1)$. Hence, isoline coalescing at $V \approx \rho$ can be observed.

In the part of the isoefficiency map above $V_{\max}(m, n)$ (see Fig. 2) the single reference machine considered in $T(1, 1, V)$ uses out-of-core memory while machines $M_1, \ldots, M_m$ use out-of-core memory at least partially. In the dominating pattern of load distribution some part of the load is processed in load

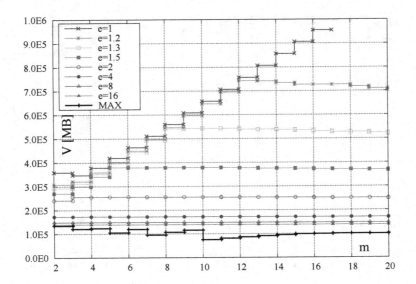

**Fig. 2.** Isoefficiency map for the load sizes $V$ above maximum efficiency.

chunks of RAM $\rho$ size while the remaining load is distributed to the machines in roughly equal sizes and processed out-of-core. Thus, for $n$ installments and $m$ machines, $n - m \geq 0$ load chunks have nearly RAM size, and the remaining chunks have size roughly $[V - (n - m)\rho]/m$. This load partitioning is intuitively effective because load as big as possible is processed in RAM, while the remaining load processed out-of-core on each machine is as small as possible. This load partitioning pattern results in the following efficiency formula

$$\mathcal{E}(m, n, V) \approx \frac{S + O + V(C + a_2) + b_2}{mS + nO + CV + (n - m)\rho a_1 + m[(V - (n - m)\rho)/ma_2 + b_2]}. \quad (14)$$

In the denominator of (14) area $mT(m, n, V)$ is calculated. It is assumed that data transfers to one machine overlap with other machines computations (latency hiding), and consequently, only $CV$ area is used on communications in all machines. $(n - m)\rho a_1$ is the area of computing in core, and $m[(V - (n - m)\rho)/ma_2 + b_2]$ out-of-core. From (14) estimation of the isoefficiency function can be derived:

$$I(e, m, n) \approx \frac{b_2(en - 1) + S(em - 1) + O(en - 1)}{(C + a_2)(1 - e)}. \quad (15)$$

In the derivation of (15) property (iii) $\rho = b_2/(a_1 - a_2)$ of (1) has been used. Note that $b_2 < 0, e > 1, n > m, |b_2| \gg S \gg O$, and $I(e, m, n) > 0$. The load size necessary for certain efficiency $e$ is almost independent of the number of machines $m$ in (15). Thus, (15) represents well the lower-right part of Fig. 2 where isoefficiency lines are nearly parallel to the horizontal axis. The top-left part of Fig. 2 can be considered an artifact. Note that with growing $V$ the time of processing load out-of-core dominates in the computation time. As a result efficiency tends to $(Va_2 + b_2)/(m[V/ma_2 + b_2]) \approx 1$ with growing $V$ and it is not possible to obtain schedules with efficiency significantly smaller than 1 without introducing artificial idle time. In other words, to construct a schedule with low efficiency, overheads are 'necessary' in the denominator of the efficiency equation like in (14). Yet, with decreasing $m$ the amount of the overheads decreases in relation to growing out-of core computation cost, and it is becoming impossible to build a schedule with some low efficiency level unless some idle time is added. Consequently, we do not show isoefficiency lines for $e < 1$ in Fig. 2 because the corresponding schedules require inserting unneeded idle time.

## 5  Conclusions

In this paper we studied time performance of divisible computations with non-linear processing time imposed by hierarchical memory. Efficiency of distributed computation has been estimated numerically by use of mixed integer linear programming. The performance has been visualized in the isoefficiency maps. It has been established that efficiency greater than 1 is possible as a result of memory hierarchy: parallel machines and multi-installment processing allow for computations in-core which is faster than if the same load were put on one machine, necessarily out-of-core. For problem sizes smaller than the maximum efficiency size, the efficiency decreases with increasing machine number. For problem sizes larger than the maximum efficiency size, the efficiency is almost independent of machine number. In the future study the idea of isoefficiency maps for systems with hierarchical memory can be extended to other pairs of system parameters than $m$ and $V$.

## References

1. Agrawal, R., Jagadish, H.V.: Partitioning techniques for large-grained parallelism. IEEE Trans. Comput. **37**, 1627–1634 (1988)
2. Akl, S.G.: The Design and Analysis of Parallel Algorithms. Prentice-Hall Int. Inc., Englewood Cliffs (1989)
3. Berlińska, J.: Communication scheduling in data gathering networks with limited memory. Appl. Math. Comput. **235**, 530–537 (2014)
4. Bharadwaj, V., Ghose, D., Mani, V., Robertazzi, T.: Scheduling Divisible Loads in Parallel and Distributed Systems. IEEE Computer Society Press, Los Alamitos (1996)
5. Drozdowski, M.: Scheduling for Parallel Processing. CCN. Springer, London (2009). https://doi.org/10.1007/978-1-84882-310-5
6. Drozdowski, M., Marszałkowski, J.M.: Divisible loads scheduling in hierarchical memory systems with time and energy constraints. In: Wyrzykowski, R., Deelman, E., Dongarra, J., Karczewski, K., Kitowski, J., Wiatr, K. (eds.) PPAM 2015. LNCS, vol. 9574, pp. 111–120. Springer, Cham (2016). https://doi.org/10.1007/978-3-319-32152-3_11

7. Drozdowski, M., Marszałkowski, J.M., Marszałkowski, J.: Energy trade-offs analysis using equal-energy maps. Future Gener. Comput. Syst. **36**, 311–321 (2014)
8. Drozdowski, M., Wielebski, Ł.: Isoefficiency maps for divisible computations. IEEE Trans. Parallel Distrib. Syst. **21**, 872–880 (2010)
9. Drozdowski, M., Wolniewicz, P.: Out-of-core divisible load processing. IEEE Trans. Parallel Distrib. Syst. **14**, 1048–1056 (2003)
10. Ghanbari, S., Othman, M.: Comprehensive review on divisible load theory: concepts, strategies, and approaches. Math. Prob. Eng. (2014). http://dx.doi.org/10.1155/2014/460354. Article ID 460354, 13 Pages
11. Ghose, D., Kim, H.J., Kim, T.H.: Adaptive divisible load scheduling strategies for workstation clusters with unknown network resources. IEEE Trans. Parallel Distrib. Syst. **16**, 897–907 (2005)
12. Gupta, A., Kumar, V.: Performance properties of large scale parallel systems. J. Parallel Distrib. Comput. **19**, 234–244 (1993)
13. Li, X., Bharadwaj, V., Ko, C.C.: Divisible load scheduling on single-level tree networks with buffer constraints. IEEE Trans. Aerosp. Electron. Syst. **36**, 1298–1308 (2000)
14. Marszałkowski, J.M., Drozdowski, M., Marszałkowski, J.: Time and energy performance of parallel systems with hierarchical memory. J. Grid Comput. **14**, 153–170 (2016)
15. Robertazzi, T.: Ten reasons to use divisible load theory. IEEE Comput. **36**, 63–68 (2003)
16. Swanson, S., Caulfield, A.M.: Refactor, reduce, recycle: restructuring the I/O stack for the future of storage. IEEE Comput. **46**, 52–59 (2013)

# Environments and Frameworks for Parallel/Distributed/Cloud Computing

# OpenMP Target Device Offloading for the SX-Aurora TSUBASA Vector Engine

Tim Cramer[1]($^{(\boxtimes)}$), Manoel Römmer[1], Boris Kosmynin[1], Erich Focht[2],
and Matthias S. Müller[1]

[1] IT Center, RWTH Aachen University, Aachen, Germany
{cramer,roemmer,kosmynin,mueller}@itc.rwth-aachen.de
[2] NEC Cooperation, Stuttgart, Germany
erich.focht@emea.nec.com

**Abstract.** Driven by the heterogeneity trend in modern supercomputers, OpenMP provides support for heterogeneous systems since 2013. Having a single programming model for all kinds of accelerator-based systems decreases the burden of code porting to different device types. The acceptance of this heterogeneous paradigm requires the availability of corresponding OpenMP compiler and runtime environments supporting different target device architectures. The LLVM/Clang infrastructure is designated to extend the offloading features for any new target platform. However, this supposes a compatible compiler backend for the target architecture. In order to overcome this limitation we present a source-to-source code transformation technique which outlines the OpenMP code regions for the target device. By combining this technique with a corresponding communication layer, we enable OpenMP target offloading to the NEC SX-Aurora TSUBASA vector engine, which represents the new generation of vector computing.

**Keywords:** HPC · OpenMP · Offloading · Vector computing · SIMD

## 1 Introduction

The requirement for large compute capabilities led to a wide use and acceptance of new computer architectures during the last decade. The fact that about 40 % of the performance-share of the TOP500 HPC systems is accelerator-based (e.g., GPGPUs) underlines this development. A new kind of accelerator is the NEC SX-Aurora TSUBASA vector engine (VE). This new architecture continues the long vector computing tradition in HPC and integrates this technology into a x86 environment. The VE is provided as a PCIe card and can be either used for executing programs which are running entirely on the device (refer to Fig. 1a) or for offloaded execution (refer to Fig. 1b), where only parts of the program run on the VE. The first method fits perfectly for highly vectorizable applications, while the latter supports programs with higher scalar parts. For the native execution, plain C/C++ or Fortran code can be cross-compiled for the device by using the NEC compiler. However, the offloading execution model requires an additional programming model including corresponding code modifications.

© Springer Nature Switzerland AG 2020
R. Wyrzykowski et al. (Eds.): PPAM 2019, LNCS 12043, pp. 237–249, 2020.
https://doi.org/10.1007/978-3-030-43229-4_21

One of the solutions are the device constructs of OpenMP [11]. The OpenMP target directives and API provide mechanisms to move the execution from the host to one or more target devices which might or might not share the same memory system. In OpenMP a target device is a logical execution unit which can be for instance a coprocesser, a GPU, a DSP, a FPGA or – as in our case – a VE. Within a target region almost all OpenMP constructs (e.g., parallel regions) can be used. Thus, performance portability is addressed, because one single and architecture-independent interface is available.

However, enabling OpenMP Offloading for new devices like the VE from a x86 environment requires a compiler with a backend for both – x86 and the VE. Since the only available compiler for the VE (i.e., the NEC compiler) lacks a x86 backend, there is no direct support for OpenMP offloading. Antao et al. [3] developed a basic infrastructure for target device offloading for the Clang compiler. However, this approach implies the availability of a LLVM IR compatible compiler backend which is not available, yet. We developed a source transformation technique to extract all target related regions which can be cross-compiled by any OpenMP compiler. This cross-compiled code is integrated as separated ELF section in the x86 binary (i.e., fat binary). As low-level offloading mechanism that controls, loads and runs the device specific kernels we use VEO[1]. Our extensions to LLVM, Clang and the OpenMP runtime are available as open source[2,3].

The option to create hybrid scalar-vector code by just adding OpenMP directives instead of doing OpenCL-alike code changes as required by VEO is hugely reducing the barrier to use the VE as an accelerator. Programs can be ported gradually, adding and tuning offloaded kernels easily while keeping the entire application functional instead of doing a huge step in porting the application to native VE code.

In this paper we describe the following contributions:

- We leverage the LLVM/Clang infrastructure in order to enable OpenMP target device offloading to the VE.
- We present a powerful (compiler-independent) source transformation technique for OpenMP target region outlining.
- We implemented a plugin for the communication between the host and the vector engine.
- We present a detailed evaluation for our approach.

The paper is organized as follows: Sect. 2 gives a brief overview on the VE architecture. In Sect. 3 we present our overall design for integrating OpenMP offloading for VEs into the LLVM toolchain, while Sect. 4 explains our source transformation technique in detail. Section 5 gives an overview on the runtime support. After evaluating our approach with different benchmarks (Sect. 6) and an overview on related work (Sect. 7), we conclude in Sect. 8.

---

[1] https://github.com/veos-sxarr-NEC/veoffload.

[2] https://github.com/RWTH-HPC.

[3] https://rwth-hpc.github.io/sx-aurora-offloading.

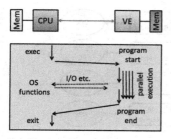
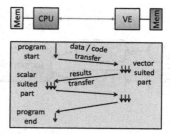

(a) Native OpenMP execution: The entire program executes on the VE.

(b) Offloaded OpenMP execution: Only vector-suited parts execute on the VE.

**Fig. 1.** SX-Aurora TSUBASA execution models.

## 2  SX-Aurora TSUBASA Vector Engine

The VE follows the tradition of long vector processors of NEC which combines SIMD and pipelining. The architecture has vector registers with a length of 256 * 64 bits = 16384 bits, with 256*32 SIMD width and pipeline depth of 8. Each vector processor of the first generation VE has eight cores with a RISC scalar unit and a vector processing unit featuring 64 architectural vector registers, three FMA, two integer ALU and a SQRT/DIV vector pipeline. The clock frequency of the VE is either 1400 or 1600 MHz, the higher frequency variant being water cooled. Each core has up to 307 GFLOPS (double precision) / 614 GFLOPS (single precision, packed) and a memory bandwidth to the shared 16 MB last level cache of 409 GB/s. The on-chip memory consists of six HBM2 stacks, four or eight layers high, with a total of 24 or 48 GB RAM and a total memory bandwidth of either 750 GB/s (24 GB model) or 1.2 TB/s [13].

The VE comes on a PCIe card with TDP of 300 Watts. Servers with x86_64 host processor(s) and up to eight such cards are available. They can be linked to large clusters by EDR Infiniband interconnects which are used by the VEs in a PeerDirect manner. The vector host (VH) server runs Linux and the VE operating system (VEOS) functionality is offloaded to the VH and implemented as user space daemons supported by kernel driver modules. Native VE programs can be built with the NEC C, C++ Fortran compilers or the ve-binutils assembler. The native VE programs offload their system call execution also to the VH, thus getting a Linux look-and-feel. NEC has released all VEOS components as open source with mainly GPL2 or LGPL licensing.

## 3  Offloading Design for SX-Aurora TSUBASA

For an architecture with a LLVM backend available, the Clang driver is designed to feed the source code of the base language (e.g. Cuda or C/C++) to its compiler frontend, which parses the source code and generates LLVM Intermediate Representation (LLVM IR) code. The Clang driver then employs LLVM to optimize the IR code and generates actual machine code for the target architecture

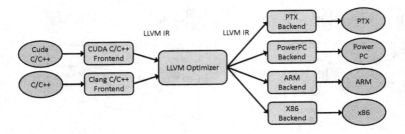

**Fig. 2.** High level view of the LLVM toolchain.

(e.g. for PowerPC, ARM, x86 or PTX) with an architecture specific LLVM backend (refer to Fig. 2). The Clang driver can then call a linker to generate a single binary from this machine code.

When compiling a program with OpenMP offloading, the driver feeds the input source code to the frontend twice – for the host and the offloading target. The frontend generates LLVM IR code for the host and specifically for offloading, which are then used to generate separate binaries. When linking the host binary, the Clang driver embeds the offloading binary into the host binary as a seperate ELF section, by employing a specially generated linker script. Thus, an adaption of the LLVM/Clang toolchain for the VE requires a backend which can transform LLVM IR code to vector engine assembly.

However, at the time writing this paper no released LLVM backend for the VE is available. To be able to compile the code necessary for offloading, we employ a source code transformation tool (refer to Sect. 4), which takes a C source code file as input and generates C code for the target regions. Instead of using a LLVM backend, we use the NEC compiler as external compiler suite for compiling and linking.

To integrate this workflow into Clang, a custom driver toolchain is employed which replaces Clang's frontend, LLVM optimizer, code generation and linking steps by calls to a set of wrapper tools. These wrapper tools feed the C code to the source code transformation tool and forwards the output to the NEC compiler to generate an object file and link the offloading binary. For generating the host binary, the Clang driver employs it's regular compilation pipeline, with the host linker integrating the machine code for the VE into a fat binary, as we will show in Sect. 5.

## 4   Source Code Transformation

As outlined in Sect. 3 we rely on an external compiler, if no LLVM backend is available for the given target device. If a target device compiler (e.g., the NEC compiler) does not natively support OpenMP target constructs, we have to extract all OpenMP target regions as well as all functions, types and global variables required on the target device.

```
#pragma omp declare target
int n = 10240;
#pragma omp end declare target
void saxpy(){
  float a = 42.0f;
  float b = 23.0f;
  float *x, *y;
  // Allocate and init x, y
  // ...
  #pragma omp target      \
    map(to:x[0:n], a)     \
    map(tofrom:y[0:n])
  #pragma omp parallel for
  for (int i = 0; i < n; ++i){
    y[i] = a*x[i] + y[i];
  }
}
```

sotoc

```
int n = 10240;

void __omp_ofld_b73b_saxpy_110( \
    int n, float * y,           \
    float *__sotoc_var_a,       \
    float * x) {

  float a = *__sotoc_var_a;

  #pragma omp parallel for
  for (int i = 0; i < n; ++i) {
    y[i] = a * x[i] + y[i];
  }

  *__sotoc_var_a = a;

}
```

**Fig. 3.** Basic function outlining with our source code transformation technique.

To this end, we employ a source-to-source transformation technique, outlining all required code parts of an OpenMP application. We developed the Clang-based tool *sotoc* and integrated it into the offloading compilation workflow. It uses the Clang compiler infrastructure to read the input source file, parse it into an abstract syntax tree (AST), and traverse this tree to search for OpenMP target regions and dependent code out of which it generates C source code to be fed into an external compiler.

This approach has the advantage that it is completely vendor-independent, because we use plain C code, thus, any kind of target device supporting C can be used, even if no LLVM IR compatible backend is available. The code optimization for the target architecture is done in the internal backend of the external compiler. However, the compilation time might be slightly higher, because we add some additional overhead for the source code transformation and the code analysis in two different compiler frontends. Here, the overhead for the latter one is less than a factor of two, because only parts of the code will be compiled for the host and the target device.

Figure 3 shows a basic example for our source code transformation technique using a scaled vector-vector multiplication. All local variables used in the target region are passed as parameter to the outlined function. The arrays x and y are passed as pointers, because of the explicit OpenMP mapping. The scalar variable a is also passed as an address pointer, but renamed in the function signature and dereferenced in the function prologue. In the epilogue the value will be updated in order to enable a data movement back to the host (not required in this example). Since no explicit mapping for a exists the variable will be implicitly firstprivate. Although an additional scalar variable b is declared and initialized, it will not be passed to the outlined function, because it is not used in the target region. The global variable n is declared in a declare target scope and thus will stay global in the target device data environment. *sotoc* will not generate any other code outside the target region (e.g., for the allocation and initialization).

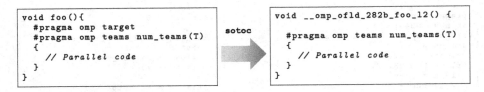

**Fig. 4.** One solution for the handling of nested `teams` constructs (works only for OpenMP 5.0). The solution `sotoc` implements at the moment is to omit all `teams` constructs and thus only fork one team.

***Teams Construct.*** In OpenMP a `parallel` construct forks a team of multiple threads, which can be synchronized by an explicit or implicit barrier. Since some target devices do not support global synchronization (e.g., between multiple streaming multiprocessors (SM) on Nvidia GPUs), it is possible to fork a league of teams with multiple threads each. Threads in different teams cannot synchronize with each other.

Until OpenMP 4.5 it was a prerequisite that a `teams` construct has to be perfectly nested into a `target` construct, which means that (1) no `teams` construct is allowed outside a `target` construct and (2) there must not be any other statements or directives outside the `teams` constructs. Due to restriction (1) it is not possible to simply outline the body of the `target` region by just keeping the `teams` construct in the outlined function as shown in Fig. 4, because the NEC compiler cannot produce code for the orphaned construct in the outlined function. For that reason we decided to just omit all `teams` constructs, which means that always only one team will be created on the target device. For an VE as a target device this does not limit the performance or functionality at all, because we have eight powerful and synchronizable cores instead of multiple SMs as on a GPU. Furthermore, this behavior is standard-compliant, because OpenMP specifies that the number of created `teams` is implementation defined, but less or equal to the value specified in the `num_teams` clause. Due to restriction (2) it would be enough omit a `teams` construct that is following a `target` construct. However, both restrictions have been relaxed in OpenMP 5.0. As a consequence, *sotoc* traverses the complete AST in order to omit all orphaned `teams` constructs. This is still required, because at the moment the NEC compiler does not fully support OpenMP 5.0.

***Combined Constructs.*** For convenience, OpenMP defines a shortcut for constructs that are immediately nested into another. The semantics for these combined constructs are identical to the semantics of the corresponding sequence of non-combined constructs. In order to be standard-compliant, our source-to-source transformation considers all target-related combined constructs defined in the specification (i.e., combinations of the `target` construct with `parallel`, `for`, `simd`, `teams` and `distribute`).

For the handling of these combined constructs it is not enough to just outline the complete structured block, because this would contradict the semantics. For instance just outlining the body of a `#pragma omp target parallel`, would

```
#pragma omp declare target
int n = 10240;
#pragma omp end declare target
void saxpy()
{
  float a = 42.0f;
  float *x, *y;
  int i;
  // Allocate and init x, y
  // ...
  #pragma omp target teams        \
    distribute parallel for       \
    map(to:n,a,x[:n]) map(y[:n])\
    device(0) private(i)
    for (i = 0; i < n; ++i)
      y[i] = a*x[i] + y[i];
}
```

sotoc

```
int n = 10240;

void __omp_ofld_3b30_saxpy_112( \
       float * y ,               \
       float *__sotoc_var_a,     \
       float * x)
{
  float a = *__sotoc_var_a;
  int i;

  #pragma omp parallel for
     private(i)
  for (i = 0; i < n; ++i)
    y[i] = a * x[i] + y[i];

  *__sotoc_var_a = a;
}
```

**Fig. 5.** Example for the source transformation of a combined construct.

mean that the target device compiler will not generate a parallel region. Thus, our source-to-source transformation has to outline the structured block and add a #pragma omp parallel at the beginning of the function.

Here, one has to consider that each directive can have different clauses (e.g., private, firstprivate, map, reduction, etc.). Some of these directives are only applicable to one of the constructs, others to multiple. Our source-to-source transformation analyzes which of the clauses belongs to which construct regarding the definition of the OpenMP specification and adds the clauses to the additional directive if necessary.

Figure 5 shows an example for the transformation of a combined construct. The target teams distribute parallel for directive is transformed into the additional parallel for directive in order to enable the parallel execution of the for loop on the target device. The teams and distribute constructs are omitted, because we only need one team for the VE, as described in the previous section. Since the map and device clauses only apply to the target construct, they are omitted as well. The private clause applies for the parallel for construct and thus is added to the additional directive. Furthermore, the declaration for the private variable i is added before the construct in order to avoid compiler errors.

*Array Handling.* The map clause in OpenMP allows to map array sections (i.e., an array or a subset of elements of an array) and structured elements into the target device data environment (e.g., the variables x and y in Listing 3). In the case of an array section the OpenMP specification demands contiguous storage. This requirement allows us to pass each array section as a pointer to the outlined target device function. This holds for dynamically allocated arrays as well as for a statically allocated (stack) array. However, in the latter case we need to cast the void pointer back the origin data type. In order to allow multidimensional stack arrays, we do this with respect to the original dimension. Otherwise we would need to transform all operator-based accesses to this array

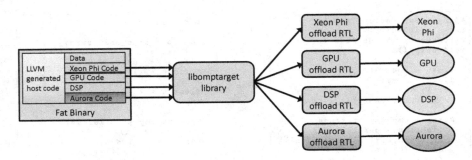

**Fig. 6.** The LLVM offloading infrastructure, based on [3].

to index-based accesses by traversing the AST. For instance an 3-dimensional array (with `size` as integer variable) defined by

    float A[2][size][5];

will be passed as a void pointer named `__sotoc_var_A` to the target device function. The pointer will be casted back to the original type by using

    float (*A)[__sotoc_vla_dim1_A][5] = __sotoc_var_A;

in the prologue of the outlined function. This works because the specification guarantees contiguous storage. Thus, every access to an array element with multiple access operators (e.g., `A[i][j][k]=42;`) works without any further code modifications within the corresponding target region. The variable `__sotoc_vla_dim1` is passed to the outlined function by the runtime system for all variable-length arrays (VLA). A VLA is an array data structure whose length is determined at runtime and not at compile time. In contrast, the size of constant-length arrays (CLA) can be determined at compile time and does not have to be passed to the target device environment. Since a combination of CLAs and VLAs is possible for every dimension of the array (see example above), our source-to-source transformer analyses the required data type for the function prologue recursively for every dimension and generates the corresponding type cast and the signature of the outlined function.

## 5  Runtime Support

OpenMP target device offloading requires support by the runtime system. In a multi-company design document[4] the principles of the offloading mechanism as well of the interface specification for the LLVM offloading runtime support is provided. The implementation of the corresponding library (libomptarget) was presented in [3]. Figure 6 show this infrastructure including our modifications. The compiler generates one fat binary including the LLVM-generated host code for the host execution and the target device codes. The infrastructure supports different kinds of target device (e.g., Xeon Phi code, GPU code or DSP code),

---

[4] https://github.com/clang-omp/OffloadingDesign.

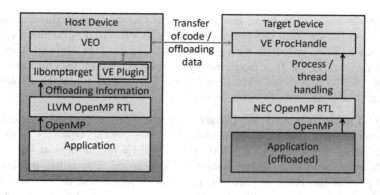

**Fig. 7.** Communication between the host and the target device.

where each code is stored in a separated ELF section. For the VE we embed the cross-compiled assembler code which was generated by our source-to-source transformation technique and the NEC compiler. The runtime system dynamically detects which target device architectures are available and if a compatible code section is included in the binary. If this is the case, the runtime system calls a provided plugin which is responsible for the setup off the target device and the communication between host and target device. We implemented the plugin for the VE by using the VEO infrastructure from NEC.

Figure 7 shows this communication between the host and the target device. As soon as an OpenMP directive is encountered in the application, the LLVM OpenMP runtime is called. If this directive is a target-related region, the offloading information is passed to libomptarget and to the VE plugin in our case. The VEO layer transfers the code and the offloading data to or from the target device and starts executing the code region. If an OpenMP region is encountered on the target device, the NEC OpenMP runtime implementation is called. Thus, two different OpenMP runtime implementations are used in our approach, which is in general not an issue (in fact there are always two instances of the runtime in distributed memory systems). However, one has to consider that changing internal OpenMP variables (e.g., for controlling the number of threads) has to be communicated between the two implementations in order to avoid side effects.

As mentioned before, our source transformation approach works for any target device architecture which has a C compiler available. However, the implementation of a device-specific plugin still has to be done for each kind of device, because no standard communication layer is defined or available. Due to the well defined interface and the provided data mapping mechanisms in libomptarget, the amount of work for the implementation of new plugins is quite low.

## 6    Evaluation

In this section we use the Himeno Benchmark [1] and a verification suite in order evaluate our approach. Since this work is intended to be a general approach for any target device architecture, the focus is on the success of the compilation and execution rather than the performance of the VE. All VE results have been executed on a VE Type 10B with 8 cores running at 1.4 GHz and a memory bandwidth of 1.2 TB/s. If a performance evaluation to an x86 architecture is given, we refer to a 2-socket system with two Intel Xeon Platinum 8160 CPUs running at 2.10 GHz with 48 cores in total.

***OpenMP Validation and Verification Suite.*** For the general verification of our source-to-source transformation technique, we use the OpenMP Validation and Verification Suite (SOLLVE) [5,6], developed by researchers at the University of Delaware and Oak Ridge National Laboratory. The suite, published in 2018 and part of the Exascale project, provides different test cases written in C and Fortran using the OpenMP offloading concepts specified in version 4.5. We configured the suite to offload to the VE using the new VE target triple and x86 for comparison[5]. In both cases, we only use the test cases written in C code as we do not support C++ or Fortran code in our implementation at the moment.

In summary, all but 6 of the 65 used tests compile correctly with the VE target and another 10 tests fail at runtime. The main reason for the compile errors is the use of anonymous (unnamed) structs or enums which causes a complex replacement in our source code transformation. Another issue in the transformation phase is the missing evaluation of the `if` clause on combined constructs like `target teams parallel for`. Most of the runtime failures are cause by missing multiple device support. At the moment this is a limitation in the VEO/VEOS infrastructure and cannot be fixed in the LLVM/Clang implementation. However, this will be fixed by NEC in some future release version and is only a minor limitation and not a general problem in our approach. Using x86 offloading, all of the tests compile correctly and only nine tests fail at runtime. The main causes for failure here are also related to arrays and the `defaultmap` clause.

***Himeno Benchmark.*** The Himeno benchmark [1] was developed by Ryutaro Himeno in 1996 at the RIKEN Institute. It is a well-known benchmark solving a 3D Poisson equation on a structured curvilinear mesh. In this part of the evaluation we use a modified version of the benchmark based on an implementation by Hart [7]. This version uses OpenMP to offload the compute kernel to the target device. For our evaluation we used the large grid size ($1024 \times 512 \times 512$) with double precision values.

With our compiler and runtime modifications we were able to successfully build the benchmark including the target offloading code for the VE. Table 1 shows the performance results compared to the given x86 system. We compare the results on both architectures using the systems with the reference version of the HIMENO benchmark without any OpenMP offloading (native execution) and with using the corresponding offloading targets. The results show that the offloading version has a maximum overhead of up to 12.3 % compared to native execution when using 8 threads on VE. The slight performance improvement

---

[5] Clang allows to define x86 as target device for testing purpose, where the target regions are executed on the host, but using the corresponding plugin in libomptarget.

**Table 1.** Performance of the Himeno Benchmark on x86 and the VE.

| Threads | SX (ofld) [GFLOPS] | SX (native) [GFLOPS] | x86 (ofld) [GFLOPS] | x86 (native) [GFLOPS] |
|---|---|---|---|---|
| 1 | 40.02 | 41.06 | 0.64 | 0.64 |
| 2 | 76.2 | 80.34 | 1.23 | 1.19 |
| 4 | 140.87 | 155.95 | 2.37 | 2.39 |
| 7 | 222.03 | 252.38 | | |
| 8 | 230.19 | 283.68 | 4.79 | 4.86 |
| 16 | | | 8.41 | 8.74 |
| 48 | | | 22.75 | 21.92 |

for the offloading version when increasing the threads from 7 to 8 can only be reached by decreasing the VEOS scheduler timer intervals and time slices due to a peculiarity of the current implementation of VEO and its interplay with the OpenMP threads. However, this issue was already identified and is being worked on while writing this paper. The results show that the VE outperforms the two Xeon CPUs by a factor of about 10 (olfd) to 12 (native), where the maxium is 283 GFLOPS on the VE.

In addition, we measured the performance of the offloading version on a Nvidia V100-SXM2 GPU by using the release versions of the GNU Compiler 8 and Clang 8 (now with Nvidia as target device). With the GNU compiler we reach a maximum performance of 33 GFLOPS and with the release Clang compiler 7.78 GFLOPS. The fact that both results are an order of magnitude lower than in our implementation does not mean that the V100 has a worse performance than a VE in general. In fact, the native execution of Himeno is only about 10 % lower [13]. However, it shows that our offloading implementation already reaches good performance results compared to other released implementations for other architectures.

## 7    Related Work

To the best of our knowledge this work is the first approach for OpenMP offloading to Aurora. However, offloading to other innovative target device architectures have been done before [4,9,10,12] and parts of our work are based on these results. In [12] Sommer et al. presented an implementation for OpenMP offloading to FPGA accelerators. Their proof-of-concept implementation is similar to our approach, although the technical realization slightly differs. Furthermore, we also support all kinds of target-related constructs (e.g., `teams` and combined directives), while their prototype focuses on the `target` construct only. In a recent publication, Álvarez et al. [2] present an infrastructure which allows to embed the source code in addition to the device-specific code in the fat binary. This work describes an alternative offloading methodology, which only requires little support from the host compiler similar to our approach.

In [8] NEC and the University Saarland present a VE backend for LLVM. The use of LLVM IR compatible backend for the VE is a good alternative to our source transformation technique, because it perfectly fits into the design of the LLVM toolchains. However, this has the disadvantage of not being usable generically on arbitrary target device architectures.

## 8    Conclusion

For the performance portability of HPC applications between different heterogeneous systems, standard-compliant programming models are required. Since OpenMP is the de-facto standard for shared memory programming and also offers support for target device offloading, it is a very good candidate for such a portable paradigm. However, for the usability and acceptance, corresponding infrastructures and implementations are required. In our work we presented a LLVM-based source-to-source transformation technique which enables support for target device offloading, even if the available OpenMP compiler and runtime for the target architecture do not support this feature natively. With our modifications to the LLVM infrastructure only the communication layer for the runtime support needs to be implemented. The availability of a LLVM IR compatible compiler backend is not required. Although our implementation is limited to C code, we showed that our approach works on the NEC SX-Aurora TSUBASA and delivers the expected performance for the Himeno Benchmark.

Since this work focus the functionality of our source transformation technique for the target region outlining and the implementation for the VE, we did not present further performance evaluations. However, as future steps we plan to analyze the performance for additional typical HPC workloads and to improve our prototype implementation of the complete offloading infrastructure in order to make it even more reliable, efficient and portable to arbitrary target device architectures.

## References

1. The Riken Himeno CFD Benchmark. http://accc.riken.jp/en/supercom/documents/himenobmt
2. Álvarez, Á., Ugarte, Í., Fernández, V., Sánchez, P.: OpenMP dynamic device offloading in heterogeneous platforms. In: Fan, X., de Supinski, B.R., Sinnen, O., Giacaman, N. (eds.) IWOMP 2019. LNCS, vol. 11718, pp. 109–122. Springer, Cham (2019). https://doi.org/10.1007/978-3-030-28596-8_8
3. Antao, S.F., et al.: Offloading support for OpenMP in Clang and LLVM. In: Proceedings of the Third Workshop on LLVM Compiler Infrastructure in HPC, LLVM-HPC 2016, pp. 1–11. IEEE Press, Piscataway (2016)
4. Bertolli, C., et al.: Integrating GPU support for OpenMP offloading directives into Clang. In: Proceedings of the Second Workshop on the LLVM Compiler Infrastructure in HPC. ACM, New York (2015)

5. Diaz, J.M., Pophale, S., Friedline, K., Hernandez, O., Bernholdt, D.E., Chandrasekaran, S.: Evaluating support for OpenMP offload features. In: Proceedings of the 47th International Conference on Parallel Processing Companion, ICPP 2018, pp. 31:1–31:10. ACM, New York (2018)
6. Diaz, J.M., Pophale, S., Hernandez, O., Bernholdt, D.E., Chandrasekaran, S.: OpenMP 4.5 validation and verification suite for device offload. In: de Supinski, B.R., Valero-Lara, P., Martorell, X., Mateo Bellido, S., Labarta, J. (eds.) IWOMP 2018. LNCS, vol. 11128, pp. 82–95. Springer, Cham (2018). https://doi.org/10.1007/978-3-319-98521-3_6
7. Hart, A.: First experiences porting a parallel application to a hybrid supercomputer with OpenMP4.0 device constructs. In: Terboven, C., de Supinski, B.R., Reble, P., Chapman, B.M., Müller, M.S. (eds.) IWOMP 2015. LNCS, vol. 9342, pp. 73–85. Springer, Cham (2015). https://doi.org/10.1007/978-3-319-24595-9_6
8. Ishizaka, K., Marukawa, K., Focht, E., Moll, S., Kurtenacker, M., Hack, S.: NEC SX-Aurora - A Scalable Vector Architecture. LLVM Developers' Meeting (2018)
9. Mitra, G., Stotzer, E., Jayaraj, A., Rendell, A.P.: Implementation and optimization of the OpenMP accelerator model for the TI keystone II architecture. In: DeRose, L., de Supinski, B.R., Olivier, S.L., Chapman, B.M., Müller, M.S. (eds.) IWOMP 2014. LNCS, vol. 8766, pp. 202–214. Springer, Cham (2014). https://doi.org/10.1007/978-3-319-11454-5_15
10. Newburn, C.J., et al.: Offload compiler runtime for the Intel® Xeon Phi coprocessor. In: 2013 IEEE International Symposium on Parallel Distributed Processing, Workshops and Phd Forum, pp. 1213–1225, May 2013
11. OpenMP Architecture Review Board: OpenMP Application Program Interface, Version 5.0, November 2018
12. Sommer, L., Korinth, J., Koch, A.: OpenMP device offloading to FPGA accelerators. In: 2017 IEEE 28th International Conference on Application-specific Systems, Architectures and Processors (ASAP), pp. 201–205, July 2017
13. Yamada, Y., Momose, S.: Vector Engine Processor of NEC's Brand-New Supercomputer SX-Aurora TSUBASA. Hot Chips Symposium on High Performance Chips, August 2018. https://www.hotchips.org. Accessed 05/19

# On the Road to DiPOSH: Adventures in High-Performance OpenSHMEM

Camille Coti[1]($\boxtimes$)(iD) and Allen D. Malony[2]

[1] LIPN, CNRS UMR 7030, Université Paris 13, Sorbonne Paris Cité,
Villetaneuse, France
camille.coti@lipn.univ-paris13.fr
[2] University of Oregon, Eugene, USA
malony@cs.uoregon.edu

**Abstract.** Future HPC programming systems must address the challenge of how to integrate shared and distributed memory parallelism. The growing number of server cores argues in favor of shared memory multithreading at the node level, but makes interfacing with distributed communication libraries more problematic. Alternatively, implementing rich message passing libraries to run across codes can be cumbersome and inefficient. The paper describes an attempt to address the challenge with OpenSHMEM, where a lean API makes for a high-performance shared memory operation and communication semantics maps directly to fast networking hardware. DiPOSH is our initial attempt to implement OpenSHMEM with these objectives. Starting with our node-level POSH design, we leveraged MPI one-sided support to get initial internode functionality. The paper reports our progress. To our pleasant surprise, we discovered a natural and compatible integration of OpenSHMEM and MPI, in contrast to what is found in MPI+X hybrids today.

**Keywords:** OpenSHMEM · Distributed run-time system · One-sided communication

## 1 Introduction

The trend of increasing core counts of shared memory servers that make up the nodes of scalable high-performance computing (HPC) systems raise questions of how parallel applications should be programmed in the future. Distributed programming models based on message passing are effective for internode parallelism, but their runtime implementation can be less efficient for intranode parallelism. This put pressure on these programming paradigms to be used in hybrid forms. For instance, while *MPI everywhere* programs are perfectly reasonable for programming HPC machines, concerns for node-level performance argues for an *MPI+X* approach, where $X$ is a shared memory programming methodology of choice. In doing so, conflicts can arise between the MPI and X runtime support, especially with respect to managing higher degrees of node-level parallelism.

© Springer Nature Switzerland AG 2020
R. Wyrzykowski et al. (Eds.): PPAM 2019, LNCS 12043, pp. 250–260, 2020.
https://doi.org/10.1007/978-3-030-43229-4_22

Alternatively, shared memory programming models are effective for intranode parallelism, but must be adapted to maintain a shared memory abstraction on distributed memories. Advances in low-latency, RDMA communication hardware make it possible to support (partitioned) global address space (P)GAS semantics with high-efficiency data transfer between nodes. For instance, the SHMEM interface and its OpenSHMEM standardization embody peer-to-peer one-sided *put* and *get* operations on distributed "shared" memory. While the abstraction is more compatible with shared memory programming, it was created originally from a perspective of internode interaction.

In this paper we consider the viability of OpenSHMEM as a unified parallel programming model for both intranode and internode parallelism. Our starting point is Coti's high-performance OpenSHMEM implementation for shared memory system called *Paris OpenSHMEM* (POSH for short) [6]. The goal of POSH is to deliver an OpenSHMEM implementation on a shared memory system that is both fast and lightweight as possible. Now we look to extend POSH for distributed HPC systems. The challenge in creating distributed POSH (DiPOSH for short) is first to support the OpenSHMEM API and second to optimize performance. For this reason, we take the strategy of layering DiPOSH on MPI one-side communication. This will give us a baseline to evaluate future enhancements. More importantly, it will expose any critical factors at the nexus between intranode and internode operation.

## 2  Related Works

Parallel programming approaches for evolving HPC systems must address the challenges of greater intranode concurrency, while connecting to powerful internode communication infrastructure. The MPI interface has dominated the message passing paradigm with important high-performance implementations available, including OpenMPI, MPICH, MVAPICH. However, MPI's generic semantics makes it more complex to implement for distributed communication and unnecessarily complicated at the node level. Efforts to integrate multithreaded shared memory programming (e.g., OpenMP) with MPI can suffer from mismatches in the runtime systems.

In contrast, the OpenSHMEM interface is very simple and straightforward to implement. Its remote memory access semantics is equally natural for targeting intranode and internode parallelism. In fact, MPI's one-sided communication support [8] is all that we needed to develop DiPOSH's remote functionality. This approach is attractive, because it takes advantage of already existing communication routines, along with the rest of MPI's infrastructure, such as optimized collective communications. However, it involves a thick software stack and required some additional synchronization to implement OpenSHMEM's communication model.

Certainly, there are several other (P)GAS programming systems actively being pursued, such as CoArray Fortran, Chapel, UPC/UPC++, Global Arrays, and HPX. As a library, DiPOSH target a lower-level OpenSHMEM API, increasing its portability and interoperability with other software.

## 3    Architecture

The core idea behind POSH is to implement the OpenSHMEM communication interface with a minimal API-to-network software stack, thereby minimizing the software overhead. As shown in [11,12], traversing the software stack can have a significant overhead on performance-critical communication networks. However, in order to be portable across parallel machines, the communication library needs to be able to use several types of networks. In this section, we are describing how DiPOSH handles the different types of communication channels.

### 3.1    Shared Heap

In the OpenSHMEM memory model, the memory of each process is made of two parts: its private memory, which only it can access, and a shared heap, that can be accessed in read and write mode by all the other processes. In POSH, this shared heap is implemented by a segment of shared memory. Each process on a node owns a segment of shared memory, which is accessed by all the other processes on the node, enabling straightforward communications locally.

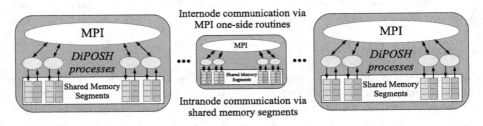

**Fig. 1.** High-level DiPOSH design illustrating intranode communication through shared segments and internode communication via MPI.

In DiPOSH, communications are also performed using this segment of shared memory: the remote memory access routines read and write data from and into the remote process's heap. Since the shared heaps are *symmetric* (*i.e.,* processes allocate the same space on their own heap), memory locations can be addressed within this segment of shared memory by using their offset from the beginning of the segment.

An example is presented in Fig. 1. Processes within a given node can communicate with each other using these segments of shared memory and processes located on different nodes use another communication channel, for instance, MPI one-sided routines.

In the particular case of MPI's one-sided communications, these routines use a *window* to perform one-sided communications. When the application is initialized, each process creates a window and associates the beginning of the segment of shared memory as its base address. Therefore, one-sided communications handle directly data in the shared heap, making both communication channels (local and MPI) compatible. Moreover, since, unlike OpenSHMEM, MPI's one-sided communications are asynchronous and non-blocking, completion of the communication is ensured by lock and unlock operations on the window.

## 3.2  Network Portability

Using the appropriate communication channel to reach another process is a highly critical point of the design of DiPOSH. Indeed, choosing the right function and calling it is in the critical path of the software stack and therefore, needs to be handled carefully in the aim of minimizing the software overhead of the communication library.

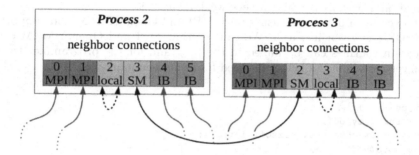

**Fig. 2.** Data structure handling how processes select the communication channel to be used with each other process. In the table of neighbors, for each other process there is a set of function pointers pointing to the appropriate communication routines.

We chose to make this decision once, when the communication library is initialized. All the processes exchange their contact information in an *allgather-like* collective operation. Then, each process knows how it can reach every other process and determine which communication channel it will use when they need to communicate with the other process.

Each process maintains a data structure that keeps information on their *neighbors* (*i.e.,* the other processes). This data structure also contains function pointers to the communication routines that correspond to the communication channel that will be used to communicate with this process. This organization is represented in Fig. 2.

It was measured in [1] that calling a function from a function pointer has an acceptable cost, compared to other call implementations (close to a direct call). Moreover, data structure requires about 1 kB per neighbor on a 64 bit architecture, which is not significant regarding the memory footprint of parallel scientific applications, even at large scale.

### 3.3   Cohabitation with Other Models

As mentioned earlier in this paper, combinations of programming models is an attractive solution to program extreme-scale machines. Therefore, it is necessary for parallel execution environment to be compatible with each other. The Open-SHMEM specification v1.4 mentions a few examples of OpenSHMEM and MPI compatibility in some implementations (annex D).

DiPOSH is *fully compatible* with MPI. In its current implementation, its run-time environment is written in MPI, which means that the coordination between the OpenSHMEM processing units (*e.g.,* communication of their contact information) is made using MPI calls. As a consequence, when the OpenSH-MEM application is initialized by start_pes(), the OpenSHMEM library calls MPI_Init. In future versions, DiPOSH might use another run-time environment in order to be independent of MPI and avoid requiring having an MPI implementation on the system. However, MPI is common enough to make having it installed on a parallel machine a very weak assumption.

An excerpt of a program using both MPI and OpenSHMEM and supported by DiPOSH is given in Fig. 3. We can see that MPI and OpenSHMEM communication routines can be mixed in the program, for the programmer to use whichever paradigm fits better its need for each communication.

```
start_pes( 0 );
rank = shmem_my_pe();
value = (int*)shmalloc( sizeof( int ) );
/* ... do stuff ... */
if( 0 == rank )
    shmem_int_put( value, &result, 1, 1 );
MPI_Barrier( MPI_COMM_WORLD );
/* ... do stuff ... */
if( 0 == rank )
    MPI_Send( &number, 1, MPI_INT, 1, 0, MPI_COMM_WORLD );
if( 1 == rank )
    MPI_Recv( &number, 1, MPI_INT, 0, 0, MPI_COMM_WORLD, &stat );
/* ... do stuff ... */
MPI_Allgather( &number, &result, 1, MPI_INT, MPI_SUM, MPI_COMM_WORLD );
```

**Fig. 3.** Excerpt of a program using both OpenSHMEM and MPI.

In addition to being an interesting feature for application developers, being able to use MPI in OpenSHMEM programs can also be useful for supporting tools, that can take advantage of some MPI-specific collective operations (such as reductions using user-defined operations and datatypes) to aggregate data.

### 3.4   Profiling

Profiling OpenSHMEM applications can be done by TAU without any specific tool interface [10]. DiPOSH can provide more low-level information, such as profiling information on the communication channels, for the user to be able to tune the communication library.

TAU's measurement model give access to a multitude of possible performance data about the execution, from hardware counters to MPI information from MPI_T. Interestingly, MPI_T also gives the possibility to modify some parameters. For instance, OpenMPI provides several *levels* of parameters: end-user, application user, developer. Some of these parameters can be used in *write* mode by the application in order to tune the library at run-time. For instance, the maximum size of a message sent in eager mode, the size of a shared memory segment... can be modified through the MPI_T interface. Moreover, TAU can keep track of the memory usage in the shared heaps.

An example of profiling information about NUMA (NUMA hits, misses, and so on) is given in the performance section of this paper, in Subsect. 4.3.

## 4    Performance

We have run preliminary performance evaluations of DiPOSH on the Grid'5000 platform [4], using the Grimoire cluster in Nancy. It is made of 8 nodes, each of which featuring two 8-core Intel Xeon E5-2630v3 CPUs, 4 10 Gb Ethernet NICs and a 56 Gbps Infiniband network interconnection. The operating system deployed on the nodes is a Debian 9.8 with a Linux kernel 4.9.0. All the code was compiled using g++ 6.3.0 with -O3 optimization flag, and DiPOSH was linked against some Boost's libraries version 1.62 and OpenMPI 2.0.1.

### 4.1    Communication Performance

We evaluated the communication time using `shmem_char_put()` and `shmem_char_get()` operations on buffers of variable sizes. Each communication time was measured using `clock_gettime()` and run 20 times, except for small buffers (less than $10^4$ bytes), that were run 200 000 times, since they are more subject to various noises on the system.

We measured the latency and the throughput on a single node and compared between when processes are bound to the same socket (`--map-by core` in OpenMPI) and on two different sockets (`--map-by socket` in OpenMPI). The latencies are given in Fig. 4 and the throughputs are given in Fig. 5.

As expected, communications are slightly faster when both processes are executed on the same socket. Also as expected, the performance of the direct shared memory implementation (called POSH SM in the captions) is significantly faster than the implementation using MPI-3 RDMA calls. This can be explained by the fact that implementing on top of MPI involves communicating through a thick software stack (see the aforementioned notion of software overhead), but more importantly, implementing the semantics of OpenSHMEM's one-sided communications using MPI-3's one-sided operations involves additional operations (`MPI_Win_lock()` and `MPI_Win_unlock()`).

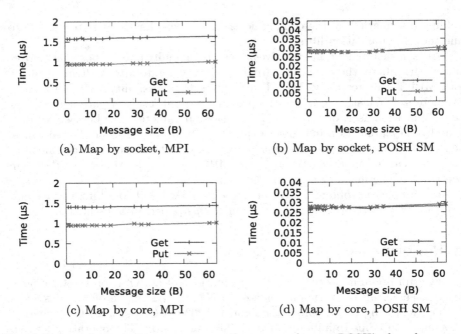

**Fig. 4.** Latency between two processes on the same node, over POSH's shared memory communication channel vs over MPI one-sided communications.

We have also measured the communication performance between two nodes, here using the implementation using MPI one-sided communications. The latency and the bandwidth are given in Fig. 6. It is interesting to see that, in spite of the harmfulness of the transposition between MPI-3's RDMA communication model and OpenSHMEM's communication model, the throughput is close to the announced bandwidth of the network used (56 Gbps).

## 4.2  Parallel Matrix-Matrix Multiplication

We implemented a parallel matrix-matrix multiplication using Cannon's algorithm. The initial local submatrices are placed in the shared heap, for other processes to fetch them using *get()* communications, hence placing their local copy in the private memory. Therefore, the local computations themselves are made on private memory. We used the DiPOSH communication channel on top of MPI-3 RDMA calls. The performance is represented in Fig. 7 and we can see that the OpenSHMEM model allowed us to write a straightforward implementation of the algorithm and DiPOSH provided good parallel performance.

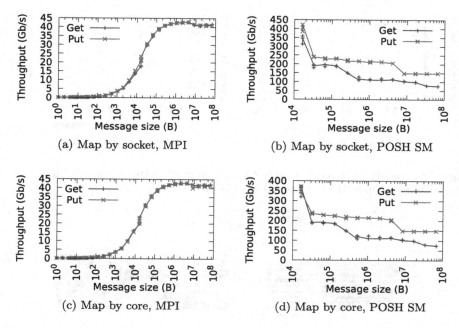

**Fig. 5.** Throughput between two processes on the same node, over POSH's shared memory communication channel vs over MPI one-sided communications.

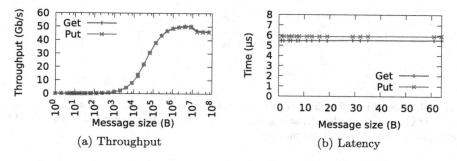

**Fig. 6.** Communication performance between two nodes, DiPOSH over MPI.

## 4.3  Some Profiling Information

As described in Sect. 3.4, DiPOSH can extract low-level profiling information from the communication channels and provide them to TAU through user-defined events. We executed the parallel matrix-matrix multiplication on a single node using the POSH shared memory communication channel, and obtained NUMA statistics during the execution. The result displayed by TAU is given in Fig. 8. For instance, we can see, for this particular shmem_get call, how much time was spent reaching data on another NUMA node (numa_foreign). Such

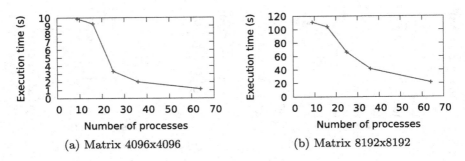

**Fig. 7.** Scalability of a parallel matrix-matrix multiplication (Cannon's algorithm) implemented in OpenSHMEM, using DiPOSH over MPI.

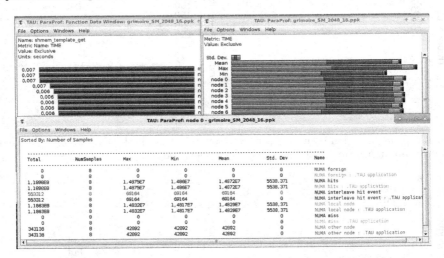

**Fig. 8.** Low-level profiling information: in addition to the usual function calls (in red, `MPI_Init()` and in purple, `shmem_*_get`), TAU displays information about the NUMA communications. (Color figure online)

information can be useful to understand how time is spent in communications and optimize data locality in order to minimize communication time.

## 5    Conclusion and Perspective

In this paper, we have presented the core design philosophy of DiPOSH, a high performance, distributed run-time environment and communication library implementing the OpenSHMEM specification. More specifically, DiPOSH focuses on taking advantage of the simple communication patterns of OpenSH-MEM in order to implement a very thin software stack between the API and the network. Furthermore, it aims at being highly portable, able to take advantage of high performance communication drivers, and delivering exceptional node-level efficiency with its POSH core.

In contrast to MPI+X hybrids, DiPOSH is completely compatible with MPI, making it possible to implement hybrid-like MPI+OpenSHMEM applications that can take advantage of the communication models of both libraries. Moreover, tools that support the parallel execution of an OpenSHMEM application can use these MPI communications, for instance, for global performance data aggregation and *in situ* analytic.

We have seen that DiPOSH can interface with the TAU measurement library in order to provide low-level profiling information about the communication channel. This can be used by application users in order to tune the library they are using, but it also opens perspective for a tooling interface that would exploit this information.

Future developments with DiPOSH include experimenting with high performance communication channels, for instance UCX [12], and KNEM for intranode communications [7], in order to be able to support a large number of networks while keeping the API-to-network path short. Moreover, we will work on providing fault tolerance capabilities, which is both necessary and challenging on exascale machines [5], particularly, for OpenSHMEM [9]. Fault tolerance can be achieved at system-level, for automatic fault-tolerance such as transparent checkpoint-restart [3] and at user-level, in order to provide the programmer with features that allow them to implement fault tolerant parallel applications, such as ULFM for MPI [2].

In addition to supporting a broad variety of networks, we are looking at a set of benchmarks and mini applications that can emphasize and stress the characteristics and choices of OpenSHMEM implementations. The next step will be to evaluate DiPOSH on large-scale supercomputers on these applications and benchmark its performance at fine grain.

We are currently working on supporting the complete OpenSHMEM interface and preparing a release. In the meantime, a partial support of the standard (providing at least point-to-point communications) can be found on a famous Git platform at the following URL: https://github.com/coti/POSH.

**Acknowledgment.** Experiments presented in this paper were carried out using the Grid'5000 testbed, supported by a scientific interest group hosted by Inria and including CNRS, RENATER and several Universities as well as other organizations (see https://www.grid5000.fr).

# References

1. Barrett, B., Squyres, J.M., Lumsdaine, A., Graham, R.L., Bosilca, G.: Analysis of the component architecture overhead in open MPI. In: Di Martino, B., Kranzlmüller, D., Dongarra, J. (eds.) EuroPVM/MPI 2005. LNCS, vol. 3666, pp. 175–182. Springer, Heidelberg (2005). https://doi.org/10.1007/11557265_25
2. Bland, W., Bouteiller, A., Hérault, T., Hursey, J., Bosilca, G., Dongarra, J.J.: An evaluation of user-level failure mitigation support in MPI. Computing **95**(12), 1171–1184 (2013)

3. Butelle, F., Coti, C.: Distributed snapshot for rollback-recovery with one-sided communications. In: 2018 International Conference on High Performance Computing & Simulation (HPCS), pp. 614–620. IEEE (2018)
4. Cappello, F., et al.: Grid'5000: a large scale and highly reconfigurable grid experimental testbed. In: SC 2005: Proceedings of the 6th IEEE/ACM International Workshop on Grid Computing CD, Seattle, Washington, USA, pp. 99–106. IEEE/ACM, November 2005
5. Cappello, F., Geist, A., Gropp, W., Kale, S., Kramer, B., Snir, M.: Toward exascale resilience: 2014 update. Supercomput. Front. Innov. **1**(1), 5–28 (2014)
6. Coti, C.: POSH: Paris OpenSHMEM: a high-performance OpenSHMEM implementation for shared memory systems. Procedia Comput. Sci. **29**, 2422–2431 (2014). 2014 International Conference on Computational Science (ICCS 2014)
7. Goglin, B., Moreaud, S.: KNEM: a generic and scalable kernel-assisted intra-node MPI communication framework. J. Parallel Distrib. Comput. **73**(2), 176–188 (2013)
8. Hammond, J.R., Ghosh, S., Chapman, B.M.: Implementing OpenSHMEM using MPI-3 one-sided communication. In: Poole, S., Hernandez, O., Shamis, P. (eds.) OpenSHMEM 2014. LNCS, vol. 8356, pp. 44–58. Springer, Cham (2014). https://doi.org/10.1007/978-3-319-05215-1_4
9. Hao, P., et al.: Fault tolerance for OpenSHMEM. In: Proceedings of the 8th International Conference on Partitioned Global Address Space Programming Models, PGAS 2014, pp. 23:1–23:3. ACM, New York (2014)
10. Linford, J.C., Khuvis, S., Shende, S., Malony, A., Imam, N., Venkata, M.G.: Performance analysis of OpenSHMEM applications with TAU commander. In: Gorentla Venkata, M., Imam, N., Pophale, S. (eds.) OpenSHMEM 2017. LNCS, vol. 10679, pp. 161–179. Springer, Cham (2018). https://doi.org/10.1007/978-3-319-73814-7_11
11. Luo, M., Seager, K., Murthy, K.S., Archer, C.J., Sur, S., Hefty, S.: Early evaluation of scalable fabric interface for PGAS programming models. In: Proceedings of the 8th International Conference on Partitioned Global Address Space Programming Models, p. 1. ACM (2014)
12. Shamis, P., et al.: UCX: an open source framework for HPC network APIs and beyond. In: 2015 IEEE 23rd Annual Symposium on High-Performance Interconnects, pp. 40–43. IEEE (2015)

# Click-Fraud Detection for Online Advertising

Roman Wiatr[1,2(✉)], Vladyslav Lyutenko[1], Miłosz Demczuk[1], Renata Słota[2], and Jacek Kitowski[2,3]

[1] Codewise, 31-503 Cracow, Poland
rwiatr@gmail.com, vlyutenko@gmail.com, milosz.demczuk@codewise.com
[2] Department of Computer Science, AGH University, 30-059 Cracow, Poland
[3] ACK Cyfronet AGH, 30-950 Cracow, Poland
{roman.wiatr,renata.slota,jacek.kitowski}@agh.edu.pl

**Abstract.** In affiliate marketing, an affiliate offers to handle the marketing effort selling products of other companies. Click-fraud is damaging to affiliate marketers as they increase the cost of internet traffic. There is a need for a solution that has an economic incentive to protect marketers while providing them with data they need to reason about the traffic quality. In our solution, we propose a set of interpretable flags explainable ones to describe the traffic. Given the different needs of marketers, differences in traffic quality across campaigns and the noisy nature of internet traffic, we propose the use of equality testing of two proportions to highlight flags which are important in certain situations. We present measurements of real-world traffic using these flags.

**Keywords:** Fraud detection · Click fraud · Real-time event processing · Online advertising

## 1 Introduction

Affiliate marketing [9] is performance-based marketing where an affiliate offers to handle the marketing effort selling products of other companies. Affiliate marketer takes the financial risk related to running the marketing campaign in return for a commission. Affiliate marketer may buy internet traffic from a website (publisher) or an advertising network (ad-network) [24] for the purpose of displaying advertisements (ads). Whatever the source is, it is crucial to capture web traffic properties for booking and analytical purposes. Knowing how the traffic behaves lets affiliate marketers optimise their strategy lowering the costs and increasing revenue.

Affiliate marketers (from now on referred to as marketers) can be subject to fraudulent traffic. One type of fraudulent traffic is click-fraud [23]. A publisher or ad-network can attempt to inflate [16,17] the number of visitors of his site by generating artificial traffic. The artificial visitor will click on an advertisement to increase the credibility of the traffic. Another way to increase the number

© Springer Nature Switzerland AG 2020
R. Wyrzykowski et al. (Eds.): PPAM 2019, LNCS 12043, pp. 261–271, 2020.
https://doi.org/10.1007/978-3-030-43229-4_23

of clicks, a site or mobile application generates, can be achieved by placing ads that a real user can click without intent [7]. A fraudster may use a botnet [20] to generate a low-frequency click attack [6] targeting a site or a specific campaign. In [12] Kim et al. propose a budget-draining attack directed at retargeting campaigns. Fraudulent traffic does not only include attacks based on clicks. Users of Voluum Tracker (Voluum TRK, Subsect. 4.1) also report fake mobile application instals. Fake app installs can be generated by mobile device farms or by mobile device emulators. Presence of fraud in the ad-networks puts additional pressure on the infrastructure serving data with low latency such as Demand Side Platforms (DSP) [22], in turn generating additional costs for the advertiser.

As stated in [14] selling any clicks might be beneficial for DSPs or ad-networks but click-fraud is damaging for the marketers as they increase the cost of traffic. Taken this into account there is a need for a solution that has an economic incentive to protect marketers while providing them with data they need to reason about the traffic quality.

This paper is organized as follows. Section 2 describes the state of the art and its shortcomings. Section 3 describes how we intend to improve the state of the art with our solution. Section 4 describes Anti-Fraud System extension of Voluum Tracker, data used for the analysis and experimental settings. Section 5 describes results of our work. Finally, Sect. 6 presents the summary of our work and the conclusions.

## 2    State of the Art

In [7] authors estimate the amount of possible fraudulent clicks by using a control ad designed in a way that it is unlikely for a human to be interested in it. The technique requires the advertiser to actively measure click-fraud thus required to pay for extra traffic used on the control ad. After manually analyzing the clicks from the control ad the authors present fraud signatures that can be used to identify individual clicks. The work shows that sophisticated ad fraud is a problem that even major ad-networks cannot handle. It does not provide a solution for a user to reason about traffic quality without setting up the control ad and it does not provide a tool to fingerprint individual clicks without investigation. In [8] ViceROI method based on observing finding publishers with anomalous revenue per user distributions is presented. It requires a baseline with examples of trusted publishers for comparison. While ViceROI is a disclosed system. It serves as an addition to a system which insights are not revealed to the public. Given we have access to the data required by ViceROI we could incorporate it into our system as an additional flag. In [17] an algorithm named SLEUTH is presented. It is a method for capturing hit inflation attacks driven by a single publisher. In [16] a complementary method named DETECTIVES is proposed for detecting coalitions of fraudulent publishers that lunch hit inflation attacks. The methods are applicable only to fraud generated by botnets. In [21] authors present a framework for detecting machine generated traffic by looking at the estimated number of users sharing the same IP. The method uses an external component to estimate the number of users per IP [15] to which marketers will not have access. Kitts et al. in [13] describe a rule-based system for click fraud detection where the rules are hidden from the public to protect them from reverse engineering. The described systems processes 1 billion events per hour which is equal to about 10% of US advertisement traffic. No marketer is able to buy, and let alone analyze that amount of traffic. This creates a market niche for an open solution with a incentive to protect marketers. In [10], authors use a Complex Event Processing engine as a tool for rule-based anomaly detection. The lack of labelled data can be partially addressed by using an unsupervised anomaly detection approach as suggested in [4]. There are also attempts to use machine learning to tag click-fraud. In [18] authors are combining information from ad-networks and advertisers about click-fraud to train three types of models achieving accuracy greater than 93%. The approach requires labelled data. In [19] work done by several teams during FDMA 2012 Competition is presented. The organizer provided mobile click database and a list of labelled publishers. The aim of the competition was to detect fraudulent publishers. The publication provides useful insights on how the models were built and what features were used.

## 3    Research Goals and Approach

Our goal is to tag fraudulent clicks in a way that enables a marketer to take action upon. The marketer tracking the traffic should be able to verify if the campaigns he is running are affected by fraud and if so he has to be able to provide a clear explanation on which clicks were fraudulent. Having a clear description of the traffic he can now either block the subset of fraudulent traffic or be entitled to a refund after providing the explanation to the publisher or ad-network.

For this purpose, we propose a set of online computed and interpretable flags describing each click such as data centre traffic, frequent clickers, etc. as well as more sophisticated yet explainable ones to describe the traffic. Given the different needs of marketers, differences in traffic quality across campaigns and the noisy nature of internet traffic, we propose the use of equality testing of two proportions [11] to highlight flags which are important in certain situations. We present measurements of real-world traffic using these flags.

## 4    Anti-Fraud System

The goal of the Anti-Fraud System (AFS) is to produce results that describe suspicious traffic, so the marketers are able to take actions upon the information feedback of the system. Hence the solution has to be opened to the marketers.

### 4.1    Voluum TRK Overview

Voluum TRK is an advertisement tracking software designed by Codewise to help marketers run and track their campaigns. In Voluum TRK campaign is a

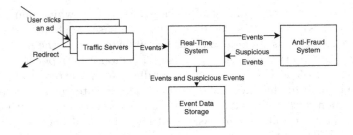

**Fig. 1.** Voluum TRK architecture overview.

set of traffic sources (publishers, ad-networks, etc.) from which the marketer buys traffic and a set of destinations to which the traffic will be redirected. The system architecture overview is presented in Fig. 1. An event is created every time by the Traffic Servers when a user is redirected by clicking on an advertisement. Next the event is sent to the Real-Time System (RTS). After processing it is accessible for the marketer in a time series aggregated form and an event log. In the same time, each event is processed by the AFS. The AFS flags each event depending on user behaviour, source, or event properties and send it back to the RTS. The AFS uses Flink [5], a streaming processing framework based on window functions [3], serving as an extension to the RTS and the data produced by it are simply visible with a delay compared to regular events. The RTS may be considered a stream processing system without the usage of window functions.

## 4.2  Data

Figure 2 illustrates a time line of tracking ads using Voluum TRK. In essence Voluum TRK may be used to track four types of consecutive events: Impressions, Visits, Clicks and Conversions. In Voluum TRK nomenclature Impression is an event created when an ad is displayed; Visit is a redirect to a Landing Page (or Offer Page) after the ad was clicked; Click is an action taken on the Landing Page; Conversion is a target action taken by the user on the offer page i.e. making a purchase, subscription or install. Figure 3 shows common state transitions in the system. The most complete path is as follows:

1. User visits a site and advertisement is displayed, an Impression is generated;
2. User clicks on the advertisement and is redirected to the Landing Page, a Visit is generated;
3. User clicks through the Landing Page, a Click is generated;
4. User takes action on the Offer Page, a Conversion is generated.

Usually marketers do not track Impressions because of the amount of events it would generate. Marketers can also resign from tracking Conversions or Clicks. There are special cases like mobile ads where the transition from the Landing Page to the Offer Page is not taken into account. The differences in how the system is used for different purposes, causes problems with evaluating the proposed traffic tags. If a marketer configures the campaign to not track Conversions, as in Fig. 3C, we will not be able to apply our method.

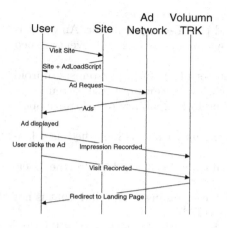

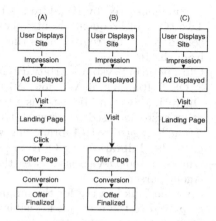

**Fig. 2.** Time line for tracking ads in Voluum TRK

**Fig. 3.** Common state transitions in Voluum TRK

For our studies, Codewise provided a sample of anonymized production data containing $2.4 \times 10^9$ events divided into two data sets: Visit to Conversion transitions (Fig. 3A, B) and Click to Conversion transitions (Fig. 3A). Next, we grouped the data by campaigns (see Subsect. 4.1) and create a summary per proposed tag.

Clicks are subsequent events to Visits. This has two consequences for our method:

- Many clicks produced by a user interacting with a Landing Page may come from one visit (we account for this by using only the last click associated with a single visit).
- If the advertiser is charged per Visits or Impression, the "Click to Conversion" data-set will not show the accurate result for campaign cost as most of the Visits don't produce Clicks. We still find this information useful as it shows how users interact with the Landing Page and can point to problematic segments of the traffic.

### 4.3   Experimental Settings

As mentioned in Sect. 4 we design our flags to make them interpretable for marketers, giving them insights on the traffic they buy from publishers or ad-networks and to decide whenever to buy the traffic or to take actions to block or attempt to get a refund. When an event such as a Click or Visit enters the system, the AFS will attempt to tag it with one of the flags. If an event is flagged it will then returns to the main processing and will be displayed to the marketer as suspicious. The set of proposed flags contains:

- Events generated by IP addresses marked as data centres.
- Two types of Frequent Events: per IP and per IP-campaign pair. An event is considered to be frequent if in the previous minute at least 100 events issued from the IP.
- Old operating system. Arbitrarily chosen set of operating systems: Android 3, OS X 10.6, iOS 6, Windows Vista or earlier versions of these systems.
- Operating System Mismatch is a tag basing P0F [2] ability to recognize operating systems based on network traffic.
- Event is tagged with Library flag when the user agent is identified as a language level libraries (e.g OkHttp [1]).
- Unrecognized Device is devices that could not be recognized by the User-Agent parser.
- Invisible Link is a tag for the number of clicks coming from a link that is not displayed to the user but is visible in the HTML code.
- Fast clickers. Event is tagged when this flag when the click on the Landing Page was executed in less than 800 ms. The threshold was determined experimentally. In some cases, it is better to change the threshold per campaign.

In addition, we implemented an offline version of SLEUTH [17]. It is an algorithm that can be used to find single publisher hit inflation attacks. The algorithm can be configured using two variables: $\phi$ defining the $\phi$-*frequent* IPs for every publisher and $\psi$ defining the $\psi$-*frequent* publishers for every IP. We used SLEUTH for $p = \phi = \psi$ to compare the results with our findings. In the system, we have the same publishers under different traffic source IDs between marketers. This means a single publisher can be found multiple times under different IDs. This can be overcome by lowering $p$ value compared to the original paper.

The goal is to pinpoint flags with high importance for a specific campaign. Given the noisy nature of internet traffic, diverse needs of marketers, different techniques of attacks and the simplicity of the tags it is important to reduce the number of false positives.

$$Z = \frac{(p_1 - p_2)}{\{P(P-1)(\frac{1}{n_1} + \frac{1}{n_2})\}^{\frac{1}{2}}} \tag{1}$$

$$P = \frac{p_1 n_1 + p_2 n_2}{n_1 + n_2} \tag{2}$$

As the next step, we propose the use of z-test (Eqs. (1) and (2)) for the equality of two proportions $p_1$ and $p_2$ [11] to highlight what tags are important in certain situations. In our case the proportions are conversion rates calculated based on Conversions and amount regular events $n_1$ and tagged ones $n_2$. Using the test we find flag and campaign pairs for which we can reject the null hypothesis that there is no difference in the conversion rate of flagged and not flagged traffic.

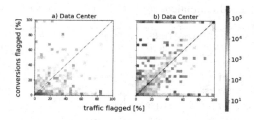

**Fig. 4.** Click to conversion data for Data Center flag.

# 5   Results

We present the experimental results as a 2D histogram with the % of flagged not converting traffic on the x-axis and the % of flagged conversions on the y-axis per campaign. The histogram is weighted by the amount of flagged traffic in the campaign. We think that this method of presenting the data is useful for the marketer because he can easily reason about what would be the potential impact of blocking flagged traffic. Looking at Figs. 4, 5 and 6 we are interested in the region under the diagonal and near the x-axis. When a campaign falls into this region it means that the flag events do not bring conversions thus they will not give the payout to the marketer. If a campaign lies near the diagonal it means that it does not differ from regular traffic in terms of conversion rate. As we show in Fig. 4 the campaigns near the diagonal will be most likely rejected by the z-test. When a campaign is above the diagonal and near the y-axis it means that flagged traffic is converting more than expected. It can also be fraudulent behaviour but our work does not focus on this type of fraud.

## 5.1   Null Hypothesis Rejection

Campaign is considered flagged by *flag X* if it contains at least one event flagged by *flag X*. A suspicious campaign is a flagged campaign where the null hypothesis was rejected. For the experiment the confidence interval was set to 95%. Rejecting flagged campaigns for whom the null hypothesis could not be rejected reduces the number of campaigns significantly.

Figure 4 shows regions which were affected by rejecting the null hypothesis rejection. Figure 4a shows only traffic from campaigns where the null hypothesis was rejected. Figure 4b shows only traffic from campaigns where the test was not able to reject the null hypothesis. As expected campaigns with a low amount of events were rejected as well as campaigns near the diagonal where differences in conversion rates were minimal. Table 1 shows the per cent of flagged campaigns rejected by the null hypothesis and the amount of flagged traffic that is a part of these campaigns. For example for Data Center 95.7% flagged campaigns containing 60.9% flagged events were rejected. This leaves only 4.3% of campaigns where a difference of conversion rate was detected and should be investigated further. Similar as in recent work [7], the flags point to a certain set of events

**Table 1.** Rejected flagged campaigns and traffic for Clicks to Conversions transition

| Flag | Campaigns rejection | Traffic rejected | Flag | Campaigns rejection | Traffic rejected |
|---|---|---|---|---|---|
| Frequent events per IP | 94.9% | 35.7% | Frequent events per IP and campaign | 81.7% | 49.1% |
| Invisible link | 91.7% | 61.4% | | | |
| Library | 97.7% | 96.7% | Data center | 95.7% | 60.9% |
| Operating system mismatch | 97.9% | 74.1% | Fast clicker | 94.1% | 58.2% |
| Old operating system | 98.6% | 71.4% | Unrecognized user agent | 98.0% | 14.9% |
| Unrecognized device | 98.6% | 99.9% | SLEUTH p = 0.001 | 90.0% | 76.7% |
| SLEUTH p = 0.005 | 92.6% | 76.6% | SLEUTH p = 0.01 | 93.4% | 74.6% |
| SLEUTH p = 0.05 | 96.4% | 76.0% | SLEUTH p = 0.1 | 97.5% | 77.4% |

and manual investigation or campaign settings adjustment may be required to reduce the number of flagged events. It must be noted that in certain situations the flags can be the proof of bad traffic quality or fraud on their own.

## 5.2   Flag Details

Figure 5 depicting transitions from Clicks to Conversions shows higher concentrations of points near the x-axis for every flag, meaning that there is a set of campaigns where flagged traffic does not convert or rarely converts. There is an interesting difference between Fig. 5b and its counterpart calculated for every campaign Fig. 5c. While the first enables sharing information about the IP click frequency between campaigns and marketers and marks more traffic as suspicious it is also less accurate. It might be due to the fact that Fig. 5c for campaigns with lower amount of traffic is less likely to capture users with shared IP. Figure 5c has a cluster of points near the x-axis while Fig. 5b has a bigger cluster near the y-axis meaning a higher amount of false positives. Table 1 also shows that rejection rate for Fig. 5c is lower than for Fig. 5b. Figure 5i tags are very rare compared to other tags. This is caused by the fact that it requires the active participation of the marketer to deploy a Landing Page with an additional script. Flags based on User Agent: Fig. 5d, f, g, h have the best accuracy when it comes to pinpointing expendable traffic. Clicks flagged by Fig. 5g are rejected as insignificant 99% of the time and clicks flagged by Fig. 5h are rejected only 14.9%. Figure 5j shows a large portion of campaigns is below the diagonal. This indicates that using Landing Page behaviour denoted by Clicks as an indication of low-quality traffic is possible. SLEUTH flag reduces its sensitivity when increasing $p$ parameter as described in the original paper. The larger the parameter the less likely it marks a false positive but also reduces the number of true positives found. Comparing SLEUTH to our tags we can see similar clustering near the x-axis meaning that traffic tagged by this flag has an abnormally low conversion rate compared to the normal traffic. It has to be noted that SLEUTH does not discover the same group of traffic that the other tags do, thus it can be used as an supplement for our method.

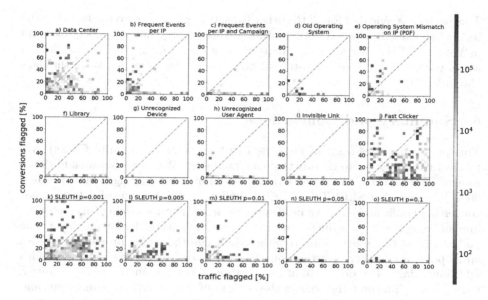

**Fig. 5.** Transitions from Clicks to Conversions

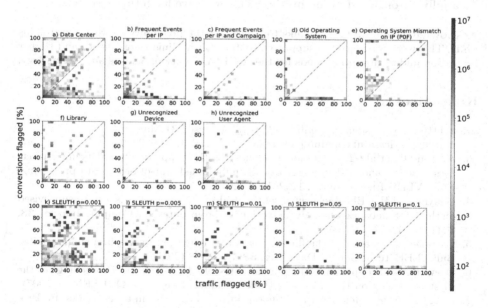

**Fig. 6.** Transitions from Visits to Conversions

Figure 6 shows less satisfaction results of the traffic flags. It does not contain Fig. 5i and j as these flag work only for clicks on a Landing Page. Comparing Fig. 6a flags with Fig. 5a it is noisier and has a large cluster just over the diagonal.

This pattern is repeated for all the other flags. It might be due to the fact that the test is less likely to reject an observation if it has a large number of samples and in Voluum TRK visits are more frequent than clicks. Even with the additional noise, there are still visible clusters of points near the x-axis for all of the flags like in Fig. 5.

## 6    Summary and Conclusions

This paper discusses a set of flags used to tag events (Visits and Clicks) in a traffic stream in order to classify them as fraudulent. We show that an additional level of filtering using to a z-test reduces the number of campaigns by more than 90%. We show that SLEUTH [17] gives similar results in terms of detecting not converting traffic and can be a used supplementary method. Using this method a human being is able to deduct which campaigns are affected by low-quality traffic and can treat this information to either challenge the traffic providers or as an entry point for further investigation. This method can be easily extended with additional flags. The downside of this method is that, as mentioned previously, a user has to interpret the flags in the context of the advertising campaigns and the traffic source he is working with and it is challenging to use this information for a fully automatic decision making. This we leave for future research.

**Acknowledgements.** R.W. thanks Codewise for the possibility to work towards his Ph.D. This work was partially supported by the Polish Ministry of Science and Higher Education under subvention funds for the AGH University of Science and Technology.

## References

1. okhttp. https://square.github.io/okhttp/. Accessed 31 Mar 2019
2. p0f. http://lcamtuf.coredump.cx/p0f3/. Accessed 31 Mar 2019
3. Akidau, T., et al.: The dataflow model: a practical approach to balancing correctness, latency, and cost in massive-scale, unbounded, out-of-order data processing. Proc. VLDB Endow. **8**(12), 1792–1803 (2015)
4. Bhuyan, M.H., Bhattacharyya, D.K., Kalita, J.K.: Towards an unsupervised method for network anomaly detection in large datasets. Comput. Inform. **33**, 1–34 (2014)
5. Carbone, P., et al.: Apache flink: Stream and batch processing in a single engine. Bull. IEEE Tech. Comm. Data Eng. **38**(4), 28–38 (2015)
6. Daswani, N., Stoppelman, M.: The anatomy of Clickbot.A. In: Proceedings of the First Workshop on Hot Topics in Understanding Botnets, p. 11. USENIX (2007)
7. Dave, V., et al.: Measuring and fingerprinting click-spam in ad networks. In: Proceedings of the ACM SIGCOMM 2012, pp. 175–186. ACM (2012)
8. Dave, V., et al.: ViceROI: catching click-spam in search ad networks. In: Proceedings of the 2013 ACM SIGSAC, pp. 765–776. ACM (2013)
9. Duffy, D.L.: Affiliate marketing and its impact on e-commerce. J. Consum. Mark. **22**(3), 161–163 (2005)
10. Frankowski, G., et al.: Application of the complex event processing system for anomaly detection and network monitoring. Comput. Sci. **16**(4), 351–371 (2015)

11. Kanji, G.K.: 100 Statistical Tests, p. 27. Thousand Oaks, SAGE (2006)
12. Kim, I.L., et al.: AdBudgetKiller: online advertising budget draining attack. In: Proceedings of the 2018 World Wide Web Conference on World Wide Web, pp. 297–307. International World Wide Web Conferences Steering Committee (2018)
13. Kitts, B., et al.: Click fraud detection: adversarial pattern recognition over 5 years at microsoft. In: Abou-Nasr, M., Lessmann, S., Stahlbock, R., Weiss, G.M. (eds.) Real World Data Mining Applications. AIS, vol. 17, pp. 181–201. Springer, Cham (2015). https://doi.org/10.1007/978-3-319-07812-0_10
14. Kshetri, N., Voas, J.: Online advertising fraud. Computer **52**(1), 58–61 (2019)
15. Metwally, A., Paduano, M.: Estimating the number of users behind IP addresses for combating abusive traffic. In: Proceedings of ACM SIGKDD, pp. 249–257. ACM (2011)
16. Metwally, A., et al.: Detectives: detecting coalition hit inflation attacks in advertising networks streams. In: Proceedings of the WWW Conference, pp. 241–250. ACM (2007)
17. Metwally, A., et al.: SLEUTH: single-publisher attack detection using correlation hunting. Proc. VLDB Endow. **1**(2), 1217–1228 (2008)
18. Mouawi, R., et al.: Towards a machine learning approach for detecting click fraud in mobile advertizing. In: 2018 IEEE IIT Conference, pp. 88–92. IEEE (2018)
19. Oentaryo, R., et al.: Detecting click fraud in online advertising: a data mining approach. J. Mach. Learn. Res. **15**(1), 99–140 (2014)
20. Silva, S.S., et al.: Botnets: a survey. Comput. Netw. **57**(2), 378–403 (2013)
21. Soldo, F., Metwally, A.: Traffic anomaly detection based on the IP size distribution. In: 2012 IEEE INFOCOM, pp. 2005–2013. IEEE (2012)
22. Wiatr, R., Słota, R., Kitowski, J.: Optimising Kafka for stream processing in latency sensitive systems. Procedia Comput. Sci. **136**, 99–108 (2018)
23. Wilbur, K.C., Zhu, Y.: Click fraud. Mark. Sci. **28**(2), 293–308 (2009)
24. Yuan, Y., et al.: A survey on real time bidding advertising. In: Proceedings of 2014 IEEE SOLI Conference, pp. 418–423. IEEE (2014)

# Parallel Graph Partitioning Optimization Under PEGASUS DA Application Global State Monitoring

Adam Smyk[1]([✉]), Marek Tudruj[1,2], and Lukasz Grochal[1]

[1] Polish-Japanese Academy of Information Technology,
86 Koszykowa Street, 02-008 Warsaw, Poland
{asmyk,tudruj}@pjwstk.edu.pl
[2] Institute of Computer Science, Polish Academy of Sciences,
5 Jana Kazimierza Street, 01-248 Warsaw, Poland

**Abstract.** The paper concerns the use of global application states monitoring in distributed programs for advanced graph partitioning optimization. Two strategies for the control design of advanced parallel/distributed graph partitioning algorithms are presented and discussed. In the first one, the parallel algorithm control runs on top of the ready to use basic graph partitioning functions available inside an existing graph partitioning METIS tool. The second control strategy is based on a genetic programing algorithm in which the applied basic graph partitioning primitives and the overall algorithmic parallel/distributed control can be freely designed by the user. In these strategies, the graph partitioning control is executed by processes/threads supervised by the application global states monitoring facilities provided inside a novel distributed program design framework PEGASUS DA. This framework provides system support to construct user-defined strongly consistent global application states and an API to define corresponding execution control. In particular, it concerns computing global control predicates on the constructed global states, the predicates evaluation and asynchronous execution control handling to obtain application global state-driven reactions. Based on such implementation, different features of the graph partitioning optimization strategies have been designed and tested. The experimental results have shown benefits of the new graph partitioning control methods designed with the use of the application global states monitoring.

**Keywords:** Parallel graph/mesh partitioning · Distributed program design tools · Application global states monitoring

## 1 Introduction

The paper concerns the methodology for advanced parallel/distributed graph partitioning optimization with the support of application global states monitoring. Efficient graph partitioning algorithms are required for solving many

© Springer Nature Switzerland AG 2020
R. Wyrzykowski et al. (Eds.): PPAM 2019, LNCS 12043, pp. 272–286, 2020.
https://doi.org/10.1007/978-3-030-43229-4_24

optimization problems in such domains as physical simulations, economics, social networks. To obtain optimal graph partitioning (NP-complete problem [5]) some direct (based on the cut-min optimization) [4] or iterative techniques [1–6] can be used. Such techniques are implemented inside some optimization frameworks as METIS [2], ParMETIS [3], Scotch project [8], Chaco [9], Zoltan [9] or PaToH [10]. The frameworks can be controlled directly from the command line or by the API (usually from C language). In order to find optimal graph partitions some control parameters must be correctly selected and specified.

Important research interest in the design of graph partitioning tools is focused on solvers for large-scale graph processing like Giraph++ [16]. It supports vertex-centric and graph-centric models. Its local nature of computations (especially in the graph-centric model) enables asynchronous work based on subgraphs partitioning. An original graph partitioning method based on the multilevel approach is embedded inside the KaFFPa framework [17]. It is based on the max-flow and min-cut computations joined with global search used in multi-grid linear solvers. KaFFPaE [18] assumes a distributed evolutionary approach to the graph partitioning using different searching strategies like F-cycle or V-cycle [21]. In this approach, modified crossover and mutation increase the diversity of individuals. It enables more efficient exploration of the solution space. KaFFPaE does not take into consideration the architectural features of the graph partitioning target system. A global result is received as the local partition improvement in one cycle, but the cycle length depends on the chosen implementation.

There are many other methods based on the local search that work in a similar way to JA-BE-JA-VC [20] which in stochastic way improves the initial partition. It works in a strongly local way but to avoid falling into local optima the simulated annealing techniques are used. Another method – different from the multilevel approach - is based on the analytical approach joining a label propagation technique with the linear programming method [19]. Some other tools for graph partitioning provide the streaming [11] or sorted [12] approaches especially for processing large graphs.

The described above methods use many different algorithmic solutions and mechanisms for graph partitioning optimization. Their main interest is focused on the essential partitioning methods with relatively smaller attention paid to the optimization of the internal control embedded in the algorithms. The aim of our paper is to pay more attention on the methods for global control embedded in the algorithms implemented in distributed way. We propose here to use the global graph partitioning execution control based on the distributed algorithm global states monitoring using a novel distributed program design and execution framework. This approach enables easy design of the partitioning algorithms to account for the influence of particular options of graph division on the quality of obtained results and for dynamic control of the optimizing process.

In this paper, we present, study and compare two general (generic) control strategies for the design of advanced parallel/distributed graph partitioning algorithms. In the first strategy, the parallel algorithm control runs on top of the ready to use basic graph partitioning functions available inside an existing graph partitioning tool. The second control strategy assumes a general algorithmic basis (a framework)) inside which the basic graph partitioning primitives and the overall algorithm control can be designed by the user.

As a means for integration of the performed studies we have introduced a Parallel Graph Partitioning Model (PGPM). It is an iterative optimization parallel algorithmic skeleton enabling testing different methods and parameter options to find the best graph partitions for some established criteria. This model additionally enables partitioning subgraphs of a given graph using the "divide and conquer" strategy. The proposed PGPM model has been implemented using two general control strategies mentioned above, however, embedded inside some realistic application environments. The first strategy assumes that the general graphs partitioning control is built on top of the popular graph processing METIS tool [4,5]. Under this control, the METIS partitioning functions are executed iteratively and in parallel to find the best graph partitions which satisfy required graph theory specific optimality criteria. The second control strategy for implementation of the PGPM model assumes building graph partitioning control on top of the genetic programming (GP) [14]. GP was used to find the sequence of graph partitioning steps which transform a computation initial data flow graph into the final macro data flow graph which has to satisfy some application-specific optimality criteria. The partitioned graph represents distributed computations used for simulation of electromagnetic wave propagation solved by the FDTD (Finite Difference Time Domain) approach [15] in a rectangular computational mesh. The resulting partitioned graph (a macro data flow graph) is mapped onto a distributed executive system to provide the minimal execution time of the FDTD-based simulation for given wave propagation areas.

In the research reported in this paper, the parallel graph partitioning is based on task distribution among co-operating computational threads. Threads produce many partitioning solutions of the same graph and various control parameters direct the flow of the iterative graph partitioning. In our case, three partition assessment criteria can be used: minimal edge-cut, minimal total communication volume and balanced computational load [2]. Depending on them, the best solution is chosen or the optimization process can be repeated for selected subgraphs.

The graph partitioning control is based on parallel multithreaded processes governed with the use of the application global states monitoring facilities provided in a novel distributed program design framework PEGASUS DA (Program Execution Governed by Asynchronous Supervision of States in Distributed Applications) [7]. It is a first runnable contemporary system of this kind which enables a fully system supported design of distributed program global execution control in the C/C++ languages. The resulting distributed applications are built using two kinds of program elements: computational processes based on threads (for application program computations) and special control processes/threads (called synchronizers) which are responsible for application program global states construction and use. The synchronizers collect local state information from application processes/threads, construct useful global application states, verify predicates specified on these states and, based on these predicates assessment, send control signals to computational processes/threads to influence their execution. All program elements, including the composition of the monitored global states and their use, are programmer definable with the use of the API and the control flow representation provided in PEGASUS DA.

The use of PEGASUS DA enables an easy identification of the algorithm control part providing clarity in the design of the graph optimization process. Many graph partitions can be found simultaneously by process threads and optimization control can be easily designed based on the analysis of the global states of the graph partitioning distributed application. Comparative studies of the multi-variant graph partitioning can be easily done by the experiment control flow restructuring. It allowed carrying out tests using the PEGASUS DA system environment for many combinations of algorithmic input options.

The paper contains 4 main sections which follow the Introduction (Sect. 1). In Sect. 2, the basic general features of the PEGASUS DA framework are described. In Sect. 3, the applied parallel graph partitioning model (PGPM) is presented. In Sects. 4 and 5, implementation of the two studied graph partitioning algorithms is described. In the last section, the experimental results are presented and discussed.

## 2 Distributed Application Global State Monitoring in PEGASUS DA

The PEGASUS DA distributed program design framework enables a programmer to design a system-supported parallel/distributed program execution control which can be optimized based on automatic monitoring of the program global states. In distributed programs designed and executed under the PEGASUS DA framework an application consists of multithreaded distributed processes programmed in C/C++. In order to design a distributed program global control a special control infrastructure is delivered, in which a programmer is supported by a communication library for acting on control data relevant for global state monitoring and an API for respective global control design. Local communication and synchronization between threads/processes inside processors is done by shared memory primitives. For communication between processors the MPI library or sockets are used. The main element of the control infrastructure for global state monitoring is called a synchronizer, Fig. 1. We have global synchronizers and local synchronizers which care about the application global states at the levels of sets of processes and threads, respectively. Global application states are defined by a programmer to contain local states of selected processes or threads in a distributed application A local synchronizer is a control thread which collects the local state messages from threads of a process to determine if the strongly consistent global state (SCGS) of these threads has been reached. A global synchronizer is a process which detects the strongly consistent global states at the multiple process level. A PEGASUS DA SCGS is defined as a set of fully parallel local states detected based on local state messages with timestamps from processor clocks synchronized with known accuracy. A user defines a distributed application control flow graph with control flow switches sensitive to global control signals representing the application global states monitored by synchronizers using predicates defined on the global states. The switches implement global state driven control constructs which direct control flow to processes/threads selected by synchronizer predicates. The C++ code is automatically generated by the compiler based on the global control flow graph, control data communication library and user specified global control attributes: monitored global states based on local states, synchronizer predicates, state reports, control signals, control communication parameters. For a detailed description of the PEGASUS DA framework, including its application to the Travelling Salesman Problem by the Branch and Bound method see [7]. For the application to parallel event-driven simulation see [22].

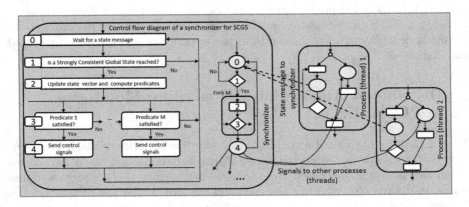

**Fig. 1.** Synchronizer co-operation with processes or threads

## 3    Parallel Graph Partitioning Model - PGPM

The general scheme of our Parallel Graph Partitioning Model (PGPM) is presented in Fig. 2. We propose here an iterative partitioning optimization algorithm which enables testing different options to select such which produce the best graph partitions. The algorithm can also partition subgraphs of graphs using the "divide and conquer" strategy. To analyze the subgraphs, different sets of partitioning options can be used. Partitioning can be recursively repeated until a satisfactory partition is obtained or can be stopped when no improvement is observed.

The proposed general partitioning algorithm can be easily implemented and controlled using the PEGASUS DA framework. It allows applying many combinations of possible local options and global control criteria at the same time.

This mainly applies to the parallel partitioning phase. The only limitation is the amount of the available main memory for the partitioning processes. However, our approach enables partitioning big graphs which cannot be stored in whole in the main memory by acting on subgraphs processed separately.

The general scheme described above can be implemented in many ways. We have implemented as a model for hierarchical graph partitioning with static assignment of the partitioning requirements and also for dynamic graph partitioning with a GP optimization support. As we will see in next sections this model can be implemented using ready to use graph partitioning tools (METIS) or it can be programmed by users from scratch (GP) to perform globally supervised dynamic graph partitioning.

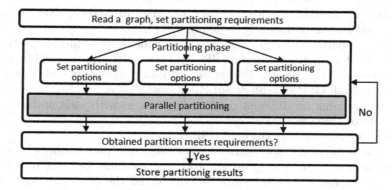

**Fig. 2.** Parallel graph partitioning model (PGPM)

## 4 Parallel Graph Partitioning Based on METIS Framework

As the first implementation of the PGPM model we studied algorithms aiming at general graphs partitioning built on top of the METIS toolkit. Its partitioning functions were executed in parallel under control driven by program global states monitoring in PEGASUS DA which used the MPI library. This made that the METIS tool was used instead of ParMETIS in which the conflicting MPI library was also used. The hierarchical partitioning enabled acting on chosen subgraphs represented as already created partition components. METIS can partition an input graph into a given number of subgraphs using different methods. Each partitioning method can be parameterized according to different options. So, we can choose different partitioning methods e.g.: multilevel recursive bi-sectioning or multilevel k-way partitioning. Also, we can specify what kind of partition optimization should be applied e.g.: edge-cut minimization or total communication volume minimization. The way of initial graph partitioning for a method also can be chosen: bisection with a greedy strategy, bisection with a random refinement, separator from an edge cut, bisection with a greedy node-based strategy. For definitions of the above notions and more information on other possible options read the METIS manual [2]. The proposed graph partitioning optimization algorithm works based on the following steps:

1. Setting partitioning requirement (partition type options and optimization criteria). In our case, such optimization criterion can be: the edge cut minimization or the total communication volume minimization.
2. Parallel graph partitioning with METIS – done for different statically defined partition types and optimization criteria by each constituent distributed process/thread.
3. Assessment of so obtained graph partitions using an assumed criterion.
4. Selection of the best partition P from obtained in step (2). Next steps are performed for the partition P chosen in step 4.
5. Selection of partition components SPs in P, which will be further partitioned recursively for a defined criterion. SPs with the worst criterion will be chosen.
6. Extraction of SPs from P – the set of partition components will be seen as a separate graph GSP (Graph of SubPartitions).
7. GSP partitioning for different partitioning options with assignment to given processes or threads.
8. Pasting the resulting partition of GSP to G.
9. Selection of the best graph partition after subgraphs partitioning (from many partition versions of the given graph). The partition for next partitioning steps is selected. The selection criterion can be the minimum edge cut or total communication volume.

The experiments for this algorithm will be described in next sections.

## 5   Parallel Graph-Partitioning by Genetic Programming

The second implemented and tested version of the PGPM model was graph partitioning by genetic programming (GP) [14] used in solving the optimized distributed FDTD problem (Finite Difference Time Domain) [15]. In this problem, a set of partial differential equations describing electromagnetic wave propagation is solved by optimized in time computations carried on in a 3-D mesh mapped onto processors of a distributed system. The optimization goal is such assignment of the computational cells of the mesh to processors of a distributed system which assures the minimal simulation time. It is obtained by node merging equivalent to the computation data flow graph partition defining the nodes of a macro data flow graph (graph node coarsening). Before the genetic programming starts, the data flow graph is initialized for partitioning. It is done in two steps. In the first step, some initial leader data flow nodes are identified in the computational cell graph (mesh) of the area. Each leader node is a prototype of a macro node being an object of further node merging to create final macro data flow nodes to be assigned for execution to separate processor nodes. In the second step, the final macro nodes are created by optimized data flow and macro data flow nodes merging. It is done by inserting data flow nodes into the initial macro nodes closest to a leader. The number of initial macro nodes is usually much bigger than the assumed number of processors in a given computational system. The genetic programming will be used to find a sequence of steps which transforms initial macro nodes into final macro nodes creating the optimal macro data flow graph (OMDFG). OMDFG corresponds to the optimal node partition for a given wave propagation area which provides the minimal execution time of the FDTD simulation program on a given number of processors. To reduce the number of nodes, ten node merging rules (MRx) have been proposed based on three criteria: computational load balancing (CLB), communication optimization: edge-cut reduction (COECR) and computational load balancing with edge cut reduction (CLBECR).

**Table 1.** Node merging rules

| Id | Objective | Merging operation |
|----|-----------|-------------------|
| MR0 | CLB | The two least loaded adjacent nodes will be merged |
| MR1 | CLB | The most loaded node will be merged with the least loaded adjacent node |
| MR2 | CLB | The least loaded node will be merged with the most loaded adjacent node |
| MR3 | COECR | Two most loaded adjacent nodes will be merged |
| MR4 | COECR | The node with the biggest comm. volume will be merged with its neighbour with the biggest comm. volume |
| MR5 | COECR | The node with the smallest comm. volume will be merged with its neighbour with the biggest comm. volume |
| MR6 | COECR | The node with the smallest comm. volume will be merged with its neighbour with the smallest comm. volume |
| MR7 | COECR | The node with the biggest comm. volume will be merged with its neighbour with the biggest comm. volume |
| MR8 | CLBECR | The least loaded node will be merged with the adjacent node with the biggest communication volume |
| MR9 | CLBECR | The least loaded node will be merged with the adjacent node with the lowest comm. volume |

The genetic programming finds in a stochastic and parallel way sequences of rules with merging operations that will be executed during the partitioning process. For rule selection, some subsets of the merging rules are determined. A partitioning process is in fact a program which performs a sequence of node merging rules with merging operations. The node identifiers for each merging operation are found by the genetic programming (GP). From the GP point of view, each sequence of merging rules defines an individual. The whole population of individuals is processed under the PEGASUS DA system to find the best partition. In Table 1, the proposed rules of merging operations are presented. No individual uses all rules at the same time. Individuals use selected subsets from the whole merging rule set. When an individual partitioning phase is finished, the local state of the individual is calculated which determines the quality of its partition. It contains information about the computational load imbalance, the cut-min value and the total execution time of the FDTD problem. The individual local state is sent to a process/thread synchronizer. The synchronizer remembers the best obtained state so far. Synchronizer predicates can determine whether in next iterations an individual can produce a better partition. If a currently received local state is the best one, it will be remembered as a current globally referenced solution. Each local state contains information on how often given operations were used to gradually improve the worse individuals. If a worse individual contains the operations with the same set of objectives as the best one, the rate of using them is gradually transferred to the local state. Otherwise, the state is remembered as the best one (for a given operation set), and in the future, either it can be used as a referenced state or it will be changed to a better one. In next iteration, merging operations will be stochastically chosen in accordance to gradually corrected distribution.

## 6    Experimental Results

In this paper, we have experimentally studied three control strategies for graph partitioning algorithms for selected benchmark graphs. The first strategy, studied only for reference purposes, consisted in using the METIS tool alone without any support for the application global state monitoring (METIS only strategy). It enabled obtaining a desired graph partition simply by execution of selected METIS functions. Two main studied control strategies enabled organizing advanced iterative graph partitioning scenarios which are based on hierarchical partitioning approach, dynamically changing partitioning methods and dynamic setting of optimization criteria. Both strategies were using the experiment execution control supported by the use of global application states monitoring inside the PEGASUS DA framework. One of the main studied strategies enabled using multiple METIS partitioning functions embedded in complicated graph partitioning scenarios organized and controlled by the PEGASUS DA framework (METIS+PEGASUS DA strategy). The other main studied strategy enabled designing user-defined graph partitioning primitives by GP together with the overall algorithmic control organized under PEGASUS DA (GP+PEGASUS DA strategy).

The implementation of all three strategies has been tested on a number of benchmark graphs (see Table 2). A distributed system consisting of eight multicore processors (AMD FX(tm)-3.1 GHz CPU and 8 GB RAM per node) providing up to 8 threads was used in graph partitioning. In the tests, the quality of each benchmark graph partition obtained by a unique use of METIS has been compared with that provided by our optimization algorithm which combined METIS with PEGASUS DA. The examined graph set contained graphs of 15000 to 1.5 million nodes. Each graph has been partitioned into 125 partitions. For METIS with PEGASUS DA, we have tested two partitioning criteria: edge-cut (EC) and total communication data volume (TCV), represented by separate predicates at process and thread levels. As we can see, the obtained partition quality in almost all cases (except for the copter2 graph and TCV criterion) was better for the METIS+PEGASUS DA version in comparison to that from METIS only (the lower value of the criterion is better). For graph Mdual the PEGASUS DA assisted version was better by about 40% for TCV criterion. For EC criterion the results were better on average by 5% to 10%. Only for the copter2 graph METIS produced a better partition. For a more detailed description of these experiments under PEGASUS DA see [13].

**Table 2.** Graphs used during static tests and experimental results.

| Graph | METIS | | | PEGASUS DA + METIS | | |
|---|---|---|---|---|---|---|
| | Criterion | Results | | Criterion | Results | |
| | | Edges | Data | | Edges | Data |
| 4elt | EC | 4276 | 4576 | EC | 4004 | 4324 |
| | TCV | 4232 | 4533 | TCV | 4016 | 4319 |
| copter2 | EC | 54899 | 37308 | EC | 54839 | 37621 |
| | TCV | 55567 | 36902 | TCV | 54214 | 37254 |
| Luxemb | EC | 682 | 1364 | EC | 611 | 1219 |
| | TCV | 687 | 1352 | TCV | 634 | 1260 |
| Mdual | EC | 32361 | 61194 | EC | 32223 | 60945 |
| | TCV | 35426 | 53886 | TCV | 32799 | 32799 |
| Belgium | EC | 2769 | 5532 | EC | 2501 | 4977 |
| | TCV | 2948 | 5707 | TCV | 2438 | 4849 |

The experiments with the GP applied for graph partitioning, described in Sect. 5, included the assistance of the PEGASUS DA control. The tests were done for many combinations of such parameters as the shape of the FDTD mesh, merging rules sets, computer system architectures with various number of processors, thread numbers per processor, communication link numbers between processors. First of all, (see Fig. 3), we have observed that the execution time of each iteration of GP significantly varied and strictly depended on two factors: the shape of the propagation area and the combinations of merging rule sets. The computational and memory complexity was different for different merging rules. The fastest operations did not analyze the neighborhood of the nodes which significantly speeded up the execution of the algorithm. They concerned computational load balancing criteria - the merging rule sets 01 (ie. MR0, MR1) or 89. For communication oriented merging rule sets (like 34567) a significantly larger execution time was observed (even 7 times for very regular shapes of the propagation area). The processing time for shapes with large regular elements also was significantly larger. It means that the GP had to perform more operations because the number of individuals in regular shapes was usually larger, and at the same time, nodes had more neighbors, and hence, the analysis itself took a longer time.

Next, we have analyzed the quality of obtained mesh cell (data flow graph) partitions. We focused on the partition load imbalance and the cut-min value. As it can be seen from Fig. 4, for computation oriented merging operations the load imbalance is significantly smaller compared to communication oriented ones. The difference is even eight times larger. The reverse relation is observed in the case of the cut-min factor (Fig. 5) where the computationally oriented merging rule sets (01 and 89) produced partitions with the highest cut-min value, whereas communication oriented ones gave cut-min even 5 times smaller.

We have also analyzed the obtained FDTD simulation time for different shapes of the computational area, Fig. 6. We can see that in all cases the shortest simulation times are produced by computationally oriented merging rule sets (01, 89). In this case, the computational load balance is more important than the cut-min value – on average - because for fast computation and fast communication configurations the proper load balance is more important. We can also see that the merging rule set 01 produced almost linear speedup independently on the number of links. The rule sets 01 and 89 produced definitely the best partitions for almost all analyzed cases. We can compare the results from Fig. 6. to Figs. 4 and 5. We can see that the FDTD simulation execution time depends on the load balance and cut-min values. The experiments have revealed that the time of the optimized graph partitioning (by METIS+PEGASUS DA strategy) is longer than the time for the version without optimization (METIS only). For hierarchical algorithm an average execution time varied from 30 s to 50 s depending on the parallel or sequential optimization (execution on a cluster or on a single processor). The GP algorithm average execution time was strongly dependent on the applied set of merging rule sets. For simple sets (01, 89), the average execution time was comparable to the hierarchical version. Tests were done for different graphs with comparable number of nodes.

Comparing two control strategies (based on PEGASUS DA) in terms of the graph partitioning time, we can say that with computation-oriented criteria, the algorithm of GP is definitely comparable to parallel optimization based on the METIS tool. In both cases the result was obtained after about 30–60 s. With communication-oriented criteria, the complexity of the GP approach was significantly higher, so that the result delivery time exceeded even a few minutes (for graphs with no more then a million nodes). The situation was very similar if the optimization goal was the graph partition which evenly distributed the application.

In the case of the graph partitioning algorithm based on the METIS tool, the obtained partitions have shown imbalance on average from 1% to 3% (from 1% to 3% partition components differed in terms of the computational nodes number). The choice of any additional optimization criteria did not matter here, whether it was cut-min or communication volume optimization. In the case of the GP-based strategy, we have analyzed the whole population generated during optimization and only for computationally oriented merging rule sets (e.g. 01, 89), the load imbalance did not exceed 5% (for 2 or 4 component partitions) and did not exceed 1% (for 32 component partitions). So as we can see, both studied strategies have shown very similar features. For the rest of merging rule sets (e.g. 0134, 456789 oriented to communication optimization) load imbalance was significantly higher (see, Fig. 2) and exceed even 15%. Such a result could exclude such partitions in terms of the load balancing quality, but, on the other side,

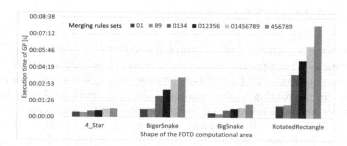

**Fig. 3.** Average GP iteration time versus area shapes and merging rule sets

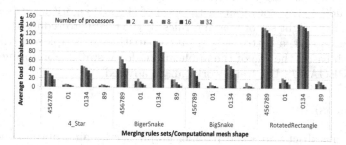

**Fig. 4.** Average computational load imbalance for different area shapes, processor numbers and merging rule sets.

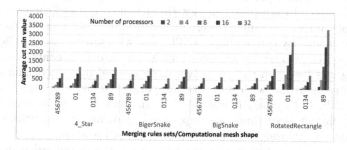

**Fig. 5.** Average cut-min for different area shapes, processor numbers and merging rule sets.

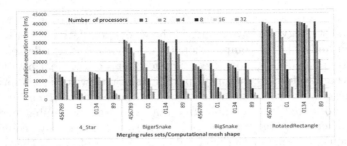

**Fig. 6.** Average execution time of the FTDT simulation for different area shapes, processor numbers and merging rule sets.

they usually have shown the best features regarding communication. The practical importance of such partition methods, ie. their practical usability strongly depends on the performance features of the executive system components. In the case of a system combining a slow communication network and very fast multi-threaded computational nodes such "degenerated" partitions, can be efficiently used in the FDTD simulation. Anyway, the computational complexity to obtain such partitions with low value of the cut-min factor (with the GP approach) is high what excludes their practical use. In summary, when the optimization goal is to achieve strongly computationally balanced graph partitions, we can use both studied strategies, but in the case of the GP algorithm it is necessary to determine the right set of merging rules. If the goal is to reduce the resulting communication cost, a better strategy is to partition the graph with the METIS tool. In both cases, the use of the application global states monitoring provided inside the PEGASUS DA framework can be advised.

## 7   Conclusions

The paper has discussed some parallel graph partitioning optimization methods designed for distributed environments controlled under the PEGASUS DA framework based on the application global states monitoring. Two strategies for control design in graph partitioning algorithms have been presented and experimentally assessed. In the first strategy, the overall control of the general partitioning algorithm was organized over the set of ready-to-use graph partitioning primitives provided by the popular METIS tool with optimization under control of graph-theoretic criteria. In the second strategy, the overall control of the graph partitioning was based on the primitives implemented by a user-defined genetic programming algorithm with optimization controlled by some application specific criteria. In both cases, it was done in an iterative way by analyzing and modifying different algorithmic assumptions. Unsatisfactory partitions could be repartitioned for different criteria and functional parameters.

In both control strategies, the overall control was organized using utilities provided inside the PEGASUS DA framework. It included user-defined synchronizers sensitive to the graph partitioning application global states, where the optimization logic was implemented using the runtime system API. Such approach has supported the design of the global control of the algorithms and decreased the programmer's effort. The partitioning methods and the applied optimizing criteria could be easily controlled depending on the observation of the global state of the partitioning processes. The application computational code was separated from the execution global control code. PEGASUS DA was used to determine the set of input parameters used in optimization. subsequent iterations. Such approach has made the design of the global control of the algorithms easier and decreased the programmer's effort. The experiments have shown that in both studied control strategies, the execution time of the optimized partitioning was longer than that without optimization. The quality of partitions obtained by METIS supported by PEGASUS DA was generally better by 5% to even 30% in comparison to that obtained by using uniquely METIS. In the case of the genetic programming algorithm, we could easily analyze the behavior of the partitioning algorithm depending on the set of node merging rules.

# References

1. Garey, M., Johnson, D., Stockmeyer, L.: Some simplified NP-complete graph problems. Theor. Comput. Sci. **1**, 237–367 (1976)
2. Karypis, D.: METIS a Software Package for Partitioning Unstructured Graphs, Partitioning Meshes, and Computing Fill-Reducing Orderings of Sparse Matrices Version 5.1.0, Department of Computer Science & Engineering, University of Minnesota, Minneapolis, March (2013)
3. https://www.glaros.dtc.umn.edu/gkhome/fetch/sw/parmetis/manual.pdf
4. Khan, M.S., Li, K.F.: Fast graph partitioning algorithms. In: IEEE Pacific Rim Conference on Communication, Computers, and Signal Processing, Victoria, B.C., Canada, May 1995, pp. 337–342 (1995)
5. Kerighan, B.W., Lin, S.: An efficient heuristic procedure for partitioning graphs. AT&T Bell Labs. Tech. J. **49**, 291–307 (1970)
6. Kirkpatrick, S., Gelatt, C.D., Vecchi, M.P.: Optimization by simulated annealing. Science **220**(4598), 671–680 (1983)
7. Tudruj, M., Borkowski, J., Kopanski, D., Laskowski, E., Masko, L., Smyk, A.: PEGASUS DA framework for distributed program execution control based on application global states monitoring. Concurr. Comput.: Pract. Exp. **27**(4), 1027–1053 (2015)
8. Pellegrini, F.: PT-Scotch and libPTScotch 6.0 User's Guide, (ver. 6) Universite Bordeaux 1 & LaBRI, UMR CNRS 5800 Bacchus team, INRIA Bordeaux Sud-Ouest December (2012)
9. http://www.cs.sandia.gov/CRF/chac_p2.html. http://www.cs.sandia.gov/zoltan/Zoltan_phil.html
10. http://bmi.osu.edu/umit/software.html
11. Roy, A., Mihailovic, I., Zwaenepoel, W.: X-stream: edge-centric graph processing using streaming partitions. In: 24th ACM Symposium on Operating Systems Principles, pp. 472–488. ACM (2013)
12. Wang, Y., et al.: GPU graph analytics. ACM Trans. Parallel Comput. (TOPC) **4**(1) (2017). Invited papers from PPoPP 2016

13. Smyk, A., Tudruj, M., Grochal, L.: Global application states monitoring applied to graph partitioning optimization. In: 17th International Symposium on Parallel and Distributed Computing, ISPDC2018, Geneva, Switzerland, June 2018, pp. 85–92. IEEE CS Press (2018)
14. Poli, R., Langdon, W.B., McPhee, N.F., Koza, J.R.: Genetic programming: an introduction and tutorial, with a survey of techniques and applications. University of Essex, UK Technical report [CES-475] (2007)
15. Wenhua, Y.: Advanced FDTD Methods: Parallelization, Acceleration, and Engineering Applications. Artech House (2011)
16. Tian, Y., et al.: From "think like a vertex" to "think like a graph". Proc. VLDB Endow. **7**(3), 193–204 (2013)
17. Sanders, P., Schulz, C.: High Quality Graph Partitioning (2013). https://doi.org/10.1090/conm/588/11700
18. Sanders, P., Schulz, C.: Distributed evolutionary graph partitioning. In: Proceedings of the 12th Workshop on Algorithm Engineering and Experimentation (ALENEX 2012), pp. 16–29 (2012)
19. Raghavan, U.N., Albert, R., Kumara, S.: Near linear time algorithm to detect community structures in large-scale networks. Phys. Rev. E **76**(3), 036106 (2007)
20. Rahimian, F., et al.: JA-BE-JA: a distributed algorithm for balanced graph partitioning. In: 7th International Conference on Self-Adaptive and Self-organizing Systems, pp. 51–60. IEEE (2013)
21. Walshaw, C.: Multilevel refinement for combinatorial optimisation problems. Ann. OR **131**(1), 325–372 (2004)
22. Masko, L., Tudruj, M.: Application global state monitoring in optimization of parallel event-driven simulation. Concurr. Comput. Pract. Exp. **31**(19), e5015 (2019)

# Cloud Infrastructure Automation
# for Scientific Workflows

Bartosz Balis[(✉)], Michal Orzechowski, Krystian Pawlik, Maciej Pawlik,
and Maciej Malawski

AGH Universtity of Science and Technology, Krakow, Poland
{balis,malawski}@agh.edu.pl

**Abstract.** We present a solution for cloud infrastructure automation
for scientific workflows. Unlike existing approaches, our solution is based
on widely adopted tools, such as Terraform, and achieves a strict separation of two concerns: infrastructure description and provisioning vs.
workflow description. At the same time it enables a *comprehensive* integration with a given cloud infrastructure, i.e. such wherein workflow execution can be managed by the cloud. The solution is integrated with our
HyperFlow workflow management system and evaluated by demonstrating its use in experiments related to auto-scaling of scientific workflows
in two types of cloud infrastructures: containerized Infrastructure-as-a-
Service (IaaS) and Function-as-a-Service (FaaS). Experimental evaluation involves deployment and execution of a test workflow in Amazon
ECS/Docker cluster and on a hybrid of Amazon ECS and AWS Lambda.
The results show that our solution not only helps in the creation of
repeatable infrastructures for scientific computing but also greatly facilitates automation of research experiments related to the execution of
scientific workflows on advanced computing infrastructures.

**Keywords:** Scientific workflows · Infrastructure automation ·
Autoscaling

## 1 Introduction

Modern computer-based scientific experiments rely on three fundamental pillars:
scientific data, scientific procedure, and computing infrastructure [11]. Scientific
workflows have emerged as a convenient paradigm for describing the scientific
procedure and automating its execution using Scientific Workflow Management
Systems (WMS) [5]. As far as the computing infrastructure is concerned, besides
the traditional HPC systems, cloud infrastructures are increasingly used for scientific computing due to their on-demand resource provisioning capabilities, flexible configurability, and a wide range of services offered in the pay-per-use business model [12]. Capabilities of the cloud infrastructures are constantly evolving,
attracting the scientific community to experiment with such advanced features
as auto-scalable container orchestration or serverless computing. However, modern WMS, usually relying on in-house infrastructure provisioning solutions, have

© Springer Nature Switzerland AG 2020
R. Wyrzykowski et al. (Eds.): PPAM 2019, LNCS 12043, pp. 287–297, 2020.
https://doi.org/10.1007/978-3-030-43229-4_25

a hard time quickly adapting to leverage these possibilities. Cloud infrastructure provisioning is a complex task due to the wide range of cloud providers and the complexity of advanced cloud services they offer. Deployment of workflow applications in such a way as to leverage advanced cloud capabilities introduces further challenge, because WMS are rather complex systems composed of many distributed components for execution management, monitoring, provenance tracking, or visualization. IT system administrators routinely utilize *infrastructure automation tools* in order to manage this complexity, speed up the infrastructure provisioning process and avoid errors [9]. Though recognized as a tool facilitating scientific workflow reproducibility [1,11], infrastructure automation has not yet been widely adopted in scientific computing.

In this paper, we present an infrastructure automation solution for scientific workflows and demonstrate its usefulness in scientific workflow research. Unlike existing solutions, we utilize standard and widely adopted infrastructure automation tools (e.g. Terraform) and strictly separate two distinct concerns: scientific workflow description vs. infrastructure description and provisioning. Using a popular tool ensures that the solution supports all capabilities of the target infrastructure, enabling *comprehensive* integration with the target infrastructure, while requiring minimum maintenance and remaining up-to-date with the latest features of ever-changing cloud offerings.

The main contributions of the paper can be summarized as follows:

- We discuss the advantages of using infrastructure automation tools for scientific workflows, and the limitations of existing solutions.
- We propose a solution for scientific workflow management in cloud infrastructures that overcomes these limitations, facilitating experimenting with various infrastructures in different configurations, and enabling execution of hybrid workflows utilizing multiple different infrastructures simultaneously.
- We show that our solution enables *comprehensive* integration of the WMS with the cloud infrastructure wherein advanced capabilities of a cloud infrastructure can be used to manage the workflow execution.
- We evaluate the feasibility of our solution by implementing it in the Hyper-Flow workflow management system [3] and using it for experimenting with auto-scaling of workflows in containerized IaaS and FaaS clouds (Amazon ECS and AWS Lambda).

The paper is organized as follows. Section 2 reviews related work. Section 3 presents the infrastructure automation solution implemented in the HyperFlow WMS. Section 4 presents experimental evaluation of the solution. Finally, Sect. 5 concludes the paper.

## 2    Related Work

The problem of infrastructure provisioning has been addressed in the context of several workflow management systems. In the Kepler system, the adopted approach is to introduce infrastructure-specific workflow nodes responsible for creation and destruction of resources, as demonstrated in [14] for the Amazon EC2 cloud. However, in such an approach the workflow description is tightly coupled to a particular infrastructure provider, leading to unwanted mixing of concerns that should be separated. Another example demonstrates integration of Kepler with the CometCloud cloud framework [13]. Also in this case the integration required development of custom Kepler director and actor, tightly coupling workflows with the framework.

Kacsuk and others propose a similar approach in the concept of infrastructure-aware workflows [7] that contain infrastructure management nodes for creating (DEPLOY) or destroying (UNDEPLOY) cloud infrastructure resources for executing the workflow application nodes. Alternatively, infrastructure-aware workflow managers can perform infrastructure provisioning tasks automatically, without explicit nodes in the workflow. The solution uses their own proprietary 'cloud orchestration' tool – Occopus.[1] Occopus supports several resource management APIs, including Amazon EC2.

PRECIP [1], a tool developed by the Pegasus WMS team [6], introduces a Python API for 'experiment management' in IaaS cloud infrastructures. The API provides functions to provision and deprovision VM instances, transfer files to and from these instances, and run commands on them. Supported cloud APIs include Amazon EC2 and several private cloud middlewares, such as Eucalyptus and OpenStack.

The goals of these solutions are similar to existing widely-used tools such as Terraform.[2] However, development and maintenance of infrastructure automation tools is a tremendous effort, so proprietary tools, such as PRECIP, Occopus and others [10,15], face the risk of becoming outdated. Moreover, one common characteristic of all the mentioned proprietary provisioning tools used by the existing WMS is that they support only a small number of infrastructure providers, and – for those which are provided – a very limited subset of their services and configuration options. Widely adopted tools, on the other hand, have the advantage of having a huge community base and being kept up-to-date.

Finally, because of the above-mentioned limitations, the existing solutions do not support *comprehensive* integration with cloud infrastructures, wherein advanced cloud capabilities are taken advantage of in the context of workflow execution. A good example of such a capability is auto-scaling and it is the subject of evaluation of our solution presented in this paper.

## 3    Cloud Workflow Execution Lifecycle in HyperFlow

The architecture of the HyperFlow workflow management system is shown in Fig. 1. The diagram presents an example deployment in two different cloud infrastructures: containerized IaaS (Amazon ECS/EC2) and FaaS (AWS Lambda). In order to run a workflow on one or more cloud infrastructures, the user needs to prepare the following: (1) Download appropriate infrastructure description

---

[1] http://occopus.lpds.sztaki.hu.
[2] https://terraform.io.

templates provided by HyperFlow for given infrastructure(s), e.g. Amazon EC2, Amazon ECS cluster, Amazon Lambda, Fargate, Google Kubernetes Engine, etc. (2) Using provided configuration files, adjust parameters of these infrastructures to ones needs. (3) Provide workflow description (graph) and input data.

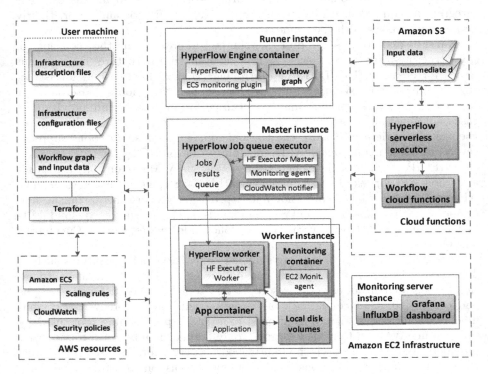

**Fig. 1.** HyperFlow workflow management system (example deployment in Amazon ECS/Docker and AWS Lambda infrastructures).

The HyperFlow infrastructure description templates specify what resources need to be created in order to provision the HyperFlow WMS components and the workflow "worker" components. Infrastructure-specific configuration files are provided wherein the user can specify security credentials and, optionally, manually set basic infrastructure parameters (such as type of the Amazon EC2 instance, autoscaling thresholds, etc.)

The workflow cloud execution life cycle implemented by HyperFlow is shown in Fig. 2. The first step is *provisioning of the Workflow Management System* which includes the monitoring server, the HyperFlow engine [3], and possibly additional executor components specific for a given infrastructure (e.g. RabbitMQ job queue) [2]. The second step is *Workflow execution planning* where the workflow activities are mapped onto resources of the chosen infrastructure(s) and the appropriate parameters of these resources are set (e.g. instance memory

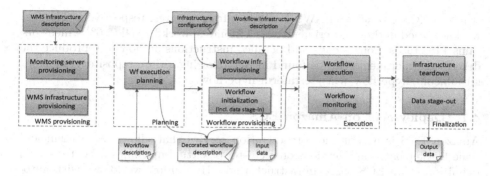

**Fig. 2.** Workflow cloud execution lifecycle.

size, cloud functions memory, or autoscaling thresholds), either by decorating the workflow description, or setting the infrastructure configuration parameters. Currently this step is done mostly manually, but work on integration of scheduling components with the HyperFlow WMS is in progress. The next step is the *Provisioning of the workflow "worker" infrastructure* – e.g. a Kubernetes cluster, and/or a set of Lambda functions – according to the execution plan. Workflow initialization is also done at this step, e.g. input data is uploaded to the cloud, Docker images are downloaded from the Docker Hub, etc. Once the infrastructure is ready, the *Execution* step may start. In this step, one or more workflow runs are executed on the provisioned infrastructure(s), leveraging their advanced capabilities, such as autoscaling. Multiple workflow monitoring components (both generic and specific ones for a given infrastructure) are also deployed and report measurements to the Monitoring server (based on Grafana and InfluxDB). More details are given in Sect. 4. In the *Finalization* step, output data is downloaded and the infrastructure is torn down.

The provisioning steps are automated on the basis of Terraform, one of the most popular and widely adopted tools, often dubbed 'de-facto standard' in infrastructure automation. In Terraform, the infrastructure to be created is defined as a set of *resources*, described in a declarative Terraform configuration language.[3] Terraform builds a graph of the resources and creates the infrastructure by executing this graph so that resources can be created in parallel where possible. When the infrastructure description is modified, Terraform detects which steps need to be re-executed.

## 4   Evaluation

In this section, we evaluate our solution by demonstrating its use in several experiments that we have conducted as part of the research on workflow autoscaling in containerized IaaS and FaaS cloud infrastructures. Here, we used the

---

[3] Infrastructure descriptions are stored in configuration files with .tf extension, and can be expressed in either JSON or a terraform-specific format.

Amazon ECS/Docker and AWS Lambda infrastructures, respectively. The test workflow used in these experiments was a Montage workflow [4] (2.0°) which contains 1482 tasks. Source code and detailed instructions on how to reproduce the described experiments can be found in the following github repository: https://github.com/hyperflow-wms/terraform-hyperflow.

### 4.1  Deployment on Amazon ECS

Amazon ECS is a container orchestration service that allows one to run and scale Docker-based applications on a cluster of virtual machines. The Terraform definition of an ECS-based infrastructure for HyperFlow workflows with auto-scaling is fairly complex, consisting of 23 different resource types, such as EC2 instances, container instances, ECS tasks and services, identities and permissions, user-defined metrics, storage volumes, auto-scaling policies, and others.

The user may configure the infrastructure by setting around 20 variables (some of which could also be set automatically by a scheduler), including the type of the EC2 instance to be used, a number of parameters specifying the auto-scaling policy, and which shared storage should be used (S3 or NFS).

The deployment of the HyperFlow WMS components in the ECS infrastructure, shown in Fig. 1, consists of a number of VM instances and Docker containers running on them. A separate *Runner instance* runs the Hyperflow engine container which orchestrates the workflow execution (both in a pure-ECS and a hybrid ECS/Lambda deployment). The workflow graph maps workflow tasks onto either the ECS workers or Lambda functions. The mapping is done via workflow functions associated with a given task – AMQPCommand for ECS cluster, and RESTServiceCommand for AWS Lambda. Tasks executed on the ECS cluster are submitted through a *HyperFlow Job queue executor* which has a master-worker architecture with the Master running on a separate instance and multiple Workers deployed the on Worker instances where the actual workflow applications

**Table 1.** Metrics collected during workflow execution on the ECS cluster.

| Metric(s) | Source monitoring component/notes |
|---|---|
| VM CPU/MEM/IO usage | Custom monitoring container deployed on cluster nodes (CloudWatch was not detailed enough) |
| Number of EC2 instances Number of containers | HyperFlow ECS monitoring plugin (via AWS API) |
| Autoscaling alarms | HyperFlow ECS monitoring plugin (via Amazon CloudWatch) |
| Job queue length | HyperFlow master (Rabbit MQ job queue)/also published to Amazon CloudWatch to trigger autoscaling rules |
| Workflow events (start, end, etc.) | HyperFlow engine |

are executed. Tasks mapped to the Lambda infrastructure are invoked via a REST *HyperFlow serverless executor*. Data is exchanged between nodes and Lambda functions through S3 (though in the case of a pure-ECS execution also NFS can be configured).

The case of automation of HyperFlow workflows in the Amazon ECS infrastructure shows that the integration of a WMS with a cloud infrastructure may be more complex than simple provisioning of nodes to run jobs on them using a custom parallel executor. A good example of this is workflow monitoring done by multiple distributed monitoring agents shown in Fig. 1. Table 1 shows metrics collected while running a workflow on the ECS cluster and their source monitoring components. The WMS *consumes* monitoring data from the cloud infrastructure in order to show them on its monitoring dashboard. Note that some charts on the dashboard will be infrastructure-specific, therefore the monitoring server is customized and provisioned on demand. However, the WMS also *publishes* metrics to Amazon CloudWatch in order to trigger autoscaling rules (*CloudWatch notifier* component). Only such a comprehensive integration allows for taking full advantage of the infrastructure's capabilities.

### 4.2   Hybrid Deployment on Amazon ECS + AWS Lambda

In the case of AWS Lambda, both servers and auto-scaling are managed by the service itself, so the description of the run time infrastructure was significantly simpler for the ECS. The code related to the function itself starts with defining a lambda function resource, where the user is expected to provide the location of the function's source code (or a ZIP file containing sources and additional dependencies). Additional parameters include: a definition of run time and code entry point, function environment parameters and timeout. The HyperFlow AWS Lambda executor logic is implemented in *JavaScript* using *Node.js* as the execution environment. Currently, the only configurable infrastructure parameter is the Lambda function's *memory size*. The final function parameter is *timeout*, i.e. its execution time limit. Usually set at maximum, the timeout value can be fine tuned in order to prevent the execution from consuming too many resources.

While both described infrastructures share a set of common characteristics, there are significant differences between them. Lambda functions free the developer from managing an infrastructure, but they also impose run time limitations. In the case of complex scientific workflows, it might be possible that some constraints, such as cost or deadline, make it beneficial to run part of the workflow on the ECS cluster, while the other part – on Lambda functions, i.e. a *hybrid infrastructure*. For a more general discussion of our earlier experiments with running scientific workflows on FaaS platforms, we refer to [8].

In the conducted experiments, the infrastructure parameters and the mappings of tasks to infrastructures were done manually. In the hybrid deployment, the mapping was decided based on individual task's requirements. Short-lived tasks with small disk space requirements were executed on AWS Lambda, while longer tasks, or those that required more disk space, were run on ECS.

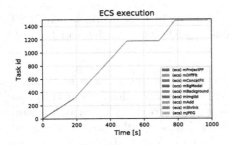 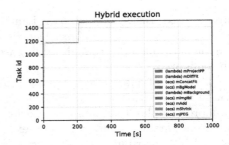

**Fig. 3.** Comparison of workflow execution on pure-ECS and hybrid (ECS+Lambda) infrastructures, executing Montage workflow (2.0°) with S3 storage. Each task is represented by a horizontal bar spanning the execution start and end time. Total execution times: 978 s (pure ECS), 423 s (hybrid).

The HyperFlow's Executor concept and asynchronous workflow execution model allowed the HyperFlow engine to manage the execution simultaneously on both infrastructures. Due to HyperFlow being agnostic when it comes to exchanging input and output files of workflows, the responsibility for managing file transfers is realized by using a single storage provider for both executors. In this case, a common storage provider is the Amazon S3 service (Fig. 3).

### 4.3    Experiments with Workflow Auto-Scaling

Having prepared infrastructure definition templates, we have performed experiments to investigate workflow auto-scaling on containerized infrastructure (managed by the client) vs. cloud functions (fully managed by the provider). The infrastructure parameters used in the runs are summarized in Table 2.

**Table 2.** Infrastructure parameters used in the experimental workflow runs.

| Infrastructure | Parameter name | Value |
|---|---|---|
| Amazon ECS | EC2 Instance type | t2.medium |
| | Number of workers added on scaling up | 1 |
| | Maximum number of workers | 3 |
| | Maximum number of EC2 instances | 3 |
| | Auto-scaling alarm threshold | 300 |
| | Auto-scaling cooldown | 500 (s) |
| Amazon EC2 + AWS Lambda | EC2 instance type | t2.medium |
| | Function memory | 2048 |
| | Maximum number of function workers | 1000 |

Here, 'workers' refer to HyperFlow Worker container instances. The *auto-scaling threshold* denotes the value of a metric which – when breached – triggers the scaling up (or down) action. In this case, the used metric was the number of jobs waiting in the queue of the HyperFlow master-worker executor. The *auto-scaling cooldown* is a period in seconds during which no scaling action may take place after a previous one was triggered.

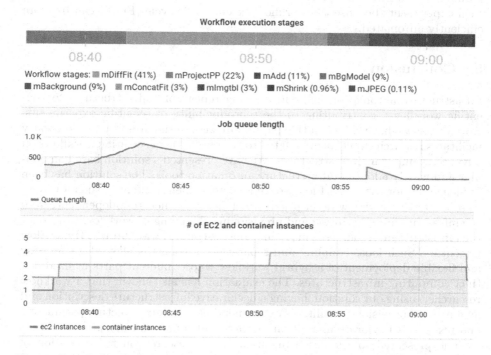

**Fig. 4.** Detailed experiment results example: Montage (2.0°) run on Amazon ECS (t2.micro instances) with auto-scaling and S3 storage.

### 4.4   Discussion of Results

Figure 4 presents Gantt charts generated based on execution traces when using pure ECS hybrid ECS/Lambda infrastructures, respectively. As one can see, automatic scaling resulted in a significant reduction of the execution time. However, the scaling is much better for the hybrid deployment than for the pure-ECS one. A closer look at details of the workflow execution visualized on the HyperFlow monitoring dashboard (Fig. 4) may help explain why this is the case.[4] As can be seen, the timing and the size of the auto-scaling actions were arguably not optimal, resulting in limited parallelism. This preliminary result

---

[4] The dashboards show a different run in which t2.micro instances were used. However, besides the longer execution time the execution patterns were the same.

proves that manual workflow autoscaling is hard and guides our further, currently ongoing research to find a scheduling heuristic which, based on the workflow graph and perhaps historic execution data, determines optimal auto-scaling parameters. Such research requires experimentation with different infrastructure parameters, and with the presented solution, it is significantly easier to perform such experiments because the workflow execution lifecycle (Fig. 2) can be more efficiently automated.

## 5    Conclusion

Infrastructure automation helps in creating repeatable infrastructures for scientific workflows, contributing to the reproducibility of scientific experiments. However, as we have shown in this paper, infrastructure automation also greatly facilitates research experiments related to the execution of scientific workflows on advanced computing infrastructures. We have presented a solution based on popular and widely adopted infrastructure automation tools. The solution has been implemented for the HyperFlow workflow management system in such a way as to separate the workflow description, which is infrastructure-independent, from the infrastructure description and provisioning. Mapping of workflow tasks onto the infrastructure resources is made in a loosely-coupled way through HyperFlow Executors. Such loose coupling and separation of concerns allowed for an easy and flexible deployment of a hybrid workflow whose different parts utilized distinct computing infrastructures. The evaluation has also shown that a workflow researcher, using our solution, having a declarative infrastructure description and an advanced provisioner facilitates experimenting with infrastructure parameters and test the behavior of an application under different conditions.

The presented solution is being used in our research on the execution of scientific workflows on elastic computing infrastructures. Future work involves the implementation of support for other computing infrastructures, and a better modularization of the implementations to increase flexibility and usability of the infrastructure automation tool.

**Acknowledgment.** This work was supported by the National Science Centre, Poland, grant 2016/21/B/ST6/01497.

## References

1. Azarnoosh, S., et al.: Introducing PRECIP: an API for managing repeatable experiments in the cloud. In: 2013 IEEE 5th International Conference on Cloud Computing Technology and Science (CloudCom), pp. 19–26. IEEE (2013)
2. Balis, B., Figiela, K., Malawski, M., Pawlik, M., Bubak, M.: A lightweight approach for deployment of scientific workflows in cloud infrastructures. In: Wyrzykowski, R., Deelman, E., Dongarra, J., Karczewski, K., Kitowski, J., Wiatr, K. (eds.) PPAM 2015. LNCS, vol. 9573, pp. 281–290. Springer, Cham (2016). https://doi.org/10.1007/978-3-319-32149-3_27

3. Balis, B.: Hyperflow: a model of computation, programming approach and enactment engine for complex distributed workflows. Future Gener. Comput. Syst. **55**, 147–162 (2016)
4. Berriman, G.B., Deelman, E., et al.: Montage: a grid-enabled engine for delivering custom science-grade mosaics on demand. In: Astronomical Telescopes and Instrumentation, pp. 221–232. International Society for Optics and Photonics (2004)
5. Deelman, E., Gannon, D., Shields, M., Taylor, I.: Workflows and e-science: an overview of workflow system features and capabilities. Future Gener. Comput. Syst. **25**(5), 528–540 (2009)
6. Deelman, E., et al.: Pegasus, a workflow management system for science automation. Future Gener. Comput. Syst. **46**, 17–35 (2014)
7. Kacsuk, P., Kecskemeti, G., Kertesz, A., Nemeth, Z., Visegradi, A., Gergely, M.: Infrastructure aware scientific workflows and their support by a science gateway. In: 7th International Workshop on Science Gateways (IWSG), pp. 22–27. IEEE (2015)
8. Malawski, M., Gajek, A., Zima, A., Balis, B., Figiela, K.: Serverless execution of scientific workflows: experiments with HyperFlow, AWS Lambda and Google Cloud Functions. Future Gener. Comput. Syst. (2017, in Press)
9. Morris, K.: Infrastructure as Code: Managing Servers in the Cloud. O'Reilly Media Inc., Newton (2016)
10. Posey, B., Gropp, C., Herzog, A., Apon, A.: Automated cluster provisioning and workflow management for parallel scientific applications in the cloud. In: Proceedings 10th Workshop on Many-Task Computing on Clouds, Grids, and Supercomputers (MTAGS) (2017)
11. Santana-Perez, I., Pérez-Hernández, M.S.: Towards reproducibility in scientific workflows: an infrastructure-based approach. Sci. Program. (2015)
12. Varghese, B., Buyya, R.: Next generation cloud computing: new trends and research directions. Future Gener. Comput. Syst. **79**, 849–861 (2018)
13. Wang, J., AbdelBaky, M., Diaz-Montes, J., Purawat, S., Parashar, M., Altintas, I.: Kepler+ cometcloud: dynamic scientific workflow execution on federated cloud resources. Procedia Comput. Sci. **80**, 700–711 (2016)
14. Wang, J., Altintas, I.: Early cloud experiences with the kepler scientific workflow system. Procedia Comput. Sci. **9**, 1630–1634 (2012)
15. Wilde, M., Hategan, M., Wozniak, J.M., Clifford, B., Katz, D.S., Foster, I.T.: Swift: a language for distributed parallel scripting. Parallel Comput. **37**(9), 633–652 (2011)

# Applications of Parallel Computing

# Posit NPB: Assessing the Precision Improvement in HPC Scientific Applications

Steven W. D. Chien[1]([⊠]), Ivy B. Peng[2], and Stefano Markidis[1]

[1] KTH Royal Institute of Technology, Stockholm, Sweden
wdchien@kth.se
[2] Lawrence Livermore National Laboratory, Livermore, CA, USA

**Abstract.** Floating-point operations can significantly impact the accuracy and performance of scientific applications on large-scale parallel systems. Recently, an emerging floating-point format called Posit has attracted attention as an alternative to the standard IEEE floating-point formats because it could enable higher precision than IEEE formats using the same number of bits. In this work, we first explored the feasibility of Posit encoding in representative HPC applications by providing a 32-bit Posit NAS Parallel Benchmark (NPB) suite. Then, we evaluate the accuracy improvement in different HPC kernels compared to the IEEE 754 format. Our results indicate that using Posit encoding achieves optimized precision, ranging from 0.6 to 1.4 decimal digit, for all tested kernels and proxy-applications. Also, we quantified the overhead of the current software implementation of Posit encoding as 4×–19× that of IEEE 754 hardware implementation. Our study highlights the potential of hardware implementations of Posit to benefit a broad range of HPC applications.

**Keywords:** HPC · Floating point precision · Posit · NPB

## 1 Introduction

Floating-point operations are indispensable for many scientific applications. Their precision formats can significantly impact the power, energy consumption, memory footprint, performance, and accuracy of applications. Moving towards exascale, optimizing precision formats in HPC scientific applications could address some key challenges identified on exascale systems [4]. Recent works in hardware-supported half-precision, software-guided mixed-precision, and adaptive precision have highlighted the importance of reconsidering precision formats [2,9,10]. *Posit* [6], an alternative to IEEE 754 floating-point format, has gained increasing attention in the HPC community because its *tapered precision* can achieve higher precision than IEEE 754 format using the same number of bits. Posit has been explored in Deep Learning applications [7], Euler and eigenvalue solvers [8]. Still, its precision improvements in general HPC scientific

© Springer Nature Switzerland AG 2020
R. Wyrzykowski et al. (Eds.): PPAM 2019, LNCS 12043, pp. 301–310, 2020.
https://doi.org/10.1007/978-3-030-43229-4_26

applications require systematic efforts to understand and quantify, which motivates our study in this paper. Our work provides a 32-bit Posit implementation of the popular NAS Parallel Benchmark (NPB) suite [1,3], called *Posit NPB* to quantify the improved precision using Posit formats compared to 32-bit IEEE 754 format in representative HPC kernels. Our main contributions are as follows:

- We provide a publicly available 32-bit Posit implementation of the NPB benchmark suite
- We define the metric for accuracy and use it to quantify the precision improvements using Posit formats in five kernels and proxy-applications compared to 32-bit IEEE 754 format
- We also provide a 128-bit IEEE 754 floating-point (Quad) implementation of the NPB benchmark suite as a high-precision solution reference
- Our Posit implementation exhibit 0.4 to 1.6 decimal digit precision improvement in all tested kernels and proxy-applications compared to the baseline
- We quantified the overhead of software-based Posit implementation as $4\times$–$19\times$ that of IEEE 754 hardware implementation
- We show that Posit could benefit a broad range of HPC applications but requires low-overhead hardware implementation.

## 2   Floating-Point Formats

Fractional real numbers are represented as floating-point numbers, and their operations are defined by floating-point operations in computer arithmetics. Instead of representing the number in its original form, a number is represented as an approximation where the trade-off between precision and range is defined. Given the same amount of memory space, a larger range of numbers can be represented if numbers in that range use a less accurate approximation.

IEEE floating-point numbers are often represented by three components, i.e., a sign bit, an exponent, and a significant. A sign bit represents whether the number is positive or negative. An exponent represents the shifting that is required to acquire the non-fraction part of a number. Finally, a significand represents the actual number after shifting. Currently, IEEE 754 format is the most broadly adopted standard.

Posit format uses four components, i.e., a sign bit, a regime, an exponent and a fraction. Different from IEEE Float, these components could have variable sizes. The first component after the sign bit is *regime*, which is used to compute a scaling factor $useed^k$ where $useed = 2^{2^{es}}$. The regime component encodes a number $k$ through a prefix code scheme. The regime contains several consecutive 1s or 0s, which is terminated if the next bit is the opposite. $k$ is defined by $k = -m$, where $m$ is the length of the bit string before the opposite bit when the bits are all zero. For instance, if the bit string is all 1s and terminated by a 0, $k$ is defined by $k = m - 1$. After the regime, depending on the number of bits left, the *exponent* begins and runs for a maximum length of *es*. The exponent encodes an unsigned integer, which represents another scaling factor $2^{exponent}$. Finally, the *fraction* has the same functionality as the significand in IEEE Float.

Figure 1 illustrates a number encoded in IEEE Float and 32-bit Posit with $es = 2$. The string begins with a zero that indicates the number is positive. The *regime* bit runs for 01, which means that $m = 1$ and thus $k = -1$. Since $es = 2$, $useed = 2^{2^{es}} = 16$, the scaling factor $useed^k = 1/16$. After the termination of *regime*, the exponent begins and has a length of $es = 2$. The exponent 00 is represented by $2^{exponent} = 2^0 = 1$. The remaining bits are used for the fraction, which encodes $1 + 130903708/2^{27}$, where the one is implicit, and the size of the fraction is 27. Since the scheme has a smaller exponent than IEEE Float, more bits can be used in the fraction, which attributes to higher accuracy.

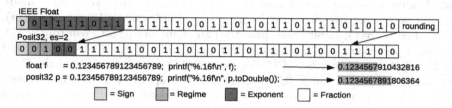

**Fig. 1.** Binary formatting of IEEE Float and 32 bit Posit with $es = 2$ when representing an arbitrary fractional number.

In this work, we also use a special type in Posit, called *Quire*, to facilitate high accuracy Fused Multiply-Add (FMA). Quire can be considered as a large scratch area for performing high precision accumulation and deferred rounding during FMA. Quire requires a large number of bits. For instance, it requires 32 bits for 8-bit Posit, 128 bits for 16-bit Posit, and 512 bits for 32-bit Posit.

# 3 Methodology

Our work aims to assess precision optimization by Posit in HPC scientific applications. To achieve this, we choose a widely-adopted parallel benchmark suite on HPC systems, the NAS Parallel Benchmark (NPB) suite [3]. The NPB suite was originally derived from Computational Fluid Dynamics (CFD) applications, and closely mimic real-world HPC applications. The suite includes five kernels (IS, EP, CG, MG, and FT) and three proxy-applications (BT, LU, and SP). In this work, we extend a subset of the suite that uses floating-point operations to evaluate the impact of Posit arithmetic in typical HPC applications. We based our implementation on the C version of the suite [1]. Our Posit NPB suite includes CG (Conjugate Gradient), MG (Multigrid), LU (Lower-Upper decomposition solver), BT (Block Tridiagonal solver) and FT (FFT solver). The benchmark suite is publicly available in a code repository[1].

---

[1] https://github.com/steven-chien/NAS-Posit-benchmark.

The original NPB implementation uses only the 64-bit IEEE 754 floating-point format. We provide a 32-bit Posit implementation and IEEE 754 floating-point implementation of the suite. To compare with a high accuracy IEEE format, we additionally provide a 128-bit IEEE floating point (quad) implementation. The NPB suite predefines problem sizes into different classes: S class for small tests, W class for workstation-size tests; A, B, C classes for standard test on supercomputer and E, D and F for large test problems on parallel machines. Our evaluation includes experiments using various problem classes for understanding the impact of Posit arithmetics.

We define the metric for accuracy as the difference between the approximated value in various precision formats and the exact value. We then evaluate the accuracy using five precision formats, i.e., Quad (`quad`), Double (`double`), IEEE Float (`float`), 32 bit Posit (`posit32`) and Quire for 32 bit Posit (`quire32`). For Quad precision, as it is not natively supported by C and C++, we adopted the `libquadmath` library by GCC. For Posit and Quire types, we used the C++ interface provided by the *SoftPosit* library where operator overloading is supported. For all evaluation, we use single-thread executions of kernels to avoid interference from multiple threads. We validate the solution from each kernel in the highest accuracy, i.e., Quad precision. We cast the generated results to the selected precision to control error propagation resulted only from the computation other than problem generation.

# 4    Results

We evaluate our NPB extension on a workstation with an Intel Core i7-7820X CPU with eight cores and 16 threads. The system has 32 GB of RAM with a 480 GB SSD. The operating system is Fedora 29 running on Kernel version 4.19.10-300. The compiler used is GCC 8.2.1 and the latest SoftPosit library from main development branch is used[2]. We compute the machine epsilon ($\epsilon$) of different formats using linear search method for reference. On this workstation, the measured $\epsilon$ values are $1.92E-34$, $2.22E-16$, $1.19E-7$, $7.45E-9$ and for `quad`, `double`, `float` and `posit32` respectively.

Reproducing floating-point computation results across platforms and optimization is a difficult task. When compiling the benchmarks, we have used the flags `-ffp-contract=off` for turning off possible Fused Multiply Add (FMA) and `-ffloat-store` to avoid storage of floating-point variables in registers thus avoiding the effect of storing intermediate value in extended precision by the processor. We select three problem sizes for the precision evaluation: class S, W and A.

**CG: Conjugate Gradient.** The CG benchmark computes the smallest eigenvalue of a sparse matrix with the Inverse Power method. At each Inverse Power iteration, a linear system is solved with the Conjugate Gradient (CG) method. The CG kernel consists of three main steps: first, the vector $x$ is initialized

---

[2] SoftPosit commit 688cfe5d.

randomly; second, the CG method solves the linear system $Az = x$ where $A$ is symmetric, definite-positive matrix, and $x$ in the known term of the linear system; third, the inverse power method uses the $||r||$ that is the norm of the residual vector $r$ to calculate $\zeta = \lambda + 1/x^T z$ where $\lambda$ is the shift factor for different problem size. In our set-up, each Inverse Power iterations includes 25 CG iterations.

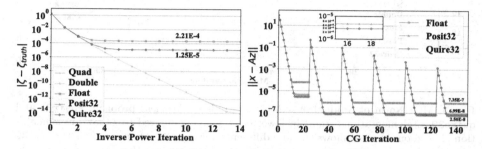

**Fig. 2.** CG: The left panel displays the error in evaluating $\zeta$ for the first 15 Inverse Power iterations and problem class W using different floating-point formats. The error decreases until reaching a minimum value approximately after five iterations for `posit32` and `float`. The right panel displays residual error of CG iterations. At the end of each Inverse Power iteration (25 CG iterations), the norm of the residual increases by more of two orders. The zoom-in view of on CG residual shows the difference in error between `posit32` and `quire32`.

The left panel of Fig. 2 shows the error calculated as the difference of the $\zeta$ estimate and its exact value for the first 15 iterations of the Inverse Power method using the problem size W. The error decreases over successive iterations until reaching a minimum value that cannot be further decreased: `float` and `posit32` reach a minimum error value of 2.21E−4 and 1.25E−5 respectively. Quire32 has the same $\zeta$ error value as `posit32` and for this reason `posit32` and Quire32 lines are superimposed. The CG benchmark using `posit32` provides a final solution that is one more digit $(\log_{10}(2.21E{-}4/1.25E{-}5) = 1.25)$ accurate than the CG benchmark using `float`.

The right panel of Fig. 2 shows the norm of the residual $r = Az - x$ for the first 150 iterations. Since each inverse power iteration consists of 25 iterations, iteration 150 of CG refers to the sixth inverse power iteration. At the last CG iteration, the norm of the residual reaches a minimum value of 7.35E−7, 6.99E−8 and 2.50E−8 for `float`, `posit32` and `quire32` implementations respectively. We note that in the case of error calculated as residual norm (right panel of Fig. 2), the error is different for `posit32` and `quire32` and it is close in value to the machine epsilon for `float` and `posit32`: 1.19E−7 and 7.45E−9.

**MG: Multigrid.** The NPB MG kernel implements the V-cycle multigrid algorithm to compute the solution of a discrete Poisson problem $(\nabla^2 u = v)$ on a

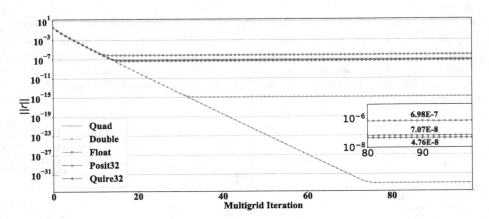

**Fig. 3.** MG: Norm of residual over MG iterations for W class problem. `Posit32` and `quire32` implementations result in lower error when compared to `float` implementation. The zoom-in view shows the fine difference between `float`, `posit32` and `quire32` when precision improvements cannot be made with more iterations.

3D Cartesian grid. After each iteration, the $L_2$ norm of residual $r = v - \nabla^2 u$ is computed for measuring the error.

As for the CG benchmark, the norm of the residual in the MG application decreases during the iterations until it cannot be further reduced as shown in Fig. 3. The norm of the residual for the `float`, `posit32` and `quire32` MG implementations at their last iteration are 6.98E−7, 7.07E−8 and 4.76E−8 respectively. These values are close to the machine $\epsilon$ values for the different floating-point formats. Also for MG, the `posit32` implementation provides a final solution that is one digit more accurate than the results that have been obtained with `float`. The `quire32` implementation is roughly 1.16 digit more accurate than the float implementation.

**LU: Lower-Upper Decomposition Solver.** LU is a CFD pseudo-application solving the Navier-Stokes system comprising a system of 5 PDEs using the Symmetric Successive Over-Relaxation (SSOR) technique. In this case, we compute the error as the difference between the estimate and the analytical solution and taking its norm over the five PDEs. This error over several iterations is shown in Fig. 4. The LU `float` implementation reaches a minimum error of 5.68E−4 while the LU `posit32` implementation reaches an error of 2.35E−5. The LU `posit32` implementation is 1.38 digit more accurate than the LU `float` implementation.

**BT: Block Tridiagonal Solver.** BT is also a CFD pseudo-application. BT employs an implicit discretization of compressible Navier-Stokes equations in 3D Cartesian geometry. The discrete solution of these equations is based on an Alternating Direction Implicit (ADI) approximate factorization. ADI factorization decouples the $x$, $y$, and $z$ directions, resulting in a linear system with block-tridiagonal of $5 \times 5$ blocks structure that is solved sequentially along each direction.

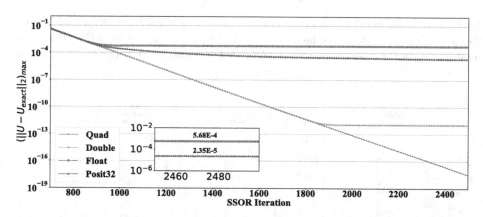

**Fig. 4.** LU: The maximum error norm over the five PDEs from iteration 700 to 2499 for LU. The error decreases reaching a minimum error. The LU `posit32` implementation leads to a lower error with respect to the LU `float` implementation.

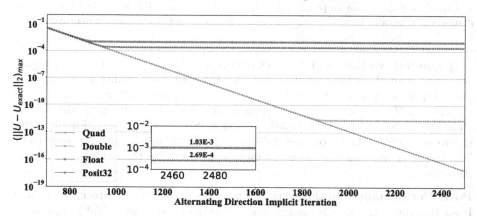

**Fig. 5.** BT: The maximum error norm over the five PDEs at each iteration from iteration 700 to 2499 for BT. The `posit32` implementation results in a lower error when compared to the `float` implementations.

We evaluate the error in the same way we evaluated the error for LU and show it in Fig. 5.

The `float` version of BT application has an error of 1.03E−3 at the last iteration while the `posit32` implementation has an error of 2.69E−4. The `posit32` implementation is 0.6 digit more accurate than the `float` version.

**FT: Fast Fourier Transform.** The NPB FT kernel solves a 3D Partial Differential Equation $\partial u(x,t)/\partial t = \alpha \nabla^2 u(x,t)$ in the spectral space using forward and inverse FFT. The application consists of three steps. Forward FFT, Evolve by multiplying a factor and inverse FFT. The solver computes the FFT of the state array $\tilde{u}(k,0) = FFT(u(x,0))$ at the initial step. The solution of the PDE is then advanced in the spectral space through an *Evolve* process by multiplying $\tilde{u}(k, 0)$ by an exponential complex factor. At each iteration, an Inverse FFT (IFFT) is computed on the result state array to move the solution from the spectral space to real space.

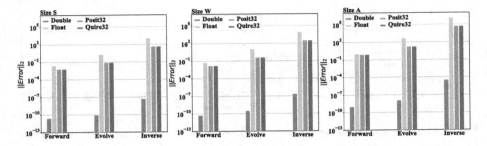

**Fig. 6.** FT: Error norm of FT for problem classes S, W and A. The ground truth is the result for the same step computed in Quad precision. FT Float implementation displays a generally higher error when compared to the `posit32` implementation.

Figure 6 shows that the FT `posit32` implementation gives a generally higher accuracy than the `float` implementation. Computation results using `quad` at each step is used as the truth. For Size W, `float`, `posit32` and `quire32` give 5.85E−2, 1.60E−2 and 1.59E−2 errors in the Forward step. After the Evolve step, the errors are 1.15E+1, 4.38E−1 and 4.38E−1 respectively. Finally, after the reverse FFT, the errors are 8.30E+3, 3.19E+2 and 3.18E+2 respectively. The `posit32` implementation is 1.4 digit more accurate than the `float` implementation.

**Posit Performance.** In this work, we have used a software implementation of Posit floating-point format resulting in performance overhead. Table 1 presents the average execution time and standard deviation of for different implementations of NPB with size S. The execution time includes also time for the validation test and possibly I/O. However, it gives an idea of how execution time differs between the different implementations. Quire implementations for LU and BT are not available so their execution time is not reported in the table. The use of Posit and Quire results in lower performance in terms of execution time and most cases are similar to results from Quad, which is also software based. For example, our CG benchmark in Posit executes 4× slower than its counterpart using IEEE Float.

## 5    Related Work

Taper precision floating-point format has a long history of development. The concept was originally proposed in the 1970s, where a larger size exponent can be used to represent a larger number range with diminishing accuracy due to

**Table 1.** Average execution time and standard deviation of different implementation of benchmark with size S in seconds over five executions.

|            | CG              | FT              | MG              | LU              | BT              |
|------------|-----------------|-----------------|-----------------|-----------------|-----------------|
| Iter/Steps | 30              | 6               | 100             | 100             | 100             |
| quad       | 5.37 ± 0.014    | 5.35 ± 0.032    | 4.30 ± 0.006    | 4.52 ± 0.040    | 7.96 ± 0.065    |
| double     | 3.16 ± 0.006    | 0.70 ± 0.002    | 0.27 ± 0.003    | 0.42 ± 0.011    | 0.56 ± 0.003    |
| float      | 3.10 ± 0.053    | 0.67 ± 0.009    | 0.24 ± 0.003    | 0.43 ± 0.003    | 0.54 ± 0.001    |
| posit32    | 12.78 ± 0.070   | 5.73 ± 0.038    | 4.59 ± 0.018    | 5.74 ± 0.023    | 7.81 ± 0.040    |
| quire32    | 12.29 ± 0.143   | 7.94 ± 0.009    | 5.41 ± 0.016    | N/A             | N/A             |

reduction in fraction bits. Universal number (unum) is another floating-point format that embraces a similar concept. Unum follows the same structure of IEEE 754 floating-point format but specifies the sizes of each component by encoding them at the end of the binary string respectively. The sizes of each component vary automatically [5]. Among other features, it specifies a special "ubit" which indicates if the number represented is exact or lies in an open number interval. Another radically different representation of fraction number is fix point representation. Fix point represents a real number as an integer and the fractional part can be identified through shifting of digit. An integer can be conceived as a subset of fix point number system where shifting is zero.

Standardization of Posit is currently underway[3] and several implementations are available. Among those, the most complete implementation is *SoftPosit. Soft-Posit*[4] is implemented as a C library and is endorsed by the Next Generation Arithmetic (NGA) team[5].

## 6   Discussion and Conclusion

In this work, we assessed the precision optimization in HPC applications using an emerging precision format called Posit. Our results showed that typical HPC kernels as in the NPB suite could improve their accuracy by using 32-bit Posit instead of 32-bit IEEE float. All tested kernels in our Posit NPB suite achieved higher precision, ranging from 0.6 to 1.4 decimal digit, compared to the IEEE Float baseline. However, a major obstacle that hinders the adoption of Posit is the overhead of software implementation. Our Posit NPB suite quantifies 4–19× overhead that of IEEE formats. This high overhead can be partially attributed to the operator overloading in C++, but more importantly, to the lack of hardware support. For the adoption of Posit by HPC applications, hardware implementations are necessary to achieve acceptable performance [11]. Overall, our results indicate Posit as a promising drop-in replacement for IEEE Float in HPC applications for precision optimization.

---

[3] https://posithub.org/docs/posit_standard.pdf.
[4] https://gitlab.com/cerlane/SoftPosit.
[5] https://posithub.org/docs/PDS/PositEffortsSurvey.html.

**Acknowledgments.** Funding for the work is received from the European Commission H2020 program, Grant Agreement No. 801039 (EPiGRAM-HS). LLNL release: LLNL-PROC-779741.

# References

1. An unofficial C version of the NAS parallel Benchmarks OpenMP 3.0 (2014). https://github.com/benchmark-subsetting/NPB3.0-omp-C
2. Anzt, H., Dongarra, J., Flegar, G., Higham, N.J., Quintana-Ortí, E.S.: Adaptive precision in block-jacobi preconditioning for iterative sparse linear system solvers. Concurr. Comput. Pract. Exp. **31**(6), e4460 (2019)
3. Bailey, D.H., et al.: The NAS parallel benchmarks. Int. J. Supercomput. Appl. **5**(3), 63–73 (1991)
4. Dongarra, J., et al.: The international exascale software project roadmap. Int. J. High Perform. Comput. Appl. **25**(1), 3–60 (2011)
5. Gustafson, J.L.: The End of Error: Unum Computing. Chapman and Hall/CRC, Boca Raton (2015)
6. Gustafson, J.L., Yonemoto, I.T.: Beating floating point at its own game: posit arithmetic. Supercomput. Front. Innov. **4**(2), 71–86 (2017)
7. Johnson, J.: Rethinking floating point for deep learning. arXiv preprint arXiv:1811.01721 (2018)
8. Lindstrom, P., Lloyd, S., Hittinger, J.: Universal coding of the reals: alternatives to IEEE floating point. In: Proceedings of the Conference for Next Generation Arithmetic (2018). https://doi.org/10.1145/3190339.3190344
9. Markidis, S., Chien, S.W.D., Laure, E., Peng, I.B., Vetter, J.S.: Nvidia tensor core programmability, performance & precision. In: 2018 IEEE International Parallel and Distributed Processing Symposium Workshops (IPDPSW), pp. 522–531. IEEE (2018)
10. Menon, H., et al.: ADAPT: algorithmic differentiation applied to floating-point precision tuning. In: Proceedings of the International Conference for High Performance Computing, Networking, Storage, and Analysis (2018)
11. Podobas, A., Matsuoka, S.: Hardware implementation of POSITs and their application in FPGAs. In: 2018 IEEE International Parallel and Distributed Processing Symposium Workshops (IPDPSW), pp. 138–145, May 2018. https://doi.org/10.1109/IPDPSW.2018.00029

# A High-Order Discontinuous Galerkin Solver with Dynamic Adaptive Mesh Refinement to Simulate Cloud Formation Processes

Lukas Krenz[(✉)] [iD], Leonhard Rannabauer, and Michael Bader

Department of Informatics, Technical University of Munich, Munich, Germany
{lukas.krenz,rannabau,bader}@in.tum.de

**Abstract.** We present a high-order discontinuous Galerkin (DG) solver of the compressible Navier-Stokes equations for cloud formation processes. The scheme exploits an underlying parallelized implementation of the ADER-DG method with dynamic adaptive mesh refinement. We improve our method by a PDE-independent general refinement criterion, based on the local total variation of the numerical solution. While established methods use numerics tailored towards the specific simulation, our scheme works scenario independent. Our generic scheme shows competitive results for both classical CFD and stratified scenarios. We focus on two dimensional simulations of two bubble convection scenarios over a background atmosphere. The largest simulation here uses order 6 and 6561 cells which were reduced to 1953 cells by our refinement criterion.

**Keywords:** ADER-DG · Navier-Stokes · Adaptive mesh refinement

## 1 Introduction

In this paper we address the resolution of basic cloud formation processes on modern super computer systems. The simulation of cloud formations, as part of convective processes, is expected to play an important role in future numerical weather prediction [1]. This requires both suitable physical models and effective computational realizations. Here we focus on the simulation of simple benchmark scenarios [10]. They contain relatively small scale effects which are well approximated with the compressible Navier-Stokes equations. We use the ADER-DG method of [5], which allows us to simulate the Navier-Stokes equations with a space-time-discretization of arbitrary high order. In contrast to Runge-Kutta time integrators or semi-implicit methods, an increase of the order of ADER-DG only results in larger computational kernels and does not affect the complexity of the scheme. Additionally, ADER-DG is a communication avoiding scheme and reduces the overhead on larger scale. We see our scheme in the regime of already established methods for cloud simulations, as seen for example in [10,12,13].

© The Author(s)
R. Wyrzykowski et al. (Eds.): PPAM 2019, LNCS 12043, pp. 311–323, 2020.
https://doi.org/10.1007/978-3-030-43229-4_27

Due to the viscous components of the Navier-Stokes equations, it is not straightforward to apply the ADER-DG formalism of [5], which addresses hyperbolic systems of partial differentials equations (PDEs) in first-order formulation. To include viscosity, we use the numerical flux for the compressible Navier-Stokes equations of Gassner et al. [8]. This flux has already been applied to the ADER-DG method in [4]. In contrast to this paper, we focus on the simulation of complex flows with a gravitational source term and a realistic background atmosphere. Additionally, we use adaptive mesh refinement (AMR) to increase the spatial resolution in areas of interest. This has been shown to work well for the simulation of cloud dynamics [12]. Regarding the issue of limiting in high-order DG methods, we note that viscosity not only models the correct physics of the problem but also smooths oscillations and discontinuities, thus stabilizing the simulation.

We base our work on the ExaHyPE Engine (www.exahype.eu), which is a framework that can solve arbitrary hyperbolic PDE systems. A user of the engine is provided with a simple code interface which mirrors the parts required to formulate a well-posed Cauchy problem for a system of hyperbolic PDEs of first order. The underlying ADER-DG method, parallelization techniques and dynamic adaptive mesh refinement are available for simulations while the implementations are left as a black box to the user. An introduction to the communication-avoiding implementation of the whole numerical scheme can be found in [3].

To summarize, we make the following contributions in this paper:

- We extend the ExaHyPE Engine to allow viscous terms.
- We thus provide an implementation of the compressible Navier-Stokes equations. In addition, we tailor the equation set to stratified flows with gravitational source term. We emphasize that we use a standard formulation of the Navier-Stokes equations as seen in the field of computational fluid mechanics and only use small modifications of the governing equations, in contrast to a equation set that is tailored exactly to the application area.
- We present a general AMR-criterion that is based on the detection of outlier cells w.r.t. their total variation. Furthermore, we show how to utilize this criterion for stratified flows.
- We evaluate our implementation with standard CFD scenarios and atmospheric flows and inspect the effectiveness of our proposed AMR-criterion. We thus inspect, whether our proposed general implementation can achieve results that are competitive with the state-of-the-art models that rely on heavily specified equations and numerics.

## 2   Equation Set

The compressible Navier-Stokes equations in the conservative form are given as

$$
\frac{\partial}{\partial t} \underbrace{\begin{pmatrix} \rho \\ \rho v \\ \rho E \end{pmatrix}}_{Q} + \nabla \cdot \underbrace{\left( \underbrace{\begin{pmatrix} \rho v \\ v \otimes \rho v + I p \\ v \cdot (I \rho E + I p) \end{pmatrix}}_{F^h(Q)} + \underbrace{\begin{pmatrix} 0 \\ \sigma(Q, \nabla Q) \\ v \cdot \sigma(Q, \nabla Q) - \kappa \nabla T \end{pmatrix}}_{F^v(Q, \nabla Q)} \right)}_{F(Q, \nabla Q)} = \underbrace{\begin{pmatrix} S_\rho \\ S_{\rho v} \\ S_{\rho E} \end{pmatrix}}_{S(Q, x, t)}
$$

(1)

with the vector of conserved quantities $Q$, flux $F(Q, \nabla Q)$ and source $S(Q)$. Note that the flux can be split into a hyperbolic part $F^h(Q)$, which is identical to the flux of the Euler equations, and a viscous part $F^v(Q, \nabla Q)$. The conserved quantities $Q$ are the density $\rho$, the two or three-dimensional momentum $\rho v$ and the energy density $\rho E$. The rows of Eq. (1) are the conservation of mass, the conservation of momentum and the conservation of energy.

The pressure $p$ is given by the equation of state of an ideal gas

$$
p = (\gamma - 1) \left( \rho E - \frac{1}{2} (v \cdot \rho v) - gz \right). \tag{2}
$$

The term $gz$ is the geopotential height with the gravity of Earth $g$ [10]. The temperature $T$ relates to the pressure by the thermal equation of state

$$
p = \rho R T, \tag{3}
$$

where $R$ is the specific gas constant of a fluid.

We model the diffusivity by the stress tensor

$$
\sigma(Q, \nabla Q) = \mu((2/3 \nabla \cdot v) - (\nabla v + \nabla v^{\mathsf{T}})), \tag{4}
$$

with constant viscosity $\mu$. The heat diffusion is governed by the coefficient

$$
\kappa = \frac{\mu \gamma}{\mathrm{Pr}} \frac{1}{\gamma - 1} R = \frac{\mu c_p}{\mathrm{Pr}}, \tag{5}
$$

where the ratio of specific heats $\gamma$, the heat capacity at constant pressure $c_p$ and the Prandtl number $\mathrm{Pr}$ depend on the fluid.

Many realistic atmospheric flows can be described by a perturbation over a background state that is in hydrostatic equilibrium

$$
\frac{\partial}{\partial z} \overline{p}(z) = -g \overline{\rho}(z), \tag{6}
$$

i.e. a state, where the pressure gradient is exactly in balance with the gravitational source term $S_{\rho v} = -\boldsymbol{k}\rho g$. The vector $\boldsymbol{k}$ is the unit vector pointing in $z$-direction. The momentum equation is dominated by the background flow in this case. Because this can lead to numerical instabilities, problems of this kind are challenging and require some care. To lessen the impact of this, we split the pressure $p = \overline{p} + p'$ into a sum of the background pressure $\overline{p}(z)$ and perturbation $p'(\boldsymbol{x}, t)$. We split the density $\rho = \overline{\rho} + \rho'$ in the same manner and arrive at

$$\frac{\partial \rho v}{\partial t} + \boldsymbol{\nabla} \cdot (v \otimes \rho v + \boldsymbol{I}p') + \boldsymbol{F}_\rho^v v = -g\boldsymbol{k}\rho'. \tag{7}$$

Note that a similar and more complex splitting is performed in [10, 12]. In contrast to this, we use the true compressible Navier-Stokes equations with minimal modifications.

## 3   Numerics

The ExaHyPE Engine implements an ADER-DG-scheme and a MUSCL-Hancock finite volume method. Both can be considered as instances of the more general PNPM schemes of [5]. We use a Rusanov-style flux that is adapted to PDEs with viscous terms [7,8]. The finite volume scheme is stabilized with the van Albada limiter [15]. The user can state dynamic AMR rules by supplying custom criteria that are evaluated point-wise. Our criterion uses an element-local error estimate based on the total variation of the numerical solution. We exploit the fact that the total variation of a numerical solution is a perfect indicator for edges of a wavefront. Let $\boldsymbol{f}(\boldsymbol{x}) : \mathbb{R}^{N_{\text{vars}}} \to \mathbb{R}$ be a sufficiently smooth function that maps the discrete solution at a point $\boldsymbol{x}$ to an arbitrary indicator variable. The total variation (TV) of this function is defined by

$$\text{TV}\left[f(\boldsymbol{x})\right] = \left\| \int_C |\boldsymbol{\nabla} f(\boldsymbol{x})| \, d\boldsymbol{x} \right\|_1 \tag{8}$$

for each cell. The operator $\|\cdot\|_1$ denotes the discrete $L_1$ norm in this equation. We compute the integral efficiently with Gaussian quadrature over the collocated quadrature points. How can we decide whether a cell is important or not? To resolve this conundrum, we compute the mean and the population standard deviation of the total variation of all cells. It is important that we use the method of [2] to compute the modes in a parallel and numerical stable manner. A cell is then considered to contain significant information if its deviates from the mean more than a given threshold. This criterion can be described formally by

$$\text{evaluate-refinement}(\boldsymbol{Q}, \mu, \sigma) = \begin{cases} \text{refine} & \text{if } \text{TV}(\boldsymbol{Q}) \geq \mu + T_{\text{refine}}\sigma, \\ \text{coarsen} & \text{if } \text{TV}(\boldsymbol{Q}) < \mu + T_{\text{coarsen}}\sigma, \\ \text{keep} & \text{otherwise.} \end{cases} \tag{9}$$

The parameters $T_{\text{refine}} > T_{\text{coarsen}}$ can be chosen freely. Chebyshev's inequality

$$\mathbb{P}\big(|X - \mu| \geq c\sigma\big) \leq \frac{1}{c^2}, \tag{10}$$

with probability $\mathbb{P}$ guarantees that we neither mark all cells for refinement nor for coarsening. This inequality holds for arbitrary distributions under the weak assumption that they have a finite mean $\mu$ and a finite standard deviation $\sigma$ [16]. Note that subcells are coarsened only if all subcells belonging to the coarse cell are marked for coarsening. In contrast to already published criteria which are either designed solely for the simulation of clouds [12] or computationally expensive [7], our criterion works for arbitrary PDEs and yet, is easy to compute and intuitive.

## 4   Results

In this section, we evaluate the quality of the results of our numerical methods and the scalability of our implementation. We use a mix of various benchmarking scenarios. After investigating the numerical convergence rate, we look at three standard CFD scenarios: the Taylor-Green vortex, the three-dimensional Arnold-Beltrami-Childress flow and a lid-driven cavity flow. Finally, we evaluate the performance for stratified flow scenarios in both two and three dimensions.

### 4.1   CFD Testing Scenarios

We begin with a manufactured solution scenario which we can use for a convergence test. We use the following constants of fluids for all scenarios in this section:

$$\gamma = 1.4, \quad \text{Pr} = 0.7, \quad c_v = 1.0. \tag{11}$$

Our description of the manufactured solution follows [4]. To construct this solution, we assume that

$$\begin{aligned}
p(\boldsymbol{x}, t) &= 1/10 \cos(\boldsymbol{k}\boldsymbol{x} - 2\pi t) + 1/\gamma, \\
\rho(\boldsymbol{x}, t) &= 1/2 \sin(\boldsymbol{k}\boldsymbol{x} - 2\pi t) + 1, \\
\boldsymbol{v}(\boldsymbol{x}, t) &= \boldsymbol{v_0} \sin(\boldsymbol{k}\boldsymbol{x} - 2\pi t),
\end{aligned} \tag{12}$$

solves our PDE. We use the constants $\boldsymbol{v_0} = 1/4 \, (1,1)^\mathsf{T}$, $\boldsymbol{k} = \pi/5 \, (1,1)^\mathsf{T}$ and simulate a domain of size $[10 \times 10]$ for $0.5\,\text{s}$. The viscosity is set to $\mu = 0.1$. Note that Eq. (12) does not solve the compressible Navier-Stokes equations Eq. (1) directly. It rather solves our equation set with an added source term which can be derived with a computer algebra system. We ran this for a combination of orders $1, \ldots, 6$ and multiple grid sizes. Note that by order we mean the polynomial order throughout the entire paper and not the theoretical convergence order. For this scenario, we achieve high-order convergence (Fig. 1) but notice some diminishing returns for large orders.

After we have established that the implementation of our numerical method converges, we are going to investigate three established testing scenarios from the field of computational fluid mechanics. A simple scenario is the Taylor-Green vortex. Assuming an *incompressible* fluid, it can be written as

$$\rho(\boldsymbol{x}, t) = 1,$$
$$\boldsymbol{v}(\boldsymbol{x}, t) = \exp(-2\mu t) \begin{pmatrix} \sin(x)\cos(y) \\ -\cos(x)\sin(y) \end{pmatrix}, \tag{13}$$
$$p(\boldsymbol{x}, t) = \exp(-4\mu t) \, {}^{1}\!/_{4} \left( \cos(2x) + \cos(2y) \right) + C.$$

The constant $C = {}^{100}\!/_{\gamma}$ governs the speed of sound and thus the Mach number $\mathrm{Ma} = 0.1$ [6]. The viscosity is set to $\mu = 0.1$.

We simulate on the domain $[0, 2\pi]^2$ and impose the analytical solution at the boundary. A comparison at time $t = 10.0$ of the analytical solution for the pressure with our approximation (Fig. 2a) shows excellent agreement. Note that we only show a qualitative analysis because this is not an exact solution for our equation set as we assume compressibility of the fluid. This is nevertheless a valid comparison because for very low Mach numbers, both incompressible and compressible equations behave in a very similarly. We used an ADER-DG-scheme of order 5 with a grid of $25^2$ cells.

The Arnold-Beltrami-Childress (ABC) flow is similar to the Taylor-Green vortex but is an analytical solution for the three-dimensional *incompressible* Navier-Stokes equations [14]. It is defined in the domain $[-\pi, \pi]^3$ as

$$\rho(\boldsymbol{x}, t) = 1,$$
$$\boldsymbol{v}(\boldsymbol{x}, t) = \exp(-1\mu t) \begin{pmatrix} \sin(z) + \cos(y) \\ \sin(x) + \cos(z) \\ \sin(y) + \cos(x) \end{pmatrix}, \tag{14}$$
$$p(\boldsymbol{x}, t) = -\exp(-2\mu t) \left( \cos(x)\sin(y) + \sin(x)\cos(z) + \sin(z)\cos(y) \right) + C.$$

The constant $C = {}^{100}\!/_{\gamma}$ is chosen as before. We use a viscosity of $\mu = 0.01$ and analytical boundary conditions. Our results (Fig. 2b) show a good agreement between the analytical solution and our approximation with an ADER-DG-scheme of order 3 with a mesh consisting of $27^3$ cells at time $t = 0.1\,\mathrm{s}$. Again, we do not perform a quantitative analysis as the ABC-flow only solves our equation set approximately.

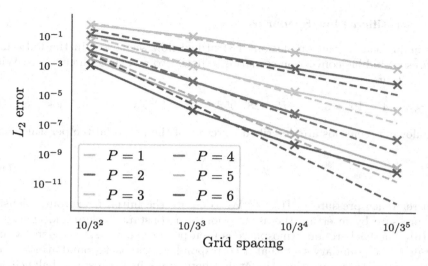

**Fig. 1.** Mesh size vs. error for various polynomial orders $P$. Dashed lines show the theoretical convergence order of $P + 1$.

As a final example of standard flow scenarios, we consider the lid-driven cavity flow where the fluid is initially at rest, with $\rho = 1$ and $p(\boldsymbol{x}) = {}^{100}/_\gamma$. We consider a domain of size $1\,\mathrm{m} \times 1\,\mathrm{m}$ which is surrounded by no-slip walls. The flow is driven entirely by the upper wall which has a velocity of $v_x = 1\,\mathrm{m/s}$. The simulation runs for $10\,\mathrm{s}$. Again, our results (Fig. 3) have an excellent agreement with the reference solution of [9]. We used an ADER-DG-method of order 3 with a mesh of size $27^2$.

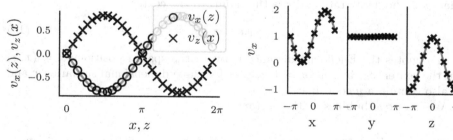

(a) Our result (markers) of the Taylor-Green vortex vs. the analytical solution (lines) Eq. (13). The plot shows two velocity slices, the respective other coordinate is held constant at a value of $\pi$.

(b) Our approximation (markers) of the ABC-flow vs. analytical solution (lines, Eq. (14)). All other axes are held constant at a value of 0. We show every 6th value.

**Fig. 2.** Two-dimensional CFD scenarios

## 4.2   Stratified Flow Scenarios

Our main focus is the simulation of stratified flow scenarios. In the following, we present bubble convection scenarios in both two and three dimensions. With the constants

$$\gamma = 1.4, \quad \mathrm{Pr} = 0.71, \quad R = 287.058, \quad p_0 = 10^5\,\mathrm{Pa}, \quad g = 9.8\,\mathrm{m/s^2}, \quad (15)$$

all following scenarios are described in terms of the potential temperature

$$\theta = T\left(\frac{p_0}{p}\right)^{R/c_p}, \tag{16}$$

with reference pressure $p_0$ [10,12]. We compute the initial background density and pressure by inserting the assumption of a constant background energy in Eq. (6). The background atmosphere is then perturbed. We set the density and energy at the boundary such that it corresponds to the background atmosphere. Furthermore, to ensure that the atmosphere stays in hydrostatic balance, we need to impose the viscous heat flux

$$F^v_{\rho E} = \kappa \frac{\partial \overline{T}}{\partial z}. \tag{17}$$

at the boundary [10]. In this equation, $\overline{T}(z)$ is the background temperature at position $z$, which can be computed from Eqs. (2) and (6).

Our first scenario is the colliding bubbles scenario [12]. We use perturbations of the form

$$\theta' = \begin{cases} A & r \leq a, \\ A\exp\left(-\frac{(r-a)^2}{s^2}\right) & r > a, \end{cases} \tag{18}$$

where $s$ is the decay rate and $r$ is the radius to the center

$$r^2 = \|\boldsymbol{x} - \boldsymbol{x}_c\|_2, \tag{19}$$

i.e., $r$ denotes the Euclidean distance between the spatial positions $\boldsymbol{x} = (x, z)$ and the center of a bubble $\boldsymbol{x}_c = (x_c, z_c)$ – for three-dimensional scenarios $\boldsymbol{x}$ and $\boldsymbol{x}_c$ also contain a $y$ coordinate.

We have two bubbles, with constants

| | | | | | |
|---|---|---|---|---|---|
| warm: | $A = 0.5\,\mathrm{K},$ | $a = 150\,\mathrm{m},$ | $s = 50\,\mathrm{m},$ | $x_c = 500\,\mathrm{m},$ | $z_c = 300\,\mathrm{m},$ |
| cold: | $A = -0.15\,\mathrm{K},$ | $a = 0\,\mathrm{m},$ | $s = 50\,\mathrm{m},$ | $x_c = 560\,\mathrm{m},$ | $z_c = 640\,\mathrm{m}.$ |

$$\tag{20}$$

Similar to [12], we use a constant viscosity of $\mu = 0.001$ to regularize the solution. Note that we use a different implementation of viscosity than [12]. Hence, it is difficult to compare the parametrization directly. We ran this scenario twice: once without AMR and a mesh of size $1000/81\,\mathrm{m} = 12.35\,\mathrm{m}$ and once with AMR with two adaptive refinement levels and parameters $T_{\mathrm{refine}} = 2.5$ and $T_{\mathrm{coarsen}} = -0.5$.

For both settings we used polynomials of order 6. We specialize the AMR-criterion (Eq. 9) to our stratified flows by using the potential temperature. This resulted in a mesh with cell-size lengths of approx. 111.1 m, 37.04 m, and 12.34 m. The resulting mesh can be seen in Fig. 5.

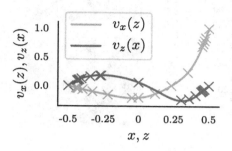

**Fig. 3.** Our approximation (solid lines) of the lid-driven cavity flow vs. reference solution (crosses) of [9]. The respective other coordinate is held constant at a value of 0.

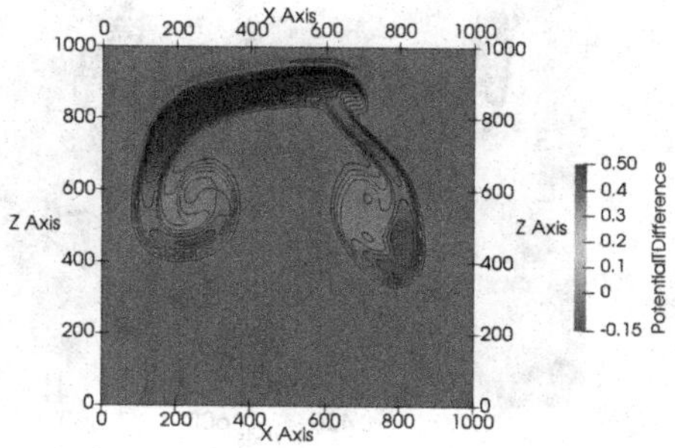

**Fig. 4.** Colliding bubbles with MUSCL-Hancock. Contour values for potential temperature perturbation are $-0.05, 0.05, 0.1, \ldots 0.45$.

We observe that the $L_2$ difference between the potential temperature of the AMR run, which uses 1953 cells, and the one of the fully refined run with 6561 cells, is only 1.87. The relative error is $2.6 \times 10^{-6}$. We further emphasize that our AMR-criterion accurately tracks the position of the edges of the cloud instead of only its position. This is the main advantage of our gradient-based method in contrast to methods working directly with the value of the solution, as for example [12]. Overall, our result for this benchmark shows an excellent agreement to the previous solutions of [12]. In addition, we ran a simulation with the same settings for a polynomial order of 3. The lower resolution leads to spurious waves (Fig. 5) and does not capture the behavior of the cloud.

Furthermore, we simulated the same scenario with our MUSCL-Hancock method, using $7^2$ patches with $90^2$ finite volume cells each. As we use limiting, we do not need any viscosity. The results of this method (Fig. 4) also agree with the reference but contain fewer details. Note that the numerical dissipativity of the finite volume scheme has a smoothing effect that is similar to the smoothing caused by viscosity.

For our second scenario, the cosine bubble, we use a perturbation of the form

$$\theta' = \begin{cases} A/2 \left[1 + \cos(\pi r)\right] & r \leq a, \\ 0 & r > a, \end{cases} \qquad (21)$$

where $A$ denotes the maximal perturbation and $a$ is the size of the bubble. We use the constants

$$A = 0.5 \, \text{K}, \quad a = 250 \, \text{m}, \quad x_c = 500 \, \text{m}, \quad z_c = 350 \, \text{m}. \qquad (22)$$

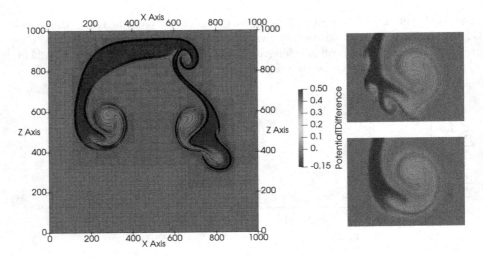

**Fig. 5.** Left: Colliding bubbles with ADER-DG. Contour values for potential temperature perturbation are $-0.05, 0.05, 0.1, \ldots 0.45$. Right: Comparison of small scale structure between order 3 (top) and order 6 (bottom).

For the three-dimensional bubble, we set $y_c = x_c = 500\,\mathrm{m}$. This corresponds to the parameters used in [11][1]. For the 2D case, we use a constant viscosity of $\mu = 0.001$ and an ADER-DG-method of order 6 with two levels of dynamic

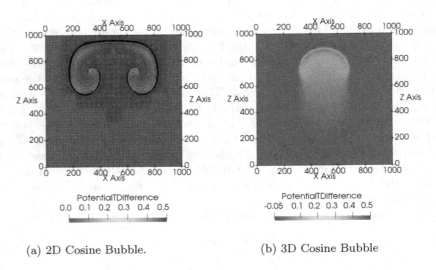

(a) 2D Cosine Bubble.          (b) 3D Cosine Bubble

**Fig. 6.** Cosine bubble scenario.

---

[1] We found that the parameters presented in the manuscript of [11] only agree with the results, if we use the same parameters as for 2D simulations.

AMR, resulting again in cell sizes of roughly $111.1\,\mathrm{m}, 37.04\,\mathrm{m}, 12.34\,\mathrm{m}$. We use slightly different AMR parameters of $T_{\mathrm{refine}} = 1.5$ and $T_{\mathrm{coarsen}} = -0.5$ and let the simulation run for 700 s. Note that, as seen in Fig. 6a, our AMR-criterion tracks the wavefront of the cloud accurately. This result shows an excellent agreement to the ones achieved in [10,12].

For the 3D case, we use an ADER-DG-scheme of order 3 with a static mesh with cell sizes of 40 m and a shorter simulation duration of 400 s. Due to the relatively coarse resolution and the hence increased aliasing errors, we need to increase the viscosity to $\mu = 0.005$. This corresponds to a larger amount of smoothing. Our results (Fig. 6b) capture the dynamics of the scenario well and agree with the reference solution of [11].

### 4.3   Scalability

All two-dimensional scenarios presented in this paper can be run on a single workstation in less than two days. Parallel scalability was thus not the primary goal of this paper. Nevertheless, our implementation allows us to scale to small to medium scale setups using a combined MPI + Thread building blocks (TBB) parallelization strategy, which works as follows: We typically first choose a number of MPI ranks that ensure an equal load balancing. ExaHyPE achieves best scalability for $1, 10, 83, \ldots$ ranks, as our underlying framework uses three-way splittings for each level and per dimension and an additional communication rank per level. For the desired number of compute nodes, we then determine the number of TBB threads per rank to match the number of total available cores.

We ran the two bubble scenario for a uniform grid with a mesh size of $729 \times 729$ with order 6, resulting in roughly 104 million degrees of freedom (DOF), for 20 timesteps and for multiple combinations of MPI ranks and TBB threads. This simulation was performed on the SuperMUC-NG system using computational kernels that are optimized for its Skylake architecture. Using a single MPI rank, we get roughly 4.9 millions DOF updates (MDOF/s) using two TBB threads and 20.2 MDOF/s using 24 threads (i.e. a half node). For a full node with 48 threads, we get a performance of 12 MDOF/s. When using 5 nodes with 10 MPI ranks, we achieve 29.3 MDOF/s for two threads and 137.3 MDOF/s for 24 threads.

We further note that for our scenarios weak scaling is more important than strong scaling, as we currently cover only a small area containing a single cloud, where in practical applications one would like to simulate more complex scenarios.

## 5   Conclusion

We presented an implementation of a MUSCL-Hancock-scheme and an ADER-DG-method with AMR for the Navier-Stokes equations, based on the ExaHyPE Engine. Our implementation is capable of simulating different scenarios: We show that our method has high order convergence and we successfully evaluated our method for standard CFD scenarios: We have competitive results for both two-dimensional scenarios (Taylor-Green vortex and lid-driven cavity) and for the three-dimensional ABC-flow.

Furthermore, our method allows us to simulate flows in hydrostatic equilibrium correctly, as our results for the cosine and colliding bubble scenarios showed. We showed that our AMR-criterion is able to vastly reduce the number of grid cells while preserving the quality of the results.

Future work should be directed towards improving the scalability. With an improved AMR scaling and some fine tuning of the parallelization strategy, the numerical method presented here might be a good candidate for the simulation of small scale convection processes that lead to cloud formation processes.

**Acknowledgments.** This work was funded by the European Union's Horizon 2020 Research and Innovation Programme under grant agreements No 671698 (project ExaHyPE, www.exahype.eu) and No 823844 (ChEESE centre of excellence, www.cheese-coe.eu). Computing resources were provided by the Leibniz Supercomputing Centre (project pr83no). Special thanks go to Dominic E. Charrier for his support with the implementation in the ExaHyPE Engine.

# References

1. Bauer, P., Thorpe, A., Brunet, G.: The quiet revolution of numerical weather prediction. Nature **525**(7567), 47 (2015). https://doi.org/10.1038/nature14956
2. Chan, T.F., Golub, G.H., LeVeque, R.J.: Updating formulae and a pairwise algorithm for computing sample variances. In: Caussinus, H., Ettinger, P., Tomassone, R. (eds.) COMPSTAT 1982, pp. 30–41. Physica, Heidelberg (1982). https://doi.org/10.1007/978-3-642-51461-6_3
3. Charrier, D.E., Weinzierl, T.: Stop talking to me-a communication-avoiding ADER-DG realisation. arXiv preprint arXiv:1801.08682 (2018)
4. Dumbser, M.: Arbitrary high order PNPM schemes on unstructured meshes for the compressible Navier-Stokes equations. Comput. Fluids **39**(1), 60–76 (2010). https://doi.org/10.1016/j.compfluid.2009.07.003
5. Dumbser, M., Balsara, D.S., Toro, E.F., Munz, C.D.: A unified framework for the construction of one-step finite volume and discontinuous Galerkin schemes on unstructured meshes. J. Comput. Phys. **227**(18), 8209–8253 (2008). https://doi.org/10.1016/j.jcp.2008.05.025
6. Dumbser, M., Peshkov, I., Romenski, E., Zanotti, O.: High order ADER schemes for a unified first order hyperbolic formulation of continuum mechanics: viscous heat-conducting fluids and elastic solids. J. Comput. Phys. **314**, 824–862 (2016). https://doi.org/10.1016/j.jcp.2016.02.015
7. Fambri, F., Dumbser, M., Zanotti, O.: Space-time adaptive ADER-DG schemes for dissipative flows: compressible navier-stokes and resistive MHD equations. Comput. Phys. Commun. **220**, 297–318 (2017). https://doi.org/10.1016/j.cpc.2017.08.001
8. Gassner, G., Lörcher, F., Munz, C.D.: A discontinuous Galerkin scheme based on a space-time expansion II. Viscous flow equations in multi dimensions. J. Sci. Comput. **34**(3), 260–286 (2008). https://doi.org/10.1007/s10915-007-9169-1
9. Ghia, U., Ghia, K., Shin, C.: High-Re solutions for incompressible flow using the Navier-Stokes equations and a multigrid method. J. Comput. Phys. **48**(3), 387–411 (1982). https://doi.org/10.1016/0021-9991(82)90058-4

10. Giraldo, F., Restelli, M.: A study of spectral element and discontinuous Galerkin methods for the Navier-Stokes equations in nonhydrostatic mesoscale atmospheric modeling: equation sets and test cases. J. Comput. Phys. **227**(8), 3849–3877 (2008). https://doi.org/10.1016/j.jcp.2007.12.009
11. Kelly, J.F., Giraldo, F.X.: Continuous and discontinuous Galerkin methods for a scalable three-dimensional nonhydrostatic atmospheric model: limited-area mode. J. Comput. Phys. **231**(24), 7988–8008 (2012). https://doi.org/10.1016/j.jcp.2012.04.042
12. Müller, A., Behrens, J., Giraldo, F.X., Wirth, V.: An adaptive discontinuous Galerkin method for modeling cumulus clouds. In: Fifth European Conference on Computational Fluid Dynamics, ECCOMAS CFD (2010)
13. Müller, A., Kopera, M.A., Marras, S., Wilcox, L.C., Isaac, T., Giraldo, F.X.: Strong scaling for numerical weather prediction at petascale with the atmospheric model NUMA. Int. J. High Perform. Comput. Appl. **33**(2), 411–426 (2018). https://doi.org/10.1177/1094342018763966
14. Tavelli, M., Dumbser, M.: A staggered space-time discontinuous Galerkin method for the three-dimensional incompressible Navier-Stokes equations on unstructured tetrahedral meshes. J. Comput. Phys. **319**, 294–323 (2016). https://doi.org/10.1016/j.jcp.2016.05.009
15. Van Albada, G., Van Leer, B., Roberts, W.: A comparative study of computational methods in cosmic gas dynamics. In: Hussaini, M.Y., van Leer, B., Van Rosendale, J. (eds.) Upwind and High-Resolution Schemes, pp. 95–103. Springer, Heidelberg (1997). https://doi.org/10.1007/978-3-642-60543-7_6
16. Wasserman, L.: All of Statistics. Springer, New York (2004). https://doi.org/10.1007/978-0-387-21736-9

# Performance and Portability
# of State-of-Art Molecular Dynamics
# Software on Modern GPUs

Evgeny Kuznetsov[1], Nikolay Kondratyuk[1,2,3], Mikhail Logunov[2,3],
Vsevolod Nikolskiy[1,2], and Vladimir Stegailov[1,2,3(✉)]

[1] National Research University Higher School of Economics, Moscow, Russia
`v.stegailov@hse.ru`
[2] Joint Institute for High Temperatures of RAS, Moscow, Russia
[3] Moscow Institute of Physics and Technology, Dolgoprudny, Russia

**Abstract.** Classical molecular dynamics (MD) calculations represent a
significant part of utilization time of high performance computing systems. As usual, efficiency of such calculations is based on an interplay of
software and hardware that are nowadays moving to hybrid GPU-based
technologies. Several well-developed MD packages focused on GPUs differ both in their data management capabilities and in performance. In
this paper, we present our results for the porting of the CUDA backend
of LAMMPS to ROCm HIP that shows considerable benefits for AMD
GPUs comparatively to the existing OpenCL backend. We consider the
efficiency of solving the same physical models using different software
and hardware combinations. We analyze the performance of LAMMPS,
HOOMD, GROMACS and OpenMM MD packages with different GPU
back-ends on modern Nvidia Volta and AMD Vega20 GPUs.

**Keywords:** LAMMPS · HOOMD · GROMACS · OpenMM ·
OpenCL · Nvidia CUDA · AMD ROCm HIP

## 1   Introduction

Classical molecular dynamics (MD) simulations method is one of the major
consumers of supercomputer resources worldwide. The development of highly
scalable MD codes that enable extreme MD system sizes is one of the important vectors of development for numerical methods [1]. Shortly after the Nvidia
CUDA technology had been introduced in 2007, hybrid MD algorithms appeared
and showed their promising performance (e.g. see [2]).

Porting MD algorithms to hybrid architectures is a cumbersome endeavor
that requires a lot of efforts. Recently, such experience has been described for
the HPC systems based on Sunway CPUs [3–6]. Another example is porting MD
algorithm to Adapteva Epiphany architecture [7]. These efforts are unique since
these architectures were programmed without established software ecosystem.

© Springer Nature Switzerland AG 2020
R. Wyrzykowski et al. (Eds.): PPAM 2019, LNCS 12043, pp. 324–334, 2020.
https://doi.org/10.1007/978-3-030-43229-4_28

At the moment, the best software ecosystem for GPU computing is based on Nvidia CUDA technology. Unfortunately, CUDA is a closed-source technology that limits portability. CUDA is a platform for writing applications for general-purpose computing on graphics processing units (GPGPU). It has mature driver and runtime support, great debugging and profiling tools, detailed documentation and samples. Almost every algorithm that has high computational complexity and parallelization possibility was implemented in CUDA. GPU acceleration is very efficient for MD calculations and gives possibility to obtain highly accurate results (e.g. see [8,9]).

In contrast, AMD Radeon Open Compute platform (ROCm) is an open-source part of the AMD's "Boltzmann Initiative" announced in 2015. At the moment, ROCm is barely known platform for developing GPGPU applications but it is developing rather quickly (e.g., version 1.9 has been announced on 15 September 2018 and version 2.3 has been announced on 12 April 2019). ROCm HIP framework allows developing application that can be compiled and run both on AMD and Nvidia GPUs. In this work, we use this framework for porting the LAMMPS MD package.

The question of efficiency of a particular type of hardware for a given software and for a given class of mathematical models is an important question in the post-Moore era that is characterized by an increasing variety of hardware and software options (see e.g. [10]). MD applications demonstrate increasing attention to the question of reproducibility and portability of MD models [11–13]. Attention to the reproducibility of the results of mathematical modelling is a rising trend in all applied fields (see e.g. [14]).

In this work we consider a set of MD models, their portability and performance across several MD packages and two GPU types (Fig. 1).

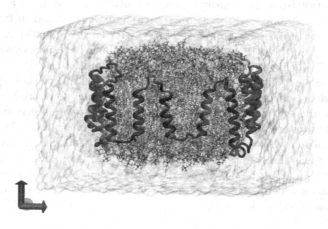

**Fig. 1.** The visual representation of the ApoA1 model in VMD [15].

## 2   Methods

ROCm contains a HIP framework that allows porting existing CUDA programs for AMD devices without deep rewriting. HIP allows creating cross-platform GPU applications. HIP provides syntactically equivalent to CUDA API and, when it comes to compiling for Nvidia devices, serves as a thin and zero-overhead shell over CUDA. From the other side, there is evidence [16] that HIP is the best performing framework from choices that ROCm support. Moreover, HIP does not prohibit the use of inline assembly to achieve maximum possible performance on both platforms.

In order to evaluate the HIP backend for LAMMPS presented in this work, we put it in the context of three other MD packages and consider two generic MD models for benchmarking.

A Lennard-Jones liquid model is an example of a model with a short-ranged interatomic potential. Systems sizes from $N = 55296$ to $4000000$ atoms with the number density $25.7444\,\text{nm}^{-1}$ are considered. The LJ parameters are $\epsilon = 0.2381$ kcal/mol and $\sigma = 0.3405$ nm. The LJ potential cut-off radius is $1.3\,\text{nm}$ that results in the average number of neighbors equal 181.

Another model is a protein-lipid system in water ApoA1 that is a popular example of a biomolecular MD system consisting of 92224 atoms (e.g. see [17]). We use the Charmm36 force field with the cut-off radius for nonbonded interactions of 1.2 nm and the relative accuracy 0.0005 for electrostatic forces (PME/PPPM). The time step of integration of equations of motion from atoms is 1 fs.

In order to check that we consider the same mathematical model in different MD packages $A$ and $B$, we calculate the divergences of atomic coordinates $< \Delta r^2(t) >= \sum_{i=1}^{N} |\mathbf{r}_i^A(t) - \mathbf{r}_i^B(t)|/N$ and velocities $< \Delta v^2(t) >= \sum_{i=1}^{N} |\mathbf{v}_i^A(t) - \mathbf{v}_i^B(t)|/N$ starting from the identical initial conditions. MD systems are subjected to exponential instability [18]. If two MD trajectories obtained in two different MD packages diverge (due to round-off errors) with the exponential rate inherent to this MD system then the numerical schemes in both MD packages are equivalent (e.g. see [7]). This approach is more illustrative and physically justified than the check of the equality of per atom forces in a given atomic configuration as it is usually done when different MD packages are being compared [13,19].

For the equivalent models in different MD packages, we determine the speed of MD calculations for two types of novel GPUs from Nvidia and AMD. For adequate comparison of GPUs, we make all the benchmarks using one and the same server. We present the results in terms of an average time per atom per step for the LJ models since in the case of short-ranged potentials the number of arithmetical operations scales as $O(N)$. For the ApoA1 model we use nanoseconds of simulated time per day (ns/day).

## 3   Hardware

The benchmark results presented in this paper have been obtained using the server based on the Supermicro H11SSL-i motherboard with AMD Epyc 7251

CPU (running in turbo mode with hyperthreading enabled) with 2 GPUs of the most novel architectures: Nvidia Titan V and AMD Radeon VII (see Table 1). Both GPU accelerators are connected to the CPU via the PCIe 3.0 × 16 interface. The server works under Ubuntu 18.04 kernel 4.15. The most important parts of software are as follows: gcc 7.3, CUDA 10.0 with 410.104 driver and ROCm 2.2 (rocm-smi–setperflevel high).

**Table 1.** The comparison of the declared peak parameters of GPUs considered

| GPU model | Architecture | SP/DP TFlops/s | Memory GB | B/w TB/s | Programming technology for HPC |
|-----------|--------------|----------------|-----------|----------|-------------------------------|
| Titan V | Volta (Nvidia) | 12.3/6.1 | 12 | 0.65 | CUDA, OpenCL 1.2, OpenACC, OpenMP |
| Radeon VII | Vega20 (AMD) | 11.1/2.8 | 16 | 1.0 | HCC/HIP, OpenCL 2.0, OpenACC*, OpenMP* |

*As announced in 2017 by "Mentor, a Siemens Business".

Nvidia GPUs support OpenACC and OpenMP technologies that are much simpler for development and result in a very portable code. The OpenACC and OpenMP support for AMD GPUs has been announced as well. The maximum efficiency of GPUs can be obtained by deep control of memory hierarchies of GPUs. This can be done using OpenCL as a standard technique for development of heterogeneous parallel programs. At the moment, Nvidia GPUs support OpenCL 1.2 and AMD ROCm support OpenCL 2.0. However the maximum efficiency of Nvidia GPU can be obtained using CUDA technology. The HIP framework is designed as an equivalent of CUDA for AMD GPUs and thus as a portability layer for a large number of existing CUDA-based software.

## 4  Software

The emergence of parallel distributed memory supercomputing systems stimulated the development of parallel codes for MD calculations. Among others, LAMMPS [20] and GROMACS [21] are two MD codes that have developed into complex simulation packages and are widely used nowadays. Since the emergence of CUDA and OpenCL, both codes have been supplemented with GPU offloading capabilities (see [22–24] and [25]). The offloading scheme uses a GPUs for non-bonded interaction and electrostatics calculations. However, the offloading efficiency depends on the CPU-GPU bandwidth that frequently poses significant limitations on performance. Moreover, the load balancing between CPU and GPU usually requires essentially more powerful CPUs for pairing with a newer generation of GPUs. From this perspective, it would be beneficial to have codes that do all the computations inside a GPU. HOOMD was among the first of such MD codes [26, 27] and is under active development today. OpenMM is another purely GPU MD code focused on biomolecular simulations [19, 28]. A comparison of key features of these packages are presented in Table 2.

**Table 2.** The comparison of the MD packages considered

| Package (version) | Supported technologies for GPUs | Supported GPU precision | Capabilities |
|---|---|---|---|
| LAMMPS (28 Feb 2019) | CUDA OpenCL* ROCm HIP** | Single Double Mixed | The widest spectrum of physical models, on-the-fly data processing, scripting language |
| HOOMD (2.5.2) | CUDA | Single Double | A wide spectrum of models, python-based scripting capabilities |
| GROMACS (2019.2) | CUDA OpenCL*** | Single Double Mixed | A limited set of force fields for biomolecules and organic compounds, a rich set of post-processing tools |
| OpenMM (7.3.1) | CUDA OpenCL | Single Double Mixed | A limited set of force fields (similar to GROMACS), python-based scripting capabilities, no MPI-parallelism for the GPU version |

*Not fully OpenCL 2.0 compliant. **Developed in this work.
***Limitations inherent to the Nvidia OpenCL runtime.

The MD algorithms allows using single precision for contributions of per-atom-forces that causes no numerical instabilities if double precision variables are used as accumulators. LAMMPS, GROMACS and OpenMM make use of this mixed precision approach that is extremely beneficial because it gives 2x performance on professional GPUs and allows using cheap consumer level GPUs for calculations [29, 30]. HOOMD, however, at the moment performs MD calculations either in pure single precision, or in pure double precision. LAMMPS, GROMACS and OpenMM have both the CUDA backend and the OpenCL backend that allows using these packages not only with Nvidia GPUs, but with other types of accelerators as well. Currently HOOMD lacks such portability.

Unfortunately, several definitions of local variables in the OpenCL backend of LAMMPS do not satisfy the OpenCL 2.0 standard that is required for ROCm platform. That is why it is not possible to run LAMMPS under ROCm without some modifications. Such modifications have been made in this work for running the LJ and ApoA1 models (lib/gpu/lal_lj.cu and lal_charmm_long.cu files).

No specific OpenCL tuning is made for Radeon VII for GROMACS and OpenMM. The AMD GPU options of the LAMMPS OpenCL backed correspond to a very old Cypress architecture. A 10% speedup on Radeon VII has been obtained from the following change BLOCK_PAIR 128 → 256, BLOCK_NBOR_BUILD 64 → 128 and BLOCK_BIO_PAIR 64 → 256 (Fig. 2).

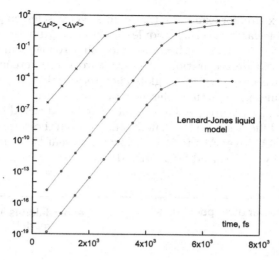

**Fig. 2.** Examples of divergences of MD trajectories for the LJ model: the closed/open circles correspond to $< \Delta r^2(t) > / < \Delta v^2(t) >$ for two MD trajectories calculated with OpenMM in double precision on Titan V (CUDA) and Radeon VII (OpenCL), the crosses correspond to two MD trajectories calculated with OpenMM and GROMACS in mixed precision.

## 5    Results and Discussion

### 5.1    Conversion of the CUDA Backend of LAMMPS to HIP

For porting to HIP framework, we use the CUDA-backend of the GPU acceleration capability of LAMMPS developed by Brown et al. [22–24]. This acceleration library is based on the *geryon* library that serves as a unification layer for CUDA and OpenCL backends.

A *hipify* tool provides a fast method to translate CUDA runtime and driver API calls and types to their HIP analogs. Not all CUDA functions have corresponding HIP wrappers and those that have, sometimes are not fully implemented. In particular, this applies to CUDA driver API calls used in the *geryon* library. Despite that *hipify*, empowered by Clang C++ parser, allows translating the most part of CUDA code in a mistake-free manner.

Due to existing architecture differences, it is essential to revise GPU kernels. This applies especially to inline assembly and warp operations. AMD and Nvidia GPUs have different warp sizes, 64 and 32 respectively, so it is necessary to make sure all warp vote and shuffle functions, _ _ballot for example, use appropriate integer types and there are no "magic" constants like 32, 31, 5 and so on. *Printf* and *assert* calls from GPU kernels are still under development, and it is worth noting that in the current ROCm version 2.2 an assert call causes kernel execution abort regardless of the condition.

Another pitfall is using texture references and objects. In the current ROCm version 2.2, AMD HSA imposes hard limits on texture dimensions (up to 4096 elements for 1D texture, for example), that makes them useless in practice. Moreover, textures are declared in kernel modules as external symbols, and they should be defined in the host scope. So, it becomes impossible to load and use binary kernel modules with texture declarations. A possible workaround to avoid significant code fixes, in this case, is to use device global pointer variables as pseudo-textures. Device pointers binding to texture references or objects turn into global variables assignments, and, like textures, global variables can be extracted by their names.

Despite these difficulties, we have successfully ported the *geryon* library backend based on CUDA driver API to HIP. Below we present the benchmark results for this new LAMMPS backend for AMD GPUs.

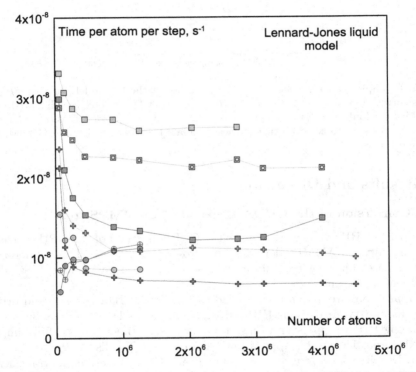

**Fig. 3.** Time per atom per step for MD calculations of the LJ liquid model with different number of atoms. The green symbols are the data from Titan V GPU with CUDA and the red symbols are the data from Radeon VII GPU with OpenCL. The squares show the data for LAMMPS (the open squares are for the ROCm HIP backend developed in this work). The crosses stand for GROMACS. And the circles stand for OpenMM. (Color figure online)

## 5.2 Benchmark Results

The results of the benchmarks for the LJ models are summarized in Fig. 3. GROMACS shows the smallest calculation times for systems larger 1 million atoms. GROMACS has the lowest requirement for the GPU memory (that is why the points for 4.5 million atoms are shown on the plot). For a still unidentified reason, OpenMM has errors for systems larger than 1.25 million atoms. At the same time, OpenMM shows the best results for small system sizes and the overall speed is the same as in GROMACS despite the fact that OpenMM uses only one CPU core (GROMACS uses all the cores with 2 threads per core).

The OpenCL backend of LAMMPS shows considerably slower execution rate than the CUDA backend. Moreover, the OpenCL backend requires more than twice GPU memory than the CUDA backend. The new HIP backend shows much better timings and uses less GPU memory.

The best results for GROMACS correspond to 8 MPI ranks with 2 threads per rank. The best results for LAMMPS on Titan V correspond to 8 MPI ranks as well. However, due to some problems with the GPU driver, the current ROCm version 2.2 does not allow using more than 4 MPI ranks for running OpenCL or HIP backends of LAMMPS (this fact lowers the measured performance values).

The benchmark results for the ApoA1 model are presented in Table 3. In this case LAMMPS uses GPU both for short-ranged forces calculations and for solving the Poisson equation for the electrostatics. In this case, using 4 MPI ranks per GPU gives the maximum speed for Titan V. Radeon VII with the HIP backend is 30% faster than with the OpenCL backend and shows the calculation speed very close to the results from Titan V with the CUDA backend.

GROMACS shows significantly higher calculation speed for the ApoA1 model as well. And OpenMM shows extraordinary speed for Titan V. For ApoA1 OpenCL backend of OpenMM is twice slower on Radeon VII than on Titan V. The reason for this is presumably connected with the electrostatic solver since the short-ranged potential solver has the same speed (see Fig. 3).

**Table 3.** The benchmark results for the ApoA1 model in mixed precision. The performance is shown in the nanoseconds of simulated time per one day of computing using the hardware and software considered (ns/day).

| Package | MPI/OMP per CPU | Titan V | | Radeon VII | |
|---------|-----------------|---------|--------|-----|--------|
| | | CUDA | OpenCL | HIP | OpenCL |
| LAMMPS | 4/1 | 11.6 | 8.5 | 10.7 | 8.0 |
| GROMACS | 8/2 | 17.0 | | | 12.9 |
| OpenMM | 1/1 | 55.2 | 51.2 | | 25.4 |

# 6   Conclusions

We have ported the LAMMPS molecular-dynamics package to ROCm HIP using the CUDA backend of the *geryon* library that implements GPU offloading for a large number of interatomic potentials and supports single, mixed and double precision calculations (available on GitHub [31]). The new ROCm HIP variant shows better calculation speed and has lower GPU memory footprint than the OpenCL backend of the *geryon* library. These results justify that the ROCm HIP framework can be used for developing portable code for hybrid application that has the same efficiency on Nvidia and AMD GPUs.

The set of models has been proposed for benchmarking classical MD codes. The method based on the divergence of trajectories has been deployed to check the identity of the mathematical models solved in different MD packages. LAMMPS, HOOMD, GROMACS and OpenMM packages have been compared using the server equipped with Titan V and Radeon VII GPUs that have quite similar peak performance and memory bandwidth. OpenCL variants of LAMMPS and GROMACS running on Radeon VII are 30–50% slower than their CUDA variants running on Titan V.

The performance of the HIP backend of LAMMPS for ApoA1 and the superiority of the OpenCL variant of OpenMM on Radeon VII for the LJ model over the CUDA variant of OpenMM show that the real-life applications can run on AMD GPUs at the same level of speed or even faster than on Nvidia GPUs of the same performance class.

**Acknowledgments.** The authors gratefully acknowledge financial support of the President grant NS-5922.2018.8.

# References

1. Tchipev, N., et al.: TweTriS: twenty trillion-atom simulation. Int. J. High Perform. Comput. Appl. **33**(5), 838–854 (2019). https://doi.org/10.1177/1094342018819741
2. Morozov, I., Kazennov, A., Bystryi, R., Norman, G., Pisarev, V., Stegailov, V.: Molecular dynamics simulations of the relaxation processes in the condensed matter on GPUs. Comput. Phys. Commun. **182**(9), 1974–1978 (2011). https://doi.org/10.1016/j.cpc.2010.12.026. Computer Physics Communications Special Edition for Conference on Computational Physics Trondheim, Norway, 23–26 June 2010
3. Dong, W., et al.: Implementing molecular dynamics simulation on Sunway TaihuLight system. In: 2016 IEEE 18th International Conference on High Performance Computing and Communications, IEEE 14th International Conference on Smart City, IEEE 2nd International Conference on Data Science and Systems (HPCC/SmartCity/DSS), pp. 443–450, December 2016. https://doi.org/10.1109/HPCC-SmartCity-DSS.2016.0070
4. Dong, W., Li, K., Kang, L., Quan, Z., Li, K.: Implementing molecular dynamics simulation on the Sunway TaihuLight system with heterogeneous many-core processors. Concurr. Comput. Pract. Experience **30**(16), e4468 (2018). https://doi.org/10.1002/cpe.4468
5. Yu, Y., An, H., Chen, J., Liang, W., Xu, Q., Chen, Y.: Pipelining computation and optimization strategies for scaling GROMACS on the sunway many-core processor. In: Ibrahim, S., Choo, K.-K.R., Yan, Z., Pedrycz, W. (eds.) ICA3PP 2017. LNCS, vol. 10393, pp. 18–32. Springer, Cham (2017). https://doi.org/10.1007/978-3-319-65482-9_2
6. Duan, X., et al.: Redesigning LAMMPS for peta-scale and hundred-billion-atom simulation on Sunway TaihuLight. In: SC18: International Conference for High Performance Computing, Networking, Storage and Analysis, pp. 148–159, November 2018. https://doi.org/10.1109/SC.2018.00015

7. Nikolskii, V., Stegailov, V.: Domain-decomposition parallelization for molecular dynamics algorithm with short-ranged potentials on Epiphany architecture. Lobachevskii J. Math. **39**(9), 1228–1238 (2018). https://doi.org/10.1134/S1995080218090159

8. Kondratyuk, N.D., Pisarev, V.V.: Calculation of viscosities of branched alkanes from 0.1 to 1000 MPa by molecular dynamics methods using COMPASS force field. Fluid Phase Equilib. **498**, 151–159 (2019). https://doi.org/10.1016/j.fluid.2019.06.023

9. Pisarev, V., Kondratyuk, N.: Prediction of viscosity-density dependence of liquid methane+n-butane+n-pentane mixtures using the molecular dynamics method and empirical correlations. Fluid Phase Equilib. **501**, 112273 (2019). https://doi.org/10.1016/j.fluid.2019.112273

10. Stegailov, V.V., Orekhov, N.D., Smirnov, G.S.: HPC hardware efficiency for quantum and classical molecular dynamics. In: Malyshkin, V. (ed.) PaCT 2015. LNCS, vol. 9251, pp. 469–473. Springer, Cham (2015). https://doi.org/10.1007/978-3-319-21909-7_45

11. Vermaas, J.V., Hardy, D.J., Stone, J.E., Tajkhorshid, E., Kohlmeyer, A.: TopoGromacs: automated topology conversion from CHARMM to GROMACS within VMD. J. Chem. Inf. Model. **56**(6), 1112–1116 (2016). https://doi.org/10.1021/acs.jcim.6b00103

12. Lee, J., et al.: CHARMM-GUI input generator for NAMD, GROMACS, AMBER, OpenMM, and CHARMM/OpenMM simulations using the CHARMM36 additive force field. J. Chem. Theory Comput. **12**(1), 405–413 (2016). https://doi.org/10.1021/acs.jctc.5b00935

13. Merz, P.T., Shirts, M.R.: Testing for physical validity in molecular simulations. PLOS ONE **13**(9), 1–22 (2018). https://doi.org/10.1371/journal.pone.0202764

14. Mesnard, O., Barba, L.A.: Reproducible and replicable computational fluid dynamics: it's harder than you think. Comput. Sci. Eng. **19**(4), 44–55 (2017). https://doi.org/10.1109/MCSE.2017.3151254

15. Humphrey, W., Dalke, A., Schulten, K.: VMD - visual molecular dynamics. J. Mol. Graph. **14**, 33–38 (1996)

16. Sun, Y., et al.: Evaluating performance tradeoffs on the radeon open compute platform. In: 2018 IEEE International Symposium on Performance Analysis of Systems and Software (ISPASS), pp. 209–218, April 2018. https://doi.org/10.1109/ISPASS.2018.00034

17. Stegailov, V., et al.: Angara interconnect makes GPU-based desmos supercomputer an efficient tool for molecular dynamics calculations. Int. J. High Perform. Comput. Appl. **33**(3), 507–521 (2019). https://doi.org/10.1177/1094342019826667

18. Norman, G.E., Stegailov, V.V.: Stochastic theory of the classical molecular dynamics method. Math. Models Comput. Simul. **5**(4), 305–333 (2013). https://doi.org/10.1134/S2070048213040108

19. Eastman, P., et al.: OpenMM 7: rapid development of high performance algorithms for molecular dynamics. PLOS Comput. Biol. **13**, 1–17 (2017). https://doi.org/10.1371/journal.pcbi.1005659

20. Plimpton, S.: Fast parallel algorithms for short-range molecular dynamics. J. Comput. Phys. **117**(1), 1–19 (1995). https://doi.org/10.1006/jcph.1995.1039

21. Berendsen, H., van der Spoel, D., van Drunen, R.: GROMACS: a message-passing parallel molecular dynamics implementation. Comput. Phys. Commun. **91**(1), 43–56 (1995). https://doi.org/10.1016/0010-4655(95)00042-E

22. Brown, W.M., Wang, P., Plimpton, S.J., Tharrington, A.N.: Implementing molecular dynamics on hybrid high performance computers – short range forces. Comput. Phys. Commun. **182**(4), 898–911 (2011). https://doi.org/10.1016/j.cpc.2010.12.021

23. Brown, W.M., Kohlmeyer, A., Plimpton, S.J., Tharrington, A.N.: Implementing molecular dynamics on hybrid high performance computers – particle-particle particle-mesh. Comput. Phys. Commun. **183**(3), 449–459 (2012). https://doi.org/10.1016/j.cpc.2011.10.012

24. Brown, W.M., Yamada, M.: Implementing molecular dynamics on hybrid high performance computers—three-body potentials. Comput. Phys. Commun. **184**(12), 2785–2793 (2013). https://doi.org/10.1016/j.cpc.2013.08.002

25. Abraham, M.J., et al.: GROMACS: high performance molecular simulations through multi-level parallelism from laptops to supercomputers. SoftwareX **1–2**, 19–25 (2015). https://doi.org/10.1016/j.softx.2015.06.001

26. Anderson, J.A., Lorenz, C.D., Travesset, A.: General purpose molecular dynamics simulations fully implemented on graphics processing units. J. Comput. Phys. **227**(10), 5342–5359 (2008). https://doi.org/10.1016/j.jcp.2008.01.047

27. Glaser, J., et al.: Strong scaling of general-purpose molecular dynamics simulations on GPUs. Comput. Phys. Commun. **192**, 97–107 (2015). https://doi.org/10.1016/j.cpc.2015.02.028

28. Eastman, P., et al.: OpenMM 4: a reusable, extensible, hardware independent library for high performance molecular simulation. J. Chem. Theory Comput. **9**(1), 461–469 (2013). https://doi.org/10.1021/ct300857j

29. Kutzner, C., Páll, S., Fechner, M., Esztermann, A., de Groot, B.L., Grubmüller, H.: Best bang for your buck: GPU nodes for GROMACS biomolecular simulations. J. Comput. Chem. **36**(26), 1990–2008 (2015)

30. Kutzner, C., Páll, S., Fechner, M., Esztermann, A., de Groot, B.L., Grubmüller, H.: More bang for your buck: improved use of GPU nodes for GROMACS 2018. CoRR abs/1903.05918 (2019). http://arxiv.org/abs/1903.05918

31. https://github.com/Vsevak/lammps

# Exploiting Parallelism on Shared Memory in the QED Particle-in-Cell Code PICADOR with Greedy Load Balancing

Iosif Meyerov[1]($\boxtimes$) , Alexander Panov[1], Sergei Bastrakov[2] ,
Aleksei Bashinov[3] , Evgeny Efimenko[3] , Elena Panova[1], Igor Surmin[1],
Valentin Volokitin[1], and Arkady Gonoskov[1,3,4]

[1] Lobachevsky State University of Nizhni Novgorod, Nizhni Novgorod, Russia
`meerov@vmk.unn.ru`
[2] Helmholtz-Zentrum Dresden-Rossendorf, Dresden, Germany
[3] Institute of Applied Physics, Russian Academy of Sciences,
Nizhni Novgorod, Russia
[4] University of Gothenburg, Gothenburg, Sweden

**Abstract.** State-of-the-art numerical simulations of laser plasma by means of the Particle-in-Cell method are often extremely computationally intensive. Therefore there is a growing need for the development of approaches for the efficient utilization of resources of modern supercomputers. In this paper, we address the problem of a substantially non-uniform and dynamically varying distribution of macroparticles in simulations of quantum electrodynamic (QED) cascades. We propose and evaluate a load balancing scheme for shared memory systems, which allows subdividing individual cells of the computational domain into work portions with the subsequent dynamic distribution of these portions among OpenMP threads. Computational experiments in 1D, 2D, and 3D QED simulations show that the proposed scheme outperforms the previously developed standard and custom schemes in the PICADOR code by 2.1 to 10 times when employing several Intel Cascade Lake CPUs.

**Keywords:** QED PIC · Load balancing · OpenMP · Multicore programming

## 1 Introduction

The numerical simulation of high-intensity laser-plasma interactions with the Particle-in-Cell (PIC) method [1] is a rapidly developing area of computational physics. The interaction of intense laser pulses with different targets provides the possibility of exciting a complex collective electron dynamics in the plasma generated by the strong laser fields. It opens up new opportunities for both

The work was funded by Russian Foundation for Basic Research and the government of the Nizhny Novgorod region of the Russian Federation, grant No. 18-47-520001.

R. Wyrzykowski et al. (Eds.): PPAM 2019, LNCS 12043, pp. 335–347, 2020.
https://doi.org/10.1007/978-3-030-43229-4_29

fundamental studies and important applications. One of the leading directions in this area is the study of effects due to quantum electrodynamics (QED) in superstrong electromagnetic fields [2–6]. Current and upcoming large-scale laser facilities are designed to provide electromagnetic fields necessary for the experimental observation of these processes. Today, an active study of future experiments is being carried out [6–9], with the central role being played by numerical simulation. Such simulations are performed with the PIC method extended with a module accounting for the QED effects, a combination referred to as the QED PIC.

Commonly the necessary QED PIC simulations are very computationally intensive and therefore are performed on supercomputers using highly optimized parallel software [10–18]. The problem of efficient implementation of the PIC method for parallel machines is well studied [18–27]. Fortunately, the method has a large potential for parallelization due to the local nature of the interactions. The spatial decomposition of the computational domain admits parallel processing on distributed memory using MPI. Shared memory parallelization is usually done either by launching an MPI process on each core, or by using a combination of MPI and OpenMP. In this case, computationally intensive loops on particles or cells (supercells) are parallelized. This parallel processing scheme is widely used for plasma simulation codes. However, when modeling the QED effects, the problem of the explosive growth of the number of particles involved in the simulation resulting from the development of electromagnetic cascades comes to the fore. The exponential increase in the number of particles in a small area requires the development of special approaches to overcome the unacceptable expenditure of RAM and computational imbalance. The proper use of special procedures for thinning and/or merging particles (see, for example, [28]) allows us to control the use of memory, but does not solve the problem of load balancing. Given the growing number of cores in modern CPUs, the problem of efficient parallelization of QED PIC codes for shared memory systems is becoming increasingly important.

Within our previous study [29] we compared five load balancing schemes for parallelizing a computational loop on cells containing substantially different numbers of particles. For the problems with a relatively small workload imbalance, these schemes enabled achieving an acceptable scaling efficiency. However, in case of a substantially non-uniform distribution, none of the considered schemes showed good results. As a unit of work all those schemes used the act of processing all particles in a cell. This is a natural choice for PIC codes using cell-based particle storage. This also simplifies parallel processing on shared memory.

In this paper we propose a new, more sophisticated, scheme of parallel processing and load balancing. As a unit of work it uses the act of processing all or some particles in a cell. It also handles the additional synchronization required when particles of the same cell are processed concurrently by different threads. Essentially, the new parallel processing scheme is a generalization of the previously considered ones in the way that the processing of some of the cells is divided into several portions of work. It increases the number of portions of work and makes it possible to avoid the typical scenario of QED PIC simulations, when few cells with a large number of particles hampers the scalability on shared memory. At the same time, the vast majority of cells are still processed as a single portion of work. We distinguish between the two types of cells during the run time and only apply the additional synchronization for the split cells, thus avoiding significant overheads.

The paper is organized as follows. In Sect. 2 we overview the QED PIC method. Section 3 introduces a baseline parallel algorithm. In Sect. 4 we give a detailed description of the new load balancing scheme. Section 5 presents numerical results. Section 6 concludes the paper.

## 2  An Overview of the Quantum Electrodynamics Particle-in-Cell Method

The Particle-in-Cell method [1] is commonly used to describe the self-consistent dynamics of charged particles in electromagnetic fields. These entities are represented numerically as the main sets of data. Electromagnetic fields and current density are set on a grid, and particles are modeled as an ensemble of macroparticles with continuous coordinates. Each macroparticle in the simulation represents a cloud of physical particles of the same kind (with equal mass and charge) closely located in the space of coordinates and momenta. According to this duality of data representation, fields and particles are updated on different stages. There are four main PIC stages.

A field solver updates the grid values of the electromagnetic fields based on the Maxwell's equations. In the PICADOR code [17] the conventional finite-difference time-domain method based on the staggered Yee grid is used [30]. On the particle update stage the position and momentum of each particle are updated according to the relativistic equation of motion, which is numerically integrated using an explicit method. Field interpolation from the grid to the particle positions is performed to compute the Lorenz force affecting the particles. Individual particle motion creates electric current, which is weighted to the grid values on the current deposition stage, to be used in the field solver, thus completing the self-consistent system of equations. For higher efficiency, it is usually convenient to combine the stages of field interpolation, Lorenz force computation and the integration of particle motion into one stage, referred to as the particle push.

This standard PIC scheme can be extended by different modules which take into account various physical processes, such as ionization and collisions [12, 19, 31]. In the case of extremely strong electromagnetic fields, the QED processes come into play. Due to the non-linear Compton scattering the charged particles accelerated in extreme laser field emit high-energy due to the Breit-Wheeler pair production photons, which in turn can decay into a pair of electron and positron [4, 18, 32]. When repeatedly occurring, these processes may lead to an avalanche-like pair density growth, similar to the avalanche gas breakdown, leading to the development of the so-called QED cascades [33]. This dynamics may lead to the formation of localized highly absorbing pair plasmas, which efficiently convert laser energy into high-energy gamma photons and charged particles. This can be treated as a source of antimatter (positrons), extremely dense electron-positron plasma, high-energy electrons and photons [6, 7, 34, 35].

The QED cascade development is a complex process which depends not only on intensity, wavelength and polarization of the electromagnetic fields, but also on their structure. This makes theoretical analysis of laser-plasma dynamics very difficult and nearly impossible when it comes to the highly non-linear stage of the interaction. This makes computer simulation extremely useful for the study of laser-plasma interactions. At high laser intensity, when radiation reaction becomes essentially stochastic, it is important to consider single particle trajectories to describe plasma dynamics correctly. A PIC code by design relies on particles' representation as an ensemble of macroparticles and allows direct modeling of particle trajectories. Moreover, the extended PIC approach allows for treating high-energy photons as particles that are generated by electrons and positrons and can later decay into pairs [32, 33]. This approach utilizes the dual treatment of the electromagnetic field: grid field values for the coherent low-frequency part and particles for the incoherent high-frequency part [18]. Photon emission and pair generation are probabilistic processes and their rates can be calculated under certain assumptions using expressions of QED based on the locally constant crossed field approximation [6]. Handling the considered QED events implies adding new particles associated with either emitted photons or produced particles.

During the development of a QED cascade the number of physical particles may rapidly increase by many orders of magnitude, so the numerical scheme should be adapted for such conditions, preserving the use of a reasonable number of macroparticles. This can be done by reweighting macroparticles in the scope of the thinout procedure. In this paper, we consider the PIC code PICADOR equipped with the adaptive event generator, which automatically locally subdivides the time step to account for several QED events in the case of high rates of the QED processes. It employs a separate thinout module for each type of particles, which allows effective processing of the rapidly increasing number of particles. Methodological and algorithmical aspects of such an extended PIC scheme have been considered in [18]. Notably, the distribution of generated electron-positron plasma can be extremely localized due to the avalanche-like character of the cascade development, its strong dependence on the field intensity, and the peculiarities of particles' motion. Particle processing in this case becomes a non-trivial problem due to a large workload imbalance, thus the technique of reducing the imbalance is of great interest.

## 3    Baseline Parallel Algorithm

Parallel processing in the PICADOR code is organized as follows. On distributed memory we use spatial domain decomposition of the simulation area, essentially standard for the PIC codes of this kind. Each MPI process handles a subarea and stores all particles and grid values, with a few layers of ghost cells. Our previous work included load balancing on the level of distributed memory based on Cartesian rectilinear partitioning [24]. Inside each MPI process we employ parallelism with OpenMP, which is the focus of this paper. Notably, this shared-memory parallelism is largely independent of the distributed-memory parallelism and thus could be considered separately.

The main computational workload of QED PIC simulations are the particle operations: both particle-grid interactions in the core PIC algorithm and QED-specific processing. These operations are spatially local, which has important implications in terms of implementation. Firstly, particles need to be stored according to their positions so that the processing of nearby particles involves compact access to grid values. Two widely used strategies are the periodical spatial sorting of particles, and storing them by cells or supercells. Secondly, processing particles sufficiently far away from one another is completely independent and thus admits a parallelization. It is particularly straightforward in case of cell- or supercell-based particle storing. The PICADOR code follows this strategy. Particles are stored and processed separately for each cell, with each particle communicating only with the closely located grid values. The radius of such communication depends on the particle form factor and numerical schemes being used, but is constant throughout a particular simulation. It allows us to separate cells into subsets, so that particles in different cells of each subset do not affect each other and can therefore be processed in parallel without any synchronization. This scheme has been presented in our earlier work [29] and is illustrated in Fig. 1. We refer to processing each of these groups, as a 'walk'. Due to the subdivision, each loop over cells in a walk has independent iterations and synchronization has to be done only between walks. The minimum unit of workload is then processing particles of a cell.

We have investigated the influence of load balancing schemes in the paper [29]. The first scheme parallelized the loop over cells in each walk using the standard OpenMP static schedule. Such a scheme worked perfectly only in relatively balanced scenarios. The second scheme used the OpenMP dynamic scheduling. This scheme substantially reduced the imbalance, but led to a large balancing overhead. In the other three schemes, the cells in each walk were occasionally sorted by decreasing the number of particles in order to improve load balancing potential. The third scheme then used the OpenMP dynamic scheduling. The fourth scheme manually distributed the cells into threads, using the greedy strategy and avoiding the overhead of dynamic balancing of OpenMP. The fifth scheme at times used the OpenMP dynamic schedule, saved the distribution obtained and used it for several subsequent iterations over time. In this way, an efficient load distribution had been achieved for systems with relatively slow dynamics.

For many PIC simulation scenarios, the particle distribution changes rather slowly, thus the standard OpenMP static schedule or the one of the custom schemes provide an excellent load balancing. For the QED PIC simulations, however, some cells can have significantly more particles than the others and the

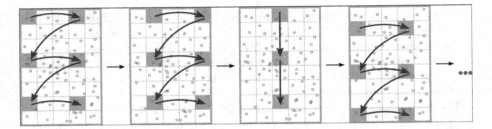

**Fig. 1.** An example of splitting cells into four walks. Particles are represented with the grey dots. Cells inside each walk are processed independently in parallel. Walks are performed sequentially with a barrier between the walks.

distribution of particles in the simulation area can vary significantly over time. In such a case a cell is too coarse of a workload unit. Therefore we developed a new load balancing scheme employing the subdivision of cells, which is described in the following section.

## 4    Dynamic Load Balancing Scheme

The main idea of the new scheme is to treat the subsets of particles in a cell as separate pieces of work. This allows balancing the workload so that each thread processes almost the same number of particles. An illustration of the algorithm is given in Fig. 2, a more detailed description is presented below.

Compared to the schemes described in the previous section, the walks play the same role, but now processing a cell can consist of several tasks, each handling a subset of particles. Importantly, the tasks of the same cell are dependent, but tasks of different cells are not. Each thread has a queue of tasks, of which no more than one task corresponds to a subdivided cell. Thus, a relatively small number of cells, not exceeding the number of threads, can be subdivided.

The pseudocode of this scheme is as follows:

```
1    updateField(Grid);
2    for Walk in Walker:
3      createTasksQueue();
4    #pragma omp parallel:
5    for Task in TasksQueue[threadIndex]:
6      for Particle in Task:
7        process(Particle);
8    for Walk in Walker:
9    #pragma omp parallel for:
10   for Task in TasksQueue[threadIndex]:
11     for Particle in Task:
12       currentDeposition(Particle);
```

We use the following greedy algorithm, which is linear-time in terms of number of cells. At the beginning, an empty queue is created for each thread. Further, the current cell with particles is completely added to the tasks queue of

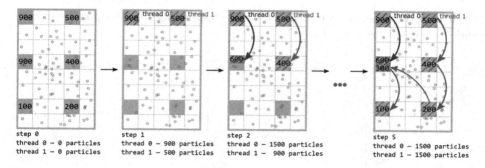

**Fig. 2.** An example of the new load balancing scheme applied to a single walk. The numbers represent the amounts of particles in cells. The blue and red arrows illustrate two threads working in parallel. One of the cells is subdivided into two tasks. (Color figure online)

the thread if the total size of the tasks in the queue does not violate the ideal balance by more than $M$ times ($M$ is a parameter of the algorithm). Otherwise, the cell is divided into two parts so that the inclusion of the first batch of work to the queue of the current thread corresponds to the ideal balance. Then, the tasks queues of the other threads are computed in a similar fashion.

The main implementation challenge is avoiding time-consuming synchronizations caused by the subdivision of cells and their distribution between threads. For example, when processing the movement of particles between cells, the algorithm remembers the numbers of the corresponding particles and processes them at the end so that each cell is processed by only one thread. Similarly, atomic operations have to be used at the current deposition stage. However, experiments have shown that these complications do not lead to a significant overhead, whereas the scheme substantially speeds up the QED PIC simulations.

## 5  Numerical Results

First, we analyze the efficiency of the load balancing schemes for a two-dimensional test problem. We use a $160 \times 160$ grid and $2.56 \times 10^6$ particles, 100 particles per cell on average. Initial sampling of particles is done with the normal distribution with the mean in the center of the simulation area and a diagonal covariance matrix with the same variance for both variables (the spatial steps are also the same). We consider three variance values: $\sigma_1{}^2 = 25\Delta x/8$, $\sigma_2{}^2 = 2\sigma_1{}^2$, $\sigma_3{}^2 = 3\sigma_1{}^2$. The smallest variance $\sigma_1{}^2$ corresponds to the most non-uniform distribution. The simulations were performed for 1000 time steps without the QED effects. Computations were carried out on a node of a supercomputer with the following parameters: $2 \times$ Intel Xeon Gold 6132 (28 cores overall), 192 GB RAM. We measured the computation time using 5 load balancing schemes implemented in [29] and the new scheme. The experiments have shown that for the first (most unbalanced) problem the new scheme outperforms

the best of the others by factor of 4.4, the corresponding speedups for the second
and third problems are 1.9 and 1.1.

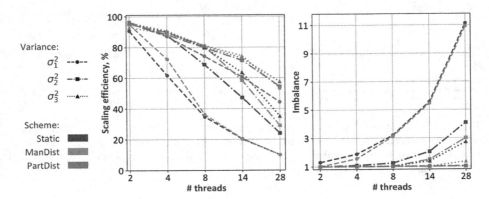

**Fig. 3.** Scaling efficiency and workload imbalance in 3 test problems with different
values of variance when employing 3 load balancing schemes on the 28-core CPU.

Figure 3 shows the scaling efficiency and the imbalance of the computational
load when using from 1 to 28 cores for each 3 test problems. Same as in [29], we
estimate workload imbalance as $I = max_w \left( \underset{i \in \{0,...,N-1\}}{mean} \left( \frac{max_t P_{wti}}{mean_t P_{wti}} \right) \right)$, where
$P_{wti}$ is a number of particles processed by the thread $t$ within walk $w$ on $i$-th
out of total $N$ iterations. In order to not overload the figure, only the data for
static balancing ('Static'), the best of the previously introduced dynamic schemes
('ManDist') and the new scheme ('PartDist'), are presented. It is shown that in
the substantially unbalanced task, the new scheme reduces the imbalance and
improves the scaling efficiency.

Secondly, we study the performance of the load balancing schemes in the
state-of-the-art simulations that take into account the QED effects. In these
applications, the processes become much more complicated, since the concentration of particles in the local regions of the computational area is not only
highly unbalanced, but also changes substantially over time. It is particularly
interesting to estimate the gain from the use of the new dynamic scheme in such
scenarios.

We consider a highly unbalanced problem of the QED cascade development
in extreme laser fields described in detail in [36]. According to recent studies
[8,32,35,37] the preferable field structure is counterpropagating laser beams.
We consider the case of circular polarization, as one of the fundamental cases.
In a wide range of intensities of the circularly polarized field, highly localized
electron-positron structures can be formed in the vicinities of electric field nodes,
antinodes or in both types of regions [36].

The maximum intensity of each of counter-propagating pulses is chosen to
be $I_0 = 10^{25}$ W/cm$^2$, which can be obtained on planned 100 PW laser systems.

The wavelength is $0.8\,\mu m$. For the sake of simplicity we consider half infinite pulses with a 1 wave period front edge. An electron-positron plasma slab with the width of one wavelength and the density of $1\,cm^{-3}$ serves as a seed and is located at the center of the simulation area. During the interaction first the incident laser pulses compress the seed plasma. Then the laser pulses overlap, a standing wave is formed and a QED cascade starts to develop. At the considered intensity during its development the plasma is highly localized in the vicinity of the antinode.

All experiments were performed on the Intel Endeavour supercomputer with high-end CPUs of the Cascade Lake generation. Performance results were collected on the following cluster nodes: 2 × Intel Xeon Platinum 8260L CPU (48 cores overall), 192 GB of RAM. In all runs we employed 1 MPI process per socket, 2 OpenMP threads per core. The code was built using the Intel Parallel Studio XE software package.

First, we consider the 1D case where the number of cells most populated by particles is minimal. The simulation box is $2\,\mu m$ long and the number of cells is 128. The time step is equal to 1/200 of the laser period. The initial number of particles and the threshold of particle thinning are $10^6$ and $2 \times 10^6$, respectively. In the 1D simulation we employ 1 node of the supercomputer (48 cores).

The experiments show that the computation time of the static scheme and two most flexible of the old dynamic schemes are roughly the same, while the new scheme is faster by almost an order of magnitude. This is due to the fact that the new scheme balances the workload much better by subdividing the cells. Figure 4 shows how imbalance of the computational load changes over time. It is calculated for every chunk of 100 consecutive iterations. The imbalance in Fig. 4 on the left is calculated using the profiler based on the computation time measurements. The imbalance in Fig. 4 on the right is estimated by the number of particles processed by the threads. Given that for the 2D and 3D simulations we obtained similar results, we can conclude that the employed model of particle imbalance works well in the case of modeling the QED cascades.

In 2D and 3D simulations the box is $2\,\mu m \times 8\,\mu m$ ($\times 8\,\mu m$), the number of cells is $64 \times 112 (\times 112)$, the initial number of macroparticles and the threshold of particle thinning are $5 \times 10^6$ in the 2D case and $2.5 \times 10^6$ in the 3D case. We consider Gaussian beams with 1 wavelength waist radius. The time step is the same as in the 1D case. The 2D and 3D cases are much more computationally intensive, therefore we employ 2 nodes of the supercomputer (4 CPUs, 96 cores overall). The PICADOR code has load balancing schemes for clusters, but in the considered case, workloads at different nodes are balanced due to the symmetry of the problem. In contrast, the uniform distribution of the work among dozens of cores on a single node is problematic.

Figure 5 shows how different the load balancing schemes work for the considered 3D simulation. In the first 300 iterations, there is no imbalance (Fig. 5, on the right) and all the schemes work for approximately the same time (Fig. 5, on the left). Further, the electromagnetic cascade begins to develop and the new scheme, unlike the others, allows keeping the imbalance under control. Similar

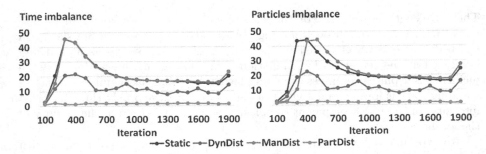

Fig. 4. 'Time imbalance' vs. 'Particles imbalance' in the 1D simulation of the QED cascades.

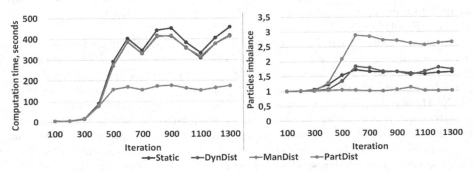

Fig. 5. Computation time of different load balancing schemes in the 3D simulation of the QED cascades. The new scheme ('PartDist') outperforms other schemes.

behavior is observed in the 2D simulation. Finally, when calculating 1300 iterations, the new scheme speeds up the simulation by 2.5 times in the 2D problem and by 2.1 times in the 3D problem.

# 6    Conclusion

In the paper, we addressed the problem of efficient utilization of new CPUs with a large number of cores in PIC laser-plasma simulations. We concentrated on simulations including the development of electromagnetic cascades, which often led to a non-uniform and time-varying concentration of particles in the computational domain. To overcome the load imbalance, we developed and implemented a special scheme in the PICADOR code that allows subdividing cells with a large number of particles. This approach substantially increased the potential for parallelization. The scheme was tested on several QED simulations. The results showed that, in the absence of a significant imbalance, the new scheme performs approximately the same as the schemes previously implemented in the PICADOR code, whereas in the other cases it outperforms them by a factor of 2.1 to 10, depending on the simulation.

Note that the current results were obtained for the problems whose internal symmetry allows using several cluster nodes in a straightforward way, or relying on the rectilinear load balancing scheme (see [24] and references therein for details). However, if a problem requires the use of dozens of nodes, there is a need for a smart combination of balancing schemes on shared and distributed memory. The development and analysis of such schemes is one of the directions of further work.

# References

1. Dawson, J.: Particle simulation of plasmas. Rev. Mod. Phys. **55**, 403 (1983). https://doi.org/10.1103/RevModPhys.55.403
2. Elkina, N.V., et al.: QED cascades induced by circularly polarized laser fields. Phys. Rev. Spec. Top. Accel. Beams **14**, 054401 (2011). https://doi.org/10.1103/PhysRevSTAB.14.054401
3. Sokolov, I., et al.: Numerical modeling of radiation-dominated and quantum electrodynamically strong regimes of laser-plasma interaction. Phys. Plasmas **18**, 093109 (2011). https://doi.org/10.1364/OE.16.002109
4. Ridgers, C.P., et al.: Modelling gamma-ray photon emission and pair production in highintensity laser-matter interactions. J. Comput. Phys. **260**, 273 (2014). https://doi.org/10.1016/j.jcp.2013.12.007
5. Grismayer, T., et al.: Laser absorption via quantum electrodynamics cascades in counter propagating laser pulses. Phys. Plasmas **23**, 056706 (2016). https://doi.org/10.1063/1.4950841
6. Gonoskov, A., et al.: Ultrabright GeV photon source via controlled electromagnetic cascades in laser-dipole waves. Phys. Rev. X **7**, 041003 (2017). https://doi.org/10.1103/PhysRevX.7.041003
7. Efimenko, E.S., et al.: Laser-driven plasma pinching in e-e+ cascade. Phys. Rev. E **99**, 031201(R) (2019). https://doi.org/10.1103/PhysRevE.99.031201
8. Tamburini, M., Di Piazza, A., Keitel, C.H.: Laser-pulse-shape control of seeded QED cascades. Sci. Rep. **7**(1), 5694 (2017)
9. Samsonov, A.S., Nerush, E.N., Kostyukov, I.Y.: QED cascade in a plane electromagnetic wave. arXiv preprint arXiv:1809.06115 (2018)
10. Brady, C.S., Arber, T.D.: An ion acceleration mechanism in laser illuminated targets with internal electron density structure. Plasma Phys. Controlled Fusion **53**(1), 015001 (2011). https://doi.org/10.1088/0741-3335/53/1/015001
11. Fonseca, R.A., et al.: OSIRIS: a three-dimensional, fully relativistic particle in cell code for modeling plasma based accelerators. In: Sloot, P.M.A., Hoekstra, A.G., Tan, C.J.K., Dongarra, J.J. (eds.) ICCS 2002. LNCS, vol. 2331, pp. 342–351. Springer, Heidelberg (2002). https://doi.org/10.1007/3-540-47789-6_36
12. Bussmann, M., et al.: Radiative signatures of the relativistic Kelvin-Helmholtz instability. In: SC 2013. ACM, New York (2013). https://doi.org/10.1145/2503210.2504564
13. Derouillat, J., et al.: SMILEI: a collaborative, open-source, multi-purpose particle-in-cell code for plasma simulation. Comput. Phys. Commun. **222**, 351–373 (2018). https://doi.org/10.1016/j.cpc.2017.09.024
14. Pukhov, A.: Three-dimensional electromagnetic relativistic particle-in-cell code VLPL (Virtual Laser Plasma Lab). J. Plasma Phys. **61**(3), 425–433 (1999). https://doi.org/10.1017/S0022377899007515

15. Bowers, K.J., et al.: Ultrahigh performance three-dimensional electromagnetic relativistic kinetic plasma simulation. Phys. Plasmas **15**(5), 055703 (2008). https://doi.org/10.1063/1.2840133

16. Friedman, A., et al.: Computational methods in the warp code framework for kinetic simulations of particle beams and plasmas. IEEE Trans. Plasma Sci. **42**(5), 1321–1334 (2014). https://doi.org/10.1109/TPS.2014.2308546

17. Surmin, I.A., et al.: Particle-in-Cell laser-plasma simulation on Xeon Phi coprocessors. Comput. Phys. Commun. **202**, 204–210 (2016). https://doi.org/10.1016/j.cpc.2016.02.004

18. Gonoskov, A., et al.: Extended particle-in-cell schemes for physics in ultrastrong laser fields: review and developments. Phys. Rev. E **92**, 023305 (2015). https://doi.org/10.1103/PhysRevE.92.023305

19. Fonseca, R.A.: Exploiting multi-scale parallelism for large scale numerical modelling of laser wakefield accelerators. Plasma Phys. Controlled Fusion **55**(12), 124011 (2013). https://doi.org/10.1088/0741-3335/55/12/124011

20. Decyk, V.K., Singh, T.V.: Particle-in-cell algorithms for emerging computer architectures. Comput. Phys. Commun. **185**(3), 708–719 (2014). https://doi.org/10.1016/j.cpc.2013.10.013

21. Germaschewski, K., et al.: The Plasma Simulation Code: a modern particle-in-cell code with patch-based load-balancing. J. Comput. Phys. **318**(1), 305–326 (2016). https://doi.org/10.1016/j.jcp.2016.05.013

22. Beck, A., et al.: Load management strategy for Particle-In-Cell simulations in high energy physics. Nucl. Instrum. Methods Phys. Res. A **829**(1), 418–421 (2016)

23. Vay, J.-L., Haber, I., Godfrey, B.B.: A domain decomposition method for pseudo-spectral electromagnetic simulations of plasmas. J. Comput. Phys. **243**(15), 260–268 (2013). https://doi.org/10.1016/j.jcp.2013.03.010

24. Surmin, I., Bashinov, A., Bastrakov, S., Efimenko, E., Gonoskov, A., Meyerov, I.: Dynamic load balancing based on rectilinear partitioning in particle-in-cell plasma simulation. In: Malyshkin, V. (ed.) PaCT 2015. LNCS, vol. 9251, pp. 107–119. Springer, Cham (2015). https://doi.org/10.1007/978-3-319-21909-7_12

25. Kraeva, M.A., Malyshkin, V.E.: Assembly technology for parallel realization of numerical models on MIMD-multicomputers. Future Gener. Comput. Syst. **17**, 755–765 (2001). https://doi.org/10.1016/S0167-739X(00)00058-3

26. Vshivkov, V.A., Kraeva, M.A., Malyshkin, V.E.: Parallel implementation of the particle-in-cell method. Program. Comput. Softw. **23**(2), 87–97 (1997)

27. Surmin, I., Bastrakov, S., Matveev, Z., Efimenko, E., Gonoskov, A., Meyerov, I.: Co-design of a particle-in-cell plasma simulation code for Intel Xeon Phi: a first look at Knights Landing. In: Carretero, J., et al. (eds.) ICA3PP 2016. LNCS, vol. 10049, pp. 319–329. Springer, Cham (2016). https://doi.org/10.1007/978-3-319-49956-7_25

28. Vranic, M., Grismayer, T., Martins, J.L., Fonseca, R.A., Silva, L.O.: Particle merging algorithm for PIC codes. Comput. Phys. Commun. **191**, 65–73 (2015). https://doi.org/10.1016/j.cpc.2015.01.020

29. Larin, A., et al.: Load balancing for particle-in-cell plasma simulation on multicore systems. In: Wyrzykowski, R., Dongarra, J., Deelman, E., Karczewski, K. (eds.) PPAM 2017. LNCS, vol. 10777, pp. 145–155. Springer, Cham (2018). https://doi.org/10.1007/978-3-319-78024-5_14

30. Taflove, A., Hagness, S.: Computational Electrodynamics: The Finite-Difference Time-Domain Method. The Artech House Antennas and Propagation Library. Artech House Inc., Boston (2005)

31. Vay, J.-L., et al.: Simulating relativistic beam and plasma systems using an optimal boosted frame. J. Phys. Conf. Ser. **180**(1), 012006 (2009). https://doi.org/10.1088/1742-6596/180/1/012006

32. Nerush, E.N., et al.: Laser field absorption in self-generated electron-positron pair plasma. Phys. Rev. Lett. **106**, 035001 (2011). https://doi.org/10.1103/PhysRevLett.106.035001

33. Bell, A.R., Kirk, J.G.: Possibility of prolific pair production with high-power lasers. Phys. Rev. Lett. **101**, 200403 (2008). https://doi.org/10.1103/PhysRevLett.101.200403

34. Vranic, M., Grismayer, T., Fonseca, R.A., Silva, L.O.: Electron-positron cascades in multiple-laser optical traps. Plasma Phys. Controlled Fusion **59**, 014040 (2016). https://doi.org/10.1088/0741-3335/59/1/014040

35. Efimenko, E.S., et al.: Extreme plasma states in laser-governed vacuum breakdown. Sci. Rep. **8**, 2329 (2018)

36. Bashinov, A.V., et al.: Particle dynamics and spatial e-e+ density structures at QED cascading in circularly polarized standing waves. Phys. Rev. A **95**, 042127 (2017). https://doi.org/10.1103/PhysRevA.95.042127

37. Jirka, M., et al.: Electron dynamics and $\gamma$ and e-e+ production by colliding laser pulses. Phys. Rev. E **93**, 023207 (2016). https://doi.org/10.1103/PhysRevE.93.023207

# Parallelized Construction of Extension Velocities for the Level-Set Method

Michael Quell[1]([✉]), Paul Manstetten[2], Andreas Hössinger[3], Siegfried Selberherr[2], and Josef Weinbub[1]

[1] Christian Doppler Laboratory for High Performance TCAD,
Institute for Microelectronics, TU Wien, Vienna, Austria
{quell,weinbub}@iue.tuwien.ac.at
[2] Institute for Microelectronics, TU Wien, Vienna, Austria
{manstetten,selberherr}@iue.tuwien.ac.at
[3] Silvaco Europe Ltd., St Ives, UK
andreas.hoessinger@silvaco.com

**Abstract.** The level-set method is widely used to track the motion of interfaces driven by a velocity field. In many applications, the underlying physical model defines the velocity field only at the interface itself. For these applications, an extension of the velocity field to the simulation domain is required. This extension has to be performed in each time step of a simulation to account for the time-dependent velocity values at the interface. Therefore, the velocity extension is critical to the overall computational performance. We introduce an accelerated and parallelized approach to overcome the computational bottlenecks of the prevailing and serial-in-nature fast marching method, in which the level-set function is used to predetermine the computational order for the velocity extension. This allows to employ alternative data structures, which results in a straightforward parallelizable approach with reduced complexity for insertion and removal as well as improved cache efficiency. Compared to the prevailing fast marching method, our approach delivers a serial speedup of at least 1.6 and a shared-memory parallel efficiency of 66% for 8 threads and 37% for 16 threads.

**Keywords:** Velocity extension · Level-set method · Parallel computing · Fast marching method

## 1 Introduction

The level-set method [9] is widely used to track moving interfaces in different fields of science, such as in computer graphics [8], fluid dynamics [7], and microelectronics [14]. The level-set method represents an interface $\Gamma$ implicitly as the zero-level-set of a higher-dimensional function, i.e., the level-set function $\phi(\boldsymbol{x}, t)$. The motion of the interface is given by the level-set equation

$$\frac{\partial \phi}{\partial t} = F_{ext}|\nabla \phi|, \tag{1}$$

© Springer Nature Switzerland AG 2020
R. Wyrzykowski et al. (Eds.): PPAM 2019, LNCS 12043, pp. 348–358, 2020.
https://doi.org/10.1007/978-3-030-43229-4_30

where $F_{ext}(\boldsymbol{x}, t)$ is the extended velocity field from the underlying model. The extended velocity field $F_{ext}$ is not unique [1], as the only formal mathematical requirement is

$$\lim_{\boldsymbol{x} \to \boldsymbol{x}_0} F_{ext}(\boldsymbol{x}) = F(\boldsymbol{x}_0), \tag{2}$$

with $F(\boldsymbol{x}, t)$ being the given continuous velocity on the interface and $\boldsymbol{x}_0$ any point on the interface, i.e., $\phi(\boldsymbol{x}_0) = 0$. The extension should not introduce new artificial zero values. An extended velocity field $F_{ext}$ fulfilling

$$\nabla F_{ext} \cdot \nabla \phi = 0 \text{ and } F_{ext}\big|_{\phi^{-1}(\boldsymbol{0})} = F \tag{3}$$

meets the requirement (2), does not introduce artificial zero values, and preserves the signed-distance property of the level-set function during the advection step, which is desirable as it leads to maximal numerical stability of the method [2]. The velocity has to be extended every time a new interface velocity is calculated (i.e., for a finite difference discretization in every time step).

A widely used approach to solve (3) is the fast marching method (FMM) [1]. Other methods such as the fast iterative method [6] and the fast sweeping method [11] are not considered here, because of their iterative nature, the computational costs to obtain an accuracy level comparable to FMM are too high. The FMM was originally developed to efficiently solve the Eikonal equation

$$\|\nabla \phi\| = 1 \quad \text{and} \quad \phi\big|_\Gamma = g(\boldsymbol{x}). \tag{4}$$

Within the level-set method, (4) and $g(x) = 0$ is solved to give the initial level-set function the signed-distance property [13]. Due to the utilization of a global heap (a single priority queue for the full domain) to track the order of the computations, the solution of (3) using the FMM has complexity $\mathcal{O}(n \log n)$, where $n$ is the number of grid points (discrete points on the computational grid) and is inherently serial. Another approach achieves complexity $\mathcal{O}(n)$ by quantization of the keys of the heap at the cost of a different error bound [17].

There have been successful attempts to parallelize the FMM through domain decomposition for distributed-memory systems [16] and for shared-memory systems [15]. Therein FMM is executed on each sub-domain with its own heap, thus enabling parallelism. An explicit synchronized data exchange via a ghost layer is used to resolve inter-domain dependencies. The decomposition approach requires knowledge about the interface position to balance the load equally [4], on the other hand the proposed algorithm employs dynamic load balancing independent of the interface position, therefore, a fair comparison is not possible. In [10], a serial approach based on fast scanning is presented, but no information is given on how the computations are ordered, which is, however, essential for cache efficiency.

In Sect. 2, we provide the original FMM algorithm (Algorithm 1) for reference and details of our approach for an accelerated velocity extension algorithm, avoiding the aforementioned difficulties when utilizing the FMM. In Sect. 3, a new serial algorithm (Algorithm 2) and a parallel algorithm (Algorithm 3) are presented. In Sect. 4, the serial and parallel run-times of our approach are presented for an application example from the field of microelectronics.

## 2    Theory

The FMM and the two proposed algorithms assign each grid point an exclusive state: *Known* means the grid point has the final velocity assigned and no further updates are required, *Unknown* means the grid point does not yet have a velocity assigned, and *Band* means that the grid point has a velocity assigned, but it is not final. The FMM, orders the grid points in the *Band* by the distance to the interface, to determine which is processed next. The ordering is achieved using a minimum heap data structure, i.e., the top element (grid point) is always the closest to the interface. This ensures that a grid point's value is *final* (i.e., it conforms to (3)), when it is removed from the *Band*. A standard implementation for the FMM is given by Algorithm 1 [12].

The *compute()* sub-routine is used to update the value of a grid point, e.g., solving the Eikonal equation, or to compute the velocity, which is described in detail in [1]. The interface and its adjacent grid points (for which the velocity is known) divide the domain in two zones (*inside* and *outside*) and the algorithm has to be applied for each zone separately as both zones are independent of each other, (see Fig. 1).

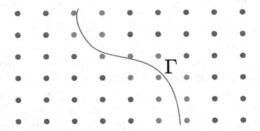

**Fig. 1.** The interface $\Gamma$ (zero-level-set) is given in blue. For the red grid points next to the interface, the velocity values are calculated. From those points the velocity is extended to the remaining domain (black grid points). (Color figure online)

The run-time contributions of the computational sub-tasks of the velocity extension are show in Fig. 2, if the FMM is used to extend the velocity. Most of

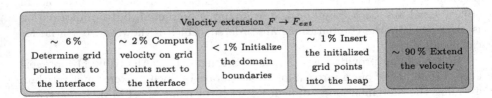

**Fig. 2.** Computational sub-tasks and their run-time contribution for the construction of the extended velocity field $F_{ext}$, utilizing the FMM. (Color figure online)

| **Algorithm 1:** FMM | **Algorithm 2:** ExtendVelocity |
|---|---|
| 1    *Known* ← ∅ <br> 2    *Band* ← initialized grid points <br> 3    *Unknown* ← all other grid points <br> 4   **while** *Band not empty* **do** <br> 5      *q* = first element of *Band* <br> 6      add *q* to *Known* <br> 7      **forall** *p neighbors of q* **do** <br> 8        **if** *p* ∈ *Unknown* **then** <br> 9          *compute(p)* <br> 10         add *p* to *Band* | 1    *Known* ← ∅ <br> 2    *Band* ← initialized grid points <br> 3    *Unknown* ← all other grid points <br> 4   **while** *Band not empty* **do** <br> 5      *q* = first element of *Band* <br> 6      add *q* to *Known* <br> 7      **forall** *p neighbors of q* **do** <br> 8        **if** *p* ∈ *Unknown and upwind* <br>            *neighbors* ∉ *Unknown* **then** <br> 9          *compute(p)* <br> 10         add *p* to *Band* |

the time is spent extending the velocity (last sub-task marked in blue), whilst the other steps require considerably less time.

For the heap data structure, the insertion or removal of a grid point triggers a sorting which results in $\mathcal{O}(\log n)$ operations. The overall complexity to populate the heap with $n$ grid points is therefore $\mathcal{O}(n \log n)$. Formulating the problem in the context of graph theory, $\mathcal{O}(n)$ is achieved.

Assume an ordered graph $G(N, E)$, the nodes $N$ are given by all the grid points and the edges $E$ are given by the direct upwind neighbors of a grid point (i.e., neighboring grid points with a smaller distance to the interface). The order in which the nodes can be computed is a topological sort problem, which is solved in linear $\mathcal{O}(|N| + |E|)$ time [5]. This is also linear in $n = |N|$, which is the number of grid points in the domain, as the number of edges $|E|$ is limited by $6|N|$ in case of a 7-point three-dimensional stencil to compute the gradient.

The topological sort problem is solved by a depth-first or breadth-first traversal over the graph [3]. These traversals can be realized by adapting the ordering of the grid points which have the status *Band* in Algorithm 1. Using a queue or a stack as data structure corresponds to a breadth-first and depth-first traversal, respectively. Adding an element to the *Band* can be done for both data structures in $\mathcal{O}(1)$; this is an advantage compared to the heap. The parallelization of these algorithms is straight forward, by processing all nodes which do not have an unresolved dependence in parallel.

## 3   Parallel Velocity Extension

Based on the findings in the previous section, we investigate – as a first step – an adapted serial algorithm (Algorithm 2) which uses different data structures to implement the *Band*. This requires a check in the *neighbors loop*, whether the upwind neighbors are not in the *Unknown* state (Line 8), only then the velocity is computed. In Algorithm 1, this check is not necessary, as the heap guarantees no unknown dependencies of the top element.

Algorithm 3 is the parallel version of Algorithm 2. The parallelization is realized by treating every grid point next to the interface as an independent starting grid point for the traversal through the graph. The grid points in the *Band* (cf. Algorithm 3 Line 6), are exclusive (i.e., OpenMP private) to each thread, but the status of a grid point is shared with all threads. In principal, explicit synchronization would be necessary in the *neighbors loop* which calls the *compute()* sub-routine – not for correctness of the algorithm, but to ensure that no grid point will be treated by two different threads (avoiding *redundant* computations). However, our approach (Algorithm 3) deliberately recomputes the values by different threads as the computational overhead is negligible compared to explicit synchronization costs. Conflicting access of two threads to the data of a grid point is resolved by enforcing atomic read and write operations (cf. Sect. 4).

In order to keep the number of *redundant* computations small, the threads start on grid points evenly distributed over the full set of interface grid points. If the *Band* of a thread is empty, the thread is dynamically assigned a new starting grid point from the initialized grid points. In case two threads operate at the same location, the check for the state *Unknown* (Line 11) reduces the *redundant* computations. To further reduce the redundant computations, it is checked again if a different thread has already processed the grid point before adding the grid point to its exclusive *Band* (Line 13). In case of a serial execution the second check is redundant to the first one (Line 11). As we present in Sect. 4, the ratio of *redundant* computations to *necessary* computations is below 1%, which plainly favors redundancy over the explicit synchronization which would limit parallel scalability.

---

**Algorithm 3:** ExtendVelocityParallel

---

```
1   Known ← ∅
2   InitPoints ← initialized grid points
3   Unknown ← all other grid points
4   forall b in InitPoints in parallel do
5   │   Band ← ∅
6   │   Band add b
7   │   while Band not empty do
8   │   │   q = first element of Band
9   │   │   add q to Known
10  │   │   forall p neighbors of q do
11  │   │   │   if p ∈ Unknown and upwind neighbors ∉ Unknown then
12  │   │   │   │   compute(p)
13  │   │   │   │   if p ∈ Unknown then
14  │   │   │   │   │   add p to Band
```

# 4   Computational Results

We evaluate the performance by benchmarking our velocity extension approach embedded in a simulation of a microelectronic fabrication process, specifically an etching simulation of a pillar-like structure (cf. Fig. 3a)[1]. This geometry provides a challenging and representative testbed as it includes flat, convex, and concave interface areas, which lead to shocks and rarefaction fans in the extended velocity field. The domain is discretized using a dense equidistant grid and the gradients are computed using first order finite differences. Symmetric boundaries are enforced by a ghost layer outside the domain.

**Table 1.** Properties of the discretization for different resolutions for the example geometry (cf. Fig. 3a).

|  | Resolution | # grid points | # initialized grid points |
|---|---|---|---|
| Low resolution case | $40 \times 40 \times 700$ | 1 235 200 | 26 168 |
| High resolution case | $160 \times 160 \times 2800$ | 73 523 200 | 411 896 |

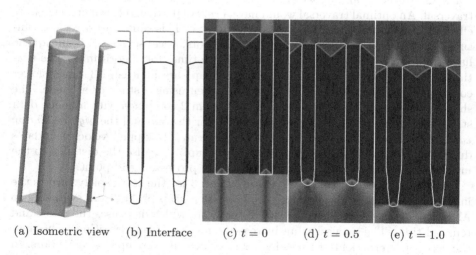

(a) Isometric view     (b) Interface        (c) $t = 0$          (d) $t = 0.5$          (e) $t = 1.0$

**Fig. 3.** (a) Initial interface in isometric view. (b)–(e) Cross-section of the simulation domain through a plane with normal $(1, 1, 0)$: (b) interface overlaid at different simulation times, (c)-(e) extended velocity (low velocity in red and high in blue) for times $t = 0$, $t = 0.5$, $t = 1.0$. (Color figure online)

---

[1] However, the presented algorithm and implementation details are not tailored or restricted to the field of microelectronics and can be applied to other fields as well.

A *low resolution case* and a *high resolution case* have been investigated. Table 1 summarizes the properties of the resulting discretization. In Fig. 3 a slice of the extended velocity is shown for three different times of the simulation.

The benchmark results are obtained on a single compute node of the Vienna Scientific Cluster[2] equipped with two Intel Xeon E5-26504 8 core processors and 64 GB of main memory. The algorithms are implemented in a GNU/Linux environment in C++11 (GCC-7.3 with optimization flag -03). OpenMP is used for parallelization. The read and write operations from and to the state (*Unknown*, *Band*, and *Known*) of a grid point and the velocity of a grid point are implemented using the atomic directives of OpenMP. C++'s standard template library (STL) containers are used for the stack and the queue. The heap is based on the STL priority queue using the distance of the grid points to the interface as key. All following results report the run-time averaged over 10 executions for all sub-tasks depicted in Fig. 2. Both of the proposed algorithms calculate the exact same results as the reference FMM, also independent to the number of threads.

## 4.1   Serial Results

Table 2 compares the serial run-times of all three algorithms for both spatial resolutions (*Run-time*). The ratio of how often an upwind neighbor is in the state *Unknown* to the total number of updates (*Un.* up.) is used as metric for optimal traversal. An optimal traversal would have a rate of 0, though this metric neglects effects of different access times (and cache misses). This causes uncorrelated run-times to the ratio of *Unknown* upwind neighbors, because the heap and queue have similar ratios but drastically different run-times for Algorithm 2. The run-times of Algorithm 1 (i.e., FMM, using a heap) are at least 1.3 times slower compared to Algorithm 2, or Algorithm 3, when using a stack or a queue. The shortest run-times are obtained using Algorithm 2 combined with a queue data structure leading to a speedup of 1.6 and 2.0 for the *low* and the *high resolution case*, respectively. The stack has the highest rate of skipped velocity updates due to an *Unknown* upwind dependence (*Un.* up.), because the distance to the interface is not used to select the subsequently processed grid point.

Switching the algorithms Algorithms 2 and 3 for the stack solely reverts the initial order of the grid points in the *Band*. The heap profits from switching to Algorithm 3, as this reduces the size of the heap, which decreases the insert and removal time. The queue has an increased run-time with Algorithm 3, because the access pattern yields 4 times higher rates of *Unknown* upwind neighbors. In conclusion, for Algorithm 3 the data structure for the *Band* is less important, because the size of the *Band* (cf. Algorithm 3, Line 6) is small (starts with a single grid point) compared to the size of the *Band* in Algorithm 2 (starts with about half the number of the initialized grid points (cf. Table 1))[3].

---

[2] http://vsc.ac.at/.

[3] The other half of the initialized grid points resides in the second zone of the domain, which is processed independently.

## 4.2   Parallel Results

Figure 4a shows the run-time of Algorithm 3 and the achieved parallel speedup for up to 16 threads. The usage of hyper-threads has been investigated, but no further speedup was measured. Therefore, the analysis focuses on the available 16 physical cores on the compute node. Each thread is pinned to its own core, e.g Thread 0 on Core 0, Thread 1 on Core 1, and so forth.

**Table 2.** Serial run-time and the ratio how often *Unknown* upwind neighbors ( *Un.* up.) were encountered compared to the total updates. Bold numbers indicate the fastest run-time for each resolution.

| Data structure | Algorithm 1 | | Algorithm 2 | | Algorithm 3 | |
|---|---|---|---|---|---|---|
| | Run-time | *Un.* up | Run-time | *Un.* up | Run-time | *Un.* up. |
| (a) *Low resolution case* | | | | | | |
| Heap | 0.265 | 0.0 | 0.258 | 0.034 | 0.190 | 0.262 |
| Stack | | | 0.196 | 0.418 | 0.200 | 0.416 |
| Queue | | | **0.162** | 0.077 | 0.177 | 0.259 |
| (b) *High resolution case* | | | | | | |
| Heap | 19.99 | 0.0 | 19.14 | 0.076 | 13.27 | 0.241 |
| Stack | | | 13.27 | 0.414 | 12.83 | 0.412 |
| Queue | | | **10.27** | 0.052 | 11.67 | 0.221 |

The serial results have already shown that for Algorithm 3 the data structure of the *Band* is less important (cf. Algorithm 3 in Table 2). The queue is also the fastest for the parallel case, because the ordering by first-in first-out avoids the sorting of the heap and reduces encountering of *Unknown* upwind neighbors compared to the stack. The shortest run-times are obtained for 8 and 16 threads for the *low resolution case* and *high resolution case*, respectively. The algorithm with the heap produces the best parallel speedup (not lowest run-time), because an increasing number of threads further reduces the size of the data structure of the *Band*. Small *Band* sizes are important for the heap, because the insertion of grid points scales with the number of grid points in the *Band* (stack and queue do not have this drawback).

The parallel efficiency for the *low resolution case* using 8 threads is 58% for the heap and the queue and 61% for the stack. For 8 threads, the *high resolution case* has a parallel efficiency of 56% for the stack, 66% for the queue, and 67% for the heap. For more than one thread, the parallel Algorithm 3 has a shorter run-time than the serial Algorithm 2. Above 8 threads, the utilization of cores on both processors induces non-uniform memory access leading to an increased run-time for the *low resolution case* (parallel efficiency of 25%) and only marginal speedup for the *high resolution case* (parallel efficiency of 37%) for all data structures. The memory is allocated by Thread 0 which resides on

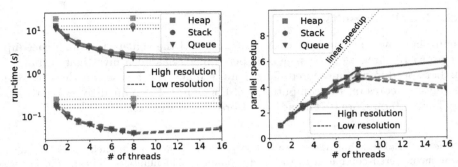

(a) Parallel run-time (left) and speedup (right) for Alg. 3. For comparison, the serial run-time of Alg. 2 is shown by the dotted lines.

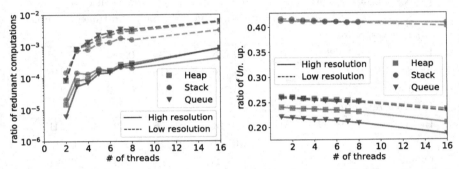

(b) On the left, the ratio of *redundant* to total calls to the *compute()* sub-routine due to implicit synchronization is shown. On the right, the ratio of how often *Unknown* upwind neighbors were encountered to the total amount of testing for *Unknown* upwind neigbors is shown.

**Fig. 4.** Parallel run-time results for Algorithm 3.

the socket of the first processor, therefore only the first 8 threads can directly access the memory. Also threads running on different sockets do not share the L3-cache, which forces communication via the main memory.

In Fig. 4b, the ratio of *redundant* computations and not performed computations due to an *Unknown* upwind neighbors are shown (cf. Sect. 3). The increase of the *redundant* computations ratio saturates with the number of threads. For 16 threads, less than 1% of the *compute()* calls are wasted (i.e., *redundant*) in the *low resolution case* and less than 0.1% in the *high resolution case*. The ratio of the *redundant* computations in the *high resolution case* is lower, because the threads process more grid points in relation to the grid points, at which threads can interfere. A similar situation is found for the ratio between the volume and the surface of a sphere. As already hinted in Sect. 3, enforcing explicit synchronization in the *neighbors loop* would lead to a significant decrease of parallel efficiency, because the explicit synchronization required to ensure that every grid point is only computed once has a higher computational cost compared to the *redundant* computations introduced otherwise. The number of redundant computations is only related to the synchronization paradigm (independent to the shared-memory approach). The ratio how often the *compute()* sub-routine is skipped, because upwind neighbors were in the *Unknown* state (cf. Sect. 3), slightly decreases with the number of threads, as the possibility increases that another thread has computed an *Unknown* upwind neighbor just in time.

In comparison with the original FMM (cf. Algorithm 1), Algorithm 3 using the queue achieved a minimal run-time of 0.038 s which is due to a serial speedup of 1.5 and a parallel speedup of 5.6 for 8 threads in the *low resolution case*. In the *high resolution case* the minimal run-time of 1.975 s is also achieved with the queue data structure when utilizing all 16 threads (Serial speed up of 1.7 and parallel speedup of 10.1).

## 5    Conclusion

A new parallel approach to accelerate the velocity extension in the level-set method has been presented and compared to the prevailing FMM. The asymptotic complexity is $\mathcal{O}(n)$ by utilizing the level-set function to determine the order of computations. Furthermore, this approach opens an attractive path for parallelization. The serial speedup compared to the FMM is at least 1.6; a speedup of 2 is observed for a high resolution test case. The proposed parallel algorithm is tailored towards a shared-memory platform. The parallel efficiency is 58% for 8 threads; 66% are achieved for a high resolution test case. Overall, we provide a straight forward parallelizable algorithm (sparing any explicit synchronization) for velocity extension in the level-set method constituting an attractive drop-in replacement for the prevailing FMM.

**Acknowledgments.** The financial support by the *Austrian Federal Ministry for Digital and Economic Affairs* and the *National Foundation for Research, Technology and Development* is gratefully acknowledged. The computational results presented have been achieved using the Vienna Scientific Cluster (VSC).

## References

1. Adalsteinsson, D., Sethian, J.A.: The fast construction of extension velocities in level set methods. J. Comput. Phys. **148**(1), 2–22 (1999). https://doi.org/10.1006/jcph.1998.6090
2. Cheng, L.T., Tsai, Y.H.: Redistancing by flow of time dependent eikonal equation. J. Comput. Phys. **227**(8), 4002–4017 (2008). https://doi.org/10.1016/j.jcp.2007.12.018
3. Cormen, T.H., Leiserson, C.E., Rivest, R.L., Stein, C.: Introduction to Algorithms. MIT Press, Cambridge (2009). ISBN: 9780262033848
4. Diamantopoulos, G., Weinbub, J., Selberherr, S., Hössinger, A.: Evaluation of the shared-memory parallel fast marching method for re-distancing problems. In: Proceedings of the 17th International Conference on Computational Science and Its Applications, pp. 1–8 (2017). https://doi.org/10.1109/ICCSA.2017.7999648

5. Hagerup, T., Maas, M.: Generalized topological sorting in linear time. Fundam. Comput. Theory **710**, 279–288 (1993). https://doi.org/10.1007/3-540-57163-9_23
6. Jeong, W.K., Whitaker, R.T.: A fast iterative method for Eikonal equations. SIAM J. Sci. Comput. **30**(5), 2512–2534 (2008). https://doi.org/10.1137/060670298
7. Losasso, F., Gibou, F., Fedkiw, R.: Simulating water and smoke with an octree data structure. ACM Trans. Graph. **23**(3), 457–462 (2004). https://doi.org/10.1145/1015706.1015745
8. Museth, K.: VDB: high-resolution sparse volumes with dynamic topology. ACM Trans. Graph. **32**(3), 1–22 (2013). https://doi.org/10.1145/2487228.2487235
9. Osher, S., Sethian, J.A.: Fronts propagating with curvature-dependent speed: algorithms based on Hamilton-Jacobi formulations. J. Comput. Phys. **79**(1), 12–49 (1988). https://doi.org/10.1016/0021-9991(88)90002-2
10. Ouyang, G.F., Kuang, Y.C., Zhang, X.M.: A fast scanning algorithm for extension velocities in level set methods. Adv. Mater. Res. **328**(1), 677–680 (2011). https://doi.org/10.4028/www.scientific.net/AMR.328-330.677
11. Qian, J., Zhang, Y., Zhao, H.: Fast sweeping methods for Eikonal equations on triangular meshes. SIAM J. Numer. Anal. **45**(1), 83–107 (2007). https://doi.org/10.1137/050627083
12. Sethian, J.A.: A fast marching level set method for monotonically advancing fronts. Proc. Natl. Acad. Sci. **93**, 1591–1595 (1996). https://doi.org/10.1073/pnas.93.4.1591
13. Sethian, J.A., Vladimirsky, A.: Fast methods for the Eikonal and related Hamilton-Jacobi equations on unstructured meshes. Proc. Natl. Acad. Sci. **97**(11), 5699–5703 (2000). https://doi.org/10.1073/pnas.090060097
14. Suvorov, V., Hössinger, A., Djurić, Z., Ljepojevic, N.: A novel approach to three-dimensional semiconductor process simulation: application to thermal oxidation. J. Comput. Electron. **5**(4), 291–295 (2006). https://doi.org/10.1007/s10825-006-0003-z
15. Weinbub, J., Hössinger, A.: Shared-memory parallelization of the fast marching method using an overlapping domain-decomposition approach. In: Proceedings of the 24th High Performance Computing Symposium, pp. 1–8. Society for Modeling and Simulation International (2016). https://doi.org/10.22360/SpringSim.2016.HPC.052
16. Yang, J., Stern, F.: A highly scalable massively parallel fast marching method for the Eikonal equation. J. Comput. Phys. **332**, 333–362 (2017). https://doi.org/10.1016/j.jcp.2016.12.012
17. Yatziv, L., Bartesaghi, A., Sapiro, G.: O(N) implementation of the fast marching algorithm. J. Comput. Phys. **212**(2), 393–399 (2006). https://doi.org/10.1016/j.jcp.2005.08.005

# Relative Expression Classification Tree. A Preliminary GPU-Based Implementation

Marcin Czajkowski$^{(\boxtimes)}$ ⓘ, Krzysztof Jurczuk ⓘ, and Marek Kretowski ⓘ

Faculty of Computer Science, Bialystok University of Technology, Wiejska 45a,
15-351 Bialystok, Poland
{m.czajkowski,k.jurczuk,m.kretowski}@pb.edu.pl

**Abstract.** The enormous amount of omics data generated from high-throughput technologies has brought an increased need for computational tools in biological analyses. Among the algorithms particularly valuable are those that enhance understanding of human health and genetic diseases. Relative eXpression Analysis (RXA) is a powerful collection of computational methods for analyzing genomic data. It finds relationships in a small group of genes by focusing on the relative ordering of their expression values. In this paper, we propose a Relative eXpression Classification Tree (RXCT) which extends major variants of RXA solutions by finding additional hierarchical connections between sub-groups of genes. In order to meet the enormous computational demands we designed and implemented a graphic processing unit (GPU)-based parallelization. The GPU is used to perform a parallel search of the gene groups in each internal node of the decision tree in order to find locally optimal splits. Experiments carried out on 8 cancer-related gene expression datasets show that the proposed approach allows exploring much larger solution space and finding more accurate interactions between the genes. Importantly, patterns in predictive structures are kept comprehensible and may have direct applicability.

**Keywords:** Relative Expression Analysis (RXA) · Decision trees · GPU · CUDA

## 1 Introduction

Rapid growth and the popularity of high-throughput technologies cause a massive amount of gene expression datasets to become publicly accessible [19]. In the literature, we may find a good number of supervised machine learning approaches for genomic classification. Among the most popular ones, we could mention the support vector machines, neural networks or random forests. Most of currently applied methods provide 'black box' classification that usually involves many genes combined in a highly complex fashion and achieves high predictive performance. However, there is a strong need for 'white box', comprehensive decision models which may actually help in understanding and identifying casual relationships between specific genes [2,5]. The popular ones, like the decision trees

© Springer Nature Switzerland AG 2020
R. Wyrzykowski et al. (Eds.): PPAM 2019, LNCS 12043, pp. 359–369, 2020.
https://doi.org/10.1007/978-3-030-43229-4_31

(DTs) which have a long history in predictive modeling [10], result in insufficient accuracy [2] when applied to gene expression data. One of the problem specific alternatives is the Relative Expression Analysis (RXA) which is a powerful collection of easily interpretable computational methods. It was designed to analyze genomic data and plays an important role in biomarker discovery and gene expression data classification. It focuses on finding interactions among a small collections of genes and studies the relative ordering of their expression rather than their raw values.

In this paper, we want to merge the strength of RXA with DTs. We propose a new hybrid solution called Relative eXpression Classification Tree (RXCT) that induces DT with the splitting rules built by the RXA methodology. In order to overcome the enormous computational complexity we designed and implemented a graphic processing unit (GPU)-based parallelization of the RXA search. Finally, we added a few changes to the RXA algorithm in order to improve speed and to enable potential multi-class prediction.

This paper is organized as follows. Section 2 provides our motivation and a brief background on RXA, DTs and GPGPU parallelization. Section 3 describes in details our hybrid RXA solution and its GPU-based implementation. Next, an experimental validation is performed on real-life datasets and in the last section, the paper is concluded and possible future works are outlined.

## 2    Background

Gene expression data is very challenging for computational tools and mathematical modelling. Traditional solutions often fail due to the high ratio of features to observations as well as genes redundancy. Therefore, there is a need for new methods to be proposed to extract significant and meaningful rules from the genomic data.

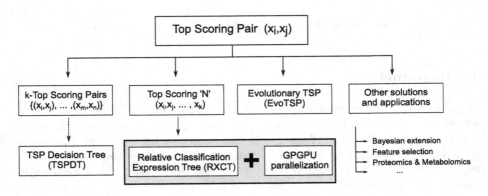

**Fig. 1.** The general taxonomy of the family of RXA

## 2.1    Algorithms for Relative Expression Analysis

Among many recent algorithms designed for the gene expression data classification, RXA methods are gaining popularity. The RXA taxonomy that includes the main development paths is illustrated in Fig. 1. A Top-Scoring Pair (TSP) is the first and the most popular RXA solution proposed by Donald Geman [8]. It uses a pairwise comparison of gene expression values and searches for a pair of genes with the highest rank. The $k$-TSP algorithm [18] is one of the first extensions of the TSP solution. It focuses on increasing the number of pairs in the prediction model and applies no more than $k$ disjoint gene pairs with the highest score, where the parameter $k$ is determined by the internal cross-validation. This method was later combined with a decision tree in algorithm called TSPDT [4]. In this system each non-terminal node of the tree divides instances according to a splitting rule that is based on TSP or $k$-TSP criteria.

Different approaches for the TSP extension focus on the relationships between more than two genes. Top Scoring Triplet (TST) [11] and Top Scoring N (TSN) [14] analyze ordering relationships between several genes, however, the general concept of TSP is retained. One of the first heuristic method applied to RXA was the evolutionary algorithm called EvoTSP [5] where the authors proposed an evolutionary search for the $k$-TSP and TSN-like rules. Performed experiments showed that evolutionary search is a good alternative to the traditional RXA algorithms. Finally, there are many variations of the TSP-family solutions that propose new ways of ranking the gene pairs.

## 2.2    Decision Trees

Decision trees [10] are one of the main techniques for discriminant analysis in knowledge discovery. The success of the tree-based approach can be explained by its ease of use, speed of classification and effectiveness. In addition, the hierarchical structure of the tree, where appropriate tests are applied successively from one node to the next, closely resembles the human way of making decisions.

However, there are not so many new solutions in the literature that focus on the classification of gene expression data with comprehensive DT models. Nowadays, much more interest is given in trees as sub-learners of an ensemble learning approach, such as Random Forests [3]. These solutions alleviate the problem of low accuracy by averaging or adaptive merging of multiple trees. However, when modeling is aimed at understanding basic processes, such methods are not so useful because they generate more complex and less understandable models.

## 2.3    GPGPU Parallelization

A general-purpose computation on GPUs (GPGPU) stands for the use of graphics hardware for generic problems. In the literature, we can find a few systems where GPU-based parallelization of the greedy induction of DTs was examined. One of the propositions was a CUDT [12] that parallelized the top-down induction process. In each internal node, in order to find the best locally optimal splits, the attributes are processed in parallel. With this approach, the authors managed to reduce the induction time of a typical decision tree from 5 to 55 times when compared with the traditional CPU version. The GPU was also introduced in GDT system which parallelizes evolutionary induction of DTs [9].

In case of RXA there exists also research considering GPU parallelization. In [13] authors managed to speed up calculations of TSP and TST by two orders of magnitude. The tests for higher number of related genes were also performed [14], but only when the total number of attributes was heavily reduced by the feature selection.

### 2.4   Motivation and Contribution

In this paper we propose a new approach that combines the strength of DTs with RXA. We are motivated by the fact that single DT represents a white-box approach, and improvements to such models have considerable potential for scientific modeling of the underlying processes in a genomic research.

Proposed contribution is inspired with the existing system called TSPDT [4] which also uses RXA concept in DT nodes. The main drawback of TSPDT as well as all traditional RXA algorithms is the enormous computational complexity that equals $O(T * k * N^Z)$, where $T$ is the size of DT, $k$ is the number of top-scoring groups, $N$ is the number of analyzed genes and $Z$ is the size of a group of genes which ordering relationships are searched. Sequential calculation of all possible gene pairs or gene groups strongly limits the number of genes and inter-relations that can be analyzed by the algorithm. This is the reason why the TSPDT focuses only on TSP and k-TSP variants and the algorithm need to be preceded by the feature selection step.

There are several major differences between the TSPDT and proposed RXCT solutions in terms of both performance and functionality:

- with the GPU parallelization the RXCT is capable of inducing the RXA-based decision tree much faster, even on entire datasets (without feature selection step);
- RXCT extends the inter-gene relations, allows testing higher number of related genes and has additional optimizations considering strict order relations;
- RXCT in contrast to TPSDT can be applied to multi-class datasets due to the different splitting rule.

## 3   Relative eXpression Classification Tree

The proposed solution's overall structure is based on a typical top-down induced [16] binary classification tree. The greedy search starts with the root node, where the locally optimal split (test) applies RXA. Then the training instances are redirected to the newly created nodes and this process is repeated for each node until the stop condition is met. Currently, we do not apply any form of pruning due to the small sample sizes, however it should be considered in the future to improve the generalizing power of the predictive model.

The general flowchart of our GPU-accelerated RXCT is illustrated in Fig. 2. It can be seen that the DT induction is run in a sequential manner on a CPU, and the most time-consuming operation (split search) is performed in parallel on a GPU. This way, the parallelization does not affect the behavior of the original algorithm.

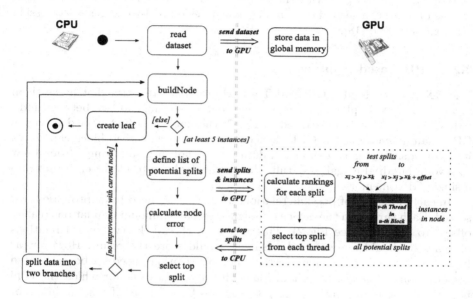

**Fig. 2.** General flowchart of a GPU-accelerated RXCT

## 3.1 RXCT Split Search

Each internal node contains information about a relation of two or three genes that is later used to constitute the split. The basic idea to analyze relations within a single instance is similar to TSP and TST solutions, however, there are some differences in strict ordering as well as in ranking of gene collections.

Let $x_i$, $x_j$ and $x_k$ be the expression values of three genes $i, j, k$ from available set of $N$ genes. However, in contrast to existing solutions, we allow triplet reduction to a pair if $j$ equals $k$. RXA can be directly applied to binary classification problems as it scores relations using probabilities of assigning instances to one of two classes. To allow application RXCT to analyze multi-class datasets, we have chosen to use Gini index [1] which is well known splitting criterion for DT. It is also slightly computationally faster than the gain ratio as there are no LOG functions.

The Gini impurity is calculated for all pairs or triplets defined by the relation: $x_i > x_j \geq x_k$ where $i \neq j$. The non-strict relation between the second and third gene allows us to test both TSP (if $j = k$) and TST ($j \neq k$) variants. However, in contrast to RXA-based solutions we do not check all possible orderings but limit them to those that meet the assumption $i > j \geq k$. This limitation does not affect the results (if one of the relation is opposite then the resulting branches are swapped) but reduces the calculations twice in case of TSP and six times in case of TST. The triplet with the highest Gini impurity becomes the split and in case of a draw, a secondary ranking that bases on real-value genes expression differences is used [18].

### 3.2   GPU-Based Approach

The RXA methods like TSP and TST exhibit characteristics that make them ideal for a GPU implementation as there is no data dependence between individual scores. As it is illustrated in Fig. 2, the dataset is first copied from the CPU main memory to the GPU device memory so each thread block can access it. Typical sizes of gene expression datasets are not large and range from a few to several hundred megabytes, thus there is no problem to fit the entire set into a single GPU memory.

In each node, $N^3$ of possible relations $x_i > x_j \geq x_k$ need to be processed and scored. Each thread on the device is assigned an equal amount of relations (called offset) to compute (see Fig. 2). This way each thread 'knows' which relations of genes it should analyze and where it should store the result. However, as it was mentioned in previous section, not all relations need to be calculated (assumption $i > j \geq k$). In addition, number of instances for which the Gini impurity is calculated varies in each tree node - from full set of samples in a root to a few instances in the lower parts of the tree.

Each launched kernel requires not only the relations that will be processed but also: an information about the instances that reach the internal node which runs the kernel; and an empty vector for the results. Within each thread there is no further parallelization: each thread simply loops over the instances that reach the node and calculates the scores to the assigned relations. After all thread blocks have finished, the top results from each threads are copied from the GPU device memory back to the CPU main memory where the top split is selected.

## 4   Experimental Validation

In this section, we present experimental analysis of RXCT predictive performance and confront its results with popular RXA extensions. In addition, we show the speedup achieved with proposed GPU parallelization.

## 4.1  Datasets and Setup

To make a proper comparison with the RXA algorithms, we use the same 8 cancer-related benchmark datasets (see Table 1) that are tested with the EvoTSP solution [5]. Datasets are deposited in NCBI's Gene Expression Omnibus and summarized in Table 1. A typical 10-fold cross-validation is applied and depending on the system, different tools are used:

- evaluation of TSP, TST, and $k$-TSP was performed with the AUERA software [7], which is an open-source system for identification of relative expression molecular signatures;
- EvoTSP results were taken from the publication [5];
- original TSPDT and RXCT implementation are used.

Due to the performance reasons concerning other approaches, the Relief-F feature selection was applied and the number of selected genes was arbitrarily limited to the top 1000. In the experiments, we provide results for the proposed RXCT solution as well as its simplified variant called $RXCT_{TSP}$ which uses only TSP-like splits.

**Table 1.** Details of gene expression datasets: abbreviation with name, number of genes and number of instances.

| Datasets | Genes | Instances | Datasets | Genes | Instances |
|---|---|---|---|---|---|
| (a) GDS2771 | 22215 | 192 | (e) GSE10072 | 22284 | 107 |
| (b) GSE17920 | 54676 | 130 | (f) GSE19804 | 54613 | 120 |
| (c) GSE25837 | 18631 | 93 | (g) GSE27272 | 24526 | 183 |
| (d) GSE3365 | 22284 | 127 | (h) GSE6613 | 22284 | 105 |

Experiments were performed on a workstation equipped with Intel Xeon CPU E5-2620 v3 (15 MB Cache, 2.40 GHz), 96 GB RAM and NVIDIA GeForce GTX Titan X GPU card (12 GB memory, 3 072 CUDA cores). We used a 64-bit Ubuntu Linux 16.04.6 LTS as an operating system. The sequential algorithm was implemented in C++ and compiled with the use of gcc version 5.4.0. The GPU-based parallelization part was implemented in CUDA-C [17] and compiled by nvcc CUDA 8.0 [15] (single-precision arithmetic was applied).

## 4.2  Accuracy Comparison to Popular RXA Algorithms

Table 2 summarizes classification performance for the proposed solution and it's competitors. From the results we can see that the proposed RXCT solution managed to outperform in average all popular RXA classifiers. Although there are no statistical differences between TSPDT and RXCT in terms of accuracy, the size of the trees generated by RXCT is significantly smaller (Friedman test and the corresponding Dunn's multiple comparison test are applied, p-value = 0.05 [6]).

There are two factors that may explain such good results achieved by the RXCT classifier. First of all, we use TST-like variant in each node instead of TSP split so more advanced relations are searched. Next, the additional experiments (not shown) revealed that in the case of DT using the original TSP rank to split the data, it returns worse results than when using one of the standard DT splitting criteria like Gini index or gain ratio. It also explains why the $RXCT_{TSP}$ using a single TSP in each node can compete with TSPDT which uses more complex split ($k$ - top pairs).

**Table 2.** Comparison of RXCT with top-scoring algorithms, including accuracy and the size of the classifier's model. The best accuracy for each dataset is bolded.

| Dataset | TSP | TST | k-TSP | | EvoTSP | | TSPDT | | $RXCT_{TSP}$ | | RXCT | |
|---|---|---|---|---|---|---|---|---|---|---|---|---|
| | acc. | acc. | acc. | size[1] | acc. | size[2] | acc. | size[3] | acc. | size[4] | acc. | size[5] |
| (a) | 57.2 | 61.9 | 62.9 | 10 | 65.6 | 4.0 | 60.1 | 16.4 | **68.2** | 15.1 | 67.2 | 10.5 |
| (b) | 88.7 | 89.4 | 90.1 | 6 | 96.5 | 2.1 | 98.2 | 2.0 | 95.1 | 2.0 | **100.0** | 2.0 |
| (c) | 64.9 | 63.7 | 67.2 | 10 | **78.1** | 2.8 | 72.3 | 6.8 | 74.6 | 6.8 | 77.7 | 5.0 |
| (d) | 93.5 | 92.8 | 94.1 | 10 | **96.2** | 2.1 | 88.3 | 3.0 | 90.0 | 2.8 | 91.6 | 2.0 |
| (e) | 56.0 | 60.5 | 58.4 | 14 | 66.9 | 3.1 | 68.1 | 5.7 | 68.6 | 6.0 | **73.2** | 4.3 |
| (f) | 47.3 | 50.1 | 56.2 | 18 | 66.2 | 2.7 | **67.2** | 11.9 | 60.6 | 12.6 | 60.0 | 9.2 |
| (g) | 81.9 | 84.2 | 87.2 | 14 | 86.1 | 4.1 | 88.6 | 4.3 | 85.0 | 4.1 | **89.7** | 3.3 |
| (h) | 49.5 | 51.7 | 55.8 | 10 | 53.6 | 6.1 | 59.6 | 8.0 | 54.3 | 7.0 | **70.4** | 5.4 |
| Average | 67.4 | 69.3 | 71.5 | 11.5 | 76.2 | 2.7 | 75.3 | 7.2 | 74.6 | 6.9 | **78.7** | 5.1 |

[1] Avg. number of k-TSP pairs (k≤18);
[2] Avg. number of unique genes;
[3] Avg. number of tree nodes, in each node no more than $k$ ($k \leq 5$) TSP;
[4] Avg. number of tree nodes, each node with single TSP;
[5] Avg. number of tree nodes, each node with single TST.

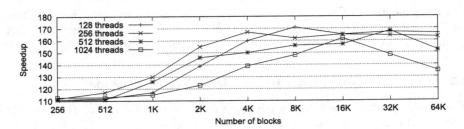

**Fig. 3.** Effect of block's and thread's number on the algorithm speedup.

## 4.3  Speed Improvement with GPGPU Approach

Even with the feature selection step, the number of possible relations for which the GPU needs to calculate score is very high. For example, for 1000 genes there is $10^9$ possible split expression rules in each tree node whereas if we take the full dataset, e.g. GSE17920, this number drastically increases to $1.63 * 10^{14}$. The high number of possible kernel tasks requires finding optimal number of threads and blocks which will perform the calculations. Figure 3 illustrates how amount of threads impacts on the GPU-accelerated RXCT speedup in comparison to its sequential version averaged on all datasets. We see that by increasing at some point the number of blocks the speedup rises. This suggests that processing too many possible relations by each thread (high load) slows down the parallelization. Decreasing the offset value improves load balancing and thus the overal RXCT speedup.

In this section we also present the times achieved by popular RXA solutions. We perform direct time comparison between the TSPDT solution and two variants of the RXCT algorithm: original RXCT and $RXCT_{TSP}$ version which is at some point similar to TSPDT. Table 3 shows the times and speedups of the sequential and parallel versions or RXCT for all the datasets. Alike in Sect. 4.2 the number of attributes in the datasets was limited to 1000. As expected, the algorithms which compare only two features perform much faster than the ones which analyze triplets. At the same time, the solutions that use hierarchical structures take much more time as they run multiple searches of top groups in each non-terminal node.

**Table 3.** Induction times of popular RXA solutions (in seconds). Impact of the GPGPU approach on the RXCT and $RXCT_{TSP}$ algorithm is also included.

| Dataset | TSP | TST | TSPDT | seqRXCT$_{TSP}$ | RXCT$_{TSP}$ | | seqRXCT | RXCT | |
|---|---|---|---|---|---|---|---|---|---|
| | Time | Time | Time | Time | Time | Speedup | Time | Time | Speedup |
| (a) | 3.01 | 999 | 325 | 15.5 | 0.067 | 232 | 4071 | 24.1 | 169 |
| (b) | 1.92 | 672 | 32.3 | 7.30 | 0.030 | 247 | 1842 | 11.9 | 155 |
| (c) | 1.49 | 1005 | 113 | 5.13 | 0.021 | 241 | 1276 | 8.04 | 158 |
| (d) | 1.95 | 590 | 76.8 | 4.87 | 0.017 | 314 | 1314 | 8.33 | 158 |
| (e) | 1.84 | 676 | 127 | 1.83 | 0.013 | 253 | 519 | 3.15 | 165 |
| (f) | 2.07 | 653 | 292 | 16.9 | 0.068 | 250 | 3444 | 23.3 | 148 |
| (g) | 2.99 | 511 | 106 | 3.33 | 0.024 | 276 | 857 | 5.50 | 156 |
| (h) | 1.78 | 592 | 151 | 6.56 | 0.031 | 250 | 1575 | 9.76 | 161 |
| Average | 2.13 | 712 | 152 | 7.68 | 0.031 | 250 | 1862 | 21.9 | 159 |

We have also performed experiments with datasets containing all attributes. Total time required to process all 8 datasets takes over 2 weeks by TSPDT, 16 h by seqRXCT$_{TSP}$ and only 3 min by RXCT$_{TSP}$ ($\sim$ x300 faster than seqRXCT$_{TSP}$ and $\sim$ x7000 faster than TSPDT).

However, titled RXCT solution is much more computationally demanding. The total time required by RXCT equals approximately 20 days which is similar to the time need by the TSPDT solution. For the seqRXCT version it would take over a decade to induce the same tree.

## 5   Conclusion

In this paper, we introduce a hybrid approach to analyze gene expression data which combines the problem-specific methodology with the popular white-box classifier. The Relative eXpression Classification Tree extends major variants of RXA solutions and is capable of finding interesting hierarchical patterns in subgroups of genes. In addition, thanks to the GPU-parallelization, we managed to induce the tree in a reasonable time.

We see many promising directions for future research. First of all, there is still a lot of ways to improve the GPU parallelization of RXCT, e.g. load-balancing of tasks based on the number of instances in each node, simultaneous analysis of two branches, better GPU hierarchical memory exploitation. Then, other variants of RXA can be used in each split like $k$-TSP, $k$-TST or even both. Next, some form of tree post-pruning could be applied, not only to limit the tree size but also to decrease the number of genes used in the splits in order to reduce overfitting and promote more accurate decisions. Finally, we are currently working with biologists and bioinformaticians to better understand the decision rules generated by RXCT and preparing the algorithm to work with protein and metabolic expression databases.

**Acknowledgments.** This work was supported by the grant S/WI/2/18 from Bialystok University of Technology founded by Polish Ministry of Science and Higher Education.

## References

1. Breiman, L., Friedman, J., Olshen, R., Stone, C.: Classification and Regression Trees. Wadsworth Int. Group, Belmont (1984)
2. Barros, R.C., Basgalupp, M.P., Freitas, A.A., Carvalho, A.C.: Evolutionary design of decision-tree algorithms tailored to microarray gene expression data sets. IEEE Trans. Evol. Comput. **18**(6), 873–892 (2014)
3. Chen, X., Wang, M., Zhang, H.: The use of classification trees in bioinformatics. Wiley Interdisc. Rev. Data Min. Knowl. **1**, 55–63 (2011)
4. Czajkowski, M., Kretowski, M.: Top scoring pair decision tree for gene expression data analysis. In: Arabnia, H., Tran, Q.N. (eds.) Software Tools and Algorithms for Biological Systems. AEMB, vol. 696, pp. 27–35. Springer, New York (2011). https://doi.org/10.1007/978-1-4419-7046-6_3
5. Czajkowski, M., Kretowski, M.: Evolutionary approach for relative gene expression algorithms. Sci. World J. **593503**, 7 (2014)
6. Demsar, J.: Statistical comparisons of classifiers over multiple data sets. J. Mach. Learn. Res. **7**, 1–30 (2006)
7. Earls, J.C., Eddy, J.A., Funk, C.C., Ko, Y., Magis, A.T., Price, N.D.: AUREA: an open-source software system for accurate and user-friendly identification of relative expression molecular signatures. BMC Bioinformatics **14**, 78 (2013). https://doi.org/10.1186/1471-2105-14-78

8. Geman, D., d'Avignon, C., Naiman, D.Q., Winslow, R.L.: Classifying gene expression profiles from pairwise mRNA comparisons. Stat. Appl. Genet. Mol. Biol. **3** (2004). Article no. 19. https://doi.org/10.2202/1544-6115.1071

9. Jurczuk, K., Czajkowski, M., Kretowski, M.: Evolutionary induction of a decision tree for large-scale data: a GPU-based approach. Soft Comput. **21**(24), 7363–7379 (2016). https://doi.org/10.1007/s00500-016-2280-1

10. Kotsiantis, S.B.: Decision trees: a recent overview. Artif. Intell. Rev. **39**(4), 261–283 (2013). https://doi.org/10.1007/s10462-011-9272-4

11. Lin, X., et al.: The ordering of expression among a few genes can provide simple cancer biomarkers and signal BRCA1 mutations. BMC Bioinformatics **10**, 256 (2009). https://doi.org/10.1186/1471-2105-10-256

12. Lo, W.T., Chang, Y.S., Sheu, R.K., Chiu, C.C., Yuan, S.M.: CUDT: a CUDA based decision tree algorithm. Sci. World J. **745640**, 12 (2014)

13. Magis, A.T., Earls, J.C., Ko, Y., Eddy, J.A., Price, N.D.: Graphics processing unit implementations of relative expression analysis algorithms enable dramatic computational speedup. Bioinformatics **27**(6), 872–873 (2011)

14. Magis, A.T., Price, N.D.: The top-scoring 'N' algorithm: a generalized relative expression classification method from small numbers of biomolecules. BMC Bioinformatics **13**, 227 (2012). https://doi.org/10.1186/1471-2105-13-227

15. NVIDIA Developer Zone - CUDA Toolkit Documentation. https://docs.nvidia.com/cuda/cuda-c-programming-guide/

16. Rokach, L., Maimon, O.Z.: Top-down induction of decision trees classifiers - a survey. IEEE Trans. SMC Part C **35**(4), 476–487 (2005)

17. Storti, D., Yurtoglu, M.: CUDA for Engineers: An Introduction to High-Performance Parallel Computing. Addison-Wesley, New York (2016)

18. Tan, A.C., Naiman, D.Q.: Simple decision rules for classifying human cancers from gene expression profiles. Bioinformatics **21**, 3896–3904 (2005)

19. Taminau, J., et al.: Unlocking the potential of publicly available microarray data using inSilicoDb and inSilicoMerging R/Bioconductor packages. BMC Bioinformatics **13**, 335 (2012). https://doi.org/10.1186/1471-2105-13-335

# Performance Optimizations for Parallel Modeling of Solidification with Dynamic Intensity of Computation

Kamil Halbiniak[1]([✉])[ID], Lukasz Szustak[1][ID], Adam Kulawik[1][ID], and Pawel Gepner[2][ID]

[1] Czestochowa University of Technology,
Dabrowskiego 69, 42-201 Czestochowa, Poland
{khalbiniak,lszustak,adam.kulawik}@icis.pcz.pl
[2] Intel Corporation, Santa Clara, USA
pawel.gepner@intel.com

**Abstract.** In our previous works, a parallel application dedicated to the numerical modeling of alloy solidification was developed and tested using various programming environments on hybrid shared-memory platforms with multicore CPUs and manycore Intel Xeon Phi accelerators. While this solution allows obtaining a reasonable good performance in the case of the static intensity of computations, the performance results achieved for the dynamic intensity of computations indicates pretty large room for further optimizations.

In this work, we focus on improving the overall performance of the application with the dynamic computational intensity. For this aim, we propose to modify the application code significantly using the *loop fusion* technique. The proposed method permits us to execute all kernels in a single nested loop, as well as reduce the number of conditional operators performed within a single time step. As a result, the proposed optimizations allows increasing the application performance for all tested configurations of computing resources. The highest performance gain is achieved for a single Intel Xeon SP CPU, where the new code yields the speedup of up to 1.78 times against the original version.

The developed method is vital for further optimizations of the application performance. It allows introducing an algorithm for the dynamic workload prediction and load balancing in successive time steps of simulation. In this work, we propose the workload prediction algorithm with 1D computational map.

**Keywords:** Numerical modeling of solidification · Phase-field method · Parallel programming · OpenMP · Workload prediction · Load balancing · Intel Xeon Phi · Intel Xeon Scalable processors

## 1 Introduction

The phase-field method is a powerful tool for solving interfacial problems in materials science [9]. It has been not only used for solidification dynamics [7,8],

© Springer Nature Switzerland AG 2020
R. Wyrzykowski et al. (Eds.): PPAM 2019, LNCS 12043, pp. 370–381, 2020.
https://doi.org/10.1007/978-3-030-43229-4_32

but has been also applied to other phenomena such as viscous fingering [3], fracture dynamics [5], and vesicle dynamics [9]. In our previous works [4,10,12], a parallel application dedicated to the numerical modeling of alloy solidification, which is based on the phase-field method, was developed and tested using various programming environments on hybrid shared-memory platforms with multicore CPUs and manycore Intel Xeon Phi accelerators. In these works, two different versions of application were considered: with the static and dynamic intensity of computations. The results of experiments shown that in the second case the performance and scalability drop significantly. In particular, for the static computational intensity, the usage of 4 KNL processors permits us to accelerate the application about 3.1 times against the configuration with a single KNL processor. At the same time, the performance results achieved in the case of the dynamic intensity of computations indicated pretty large room for further optimizations, since for example the utilization of 4 KNL devices gives only the speedup of 1.89 times.

In this work, we focus on improving the overall performance of the solidification application with the dynamic computational intensity. We propose an optimization method which is based on the *loop fusion* technique and assumes a significant modification of the application source code. This method permits us to compress the main application workload into a single nested loop, as well as reduce the number of conditional operators which have to be performed within a single time step. At the same time, the developed method is vital for further optimizations of the application performance. It allows introducing an algorithm for the dynamic workload prediction and load balancing in successive time steps of simulation. In this work, we propose the workload prediction algorithm with 1D computational map.

This paper is organized as follows. Section 2 outlines the numerical model, as well as presents the basic version of the solidification application with the dynamic intensity of computations. Section 3 outlines the approach for performance optimization of the application. The next section presents performance evaluation of the proposed method on platforms equipped with Intel Xeon Scalable CPUs and Intel KNL accelerators. Section 5 outlines the algorithm for the workload prediction and load balancing, which is based on 1D computational map. Section 6 concludes the paper.

## 2 Parallelization of Numerical Modeling of Solidification with Dynamic Intensity of Computation

### 2.1 Overview of Numerical Model

In the modeling problem studied in this paper, a binary alloy of Ni-Cu is considered as a system of the ideal metal mixture in liquid and solid phases. The numerical model refers to the dendritic solidification process [1,13] in the isothermal conditions with constant diffusivity coefficients for both phases. It allows us to use the field-phase method defined by Warren and Boettinger [14]. In the model, the growth of microstructure during the solidification is determined by solving a system of two PDEs. The first equation defines the phase content $\phi$:

$$\frac{1}{M_\phi}\frac{\partial\phi}{\partial t} = \varepsilon^2 \left[\nabla \cdot \left(\eta^2\nabla\phi\right) +\eta\eta' \left(sin(2\theta)\left(\frac{\partial^2\phi}{\partial y^2} - \frac{\partial^2\phi}{\partial x^2}\right) + 2cos(2\theta)\frac{\partial^2\phi}{\partial x\partial y}\right)\right]$$

$$-\frac{1}{2}\left(\eta'^2 + \eta\eta''\right)\left(-cos(2\theta)\left(\frac{\partial^2\phi}{\partial y^2} - \frac{\partial^2\phi}{\partial x^2}\right) + 2sin(2\theta)\frac{\partial^2\phi}{\partial x\partial y} - \frac{\partial^2\phi}{\partial x^2} - \frac{\partial^2\phi}{\partial y^2}\right) \quad (1)$$

$$-cH_B - (1-c)H_A - cor,$$

where: $M_\phi$ is defined as the solid/liquid interface mobility, $\varepsilon$ is a parameter related to the interface width, $\eta$ is the anisotropy factor, $H_A$ and $H_B$ denotes the free energy of both components, $cor$ is the stochastic factor which models thermodynamic fluctuations near the dendrite tip. The coefficient $\theta$ is calculated as follows:

$$\theta = \frac{\partial\phi}{\partial y}\bigg/\frac{\partial\phi}{\partial x}. \quad (2)$$

The second equation defines the concentration $c$ of the alloy dopant, which is one of the components of the alloy:

$$\frac{\partial c}{\partial t} = \nabla \cdot D_c \left[\nabla c + \frac{V_m}{R}c(1-c)(H_B(\phi,T) - H_A(\phi,T))\nabla\phi\right], \quad (3)$$

where: $D_c$ is the diffusion coefficient, $V_m$ is the specific volume, $R$ is the gas constant.

In this model, the generalized finite difference method [2, 6] is used to obtain the values of partial derivatives in Eqs. (1) and (2). In order to parallelize computations with a desired accuracy, the explicit scheme is applied with a small time step $\Delta t = 1e - 7s$.

The resulting computations [12] belong to the group of forward-in-time, iterative algorithms since all the calculations performed in the current time step $k$ depend on results determined in the previous step $k-1$. The application code consists of two main blocks of computations, which are responsible for determining either the phase content $\phi$ or the dopant concentration $c$. In the model, the values of $\phi$ and $c$ are determined for nodes distributed across a considered domain (Fig. 1). For this aim, the values of derivatives in all nodes have to be calculated at every time step.

In our previous work [12], two different cases were introduced – with the static and dynamic intensity of computations. In the first case, the workload of computing resources is constant during the application execution, since a constant number of equations is solved. This assumption corresponds to modeling problems in which the variability of solidification phenomena in the whole domain has to be considered. In the second case, the model is able to solve differential equations only in part of nodes, which is changing during the simulation following the growth of microstructure. The use of a suitable selection criterion allows reducing significantly the amount of computations. The consequence is a significant workload imbalance, since the selection criterion is calculated after the static partitioning of the grid nodes across computing resources.

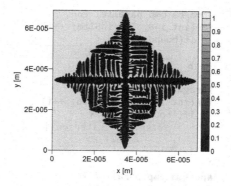

**Fig. 1.** Phase content for the simulated time $t_S = 2.75 \times 10^{-3}s$

## 2.2 Basic Version of Solidification Application with the Dynamic Intensity of Computations

Figure 2 illustrates the computational core of the solidification application with the dynamic intensity of computations, for a single time step. All the computations in the application are organized as five loops that iterate over all nodes of the grid. Two of them, with kernels $K_1$ and $K_3$, execute calculations in the boundary nodes, while the other two loops, with kernels $K_2$ and $K_4$, perform computations for the internal part of the grid. The last loop completes the execution of a single time step. The selection of the boundary and internal nodes of the grid is implemented using four conditional statements, which have to be executed within a single time step.

All kernels of the application are organized as two nested loops, where the outer and inner loops iterate over the grid nodes and neighbors of each node, respectively. The inner loop corresponds to stencil computations used for determination of partial derivatives. Figures 3 and 4 depict the code snippets corresponding to kernels $K_1$ and $K_2$ that are responsible for determining the dopant concentration $c$ for the boundary and internal nodes of the grid, respectively. The structure of kernels $K_3$ and $K_4$ that calculate the phase content $\phi$ is similar. For a given node $i$, the indices `node_e[offset+j]` of its neighbours are kept in the configuration file describing the whole grid. In consequence, the patterns of all 22 stencils are determined at runtime. The structure of kernels allows their parallelization using **omp parallel for** directives of OpenMP for the outer loops.

The application studied in this paper uses the SOA (structure of arrays) layout of memory, where computations are performed using one-dimensional arrays. For instance, `node_conc0[i]` contains value of the dopant concentration for the $i$-$th$ node, while `node_Fi0[i]` corresponds to value of the phase content for this node. The transformation to the SoA organization of memory from the original AoS (array of structures) layout, which was used in the original code, are in line with the first step of the methodology proposed in work [4].

```
#pragma omp parallel for
for(int i=0; i<gridSize; ++i) {
    if(boundary[i]) {
        Kernel K1
    }
}
#pragma omp parallel for
for(int i=0; i<gridSize; ++i) {
    if(!boundary[i]) {
        Selection criterion
        Kernel K2
    }
}
#pragma omp parallel for
for(int i=0; i<gridSize; ++i) {
    if(boundary[i]) {
        Kernel K3
    }
}
#pragma omp parallel for
for(int i=0; i<gridSize; ++i) {
    if(!boundary[i]) {
        Selection criterion
        Kernel K4
    }
}
#pragma omp parallel for
for(int i=0; i<gridSize; ++i) {
    Kernel K5
}
```

**Fig. 2.** General scheme of basic version of solidification application with the dynamic intensity of computations

The criterion for selecting grid nodes plays a vital role for the overall performance of the application. In the basic version of the solidification application with the dynamic computational intensity, the selection criterion (see Fig. 2) is checked during execution of kernels $K_2$ and $K_4$. The execution of this criterion involves two additional conditional operators. As a result, six conditional statements have to be executed within a single time step. Moreover, the execution of the selection criterion leads to the analysis of practically all nodes of the grid (excluding boundary ones), not just nodes within the area of grain growth.

```
#pragma omp parallel for
for(int i=0; i<grid_size; i++) {
    const int offset = i*max_neighbors;
    double d0(0.0), d2(0.0);
    double z[max_neighbors];
    /.../
    for(int j=0; j<neighbors_count[i]; ++j) {
        // Stencil computations used to determine partial derivatives
        z[j] = (node_g2[offset+0]*node_cosAlf[i]+
                node_g2[offset+2]*node_cosBet[i])*node_hx[offset+j]+
                (node_g2[offset+1]*node_cosAlf[i]+
                node_g2[offset+3]*node_cosBet[i])*node_hy[offset+j]
        const int idx = node_e[offset+j];
        d2 += node_conc0[idx]*z[j];
        /.../
    }
    // Computations performed within nodes
}
```

Fig. 3. Kernel $K_1$

```
#pragma omp parallel for
for(int i=0; i<grid_size; i++) {
    const int offset = i*max_neighbors;
    const int gOffset = i*25;
    double d1xCj(0.0), d1xDcj(0.0), d1xFj(0.0);
    double zx[max_neighbors];
    /.../
    for(int j=0; j<neighbors_count[i]; ++j) {
        // Stencil computations
        // 3 of all 15 stencils used in kernel K2
        zx[j] = 1/pow(node_h[offset+j],2*m)*
                (node_g[gOffset+0]*node_hx[offset+j]+
                node_g[gOffset+1]*node_hy[offset+j]+
                0.5*node_g[gOffset+2]*node_hx[offset+j]*node_hx[offset+j]+
                0.5*node_g[gOffset+3]*node_hy[offset+j]*node_hy[offset+j]+
                node_g[gOffset+4]*node_hx[offset+j]*node_hy[offset+j]);
        const int idx = node_e[offset+j];
        d1xCj += zx[j]*node_conc0[idx];
        d1xDcj += zx[j]*node_Dc[idx];
        d1xFj += zx[j]*node_Fi0[idx];
        /.../
    }
    // Computations performed within nodes
}
```

Fig. 4. Kernel $K_2$

```
#pragma omp parallel for
for(int i=0; i<gridSize; ++i) {
        Selection criterion
    if(boundary[i]) {
                Kernel K1
                Kernel K3
    }
    else {
                Kernel K2
                Kernel K4
    }
            Kernel K5
}
```

**Fig. 5.** General scheme of modified version of solidification application with the dynamic intensity of computations

## 3    Performance Optimization of the Application Using *Loop Fusion*

In this section, we present a method for optimizing the performance of solidification application with the dynamic computational intensity. This method assumes a significant modification of the application source code using the *loop fusing* technique. This optimization technique assumes merging selected loops, in order to reduce the loop overheads, as well as increase the instruction parallelism, improve the data locality, and even reduce data transfers [11].

Figure 5 presents the general scheme of executing a single time step of the application after applying the *loop fusion* technique. In contrast to the basic version of code (Fig. 2), all workloads of the modified version are executed in a single nested loop. Such a solution allows us to reduce the number of conditional statements used to selecting the boundary and internal nodes of the grid, as well as to decrease the number of conditional statements required for checking the selection criterion. In practice, the execution of a single time step requires now to perform only two conditional statements, instead of six ones in the basic version.

Implementing the *loop fusion* requires also a suitable modification of the selection criteria. The resulting criterion is obtained by merging the selection criteria used for the kernels $K_2$ and $K_4$ in the basic code. Moreover, due to removing the loop completing each time step, some additional calculations have to be performed in the new selection criterion.

In the modified scheme of the application execution (Fig. 5), the selection criterion is still calculated for all nodes of the grid. However, this scheme allows us to introduce further optimization of the application performance by providing a method for the efficient workload prediction and load balancing across resources of a computing platform (see Sect. 5).

**Table 1.** Specification of tested platforms

|  | Intel Xeon Platinum 8180 (SKL) | Intel Xeon Phi 7250F (KNL) |
| --- | --- | --- |
| Number of devices | 2 | 1 |
| Number of cores per device | 28 | 68 |
| Number of threads per device | 56 | 272 |
| Base frequency [GHz ] | 2.5 | 1.4 |
| (AVX frequency) | (1.7) | (1.2) |
| SIMD width [bits] | 512 | 512 |
| AVX peak for DP [GFlop/s] | 3046.4 | 2611.2 |
| Size of last level cache [MB] | 38.5 | 34 |
| Memory size | 512 GB DDR4 | 16 GB MCDRAM 96GB DDR4 |
| Memory bandwidth [GB/s] | 119,2 | MCDRAM: 400+ DDR4: 115.2 |

# 4 Experimental Results

In this section, we present performance results obtained for the approach proposed in the previous section, assuming the double precision floating point format. The experiments are performed for two platforms (Table 1):

1. SMP server equipped with two Intel Xeon Platinum 8180 CPUs (first generation of Intel Xeon Scalable Processor architecture);
2. single Intel Xeon Phi 7250 F processor (KNL architecture).

The KNL accelerator is utilized in the *quadrant* clustering mode with the MCDRAM memory configured in the *flat* mode. All the tests are compiled using the Intel `icpc` compiler (version 19.0.1) with `-O3` and `-xMIC-AVX512` flags for Intel KNL, and `-xCore-AVX512 -qopt-zmm-usage = high` flags for Intel Xeon Platinum CPUs. In order to ensure the reliability of performance results, the measurements of the execution time are repeated $r = 10$ times, and the median value is used finally.

Table 2 presents the total execution times and speedups achieved for the basic and optimized versions of the solidification application with the dynamic computational intensity. The tests are executed for 110 000 time steps, and two grid sizes: $2000 \times 2000$ and $3000 \times 3000$, using three configurations of computing resources:

1. single Intel Xeon Platinum 8180 CPU;
2. two Intel Xeon Platinum 8180 CPUs;
3. single Intel KNL processor.

The analysis of Table 2 permits us to conclude that the proposed method allows increasing significantly the performance of computations for all configurations of computing resources. for the grid of size $2000 \times 2000$, the highest performance gain is achieved for configuration with a single Intel Xeon Platinum CPU, when the new code yields the speedup of about 1.78 times against the basic

**Table 2.** The execution times and speedups achieved for the basic and optimized versions of the solidification application with the dynamic intensity of computational

| Size of grid | Computing resources | Execution time [s] | | Speedup |
|---|---|---|---|---|
| | | Basic $T_B$ | Optimized $T_{Op}$ | $S = T_B/T_{Op}$ |
| 2000 × 2000 | 1 × SKL | 1785 | 1001 | 1.78 |
| | 2 × SKL | 1456 | 837 | 1.74 |
| | 1 × KNL | 1661 | 1078 | 1.54 |
| 3000 × 3000 | 1 × SKL | 4061 | 2359 | 1.72 |
| | 2 × SKL | 3266 | 2005 | 1.63 |
| | 1 × KNL | 3869 | 2540 | 1.52 |

version. At the same time, the lowest speedup equal to 1.54 times is obtained on Intel KNL processor. For configuration with two Intel Xeon CPUs, the developed implementation allows accelerating the simulation by 1.74 times. For the second grid, the performance gains achieved by the proposed optimization method are similar.

## 5    Toward Dynamic Workload Prediction and Load Balancing: 1D Map Approach

Although the proposed optimization method (see Sect. 3) allows increasing the performance of computations, it does not ensure an efficient workload distribution across computing resources. Thus, the next step of performance optimization of the solidification application with the dynamic computational intensity will focus on resolving this issue. To achieve this goal, we propose an algorithm for the dynamic workload prediction.

This algorithm is responsible for predicting the computational workload in successive time steps of the simulation. In practice, it permits us to adjust the computational domain to the domain of simulation. The computational domain refers to the grid area wherein the primary computations are performed and the selection criterion is checked. The prediction of the workload for the next time step $k + 1$ is based on results of computations performed in the current step $k$. In practice, if values of variables in a grid node are computed in a given time step, this node and its neighbours are taken into consideration when predicting the computational domain for the next time step.

The workload prediction algorithm proposed in this paper is illustrated in Fig. 6. It is based on representing the grid nodes using an 1D array. The computational domain predicted for the next time step is defined by two coordinates referring respectively to the beginning (`minNode`) and end (`maxNode`) of the area wherein the computations are performed and the selection criterion is checked. For the first time step ($k = 1$), the area of checking the selection criterion includes

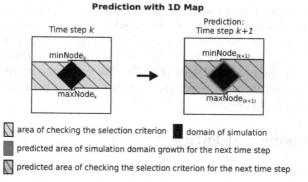

**Fig. 6.** Predicting domain of computation in successive time steps with 1D map

the whole grid. In this case, `minNode = 0` and `maxNode = grid_size`. Then, starting from the second time step ($k = 2$), the selection criterion is checked only within the area adjusted to the domain of simulation.

Predicting the computational domain plays a significant role in ensuring the efficient load balancing across computing resources (cores). In the basic version of the application, the selection criterion is checked for all nodes of the gird. It

```
int minTemp = minNode;
int maxTemp = maxNode;
#pragma omp parallel for \
    reduction(min:minTemp) reduction(max:maxTemp)
for(int i=minNode; i<maxNode; ++i) {
    Selection criterion
    minTemp = min(i, minTemp);
    maxTemp = max(i, maxTemp);
    if(boundary[i]) {
        Kernel K1
        Kernel K3
    }                          Primary
    else {                     computations
        Kernel K2
        Kernel K4
    }
    Kernel K5
}
Workload prediction: determining new values of
minNode and maxNode using minTemp and maxTemp
```

**Fig. 7.** General scheme of a single time step in solidification application with workload prediction using 1D map

leads to undesirable situations when only a part of cores perform primary computations, while the rest of cores are responsible only for checking the selection criterion. The usage of the algorithm for workload prediction permits resolving this problem. As a result, the selection criterion is not checked for the entire grid, but only for the predicted area (Fig. 7). This ensures a more efficient workload distribution across cores, since all cores will perform primary computations within the domain of simulation in successive time steps.

## 6   Conclusions and Future Works

The main challenge of this work is the performance optimization of the solidification application with the dynamic computational intensity. For this aim, we propose to modify the application code significantly using the *loop fusion* technique. The proposed approach permits us to execute all kernels of the application in a single nested loop, as well as reduce the number of conditional operators that have to be performed within a single time step.

The achieved performance results show that the proposed optimization method allows increasing the application performance for all tested configurations of computing resources. The highest performance gain is achieved for a single Intel Xeon CPU, where the new code yields the speedups of about 1.78 and 1.72 times against the basic version, respectively for grids of size $2000 \times 2000$ and $3000 \times 3000$. At the same time, the usage of the proposed method for a single Intel KNL accelerator permits to reduce the execution time of about 1.54 and 1.52 times, respectively.

The aim of our future work is to investigate the possibility of accelerating the studied application using the dynamic workload prediction and workload balancing. In particular, we are planning to incorporate the algorithm presented in this paper, which is based on 1D computational map, into the modified code which uses the *loop fusion* technique. Also, it is expected to develop another algorithm for workload prediction and load balancing which uses 2D computational map. This solution should allow adjusting the domain of computations to the domain of simulation more accurately.

**Acknowledgments.** This research was conducted with the financial support of the National Science Centre (Poland) under grants no. UMO-2017/26/D/ST6/00687. The authors are grateful to: (i) Intel Technology Poland and (ii) Czestochowa University of Technology (MICLAB project no. POIG.02.03.00.24-093/13) for granting access to HPC platforms.

## References

1. Adrian, H., Spiradek-Hahn, K.: The simulation of dendritic growth in Ni-Cu alloy using the phase field model. Arch. Mater. Sci. Eng. **40**(2), 89–93 (2009)
2. Benito, J.J., Ureñ, F., Gavete, L.: The generalized finite difference method. In: Àlvarez, M.P. (ed.) Leading-Edge Applied Mathematical Modeling Research, pp. 251–293. Nova Science Publishers, New York (2008)

3. Folch, R., Casademunt, J., Hernandez-Machado, A., Ramirez-Piscina, L.: Phase-field model for Hele-Shaw flows with arbitrary viscosity contrast. II. Numerical study. Phys. Rev. E **60**(2), 1734–1740 (1999)
4. Halbiniak, K., Wyrzykowski, R., Szustak, L., Olas, T.: Assessment of offload-based programming environments for hybrid CPU-MIC platforms in numerical modeling of solidification. Simul. Model. Pract. Theory **87**, 48–72 (2018)
5. Karma, A., Kessler, D., Levine, H.: Phase-field model of mode III dynamic fracture. Phys. Rev. Lett. **87**(4), 045501 (2001). https://doi.org/10.1103/PhysRevLett.87.045501
6. Kulawik, A.: The modeling of the phenomena of the heat treatment of the medium carbon steel. Wydawnictwo Politechnki Czestochowskiej, (281) (2013). (in Polish)
7. Provatas, N., Elder, K.: Phase-Field Methods in Materials Science and Engineering. Wiley, Weinheim (2010)
8. Shimokawabe, T., et al.: Peta-scale phase-field simulation for dendritic solidification on the TSUBAME 2.0 supercomputer. In: Proceedings of the 2011 ACM/IEEE International Conference on High Performance Computing, Networking, Storage and Analysis, SC 2011. IEEE Computer Society (2011). https://doi.org/10.1145/2063384.2063388
9. Steinbach, I.: Phase-field models in materials science. Model. Simul. Mater. Sci. Eng. **17**(7), 073001 (2009). https://doi.org/10.1088/0965-0393/17/7/073001
10. Szustak, L., Halbiniak, K., Kulawik, A., Wrobel, J., Gepner, P.: Toward parallel modeling of solidification based on the generalized finite difference method using Intel Xeon Phi. In: Wyrzykowski, R., Deelman, E., Dongarra, J., Karczewski, K., Kitowski, J., Wiatr, K. (eds.) PPAM 2015. LNCS, vol. 9573, pp. 411–422. Springer, Cham (2016). https://doi.org/10.1007/978-3-319-32149-3_39
11. Szustak, L., Rojek, K., Olas, T., Kuczynski, L., Halbiniak, K., Gepner, P.: Adaptation of MPDATA heterogeneous stencil computation to Intel Xeon Phi coprocessor. Sci. Prog. (2015). https://doi.org/10.1155/2015/642705
12. Szustak, L., Halbiniak, K., Kuczynski, L., Wrobel, J., Kulawik, A.: Porting and optimization of solidification application for CPU-MIC hybrid platforms. Int. J. High Perform. Comput. Appl. **32**(4), 523–539 (2018)
13. Takaki, T.: Phase-field modeling and simulations of dendrite growth. ISIJ Int. **54**(2), 437–444 (2014)
14. Warren, J.A., Boettinger, W.J.: Prediction of dendritic growth and microsegregation patterns in a binary alloy using the phase-field method. Acta Metall. Mater. **43**(2), 689–703 (1995)

# Parallel Non-numerical Algorithms

# SIMD-node Transformations
# for Non-blocking Data Structures

Joel Fuentes[1]([✉]), Wei-yu Chen[3], Guei-yuan Lueh[3], Arturo Garza[2],
and Isaac D. Scherson[2]

[1] Department of Computer Science and Information Technologies,
Universidad del Bío-Bío, Chillán, Chile
`jfuentes@ubiobio.cl`
[2] Department of Computer Science, University of California, Irvine, CA, USA
[3] Intel Corporation, Santa Clara, CA, USA

**Abstract.** Non-blocking data structures are commonly used in multi-threaded applications and their implementation is based on the use of atomic operations. New computing architectures have incorporated data-parallel processing through SIMD instructions on integrated GPUs, including in some cases support for atomic SIMD instructions. In this paper, a new framework is proposed, *SIMD-node Transformations*, to implement non-blocking data structures that exploit parallelism through multi-threaded and SIMD processing. We show how one- and multi-dimensional data structures can embrace SIMD processing by creating new data structures or transforming existing ones. The usefulness of this framework and the performance gains obtained when applying these transformations, is illustrated by means of SIMD-transformed skiplists, $k$-ary trees and multi-level lists.

**Keywords:** Non-blocking · Data structures · SIMD

## 1 Introduction

The advent of machines with many CPU cores has raised the need for efficient, scalable, and linearizable non-blocking data structures. New computing architectures, such as Intel's heterogeneous processors, consisting of CPU cores and an integrated GPU (iGPU) on the same die, have brought new processing capabilities such as SIMD and SIMT processing on the iGPU [10]. SIMD (Single Instruction, Multiple Data) is a term introduced by Flynn [2] and refers to a processor architecture that simultaneously executes a single instruction over multiple data items. However, to take advantage of this architecture is not straightforward as the programmer requires to manually manage the data in arrays or vectors (SIMD-friendly format), so the SIMD execution unit can act over them.

Introduced by Intel, C for Metal (CM) is a high-level programming framework based on C/C++ that provides an efficient interface to utilize the SIMD/SIMT capabilities of Intel's iGPU [9]. This programming environment was built and

© Springer Nature Switzerland AG 2020
R. Wyrzykowski et al. (Eds.): PPAM 2019, LNCS 12043, pp. 385–395, 2020.
https://doi.org/10.1007/978-3-030-43229-4_33

designed to efficiently exploit the SIMD capability of the iGPU through a syntax to facilitate data-parallel implementation. Vector operations defined at a high level by the user are converted into SIMD operations by the CM compiler, extracting the maximum possible parallelism for the underlying hardware.

There have been some proposals for using SIMD instructions to accelerate data structures and indices on CPU and GPU (discrete and integrated) [3,5]. The main focus is the use of SIMD instructions for traversals and search operations in different data structures and indices [11,13,14,16]. However, non-blocking and linearizable data structures with SIMD processing have not been explored yet.

In this paper, we propose a novel framework dubbed *SIMD-node Transformations* to implement non-blocking data structures using multi-threaded and SIMD processing. As a result, it is shown that one- and multi-dimensional data structures, such as skiplists, k-ary trees and multi-level lists, can embrace SIMD processing through these transformations and remain linearizable. Finally, we report important performance gains obtained when applying these transformations using CM and Intel's iGPU to a concrete lock-free skiplist, $k$-ary tree and multi-level list.

## 2    Non-blocking Data Structures with SIMD Processing

To correctly implement concurrent data structures on SIMD architectures, support of atomic operations is required. The goal of implementing concurrent or non-blocking data structures is two-fold: to achieve higher thread concurrency by using non-blocking techniques and to achieve data parallelism at the instruction level through SIMD processing.

Similarly to CPU programming, concurrent data structures on SIMD architectures can be classified as blocking and non-blocking. Additionally, as it is shown in this paper, non-blocking properties such as lock-free and wait-free can also be achieved on SIMD-based non-blocking data structures.

In terms of the design of SIMD-based non-blocking data structures, the organization of internal data items and nodes is the most important aspect to efficiently exploit the SIMD capabilities within concurrent operations. Therefore, to achieve correctness, non-blocking properties, together with data parallelism requires a new abstraction when implementing SIMD-based concurrent data structures from scratch or transforming existing ones.

Table 1 shows a classification of concurrent data structures based on their operation implementations using SISD (Single Instruction Single Data – classical CPU instructions) and SIMD. Atomic operations and specifically the compare-and-swap operation (CAS) are fundamental for implementing update operations in non-blocking data structures. SISD CAS is supported in most multi-core CPU and GPU architectures. However, SIMD CAS is not commonly supported on CPU architectures, nonetheless, modern iGPU on heterogeneous CPU/iGPU architectures do support it [10].

Search operations, that traverse data structures, are typically the most common operation in applications such as databases, search engines, file systems,

**Table 1.** Classification of concurrent data structures based on their implementation

| Operation | Implemented with | Concurrent behavior per thread | Suitable concurrent data structures |
|---|---|---|---|
| Search | SISD | Nodes or data items are traversed and processed one a time | Linked lists, queues, stacks |
|  | SIMD | Several nodes or data items are traversed and processed within an instruction. Improvement in performance is potentially high | $k$-ary trees, skiplists, multi-level hash tables, indices and succinct structures |
| Update | SISD CAS | Nodes or data items are updated one at a time | Linked lists, queues, stacks |
|  | SIMD CAS | Several nodes or data items are updated atomically if they are contiguous memory addresses | $k$-ary trees, skiplists, multi-level hash tables |

etc. So it is important to implement them efficiently. SIMD traversals are search operations where comparisons are performed on several data items within a single instruction [11]. For example, chunk-based nodes of a $k$-ary tree can contain several keys that serve as decision to continue the search to their children nodes and whose search is performed using data-parallel comparisons with SIMD instructions. Besides chunk-based trees, multi-dimensional data structures such as skiplists, multi-level hash tables, and index structures are good candidates to be implemented in such manner and take advantage of the SIMD processing. On the other hand, one-dimensional data structures such as linked lists, queues, and stacks are not well-suited for SIMD processing due to their operational nature, i.e. in some cases data items are obtained directly without search, or traversals are inherently sequential.

## 2.1 SIMD-node Transformations

In order to implement non-blocking data structures based on SIMD processing, a new abstraction is defined. This abstraction will serve as a guideline for designing new SIMD-friendly non-blocking data structures or transforming existing SISD non-blocking data structures to SIMD-friendly ones.

**Definition 1.** *Let* $\Theta_X$ *denote a super node encapsulating single nodes or data items of a data structure that can be processed by a SIMD instruction of size* $X$ *(SIMD width).* $\Theta_X$ *can be updated atomically by SIMD CAS operations in concurrent executions and its only restriction is to have the* $X$ *data items contiguously stored in memory.*

**Definition 2.** *Let* $S$ *be a one- or multi-dimensional data structure, then a* $\Theta$-*node transformed* $S^\Theta$ *can be obtained such that*

- *Every node in* $S$ *is in some* $\Theta_X$ *of* $S^\Theta$.

$$\forall r \in S, r \in S^\Theta \wedge |S| = \sum_{i=0}^{n} |\Theta_X^i|$$

*where* $|S|$ *is the number of single nodes or data items in* $S$.
- $S^\Theta$ *preserves the same* $S$'s *traversal structure and hence the complexity of all its operations.*
- $S^\Theta$ *can be traversed using the original traversal algorithms of* $S$ *(single node-based traversal), and new traversal algorithms using SIMD instructions (*$\Theta_X$-*based traversal).*

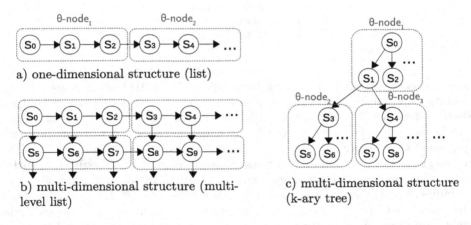

**Fig. 1.** $\Theta$-node transformation from a one-dimensional (a) and multi-dimensional (b and c) structures

Figure 1 depicts $\Theta$-node transformations from three different data structures. Figure 1a shows the layout of a simple linked list (one-dimensional structure) whose nodes have been encapsulated by two $\Theta_X$. Figure 1b and c illustrate the $\Theta$-node transformations of two multi-dimensional structures, a multi-level list and a $k$-ary tree. Notice that these examples illustrate pointer-based data structures, but $\Theta$-node transformations can also be applied on to other kinds of data structures, e.g. array structures, where several data items can be contained in $X$.

The manner that $\Theta$-node transformations are applied depends mainly on the organization of the data structure. In hierarchical data structures such as $k$-ary trees, $\Theta$-node transformations are to be performed in *parent-children* form, i.e. $\Theta_X$ encapsulates a parent and all its children. Therefore, the maximum size $X$, which is defined by hardware, is the most important factor when performing $\Theta$-node transformations; it defines whether the transformation in a hierarchical data structure is possible or not. In one-dimensional structures such as lists, queues, etc., the only restriction is to encapsulate more than one node or data item in $\Theta_X$.

**Theorem 1.** *A $\Theta$-node transformation, applied on a one- or multi-dimensional data structure $S$, produces $S^\Theta$ with the same traversal structure and complexity.*

*Proof.* By contradiction. Assume that the produced $S^\Theta$ has different traversal structure and complexity than the original $S$. Then, for this to happen, new pointers must have been added/removed between nodes or data items during the $\Theta$-node transformation. However, the previous statement leads to contradiction with Definition 2 which states that a $\Theta$-node transformation on one- or multi-dimensional structures only encapsulates contiguous nodes or data items and no new pointers are added and no existing pointers are modified outside each $\Theta_X$. □

The concept of *linearizability*, introduced in [8], is used to prove the correctness of a $\Theta$-node transformation from a concurrent data structure. Linearizability is a global property that ensures that when two processes each execute a series of method calls on a shared object there is a sequential ordering of these method calls that do not necessarily preserve the program order (the order in which the programmer wrote them down), but each method call does seem to happen instantly (i.e., invocation and response follow each other directly), whilst maintaining the result of each method call individually and consequently the object its state [7]. This definition of linearizability is equivalent to the following:

- All function calls have a linearization point at some instant between their invocation and their response.
- All functions appear to occur instantly at their linearization point, behaving as specified by the sequential definition.

*Conjecture 1.* A $\Theta$-node transformed data structure $S^\Theta$ of $S$ can be linearizable if $S$ is linearizable.

*Observation.* As any $\Theta_X$ can be updated atomically by Definition 1, $S^\Theta$ can be proved linearizable when all its functions hold the linearization properties in any $\Theta_X$ of $S^\Theta$ of $S$.

## 2.2  Restrictions and Considerations

The main restriction for implementing non-blocking data structures with SIMD processsing is the hardware support. Currently, most of the modern CPUs and GPUs support SIMD instructions, however not all of them support atomic SIMD instructions. There have been some proposals for adding multi-word CAS (CASN) support to CPUs [1,6,15], but their performance decreases considerably with high values of N.

The proposed framework also considers the availability of a memory allocator that is capable of allocating contiguous memory addresses for each SIMD node. Recall that SIMD-node transformations are based on data item encapsulations within contiguous memory addresses.

In terms of performance, defining the proper width of each SIMD node could impact the performance significantly. SIMD instructions might have many performance implications due to their width and interpretations from different compilers: under-utilization of the SIMD block, incorrect SIMD width defined by user, unmatched cache line size, and so on.

## 3  A Lock-Free Skiplist Based on $\Theta$-node Transformations

To illustrate the use of the $\Theta$-node transformations on a concrete data structure, a $\Theta$-node transformed lock-free skiplist is presented. The skiplist is a probabilistic data structure that is built upon the general idea of a linked list [12]. The skiplist uses probability to build subsequent layers of linked lists upon an original linked list, allowing search procedures to jump quickly into the desired list. It provides $O(log(n))$ complexity for all its operations. We capitalized on the fundamental of $\Theta$-node transformation to design a new skiplist, named CMSL, based on chunked lists. So instead of having nodes and individual updates when inserting or deleting keys, entire chunks are updated using SIMD atomic operations.

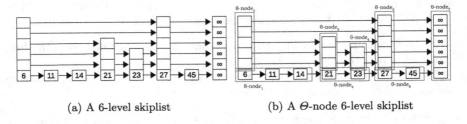

(a) A 6-level skiplist            (b) A $\Theta$-node 6-level skiplist

**Fig. 2.** Skiplists before and after $\Theta$-node transformation

Figure 2 illustrates the layout of the original and transformed skiplists. In Fig. 2b, upper (vertical) $\Theta_X$ nodes can be $\Theta_1$, $\Theta_8$, $\Theta_{16}$, which store the offsets to pointed lists. On the other side, every bottom (horizontal) $\Theta_X$ is a $\Theta_{16}$ which maintains up to 15 sorted keys, and has space reserved for a pointer to next $\Theta_X$. When more levels or keys are required to be store in a list, new $\Theta_X$ are attached. Recall that the skiplist's maximum level is directly related to the total number of keys that it can store while maintaining its logarithmic complexity. As the CMSL can keep up to 32 levels in its main list, only two $\Theta_{16}$ are enough to maintain up to $2^{32}$ sorted keys with logarithmic running times for search operations and $p = 1/2$.

CMSL provides three main operations: *searchKey*, *insertKey* and *deleteKey*. The implementation details and complete source code of the $\Theta$-based skiplist can be found in [4]. Following the original skiplist algorithm, the *searchKey*, which is the main function and called from all others, searches for the key's range first and then performs a linear search within the list found. For example, considering the skiplist structure from Fig. 2 as reference, the search operation is performed in two dimensions: *findList* performs the search within the chunk of levels horizontally, while *searchKey* performs the search within the chunks of keys vertically.

The updated operations, namely *insertKey* and *deleteKey*, are based on atomic operations on $\Theta_X$ nodes. To achieve lock-freedom, atomic updates are retried until they succeed. *deleteKey* also uses marks to label removed $\Theta_X$ nodes. The entire procedure is performed in two steps: $\Theta_X$ nodes are marked first, and then physically removed.

A detailed proof of correctness using linearization points establishes that CMSL guarantees the lock-freedom property for the insert and delete operations, and wait-free property for its search operation.

The proposed skiplist represents a set data structure. A key $k$ is in the set if there is a path from the first list to a chunk of keys that contains $k$, otherwise it is not in the set.

A valid skiplist has the following properties:

- The keys in the skiplist are sorted by their ascending order
- There are no duplicated keys
- Every key has a path to it from the first list (vertical list in Fig. 2a.)

The following fundamental lemma is derived from the properties of a valid skiplist:

**Lemma 1.** *For any list $l$, composed of one or more $\Theta_X$, key $k \in l$ and $0 < i \leq l.level$*

$$l[i].minKey \neq null \Rightarrow k < l[i].minKey$$

Where $l[i].minKey$ represents the minimum key of the pointed list from $l$ at level $i$.

*Proof.* In *insertKey* operation, it is ensured that new keys are inserted at the corresponding position (ascending order) in the $\Theta_X$ of keys. On the other hand, in *findList* procedure, it is ensured that always the right list $l$ is found such that $l.keys[0] \leq k < l[0].minKey$.

**Lemma 2.** *The successful searchKey and failed insertKey of an existing key operations on a valid skiplist are linearizable.*

*Proof.* First note that neither of these operations make modifications to the skiplist so they result in a valid skiplist. For these operations we must have a $\Theta_X$ with key $k$ at the point of linearization. The linearization point is when $\Theta_x$ is read in *searchKey* and in *insertKey*. The check in the following step ensures that this is the right $\Theta_X$ and it contains $k$. Finally, lemma 1 ensures that the skiplist is valid and it is the only list with a $\Theta_X$ containing key $k$.

**Lemma 3.** *The failed searchKey and failed delete operations on a valid list are linearizable.*

*Proof.* Firstly, note that neither of these operations make modifications to the skiplist so they result in a valid skiplist. We must have no key $k$ in the skiplist at the linearization point. After locating the list where the key could have been located (whose first key is smaller or equal than $k$), the linearization point is where the last $\Theta_X$ of keys is read in *searchKey* and *deleteKey*. Later, in both procedures, the entire $\Theta_X$ is compared with $k$ which must not have a key equal to $k$. Observe that even though $\Theta_X$ of levels maintains pointers to every next list, the *findList* operation always goes to the actual list that contains the chunk with the key. So in the case of a lists marked as deleted, they are discarded.

**Lemma 4.** *The successful insert operations on a valid list are linearizable.*

*Proof.* The precondition is that there is no key equal to $k$ in the skiplist. The post-condition is a valid skiplist with the key $k$ in some list's $\Theta_X$. There are two cases of linearization point. The first linearization point is the successful SIMD CAS operation inserting $k$ in a $\Theta$ of keys in *insertInChunk*, in this case there is available space to insert the new key in the chunk. The second case is when a $\Theta_X$ is full of keys or a new list is created due to a level greater than 0, here the linearization point is the successful SIMD CAS changing $\Theta_X$'s next pointer. Notice that for the case of creating a new list with level greater than 0, there is at least one more SIMD CAS operation on the previous list that updates the pointer to the new list. Even when the key was already inserted and the skiplist is still valid, the post-processing step to update upper levels is necessary to achieve the probabilistic assignment of levels on new lists.

**Lemma 5.** *The successful delete operations on a valid list are linearizable.*

*Proof.* This operation has a precondition that there exists a $\Theta_x$ of keys in the skiplist with key $k$. The post-condition is having the same $\Theta_X$ in the skiplist without the key $k$ or without the entire $\Theta_X$ if it only had the key $k$ at the time of linearization point. The linearization point is the successful SIMD CAS removing the key from $\Theta_X$ or removing the entire $\Theta_X$. The existence of $k$ for the precondition is ensured by the SIMD CAS itself (it would fail otherwise). The CAS ensures the post-condition of $\Theta_X$ without $k$ along with Lemma 2 and that the skiplist is still valid.

**Theorem 2.** *CMSL satisfies linearizability for the set abstraction.*

*Proof.* This follows from Lemmata 2, 3, 4 and 5.

Notice that other combinations of operations and their linearization points can be directly derived from the lemmata described above.

CMSL is lock-free. It only uses CAS operations (atomic SIMD1, SIMD8 and SIMD16) as a synchronization technique for update operations and no synchronization during search operations. The only way for an operation to not exit the main loop in *insertkey* and *deletekey* is to have its CAS interfere with another CAS executed by another thread on the same $\Theta_X$. Note that this guarantees that each time a CAS fails, another thread succeeds and thus the whole system always makes progress, satisfying the *lock-freedom* property.

## 3.1   Experimental Results

A set of experiments were carried out in order to evaluate the performance of the $\Theta_X$-based skiplist CMSL. An Intel's iGPU was used for running the experiments since it supports atomic SIMD instructions. CMSL's performance was compared with CPU state-of-the-art implementations.

Algorithms were implemented in CM for iGPU and in C++11 for CPU on a Intel(R) Core i7-8670HQ CPU, with a total of 4 physical cores and 8 logical processors running at 2.70 GHz. Also, this machine has an Intel(R) Iris Pro Graphics 580, dubbed GT4e version and Gen9 microarchitecture, with 128 MB of dedicated eDRAM memory and features 72 Execution Units (EUs) running at 350–1100 (Boost) MHz.

Use cases to test concurrent data structures are typically lists of combined concurrent operations distributed among all the threads. The set of operations was defined by ratios, i.e. the percentages that correspond to search, insert, and delete operations from the total amount of operations performed by a thread. These use cases were measured using CMSL and state-of-the-art concurrent Skiplist for CPU: *(a) Herlihy's*, update operations are optimistic, so it finds the node to update and then acquires the locks at all levels, validates the nodes, and performs the update. Searches simply traverse the multiple levels of lists; *(b) Herlihy-Optik's*, optimized version of Herlihy's skiplist using Optik patterns; *(c) Optik's*, skiplist implementation using Optik patterns; *(d) Pugh's*, optimized implementation of the first skiplist proposal by Pugh, it maintains several levels of lists where locks are used to perform update operations; *(e) Fraser's*, optimistically searches/parses the list and then does CAS at each level (for updates). A search/parse restarts if a marked element is met when switching levels. The same applies if a CAS fails.

Figure 3a shows the number of operations per second (in millions) that are achieved by different skiplists on CPU and our proposal on iGPU. The $x - axis$ depicts the ratio of [search, insert, delete] operations, i.e. the leftmost data item corresponds to insert-only operations, while the rightmost data item corresponds to search-only operations.

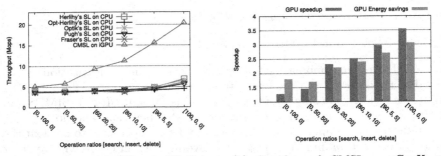

(a) Throughput of different Skiplists on GenX and CPU

(b) Speedup of CMSL on GenX over Fraser's skiplist for CPU

**Fig. 3.** Performance results of skiplists on GenX and CPU

Notice that the concurrent skiplists tested on CPU are mainly designed for many-core CPUs. They scale well for multiple threads (e.g. over 16). However, when testing up to 8-thread on a 4-core CPU their performance is not scalable. In contrast, CMSL scales well on the iGPU of the same 4-core processor.

CMSL achieves higher number of operations for all the scenarios, and its best performance is when all the operations are search with up to 3.1× speedup. When all the operations are updates, it still presents 1.2–1.5× speedup. We chose the Fraser's skiplist to perform an individual comparison in terms of GPU speedup and energy savings. Figure 3b illustrates the speedup of CMSL over Fraser's skiplist. CMSL achieves up to 3.5× speedup when all the operations are search and 1.3× speedup when all the operations are insert; the rest of mixture operations present speedups between these numbers. Similarly, it also shows the energy savings for all the experiments, achieving of up to 300% when all the operations were search.

It is noteworthy where the performance gains of CMSL come from. All the CPU implementations are based on single-node design; e.g, traversals on levels are done through node pointers one at a time, comparison operations within the bottom level are performed one by one, and so on. In contrast, CMSL performs traversals on levels through 8, 16, 24 or 32 levels at a time, and comparisons at the bottom levels are performed every 16 elements with one single SIMD operation.

## 4    Other Applications of $\Theta$-node Transformations

One- and multi-dimensional data structures, can benefit from new SIMD computing capabilities available in accelerators such as iGPUs by applying $\Theta$-node transformations. Additional non-blocking data structures such as $k$-ary trees, multi-level lists, and multi-level hash tables were implemented using the presented transformations. Similar to the skiplist CMSL, experimental results with these transformed data structures showed important performance gains when comparing with non-blocking implementations for CPU. Due to constrains in the length of this paper, their study is left for a follow-up extended paper.

# References

1. Feldman, S., Laborde, P., Dechev, D.: A wait-free multi-word compare-and-swap operation. Int. J. Parallel Program. **43**(4), 572–596 (2015)
2. Flynn, M.J.: Very high-speed computing systems. Proc. IEEE **54**(12), 1901–1909 (1966)
3. Fuentes, J.: Towards methods to exploit concurrent data structures on heterogeneous CPU/iGPU processors. Ph.D. thesis, UC Irvine (2019)
4. Fuentes, J., Chen, W.Y., Lueh, G.Y., Scherson, I.D.: A lock-free skiplist for integrated graphics processing units. In: 2019 IEEE International Parallel and Distributed Processing Symposium Workshops (IPDPSW), pp. 36–46. IEEE (2019)
5. Fuentes, J., Scherson, I.D.: Using integrated processor graphics to accelerate concurrent data and index structures (2018)
6. Harris, T.L., Fraser, K., Pratt, I.A.: A practical multi-word compare-and-swap operation. In: Malkhi, D. (ed.) DISC 2002. LNCS, vol. 2508, pp. 265–279. Springer, Heidelberg (2002). https://doi.org/10.1007/3-540-36108-1_18
7. Herlihy, M., Shavit, N.: The Art of Multiprocessor Programming. Morgan Kaufmann, Burlington (2011)
8. Herlihy, M.P., Wing, J.M.: Linearizability: a correctness condition for concurrent objects. ACM Trans. Program. Lang. Syst. (TOPLAS) **12**(3), 463–492 (1990)
9. Intel Corporation: C-for-metal compiler (2018). https://github.com/intel/cm-compiler
10. Junkins, S.: The compute architecture of Intel® Processor Graphics Gen9. Intel whitepaper v1 (2015)
11. Kim, C., et al.: Fast: fast architecture sensitive tree search on modern CPUs and GPUs. In: Proceedings of the 2010 ACM SIGMOD International Conference on Management of Data, pp. 339–350. ACM (2010)
12. Pugh, W.: Skip lists: a probabilistic alternative to balanced trees. Commun. ACM **33**(6), 668–676 (1990)
13. Ren, B., Agrawal, G., Larus, J.R., Mytkowicz, T., Poutanen, T., Schulte, W.: SIMD parallelization of applications that traverse irregular data structures. In: Proceedings of the 2013 IEEE/ACM International Symposium on Code Generation and Optimization (CGO), pp. 1–10. IEEE (2013)
14. Schlegel, B., Gemulla, R., Lehner, W.: K-ary search on modern processors. In: Proceedings of the Fifth International Workshop on Data Management on New Hardware, pp. 52–60. ACM (2009)
15. Sundell, H., Tsigas, P.: Lock-free and practical doubly linked list-based deques using single-word compare-and-swap. In: Higashino, T. (ed.) OPODIS 2004. LNCS, vol. 3544, pp. 240–255. Springer, Heidelberg (2005). https://doi.org/10.1007/11516798_18
16. Zeuch, S., Huber, F., Freytag, J.C.: Adapting tree structures for processing with SIMD instructions (2014)

# Stained Glass Image Generation Using Voronoi Diagram and Its GPU Acceleration

Hironobu Kobayashi, Yasuaki Ito$^{(\boxtimes)}$ ⓘ, and Koji Nakano ⓘ

Department of Information Engineering, Hiroshima University,
Kagamiyama 1-4-1, Higashi-Hiroshima, Hiroshima 7398527, Japan
{hironobu,yasuaki,nakano}@cs.hiroshima-u.ac.jp

**Abstract.** The main contribution of this work is to propose a stained glass image generation based on the Voronoi diagram. In this work, we use the Voronoi cells and edges of the Voronoi diagram as colored glasses and leads in the stained glass, respectively. To fit Voronoi cells to the original image, we use a local search technique. Using this technique, we can obtain a high quality stained glass image that well-represents an original image. However, considering the computing time, it is not pragmatic for most applications. Therefore, this paper also proposes a graphic processing unit (GPU) implementation for the stained glass image generation employing the local search to produce the stained glass images. Experimental result shows that the proposed GPU implementation on NVIDIA Tesla V100 attains a speed-up factor of 362 and 54 over the sequential and parallel CPU implementations, respectively.

**Keywords:** Stained glass image generation · Human visual system · GPU · Parallel processing

## 1 Introduction

*Stained glass* is colored glass used to form pictorial design by setting contrasting pieces in a lead framework. *Stained glass image generation* is known as one of the *non-photorealistic rendering* techniques which produce an image resembling artistic representation, such as oil-painting, tile art, and mosaic. Thus far, several studies of non-photorealistic rendering including stained glass image generation have been devoted [5]. These researches employ image segmentation [6,10,13,14] and Voronoi diagrams [4,7,8] to generate stained glass images. In this work, we focus on the stained glass image generation based on *the Voronoi diagram*. Let $d(p,q)$ denote the Euclidean distance of two points $p$ and $q$ in a plane. The Voronoi diagram of a set $P$ of points, called *Voronoi seeds*, in a plain is a partitioning of a plain into regions $V(p)$ ($p \in P$) called *Voronoi cells*, each of which consists of all points closet to a point $p(\in P)$ over all points in $P$, that is, $V(p) = \{q \in Q | d(p,q) \leq d(q',q)$ for all $q' \in P\}$, where $Q$ is a set of all points in a plane. The boundary of Voronoi cell $V(p)$ for each point $p$ is determined

© Springer Nature Switzerland AG 2020
R. Wyrzykowski et al. (Eds.): PPAM 2019, LNCS 12043, pp. 396–407, 2020.
https://doi.org/10.1007/978-3-030-43229-4_34

by line segments, each of which is a perpendicular bisector of $p$ and a neighbor point in $P$. We call a line segment representing the boundary of a Voronoi cell *a Voronoi edge*. In this work, we use the Voronoi cells and edges as colored glasses and leads in the stained glass, respectively. Figure 1(a) illustrates an example of Voronoi diagram.

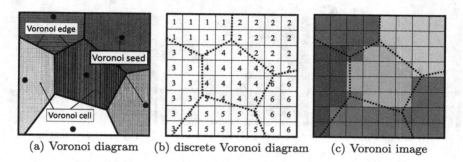

(a) Voronoi diagram    (b) discrete Voronoi diagram    (c) Voronoi image

**Fig. 1.** Voronoi diagram, discrete Voronoi diagram, and Voronoi image (Color figure online)

Using Voronoi diagrams, in [7], the method uses the adjustment of Voronoi cells by moving Voronoi seeds to one of the 8-connected pixels to decrease the error. To reduce the computing time, in spite of the race condition by the concurrent movement of Voronoi seeds, the adjustment for all Voronoi seeds is performed at the same time. After that, the adjustment for each Voronoi seed is sequentially performed one by one. Also, the size of Voronoi cells are almost the same without respect to the characteristic of local areas in the original image. In [8], a centroidal Voronoi diagram is used. By varying the size of Voronoi cells, local characteristics are well-represented. However, smooth long edges consisting of edges of multiple Voronoi cells cannot be represented. While, in [4], a weighted Voronoi diagram is used to generate images consisting of curve-shaped glasses.

The main contribution of this paper is to propose a method for generating a stained glass image which reproduces an input color image using the Voronoi diagram. Especially, we introduce the idea of the human visual system to generate high quality images. This idea has been utilized in the digital halftone [3,9]. Also, to represent fine and coarse regions, we vary the size of Voronoi cells. In addition, we employ a Voronoi seed adjustment in a similar way in [7]. The second contribution of this paper is to implement the above stained glass generation method on a *GPU* to accelerate the computation. GPUs (Graphics Processing Units) consisting of a lot of processing units, can be used for general-purpose computation. Our experimental results show that the GPU implementation can run up to 362 times faster than the sequential CPU implementation and 54 times faster than the parallel CPU implementation with 36 threads.

To compare other methods, we generated a stained glass image by GIMP [1]. The generated image by GIMP can be obtained in short time. However, all cell sizes are the same and border of objects are unclear. Also, detailed part such as eyes and hair cannot be represented well. On the other hand, the stained glass image obtained by the proposed method has clear borders and detailed parts are well represented by various size of cells (Fig. 2).

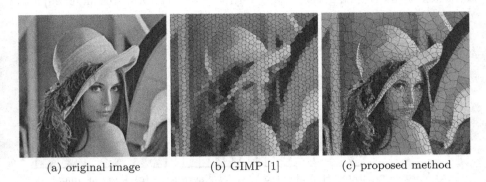

(a) original image          (b) GIMP [1]          (c) proposed method

**Fig. 2.** Comparison of stained glass image generarion between GIMP and the proposed method

This paper is organized as follows. Section 2 explains the proposed stained glass image generation based on the Voronoi diagram. We then show a GPU implementation to accelerate the computation in Sect. 3. Section 4 shows the performance evaluation of the GPU implementation compared with the CPU implementations. Section 5 concludes our work.

## 2   Proposed Stained Glass Image Generation

We first introduce *a discrete Voronoi diagram* and *a Voronoi image* using the Voronoi diagram. A Voronoi diagram is mapped to a discrete Voronoi diagram such that each Voronoi seed of the Voronoi diagram is located at the center of a pixel, and pixels in each Voronoi cell are assigned the seed ID that is a unique number assigned to each Voronoi seed (Fig. 1(b)). A Voronoi image is made from in a discrete Voronoi diagram. Pixels in each Voronoi cell are assigned to the same color. In following, a new stained glass image generation method is presented. Next, we define the goodness of a generated stained glass image.

### 2.1   The Goodness of a Generated Stained Glass Image Based on the Human Visual System

We introduce the error from an original image based on the human visual system. After that, we will show an algorithm of stained glass image generation.

A gray scale image is considered first, and then we will extend it to a color image. Consider an original image $A = (a_{i,j})$ of size $N \times N$, where $a_{i,j}$ denotes the intensity level at position $(i, j)$ $(1 \leq i, j \leq N)$ taking a real number in the range $[0, 1]$. The stained glass image generation is to find a Voronoi image $B = (b_{i,j})$

that reproduces the original image $A$. The goodness of the output image $B$ can be computed using the Gaussian filter that approximates the characteristic of the human visual system. Let $G = (g_{p,q})$ denote a Gaussian filter, that is, a two-dimensional symmetric matrix of size $(2w + 1) \times (2w + 1)$, where each non-negative real number $g_{p,q}$ $(-w \le p, q \le w)$ is determined by a two-dimensional Gaussian distribution such that their sum is 1. In other words, $g_{p,q} = s \cdot e^{-\frac{p^2+q^2}{2\sigma^2}}$, where $\sigma$ is a parameter of the Gaussian distribution and $s$ is a fixed real number satisfying $\sum_{-w \le p,q \le w} g_{p,q} = 1$. Let $R = (r_{i,j})$ be the projected gray scale image of a Voronoi image $B$ obtained by applying the Gaussian filter as follows:

$$r_{i,j} = \sum_{-w \le p,q \le w} g_{p,q} b_{i+p,j+q} \quad (1 \le i, j \le N).$$

As $\sum_{-w \le p,q \le w} g_{p,q} = 1$ and $g_{p,q}$ is non-negative, each $r_{i,j}$ takes a real number in the range $[0, 1]$. Hence, the projected image $R$ is a gray scale image. A Voronoi image $B$ is a good approximation of the original image $A$ if the difference between $A$ and $R$ is small enough. The error $e_{i,j}$ at each pixel point location $(i, j)$ of image $B$ is defined by

$$e_{i,j} = a_{i,j} - r_{i,j}, \tag{1}$$

and the total error is defined by

$$\text{Error}(A, B) = \sum_{1 \le i,j \le N} |e_{i,j}|. \tag{2}$$

Because the Gaussian filter approximates the characteristic of the human visual system, the Voronoi image $B$ reproduces the original image $A$ if $\text{Error}(A, B)$ is small enough.

Now, we extend the error computation for a gray scale image to a color image. In this work, we consider RGB colors whose value is specified with three real numbers in the range $[0, 1]$ that represents red, green, and blue. For color images, projected image $R$ and the error in Eq. (1) are computed for each color. Namely, for each color, Gaussian filter is applied and the error is computed. Let $e_{i,j}^{R}$, $e_{i,j}^{G}$, and $e_{i,j}^{B}$ denote the errors of red, green, and blue at each pixel position $(i, j)$, respectively. Equation (2) is extended to the sum of each color value as follows:

$$\text{Error}(A, B) = \sum_{1 \le i,j \le N} (|e_{i,j}^{R}| + |e_{i,j}^{G}| + |e_{i,j}^{B}|). \tag{3}$$

In the proposed stained glass image generation, the color of each glass is selected from one of the 4096 colors consisting of 16 colors of the red, green, and blue color components each.

## 2.2   Algorithm for Stained Glass Image Generation

The main purpose of this section is to propose a stained glass image generation algorithm. Briefly explaining, given an input image, the algorithm finds a Voronoi image that minimizes the total error as shown in the previous section.

The proposed algorithm consists of the following three steps:

**Step 1:** Generate an initial Voronoi image
**Step 2:** Merge neighboring Voronoi cells that have the same color
**Step 3:** Iterate adjustment of the Voronoi seeds with the local search

The resulting images in each step are shown in Fig. 3 and the details are shown in the following.

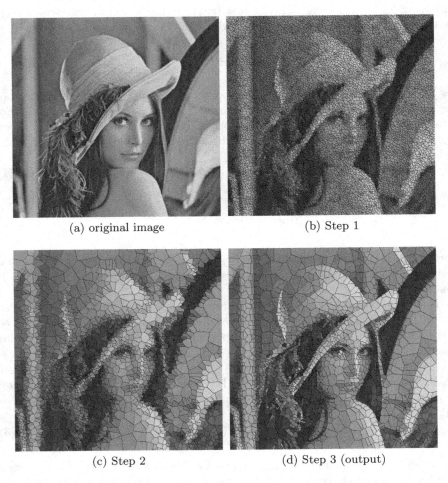

(a) original image           (b) Step 1

(c) Step 2           (d) Step 3 (output)

**Fig. 3.** Resulting images in each step for 'Lenna' of size $1024 \times 1024$

In Step 1, an initial Voronoi image to be a stained glass image is generated. First, Voronoi seeds are randomly distributed to a Voronoi image that is the same size as an input image such that each pixel in the image becomes a Voronoi seed with a probability $p$. We can control the size of Voronoi cells with $p$. If $p$ is higher, the cell size becomes smaller since the density of the Voronoi seeds is higher. According to the distribution of Voronoi seeds, the Voronoi image is generated such that pixels in each Voronoi cell are assigned a color that is close to pixels in corresponding to the Voronoi cell in the input image. Figure 3(b) shows an initial Voronoi image for the input original image Lenna [12] of size $1024 \times 1024$ in Fig. 3(a) with $p = \frac{1}{100}$. The Voronoi image has 10433 Voronoi cells. Figure 4 shows the comparison of initial Voronoi diagrams for various $p$. From the figure, when $p$ is large, the size of Voronoi cells is small (Fig. 4(a)). Therefore, to make the size appropriate, the merging procedure is iterated many times. On the other hand, when $p$ is small, the size of cells is large (Fig. 4(c)). Large cells cannot represent detailed parts. Thus, the probability $p$ should be set so as to make the size of cells appropriate.

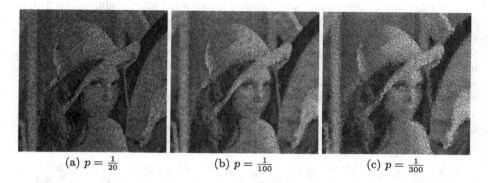

(a) $p = \frac{1}{20}$          (b) $p = \frac{1}{100}$          (c) $p = \frac{1}{300}$

**Fig. 4.** Comparison of initial Voronoi diagrams for various $p$

In Step 2, neighboring Voronoi cells that have the same color are merged into one Voronoi cell. Specifically, for each Voronoi cell, if the neighboring cells have the same color, the neighboring seeds are removed and the seed is moved to the centroid of them. After that, the Voronoi image is reconstructed and color of each cell is updated as illustrated in Fig. 5. The above merging process for all cells is repeatedly preformed several times. We can control the size of Voronoi cells with the times of the iteration. In this work, we repeat this merging process 5 times. Figure 3(c) shows the resulting Voronoi image in Step 2 in which the number of Voronoi cells is reduced to 2141. From the figure, at this time, outlines such as her shoulder and hat cannot be represented well.

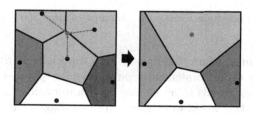

**Fig. 5.** Merging Voronoi cells

In Step 3, the Voronoi seeds are iteratively moved to minimize the total error in Eq. (3) by repeating the local search based on the greedy approach. In the local search, the movement of each Voronoi seed is limited to the adjacent 8-connected pixels and the Voronoi seed is moved to the neighboring pixel where the total error is minimized most. If the error cannot be reduced by the movement, the Voronoi seed is not moved. In Step 3, a round of the local search for all Voronoi seeds is repeated until no movement of Voronoi seeds is performed, that is, no more improvement is possible. Figure 3(d) shows the image obtained in Step 3. This image is the generated stained glass image in the proposed method. In the image, apparently, long straight lines and curves are well-represented. Also, to represent fine areas such as her hair and eyes, small cells are used. On the other hand, coarse areas in the background consist of large cells. To obtain this stained glass image, the above round was iterated 51 times. Figure 7 shows the graph of the number of moved Voronoi seeds and the total error in Step 3. According the graph, in the first several rounds, most of the Voronoi seeds are moved and the total error rapidly decreases. After round 40, almost Voronoi seeds are not moved and the total error does not change. Figures 6 shows generated stained glass images for different size that are obtained by the same parameters in the above except that to make up the number of Voronoi seeds nearly, we set $p = \frac{1}{25}$ and $\frac{1}{50}$ in Figs. 6(a) and (b), respectively. Although the resolution of these images is smaller than that of Fig. 3(d), outlines can be represented well.

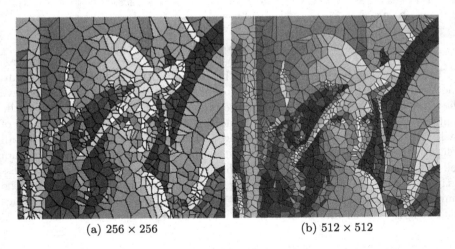

(a) $256 \times 256$         (b) $512 \times 512$

**Fig. 6.** Generated stained glass images for 'Lenna' of sizes $256 \times 256$ and $512 \times 512$

Since the computing time cost is very high especially in Step 3, in the next section, we show a GPU implementation employing parallel execution of the adjustment of Voronoi seeds to accelerate the computation.

We note that, in this work, we utilize a Voronoi diagram to segment an image into convex polygons that represent glasses. However, this work cannot limit it since our technique of Steps 2 and 3 can be applied to any image segmentation method if segmented images consist of convex polygons.

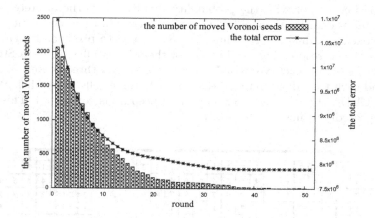

**Fig. 7.** The number of moved Voronoi cells and the total error in Step 3

## 3   GPU Implementation

This section shows a brief explanation about the CUDA on NVIDIA GPUs. Following it, we present the details of the proposed GPU implementation of the stained glass image generation algorithm shown in Sect. 2.2.

NVIDIA provides a parallel computing architecture called *CUDA* on NVIDIA GPUs. CUDA uses two types of memories: *the global memory* and *the shared memory* [11]. The global memory is implemented as an off-chip DRAM or an on-chip HBM2 of the GPU, and has large capacity. The shared memory is an extremely fast on-chip memory with lower capacity. CUDA parallel programming model has a hierarchy of thread groups called *grid*, *block*, and *thread*. A single grid is organized by multiple blocks, each of which has equal number of threads. The blocks are allocated to streaming multiprocessors such that all threads in a block are executed by the same streaming multiprocessor in parallel. All threads can access to the global memory. However, threads in a block can access to the shared memory of the streaming multiprocessor to which the block is allocated. Since blocks are arranged to multiple streaming multiprocessors, threads in different blocks cannot share data in the shared memories. CUDA C extends C language by allowing the programmer to define C functions, called *kernels*. By invoking a kernel, all blocks in the grid are allocated in streaming multiprocessors, and threads in each block are executed by processor cores in a single streaming multiprocessor.

We now show how we implement the algorithm on the GPU. We assume that an input original image of size $N \times N$ is stored in the global memory in advance, and the implementation writes the resulting stained glass image to the global memory. To implement Step 1, to distribute Voronoi seeds randomly, we assign threads to each pixel in which a thread determines that the pixel becomes a Voronoi seed or not with the probability $p$. After that, the discrete Voronoi diagram and the initial Voronoi image are generated. In these generation, a kernel is invoked, in which CUDA threads are assigned to pixels and each thread computes the corresponding pixel values of them in parallel. Also, in Step 2, a kernel is invoked for each Voronoi cells in which CUDA threads are assigned to pixels and each thread computes the merging Voronoi cells in parallel. We synchronize CUDA threads whenever merging procedure is completed. This kernel call is repeatedly 5 times in this work.

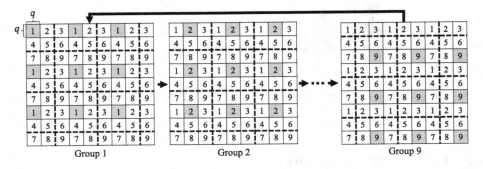

**Fig. 8.** Nine groups of subimages and parallel execution without the race condition.

In Step 3, Voronoi seeds are moved to one of the 8-connected pixels one by one. However, two or more Voronoi seeds cannot be moved at the same time due to the race condition. Therefore, it is difficult to implement the algorithm as parallel execution. To perform the computation in parallel, we split the input image into subimages of size $q \times q$ and partition subimages into nine groups Group 1, Group 2, . . . , Group 9 as illustrated in Fig. 8. Using these groups of subimages, we first perform the local search for Voronoi cells of size up to $3q \times 3q$ in parallel, and then the adjustment for the remaining Voronoi cells is performed one by one, as follows. We use $\frac{N^2}{9q^2}$ CUDA blocks, and perform the local search for the Voronoi seeds in all subimages of each group one by one. Each CUDA block first performs the local search only for Voronoi cells of which Voronoi cell does not stick out from the nine subimages of size $3q \times 3q$ separated by the broken lines (Fig. 9(a)). The remaining Voronoi seeds whose cells stick out as illustrated in Fig. 9(b) are recorded at this time. For example, in Fig. 8, to perform subimages in Group 1, we launch nine CUDA blocks, each of which performs the local search for the corresponding subimage of size $3q \times 3q$. Each CUDA block performs Step 3 for all Voronoi cells in the subimage.

Each round in Step 3 performs the local search for Group 1, Group 2, . . . , Group 9, in turn. A CUDA kernel that performs the local search for the corresponding group is invoked for each group, that is, the execution is synchronized whenever the computation of each round is finished. After that, the local search for the recorded Voronoi seeds is performed one by one. In the above local search, the input image and the Voronoi image stored in the global memory are frequently read during the computation. Therefore, to reduce the memory access time to them, we cache the elements of them to the shared memory. We note that in the above parallel computation, the reconstruction of the Voronoi diagram and the computation of the total error are performed only for the partial subimage to be updated by the movement of the seeds.

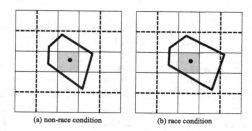

(a) non-race condition        (b) race condition

**Fig. 9.** Race condition of Voronoi cells for the parallel local search in each group

## 4   Performance Evaluation

In this section, we show the evaluation of the computing time of the proposed GPU implementation. We have used Lenna [12] in Fig. 3(a) of sizes $256 \times 256$, $512 \times 512$, and $1024 \times 1024$. In the evaluation, the Gaussian filter has been set with the parameters $\sigma = 1.3$. The probability of the distribution of Voronoi seeds in Step 1 is $p = \frac{1}{100}$. Also, the size of subimages in Step 3 of the GPU implementation is $32 \times 32$, that is $q = 32$. In order to evaluate the computing time, we have used NVIDIA Tesla V100, which has 5120 processing cores running on 1.455 GHz. For comparison purpose, Intel Core i9-7980XE with 18-cores running of 2.6 GHz has been used to evaluate the GPU implementation by CPU implementations. The processor has 18 physical cores each of which acts 2 logical cores by hyper-threading technology. Using this, a sequential CPU implementation and a parallel CPU implementation with multiple threads have been evaluated. The parallel CPU implementation has been implemented with 36 threads using OpenMP [2] and performs a similar manner with the GPU implementation to perform parallel execution. In the GPU implementation, we assign each CUDA block of threads to a subimage of size $3q \times 3q$. On the other hand, we assign each thread to a subimage of size $3q \times 3q$.

Table 1 shows the reduction of Voronoi seeds in Step 2 and the computing time for an input image stored in the main memory in the CPU until the stained glass image is stored to the main memory. In both CPU and GPU implementations, the computing time of Step 3 dominates the total computing time as shown in the table. The computation time of Step 1 and Step 2 is up to 1.2% of the total computation time. Therefore, the speed-up with the GPU is almost equivalent to accelerating the computation of Step 3. According to the table, the computing time of the GPU implementation is reduced by a factor of up to 362 over the sequential CPU implementation. On the other hand, the GPU implementation runs 5–54 times faster than the parallel CPU implementation. Since the performance of each core in the CPU is much higher than that in the GPU, the speed-up factor is small though 5120 cores are used in the GPU implementation. However, the proposed GPU implementation seems to be sufficiently fast.

**Table 1.** Computing time in seconds of the stained glass image generation for 'Lena' of sizes $256 \times 256$, $512 \times 512$, and $1024 \times 1024$

| Image size | # of seeds in Step 2 | # of rounds in Step 3 | CPU1 (1 thread) | CPU2 (36 threads) | GPU | $\frac{CPU1}{GPU}$ | $\frac{CPU2}{GPU}$ |
|---|---|---|---|---|---|---|---|
| $256 \times 256$ | $2675 \rightarrow 812$ | 15 | 360.46 | 118.98 | 21.96 | 16.41 | 5.42 |
| $512 \times 512$ | $4963 \rightarrow 1270$ | 27 | 2761.38 | 582.13 | 51.02 | 54.12 | 11.41 |
| $1024 \times 1024$ | $10433 \rightarrow 2141$ | 51 | 51232.45 | 7661.40 | 141.44 | 362.22 | 54.17 |

## 5   Conclusion

In this paper, we have proposed a stained glass image generation algorithm based on Voronoi diagram and its GPU acceleration. By introducing the error based on the human visual system, we can generate high quality stained glass images. In addition, we have shown an efficient parallel adjustment of Voronoi seeds on the GPU. We have implemented it on NVIDIA Tesla V100 GPU. The experimental result shows that our GPU implementation attains a speed-up factor of up to 362 and 54 over the sequential and parallel CPU implementation, respectively.

## References

1. GIMP - GNU image manipulation program. https://www.gimp.org/
2. OpenMP. http://www.openmp.org/
3. Analoui, M., Allebach, J.: Model-based halftoning by direct binary search. In: Proceedings of SPIE/IS&T Symposium on Electronic Imaging Science and Technology, San Jose, CA, USA, vol. 1666, pp. 96–108 (1992)
4. Ashlock, D., Karthikeyan, B., Bryden, K.M.: Non-photorealistic rendering of images as evolutionary stained glass. In: Proceedings of IEEE International Conference on Evolutionary Computation, pp. 2087–2094, July 2006
5. Battiato, S., di Blasi, G., Farinella, G.M., Gallo, G.: Digital mosaic frameworks - an overview. Comput. Graph. Forum **26**(4), 794–812 (2007)
6. Brooks, S.B.S.: Image-based stained glass. IEEE Trans. Vis. Comput. Graph. **12**(6), 1547–1558 (2006)
7. Dobashi, Y., Haga, T., Johan, H., Nishita, T.: A method for creating mosaic images using Voronoi diagrams. In: Proceedings of Eurographics 2002, pp. 341–348, September 2002

8. Faustino, G.M., de Figueiredo, L.H.: Simple adaptive mosaic effects. In: Proceedings of XVIII Brazilian Symposium on Computer Graphics and Image Processing (SIBGRAPI 2005), pp. 315–322 (2005)
9. Kouge, H., Honda, T., Fujita, T., Ito, Y., Nakano, K., Bordim, J.L.: Accelerating digital halftoning using the local exhaustive search on the GPU. Concurr. Comput. Pract. Exp. **29**(2), e3781 (2017)
10. Mould, D.: A stained glass image filter. In: Proceedings of the 14th Eurographics Workshop on Rendering, pp. 20–25 (2003)
11. NVIDIA Corporation: NVIDIA CUDA C programming guide version 9.1, January 2018
12. Po, L.M.: Lenna 97: A complete story of Lenna (2001). http://www.ee.cityu.edu.hk/~lmpo/lenna/Lenna97.html
13. Seo, S.H., Lee, H.C., Nah, H.C., Yoon, K.H.: Stained glass rendering with smooth tile boundary. In: Shi, Y., van Albada, G.D., Dongarra, J., Sloot, P.M.A. (eds.) ICCS 2007, Part II. LNCS, vol. 4488, pp. 162–165. Springer, Heidelberg (2007). https://doi.org/10.1007/978-3-540-72586-2_23
14. Setlur, V., Wilkinson, S.: Automatic stained glass rendering. In: Nishita, T., Peng, Q., Seidel, H.-P. (eds.) CGI 2006. LNCS, vol. 4035, pp. 682–691. Springer, Heidelberg (2006). https://doi.org/10.1007/11784203_66

# Modifying Queries Strategy for Graph-Based Speculative Query Execution for RDBMS

Anna Sasak-Okoń[✉] [iD]

University of Maria Curie Skłodowska in Lublin,
Pl. Marii-Curie skłodowskiej 5, 20-031 Lublin, Poland
anna.sasak@umcs.pl

**Abstract.** The paper relates to parallel speculative method that supports query execution in relational database systems. The speculative algorithm is based on a dynamic analysis of input query stream in databases serviced in SQLite. A middleware called the Speculative Layer is introduced, which based on a specific graph representation of query streams chooses the Speculative Queries to be executed. The paper briefly presents the structure of the Speculative Layer and graph modeling method. Then an extended version of speculative algorithm is presented which assumes an increased number of modifying queries in input query stream. Each modifying query present in the analysed query stream endangers already executed Speculative Queries with possibly invalid data and blocks their further use. We propose more sophisticated modifying queries analysis which aims in reducing the number of Speculative Queries which have to be deleted and thus decreases the necessary data manipulations. Experimental results are presented based on the proposed algorithms assessment using a real testbed database serviced in SQLite.

**Keywords:** Speculative query execution · Relational databases · Modifying queries

## 1 Introduction

Speculative execution is an optimization technique where a computer system performs some tasks in an anticipation of them being needed. Depending on speculation accuracy there may be gain or loss in application execution time [1]. Speculative instruction execution derives from branch prediction technology adopted in processor architecture in late 70-ties of the XIXth century [2,3]. The concept of speculative decomposition [4] was developed in the context of parallel processing, to provide more concurrency if extra resources are available. Speculative execution enables execution of several conditional code branches in parallel with computations which actually determine the branch choice. Good surveys of speculative execution are given in [5,6].

© Springer Nature Switzerland AG 2020
R. Wyrzykowski et al. (Eds.): PPAM 2019, LNCS 12043, pp. 408–418, 2020.
https://doi.org/10.1007/978-3-030-43229-4_35

Practical application of speculative execution for relational databases has gathered already a fairy numerous bibliography including [1, 7–13]. The common approach is an attempt to use the idle time of the DBMS functioning to perform some anticipated, potentially useful operations. These could be some data transformations or subqueries executed in advance and gathering data which could be useful for the final query. Also a notion of the speculative protocol has been introduced for database transactions. It enables a faster access to data locked in a transaction as soon as the blocking transaction had produced the resulting data images. Two speculative executions based on old and recent images are being performed. Finally, one of the speculative results is validated depending on the obtained blocking transaction real result.

The important fact is that none of the discussed above methods was targeting a cooperative definition of the speculative actions that would cover needs of many queries at a time. This feature together with a graph-oriented approach is essential for our method presented in this paper. The speculative execution model we proposed in our previous papers [14–16] performs additional speculative queries which are determined based on automatic analysis of the stream of queries dynamically arriving to a RDBMS. This analysis is done by an additional middleware layer between the user application and the relational database management system called the Speculative Layer. Based on an aggregated graph representation of current subset of user queries (Speculation Window) it generates speculative queries which are executed whenever resources are available. Data already gathered by the speculative queries can be used while executing original user queries, significantly shortening response time.

In our previous papers we introduced the Speculative Layer with the simplified speculation model which assumed the use of one speculative query for one input user query [14, 15]. Then we extended its functions by allowing the use of results of multiple speculative queries to support a single user query. In our previous work we assumed that data modifications are relatively rare and so the reaction to modifying queries was rather basic, simply removing endangered speculations. In this paper we propose more sophisticated algorithm of dealing with modifying queries, considering not only relations they are referring to but also attributes and conditions they use. With this new approach we managed to reduce the number of speculations which had to be deleted due to potential data modifications.

The proposed speculative support for RDBS has been implemented with C++ pthread library and SQLite 3.8.11.1. It has been positively verified by extensive tests performed using TPC benchmark generator [17] in the environment including 200000 products and 150000 customers.

## 2   A SQL DB with a Speculative Layer

In our previous papers [14–16], we have proposed a model of speculative query execution implemented as an additional middleware called the Speculative Layer, located between user database applications and the RDBMS. It analyses the current stream of arriving user queries and dynamically supports and speedups its execution. The general scheme of the RDBMS cooperating with the Speculative Layer is presented in Fig. 1.

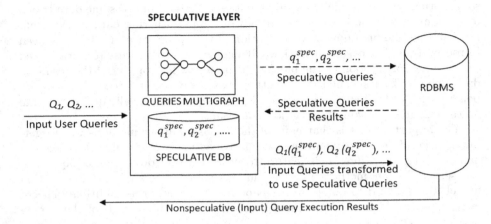

**Fig. 1.** The DBMS with an additional Speculative Layer analysing user queries.

The aim of the Speculative Layer is to identify some common constituent operations in user queries which, executed as speculative queries, can be used multiple times. In order to do that it creates the graph representation of sets of user queries called the Speculation Window. In each step that joint representation called Queries Multigraph is analysed to generate a set of speculative queries which will be executed in the RDBMS and stored in server main memory (RAM) as a subset of data called Speculative DB (Database). The data from the Speculative DB are used during execution of input user queries. They constitute ready-to-use working data and eliminate scanning larger numbers of records usually performed with the use of slow disk memory transactions as it would happen if the input queries were executed in a speculatively non-supported way. It improves system throughput and shortens users waiting time. For a detailed description of the Speculative Layer functions including the Speculative Graph Analysis and Queries Execution see [14,15].

## 3    Assumed Query Structure and Query Graphs Creation Rules

The Speculative Layer supports the CQAC (Conjunctive Queries With Arithmetic Comparisons) select queries. They are Selection-Projection-Join (SPJ) queries with one type of logical operator AND in WHERE clauses. Additionally there are two more operators allowed: IN for value sets and LIKE for string comparisons. Additionally, we allow all types of modifying queries id est. update, insert and delete.

Each CQAC query is represented by its Query Graph $G_Q(V_Q, E_Q)$ according to rules similar to these proposed in [18,19]. There are three types of Query Graph vertices: Relation, Attribute or Value. Query Graph edges represent functions performed by adjacent vertices in the represented query. Thus we have:

1. Membership Edges - $\mu$ - between a relation and each of its attributes from a SELECT clause,
2. Predicate Edges - $\theta$ - for each predicate of WHERE clause between two attributes or an attribute and a value,
3. Selection Edges - $\sigma$ - one for each predicate of WHERE clause between relation and attribute,
4. Delete - $\delta$, Insert - $\eta$ and Update - $v_S, v_W$ Edges - for each modifying query, between a relation and a modified attribute. In case of an insert query, $\eta$ edge is a loop that connects a relation vertex to itself. There are two types of update edges to distinguish between attribute vertices form update's SET and WHERE clauses.

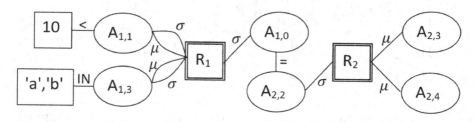

**Fig. 2.** A graph representation for a single SELECT query.

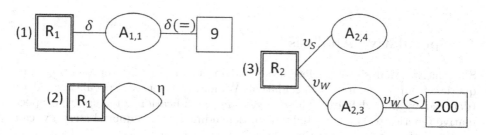

**Fig. 3.** Graph representations of the following modifying queries: (1)

Figure 2 presents graph representation of the following query: SELECT $A_{1,1}$, $A_{1,3}$, $A_{2,3}$, $A_{2,4}$ FROM $R_1, R_2$ WHERE $A_{1_0} = A_{2_2}$ AND $A_{1,1} < 10$ AND $A_{1,3}$ IN $('a', 'b')$. Figure 3 shows separate graph representations of three modifying queries: DELETE FROM $R_1$ WHERE $A_{1_1} = 9$ **(2)** INSERT INTO $R_1$ VALUES (1,'xx','b') **(3)** UPDATE $R_2$ SET $A_{2,4} = $ 'yy' WHERE $A_{2,3} < 200$.

To represent a set of queries by a single graph some additional rules have been defined. Such graph $G_S(V_S, E_S)$ will be called a Queries Multigraph or $Q_M$. $Q_M$ vertices set is an union of vertices of all component query graphs: $V_s = V_{Q_1} \cup V_{Q_2} \cup ... \cup V_{Q_n}$. $Q_M$ edges set is a multiset of all component query graph edges: $E_s = E_{q1} + E_{q2} + ... + E_{qn}$. This way multiple edges of the same type are allowed, raising the issue of some edge grouping. Figure 4 shows the $Q_M$ representing four user queries: First query is the one presented on Fig. 2, the second one is a modifying (DELETE) query from Fig. 3, third and fourth are SELECT queries also referring to the $R_1$ relation.

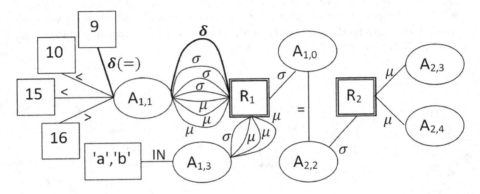

**Fig. 4.** A query multigraf of four user queries: (1) SELECT $A_{1,1}$, $A_{1,3}$, $A_{2,3}$, $A_{2,4}$ FROM $R_1, R_2$ WHERE $A_{1_0} = A_{2_2}$ AND $A_{1,1} < 10$ AND $A_{1,3}$ IN $('a','b')$. (2) DELETE FROM $R_1$ WHERE $A_{1_1} = 9$ (3) SELECT $A_{1,1}$, $A_{1,3}$, FROM $R_1$ WHERE $A_{1,1} < 15$ (4) SELECT $A_{1,1}$, $A_{1,3}$, FROM $R_1$ WHERE $A_{1,1} > 16$.

## 4   Speculative Analysis

Speculative Analysis is to determine and insert in the QM representing user queries from the current Speculation WIndow, a set of Speculative edges (denoted by dotted lines). These edges are an indication for respective Speculative Queries to be generated which, depending on the adopted strategy, can be one of three types of queries/edges that we introduce:

- Speculative Parameter Queries - the inserted edges mark selected nested queries. If a nested query has been marked to become a Speculative Query, it is possible to use its results as a parameter in its parent query.
- Speculative Data Queries - the aim of these speculative queries is to obtain and save in the Speculative DB a specific subset of records or/and attributes of relation. The main goal is to choose this subset so as it could be used in

the execution of as many input queries as possible. The starting point for the process of inserting Speculative Data Edges are value vertices which are then used to create a WHERE clause of the Speculative Data Query.
- Speculative State - the inserted edges relate to modifying queries in the Speculation Window. The presence of a modifying query means that both already executed and awaiting Speculative Queries are in danger of processing invalid data. The strategy of dealing with modifying queries to minimize the number of Speculative Queries which have to be deleted due to invalid data is the subject of interest of this paper with the broader description in the following section. The detailed algorithms for inserting speculative edges into QM and generating speculative queries are included in paper [15]. Figure 5 shows a piece of the QM from Fig. 4 with speculative edges representing on speculative query: SELECT $A_{1,1}$ FROM $R_1$ WHERE $A_{1,1} < 15$ which is also marked as Speculative State. Assuming that we manage to positively verify data state after execution of modifying query, this speculation can be used by two of the queries represented by an analysed QM.

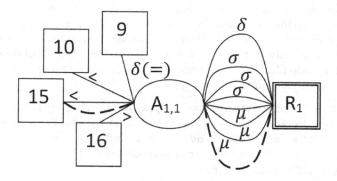

**Fig. 5.** QM with speculative data edges representing one speculative query

## 5 The Speculative Algorithm with a New Strategy for Modifying Queries

In our previous papers [14–16] we proposed and then upgraded for multiple speculation use a dual-speculation type algorithm. For each Speculation Window, based on representing it QM, 2 Speculative Queries (called Speculative Queries in Advance) are generated with the following features: (1) The first Speculative Query in Advance corresponds to the highest number of input queries which could utilize its results. (2) The second query has the highest row reduction count for all speculative queries generated for the analysed Speculation Window.

If there are idle threads they are used to execute a different type of Speculative Queries called Speculative Queries on Demand. These queries are generated only when a worker thread sends such a no-job request and are based on the analysis of the history of previously executed queries with the following rules: (1) The Speculative Query on Demand is generated only for one, the most frequently occurring in the History Structure attribute, (2) For each attribute in the History Structure, there can be only one Speculative Query on Demand stored in the Speculative DB, (3) If there already exist Speculative Queries on Demand for all attributes in the History Structure, the new one is generated for the attribute with the highest occurrence rise since the previous Speculative Query on Demand has been executed. For each relation that occurs in an input query the speculative algorithm tries to find an executed speculative query of any type whose results could be used.

Previously, we assumed that data modifications are relatively rare and thus adopted a very basic strategy for modifying queries which simply removed all Speculative Queries endangered with data modification. In this paper, the goal is to minimize the number of deleted Speculations reducing number of unnecessary data manipulations.

Thus, depending on the modification type we verify which relation and attributes its referring to. When Executed Speculative query consists of enough attributes we execute modifying query both on main speculative databases and delete speculation only if it cannot be done due to missing attributes. The following pseudo code illustrates the process of Speculative Query Verification due to executed modifying query:

```
VerifyAfterModification(Q_mod={R,[A_where],[A_set]}){
//R is a Relation, A_where, A_set - attributes from the where
//and set clauses if they occur in a modifying query
foreach ( Q_s ∈ ExecutedSpeculativeQueries ){
  if (Q_s.InDanger == Q_mod.Id){
    if (Q_mod.type == INSERT) toBeSaved = 1;
    else if (Q_mod.type == DELETE){
      if (A_where ∈ Q_s.SELECTclause) toBeSaved = 1;
      else toBeSaved = 0;}
    else if (Q_mod.type == UPDATE) {
      if (A_set ∈ Q_s.SELECTclause){
        if (A_where ∈ Q_s.SELECTclause)  toBeSaved = 1;
        else toBeSaved = 0;}
      else toBeSaved = 1;}
    if (toBeSaved == 1){
      Execute(MainDB, Q_mod); Execute(SpecDB, Q_mod);
      Q_s.RemoveDanger(); }
    else RemoveSpeculation (Q_s);
}} return;}
```

# 6   Test Environment and Queries

The Speculative Layer was implemented with C++ and Visual Studio 2013 with Pthread library and SQLite 3.8.11.1. Experimental results were obtained under Windows 8.1 64b with Intel Core i7-3930K processor and 8GB RAM. For the experiments we used the database structure and data (8 relations, 1GB of data) from the TPC benchmark described in [17]. However, a new set of 12 Query Templates was prepared and used to generate test sets of input queries. Each experiment is conducted three times on three different sets of 1000 input queries generated for the fixed templates densities.

Templates T1 - T6 are for select clauses representing different search scenarios which as a consequence generates more diverse Speculative Queries in the Speculative Layer. Templates T1 and T2 are both referring to the biggest relation in database LINEITEM but with more narrow (T1) or more general (T2) search criteria. T3 and T4 joins two relations one of which is small and also a fixed set of values is used for one of the attributes in a WHERE clause. T5 and T6 are bot referring to the second biggest relation ORDERS simulating the narrow product search.

Templates T7 - T12 are used to create modifying queries. T7 is an example of INSERT query, T8 - T10 UPDATE on Lineitem, T11, T12 DELETE on Orders. Each template has a fixed structure. If there are values used in a WHERE or SET clauses, those are proper random values for the attribute they are referring to. We have used as much as 6 modifying templates to properly test each condition branch from the algorithm presented in the previous section.

First set of generated and tested queries are SELECT only queries based on templates T1 - T6 with equal occurrence densities. Then, we started to add modifying queries based on templates T7 - T12 starting with 1% density each. So the second set of test queries consists of 94% of SELECT queries and 6% of modifying queries. Next in each test we increase the densities of modifying queries by 1% each and thus we get 88%-SELECT, 12%-modifying then 76%-SELECT, 18%-modifying and so on.

# 7   Experimental Results - Modifying Queries

In our previous papers we have presented the results of experiments in which we have determined the size of the Speculation Window (i.e. the number of input queries represented by QM), and the number of worker threads (i.e. the number of threads executing the Speculative Queries). Thus, the following set of experiments were conducted for the Speculation Window = 5 and the number of worker threads = 3. All presented results are the average values obtained for the execution of 3 input query sets with fixed templates densities for each experiment. We have conducted e series of six experiments with the density of modifying queries varying from 6% to 36% (we mark those test with labels MOD6, MOD12, ... MOD36 where the number 6, 12 and so on, states the percent of modifying queries in the test queries set). Additionally, in each experiment we compare two executions: first - using old (basic) modifying approach and second - using new (extended) strategy.

Figure 6 presents the number of executed Speculative Queries which had to be deleted due to the data modifications. We can see that, for the old strategy, the number of deleted speculative queries is significantly growing with the growing density of modifying queries in test queries sets. On the other hand, Fig. 7 shows that by adopting the new strategy, we managed to save a considerable number of Speculative Queries from being deleted and the number of those that were deleted doesn't grow so rapidly with growing data modifications.

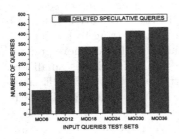

**Fig. 6.** The average number of deleted speculative queries for each input query set for old modifying queries strategy.

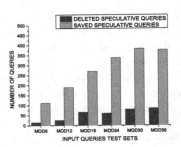

**Fig. 7.** The average number of deleted and saved speculative queries for each input query set for new modifying queries strategy.

**Table 1.** Characteristic values improvement for the new modifying queries strategy.

| The % reduction of number of executed Speculative Queries | | | | | |
|---|---|---|---|---|---|
| MOD6 | MOD12 | MOD18 | MOD24 | MOD30 | MOD36 |
| 1,07% | 3,27% | 6,41% | 6,77% | 9,92% | 10,74% |

| The % improvement o number of input test queries executed with Speculative Queries results | | | | | |
|---|---|---|---|---|---|
| MOD6 | MOD12 | MOD18 | MOD24 | MOD30 | MOD36 |
| 2,02% | 2,96% | 4,64% | 5,76% | 7,68% | 9,22% |

| The % reduction of average execution time if input test queries | | | | | |
|---|---|---|---|---|---|
| MOD6 | MOD12 | MOD18 | MOD24 | MOD30 | MOD36 |
| 5,78% | 8,05% | 4,58% | 11,31% | 12,6% | 21,86% |

| The % of Speculative Queries saved from being deleted | | | | | |
|---|---|---|---|---|---|
| MOD6 | MOD12 | MOD18 | MOD24 | MOD30 | MOD36 |
| 88,19% | 87,9% | 80,36% | 84,92% | 87,72% | 81,6% |

Table 1 presents what other benefits we managed to obtain with improving our strategy for modifying queries. Each presented value is a difference between obtained with old and new strategy. As we have saved (ILE) of Speculative Queries from being deleted the total number of executed Speculative Queries has dropped from 1,07% to 10,74%. On the same time the number of input test queries executed with the Speculative Query result improved, from 2,01% to 9,22%, also reducing the average execution time of input queries, from 5,78% to even 21,86%.

## 8    Conclusion

This paper has presented the model of speculative query execution support to sppedup SQL query execution in relational databases. The Speculative Layer, based on joint graph modelling and analysis of groups of queries executes speculative queries. However, modifying queries occurring in user queries stream, endangers executed Speculative Queries with invalid data. In this paper we have concentrated on exploring the possibility to save Speculative Queries from deleting by developing more complex analysis strategy for modifying queries. We have tested the proposed solution using a practically implemented testbed and real large experimental benchmark.

Experimental results for the test database and six different group of queries are satisfactory. For all sets we have prevented we saved on average 84,28% of Speculative Queries endangered by possible data modification. Due to that the total number of executed Speculative queries was reduced on average by 6,36% while improving the number of input queries executed with use of speculative query (5,38%) an reducing total average query execution time by 10,69% avg.

Further work on the described project will concentrate on developing more sophisticated ways of combining multiple speculative queries results, for example by joining fragments of speculative queries results to obtain one set appropriate for an input query and also combining it with new developed strategy for modifying queries.

## References

1. Hristidis, V., Papakonstantinou, Y.: Algorithms and applications for answering ranked queries using ranked views. VLDB J. **13**(1), 49–70 (2004)
2. Liles Jr., E.A., Wilner, B.: Branch prediction mechanism. IBM Tech. Discl. Bull. **22**(7), 3013–3016 (1979)
3. Smith, J.E.: A study of branch prediction strategies. In: ISCA Conference Proceedings, New York, pp. 135–148 (1981)
4. Grama, A., Gupta, A., Karypis, G., Kumar, V.: Introduction to Parallel Computing, 2nd edn. Addison Wesley, Harlow (2003)
5. Kaeli, D., Yew, P.: Speculative Execution in High Performance Computer Architectures. Chapman Hall/CRC, Boca Raton (2005)
6. Padua, D.: Encyclopedia of Parallel Computing A-D. Springer, Boston (2011)
7. Polyzotis, N., Ioannidis, Y.: Speculative query processing. In: CIDR Conference Proceedings, Asilomar, pp. 1–12 (2003)

8. Karp, R.M., Miller, R.E., Winograd, S.: The organization of computations for uniform recurrence equations. J. ACM **14**(3), 563–590 (1967)
9. Barish, G., Knoblock, C.A.: Speculative plan execution for information gathering. Artif. Intell. **172**(4–5), 413–453 (2008)
10. Barish, G., Knoblock, C.A.: Speculative execution for information gathering plans. In: AIPS Conference Proceedings, Toulouse, pp. 184–193 (2002)
11. Reddy, P.K., Kitsuregawa, M.: Speculative locking protocols to improve performance for distributed database systems. IEEE Trans. Knowl. Data Eng. **16**(2), 154–169 (2004)
12. Ragunathan, T., Krishna, R.P.: Performance Enhancement of Read-only Transactions Using Speculative Locking Protocol. IRISS, Hyderabad (2007)
13. Ragunathan, T., Krishna, R.P.: Improving the performance of read-only transactions through asynchronous speculation. In: SpringSim Conference Proceedings, Ottawa, pp. 467–474 (2008)
14. Sasak-Okoń, A.: Speculative query execution in relational databases with graph modelling. In: Proceedings of the FEDCSIS, ACSIS, vol. 8, pp. 1383–1387 (2016)
15. Sasak-Okoń, A., Tudruj, M.: Graph-Based speculative query execution in relational data-bases. In: ISPDC 2017, July 2017. Innsbruck, Austria, CPS. IEEE Explore (2017)
16. Sasak-Okoń, A., Tudruj, M.: Graph-based speculative query execution for RDBMS. In: Wyrzykowski, R., Dongarra, J., Deelman, E., Karczewski, K. (eds.) PPAM 2017, Part I. LNCS, vol. 10777, pp. 303–313. Springer, Cham (2018). https://doi.org/10.1007/978-3-319-78024-5_27
17. TPC benchmarks (2015). http://www.tpc.org/tpch/default.asp
18. Koutrika, G., Simitsis, A., Ioannidis, Y.: Conversational Databases: Explaining Structured Queries to Users, Technical Report Stanford InfoLab (2009)
19. Koutrika, G., Simitsis, A., Ioannidis, Y.: Explaining structured queries in natural language. In: ICDE Conference Proceedings, Long Beach, pp. 333–344 (2010)

# Soft Computing with Applications

# Accelerating GPU-based Evolutionary Induction of Decision Trees - Fitness Evaluation Reuse

Krzysztof Jurczuk[(✉)] [iD], Marcin Czajkowski[iD], and Marek Kretowski[iD]

Faculty of Computer Science, Bialystok University of Technology,
Wiejska 45a, 15-351 Bialystok, Poland
{k.jurczuk,m.czajkowski,m.kretowski}@pb.edu.pl

**Abstract.** The rapid development of new technologies and parallel frameworks is a chance to overcome barriers of slow evolutionary induction of decision trees (DTs). This global approach, that searches for the tree structure and tests simultaneously, is an emerging alternative to greedy top-down solutions. However, in order to be efficiently applied to big data mining, both technological and algorithmic possibilities need to be fully exploited. This paper shows how by reusing information from previously evaluated individuals, we can accelerate GPU-based evolutionary induction of DTs on large-scale datasets even further. Noting that some of the trees or their parts may reappear during the evolutionary search, we have created a so-called repository of trees (split between GPU and CPU). Experimental evaluation is carried out on the existing Global Decision Tree system where the fitness calculations are delegated to the GPU, while the core evolution is run sequentially on the CPU. Results demonstrate that reusing information about trees from the repository (classification errors, objects' locations, etc.) can accelerate the original GPU-based solution. It is especially visible on large-scale data where the cost of the trees evaluation exceeds the cost of storing and exploring the repository.

**Keywords:** Evolutionary algorithms · Decision trees · Big data mining · Graphics processing unit (GPU) · CUDA

## 1 Introduction

Decision trees (DTs) [9] are one of the most useful supervised learning methods for classification. During over 50 years of their applications [12], they were mainly induced using greedy heuristics like a top-down approach [16]. Involving evolutionary algorithms (EAs) into the trees induction [2] was a breath of fresh air. However, evolutionary tree induction is much more computationally demanding, and there were raised many questions of the time efficiency of such an approach. Nevertheless, evolutionary induced DTs have managed to gain in popularity for less demanding data mining problems for which the generation

© Springer Nature Switzerland AG 2020
R. Wyrzykowski et al. (Eds.): PPAM 2019, LNCS 12043, pp. 421–431, 2020.
https://doi.org/10.1007/978-3-030-43229-4_36

time of the prediction model was not crucial. Their main advantage is the global approach in which a tree structure, tests in internal nodes and predictions in leaves are searched simultaneously [10]. As a result, the generated trees are significantly simpler and at least as accurate as the greedy alternatives that are based on the classical divide and conquer schema.

Rapid development of new computational technologies and parallel frameworks is a chance to conquer the barriers concerning slow evolutionary DT induction [7]. To make it work, proposed new solutions need to be efficient for large-scale data and run on relatively cheap and generally available hardware/software. For these reasons, we have concentrated on using graphics processing units (GPU)s to parallelize and speed up the induction.

This paper focuses on accelerating a GPU-based evolutionary induction of classification trees [8] for the large-scale data. In the approach, the main evolutionary loop (selection, genetic operators, etc.) is performed sequentially on a CPU, while the most time-consuming operations like fitness calculation are delegated to a GPU. Noting that some of the trees or their parts may reappear during the evolutionary search, we examine if it is worth to store and reuse this information. In this paper, we introduce a concept of a repository of previously evaluated DTs in order to limit fitness recalculation of the new ones founded by EA. The search of the same (or similar) individuals is performed fully on the GPU where they are stored as a part of the repository. The second part of the repository, which gathers the corresponding fitness results, is located on the CPU in order to limit CPU/GPU memory transfers. The reuse strategy is not new in EA; however, it was studied in the context of improving the genetic diversity [3] (e.g., by chromosome revisiting) and it was not applied to DTs induction.

The next section provides a brief background on DTs and the Global Decision Tree (GDT) system which is used to test the proposed solution. Section 3 presents our approach, and Sect. 4 shows its experimental validation. In the last section, the paper is concluded and possible future work is outlined.

## 2    Evolutionary Induction of Decision Trees

Decision trees [9] (DTs) have a knowledge representation structure that is built of nodes and branches, where each internal node holds a test on one or more attributes; each branch represents the outcome of a test; and each leaf (terminal node) holds a prediction. Such a hierarchical tree structure, where appropriate tests from consecutive nodes are sequentially applied, closely resembles the human way of making decisions. The success of tree-based approaches can be explained by their ease of application, fast operation, and effectiveness. Nevertheless, the execution times and resource utilization still require improvement to meet ever-growing computational demands.

Evolutionary algorithms [14] (EAs) belong to a family of meta-heuristic methods. They represent techniques for solving a wide variety of difficult optimization problems. The framework of EA is inspired by biological mechanisms of evolution. The algorithm operates on individuals that compose a current population. Each individual represents a candidate solution to the target problem.

Individuals are assessed using a fitness function that measures their performance. Next, individuals with higher fitness usually have a higher probability of being selected for reproduction. Genetic operators such as mutation and crossover influence individuals, thereby producing new offspring(s). This guided random search (the offspring usually inherits some traits from its ancestors) is stopped when some convergence criteria is satisfied.

Typically, DTs are induced by a top-down approach [16] that realises the classical divide and conquer schema. The main consequences of locally optimal choices made in each tree node during the induction are overgrown and often less stable DT classifiers. Emerging alternatives to the top-down solutions include primarily EAs. Their global approach limits the negative effects of locally optimal decisions but it is much more computationally demanding [2, 6].

## 2.1 Parallelization of DTs Induction

Fortunately, EAs are naturally prone to parallelism and the artificial evolution can be implemented in various ways [4]. There are at least three main strategies that have been studied to parallelize EAs: master-slave model, island model and cellular model. Since EAs work on a set of independent solutions, it is relatively easy to distribute the computational load among multiple processors through a data or/and population decomposition approach. Recent research on the parallelization of various evolutionary computation methods has seemed to focus on GPUs as the implementation platform [10, 18]. The popularity of GPUs results from their general availability, relatively low cost and high computational power.

In the context of parallelization of DTs induction, not enough has been done and the GPGPU topic has hardly been studied in the literature at all. There are several GPU-based parallelizations but they consider either greedy inducers [11] or random forests [13]. As for evolutionary induced DTs, we found only a few papers that cover parallel extensions of the Global Decision Tree (GDT) data mining system: using MPI/OpenMP [5], GPU [8] and Spark [15]. One of the possible reasons why this topic has not yet been adequately explored by other systems is that the straightforward EA parallelization of DT induction may be insufficient. In order to achieve high speedup and exploit the full potential of e.g., GPU parallelization, there is a strong need to incorporate knowledge about DT specificity and its evolutionary induction.

## 2.2 Global Decision Tree System

This paper uses the GDT system [10] which supports various decision trees, and it has been already studied in terms of parallelism. Another benefit of the GDT system is that its scheme follows a typical EA framework [14] with an unstructured, fixed size population and a generational selection. In this study, we have deliberately limited the GDT description to a univariate binary classification tree version to facilitate understanding and to eliminate less important details.

Individuals in the population are not specially encoded and are represented and processed in their actual form as univariate classification trees. This is the most popular and basic variant of DTs that uses typical inequality tests with two outcomes in the internal nodes. Initialization is performed with a simplified top-down manner, which is applied to randomly selected small sub-samples of the learning data. Tests in the internal nodes are created by randomly selecting two objects from different classes (mixed dipole) that are located in the considered node [10]. Then, an effective test that separates these two objects into subtrees is randomly created, considering attributes only with different feature values.

In order to preserve population diversity, the GDT system applies two specialized genetic meta-operators corresponding to the classical mutation and crossover. For both operators framework provides several variants [10] that influence the tree structure and the splitting tests in the internal nodes, e.g.: *(i)* replace subtree/branch/node/test between two affected individuals; *(ii)* prune the internal node into the leaf; *(iii)* modify the test in internal nodes (shift threshold); *(iv)* replace existing test with a new one created on a randomly chosen dipole.

Each time, the selection of the variant and affected node (or nodes) are random. The probability distributions are different for nodes and for variants. For nodes, the location in the tree is taken into account (modification of nodes from the upper levels results in more global changes) and the quality of the subtree (less accurate nodes should be modified more frequently).

Direct minimization of the prediction error measured on the learning dataset usually leads to the over-fitting problem, especially in the case of DTs. In typical top-down induction, this is partially mitigated by defining a stopping condition and post-pruning. In the case of an evolutionary approach, the multi-objective function is required to minimize the prediction error and the tree complexity at the same time. As, in this paper, only univariate classification trees are considered, the simple weighted form of the fitness function is used:

$$Fitness(T) = Accuracy(T) - \alpha * (Size(T) - 1.0), \tag{1}$$

where $Accuracy(T)$ represents the classification quality of the tree $T$ estimated on the learning set, $Size(T)$ is the number of nodes in $T$ and $\alpha$ is a user-defined parameter that reflects the relative importance of the complexity term.

The selection mechanism is based on the linear ranking selection [14]. Additionally, in each iteration one individual with the highest fitness in the current population is copied to the next one (elitist strategy). The evolution ends when the fitness of the best individual in the population does not improve during the fixed number of generations or the maximum number of generations is reached.

## 3 Repository-Supported GPU-based Approach

The general flowchart of the proposed solution is illustrated in Fig. 1. The following operations: initialization of the population as well as selection of the individuals/trees remain unchanged compared to the original GDT system. They are

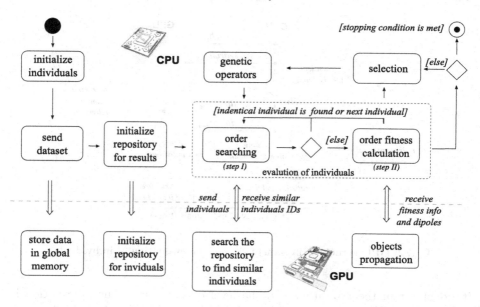

**Fig. 1.** Flowchart of a GPU-based approach supported by the repository of individuals.

run in a sequential manner on a CPU since they are relatively fast. In addition, the initial population is created only once on a small fraction of the dataset. The GPU is called after each successful application of the crossover/mutation operator, when there is a need to evaluate the individuals. It is the most time-consuming operation because all dataset objects have to be propagated from the tree root to one of the leaves. Both the dataset size and individual size influence the computational requirements. This part of the algorithm is isolated and performed in parallel on a GPU. This way, the parallelization does not affect the behavior of the original EA.

In the initialization phase (see Fig. 1), the whole dataset is sent to the GPU. This CPU to GPU transfer is done only once and the data is saved in the allocated space in the global memory. This way, all objects of the dataset are accessible for all threads at any time. In addition, the repository of the individual is created. A fixed amount of memory is allocated both on the GPU and CPU. The size of the repository is set before the evolution. Initially, the repository is empty. On the GPU, the repository is devoted to store the structure of DTs (tree nodes, branches, tests, etc.). The part of repository located on the CPU is responsible for keeping fitness results and dipoles of the trees stored on the GPU. The splitting of the repository location between CPU and GPU allows us to avoid unnecessary CPU/GPU memory transfers.

In the evolutionary loop, each time, when the genetic operator is successfully applied, the GPU is asked to help the CPU. First, the modified individual/tree is sent to the GPU. Then, the GPU searches in the repository for a similar individual (see Fig. 1, step I). If such an individual is found, the GPU returns

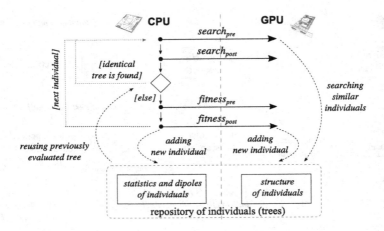

**Fig. 2.** GPU kernels arrangement during the evaluation of individuals.

its identifier in the repository and the level of similarity. This information is used by the CPU to get previously calculated (individual's) statistics from its part of the repository. If the same individual is found, then the CPU has only to copy previously calculated statistics and dipoles from the repository to the currently evaluated tree. If a similar individual is found, then information from the matched tree part is used, while for the remaining part of the tree, the GPU is called to calculate the missing information. In the current version of the algorithm, when the trees aren't the same, two types of similarity are considered. The first case concerns the situation when the tree root and its left subtree are the same as in the tree from the repository. The second case is analogical but it refers to the right subtree of the root.

To find a similar individual, two kernel functions are called (see Fig. 2). The first kernel ($search_{pre}$) is used to compare the evaluated indvidual with all individuals from the repository. A two-level decomposition is applied. Each GPU block processes one individual from the repository. Threads inside blocks are responsible for comparing various nodes inside a tree. On the GPU, trees are represented as one-dimensional arrays where the position of the left and right child of the $i$-th node equals $(2 * i + 1)$ and $(2 * i + 2)$, respectively [8]. Thanks to this, the comparison of the trees consists in checking corresponding array elements. Additionally, for two types of similar trees, one-dimensional arrays (so-called maps) are prepared to know which tree nodes have to be checked. Finally, this kernel provides the level of similarity for each tree in the repository.

The second kernel ($search_{post}$) goes through the results from the first kernel and provides the identifiers of similar individuals in the repository and their level of similarity. If no individual is found, $-1$ identifier is returned. The second kernel uses only one block to synchronize the results merging effectively. Each GPU thread is responsible for a bunch of individuals from the repository.

In case of not finding similar individuals in the repository, the GPU is called to calculate the required information as well as to search dipoles (see Fig. 1, part II) [8]. Two kernel functions are used (see Fig. 2). The first kernel ($fitness_{pre}$) is called to propagate all objects in the training dataset from the tree root to appropriate leaves. It uses the data decomposition strategy. At first, the whole dataset is spread into smaller parts over GPU blocks. Next, in each block, the assigned objects are further spread over the threads. Each GPU block makes a copy of the evaluated individual that is loaded into the shared memory. This way, the threads process the same individual in parallel but handle different chunks of the dataset. At the end of the kernel, in each tree leaf, the number of objects of each class that reach that particular leaf is stored.

The second kernel function ($fitness_{post}$) merges information from multiple copies of the individual allocated in each GPU block. This operation sums up the counters from copies of the individual, and the total number of objects of each class in each tree leaf is obtained. Finally, reclassification errors in each leaf are calculated. Then, all gathered information is propagated from the leaves towards the root. The obtained tree statistics (like coverage, errors) as well as dipoles are sent back to the CPU in order to finish the individual evaluation.

Each time when the GPU calculates new fitness results, the repository is updated. On the CPU, the obtained tree statistics and the dipoles are added to the repository. As regards the GPU, the corresponding tree structures are stored there. In the current implementation, when a new tree is added to the repository, it replaces the oldest one if the repository is full.

## 4   Experimental Validation

Experimental validation was performed on an artificially generated dataset called *chess3x3*. This dataset represents a classification problem with two classes, two real-values attributes and objects arranged on a 3×3 chessboard [10]. We used the synthetic dataset to scale it freely, unlike real-life datasets. All presented results correspond to averages of 5–10 runs and were obtained with a default set of parameters from the original GDT system. As we are focused in this paper only on size and time performance, results for the classification accuracy are not included. However, for the tested dataset variants, the GDT system managed to induce trees with optimal structures and accuracy over 99%.

Experiments were performed on a server equipped with 2 eight-core processors Intel Xeon E5-2620 v4 (20 MB Cache, 2.10 GHz), 256 GB RAM and running 64-bit Ubuntu Linux 16.04.02 LTS. We tested NVIDIA Tesla P100 GPU card (3 584 CUDA cores and 12 GB of memory). The original GDT system was implemented in C++ and compiled with the use of gcc version 5.4.2. The GPU-based parallelization was implemented in CUDA-C [17] and compiled by nvcc CUDA 8.0 [1] (single-precision arithmetic was applied).

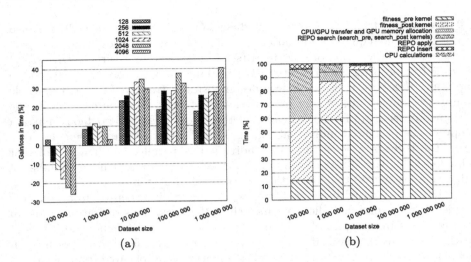

**Fig. 3.** Performance of the accelerated approach: (a) influence of the repository size (from 128 to 4096 trees) on the reduction of induction time, (b) detailed time-sharing information (mean time as a percentage) for repository size of 512 individuals.

## 4.1    Results

Figure 3(a) shows how much the reusing mechanism accelerates the evolutionary induction. The influence of both the dataset size and the repository size is presented. It is clearly visible that the gain in time is more prominent when the dataset size increases. For more than 10 millions of objects, the repository-supported version is about 30% faster than the original GPU-acceleration. An optimal number of trees stored in the repository grows when larger data is processed (for 100 000 objects 128 trees, for 1 000 000 objects 512 trees, etc.)

For smaller datasets (e.g., 100 000 objects), the induction time is even increased if too many individuals are held in the repository. When the number of trees exceeds 256, the loss in time is observed. It is caused by a repository overhead, which is illustrated in Fig. 3(b). Searching in the repository is mainly blamed for time efficiency drop ($REPO_{search}$ is the time that the algorithm spent on searching in the repository). The time spent on other repository operations, like inserting new individuals ($REPO_{insert}$) and reusing stored data ($REPO_{apply}$), is negligible and does not depend on the repository size. For the smallest dataset, the repository search overhead is not compensated by the time profit resulting from the reusage. When the number of objects increases, then the evaluation of individuals is more time demanding and the repository overhead becomes negligible.

In Table 1, we present mean execution times of all tested implementations. For the GPU-based version with the reusing mechanism, the optimal repository size is considered. The results show that the gain in time is more important when large-scale data is processed. For 1 billion of objects, the proposed solution

**Table 1.** Mean execution times of the repository-supported (GPU with REPO), GPU, OpenMP and sequential versions of the algorithm (in seconds, for larger datasets also time in hours is included).

| Dataset size | GPU with REPO | | GPU | | OpenMP | Sequential |
|---|---|---|---|---|---|---|
| 100 000 | 21.1 | | 21.8 | | 100.2 | 685 |
| 1 000 000 | 55.3 | | 62.7 | | 3 605.7 | 23 536 |
| 10 000 000 | 449.9 | | 666.7 | | 47 600.4 | 324 000 |
| 100 000 000 | 5 428.4 | (1.5 h) | 7 471.3 | (2 h) | weeks | months |
| 1 000 000 000 | 59 404.2 | (16.5 h) | 84 245.2 | (23.5 h) | months | over a year |

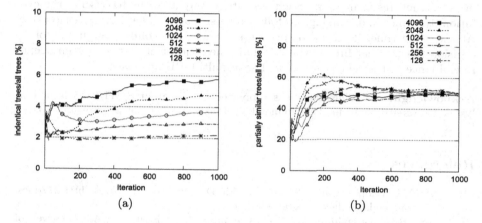

(a)                                             (b)

**Fig. 4.** Searching success ratio for different repository size (from 128 to 4096 individuals): (a) identical tree, (b) similar trees (root and its left or right child).

decreases the induction time by $\approx 7$ h (it is 40% faster). It was estimated that the sequential GDT system would need over a year to calculate such a dataset and the OpenMP parallelization [5] would decrease this time to a few months.

We have also examined the searching success ratio (how frequently the search in the repository ends with a success). Figure 4(a) shows that an identical tree is found in a few percent of cases (from 2% to 6%). Moreover, we observe that this ratio grows with the larger repository. However, we should remind that the repository size influences the search overhead which in the case of smaller datasets could be significant (as it is shown in Fig. 3). As regards partial similarity, the search ends with success much more often (in about 50% of trials) and it is not dependent on the repository size. The results suggest that similar trees contribute more to obtain an acceleration.

## 5 Conclusion

In this paper, we extend the GPU-based approach for evolutionary induction of decision trees. The concept of fitness evaluation reusage is proposed. A so-called repository of individuals is used to store previously considered trees. When a new individual is evaluated, first, a similar tree to reuse is searched in the repository. If it is found, the results are read from the repository. Otherwise, the GPU has to perform required calculations. The results show that the proposed strategy is able to speed up the induction even further, especially on large-scale data.

This research is a first step in building a so-called 'multi-tree' strategy. It assumes that similar individuals are corepresented by partially sharing fragments/structures in memory. Such a strategy may allow us to observe the evolution dynamics in detail, e.g., to follow diversity at each level of a decision tree. At the same time, it can also speed up further the evolutionary induction of decision trees. Other future works include the concept of searching even deeper resemblance in the repository as well as a multi-GPU approach.

**Acknowledgments.** This work was supported by the grant S/WI/2/18 from Bialystok University of Technology founded by Polish Ministry of Science and Higher Education.

## References

1. NVIDIA Developer Zone - CUDA Toolkit Documentation (2019). https://docs.nvidia.com/cuda/cuda-c-programming-guide/
2. Barros, R.C., Basgalupp, M.P., De Carvalho, A.C., Freitas, A.A.: A survey of evolutionary algorithms for decision-tree induction. IEEE Trans. SMC, Part C **42**(3), 291–312 (2012)
3. Charalampakis, A.E.: Registrar: a complete-memory operator to enhance performance of genetic algorithms. J. Glob. Optim. **54**(3), 449–483 (2012)
4. Chitty, D.M.: Fast parallel genetic programming: multi-core CPU versus many-core GPU. Soft Comput. **16**(10), 1795–1814 (2012)
5. Czajkowski, M., Jurczuk, K., Kretowski, M.: A parallel approach for evolutionary induced decision trees. MPI+OpenMP implementation. In: Rutkowski, L., Korytkowski, M., Scherer, R., Tadeusiewicz, R., Zadeh, L.A., Zurada, J.M. (eds.) ICAISC 2015, Part I. LNCS (LNAI), vol. 9119, pp. 340–349. Springer, Cham (2015). https://doi.org/10.1007/978-3-319-19324-3_31
6. Czajkowski, M., Kretowski, M.: Evolutionary induction of global model trees with specialized operators and memetic extensions. Inf. Sci. **288**, 153–173 (2014)
7. Franco, M.A., Bacardit, J.: Large-scale experimental evaluation of GPU strategies for evolutionary machine learning. Inf. Sci. **330**(C), 385–402 (2016)
8. Jurczuk, K., Czajkowski, M., Kretowski, M.: Evolutionary induction of a decision tree for large-scale data: a GPU-based approach. Soft Comput. **21**(24), 7363–7379 (2017)
9. Kotsiantis, S.B.: Decision trees: a recent overview. Artif. Intell. Rev. **39**(4), 261–283 (2013)
10. Kretowski, M.: Evolutionary Decision Trees in Large-Scale Data Mining. Springer, Cham (2019). https://doi.org/10.1007/978-3-030-21851-5

11. Lo, W.T., Chang, Y.S., Sheu, R.K., Chiu, C.C., Yuan, S.M.: CUDT: A CUDA based decision tree algorithm. Sci. World J. (2014)
12. Loh, W.Y.: Fifty years of classification and regression trees. Int. Stat. Rev. **82**(3), 329–348 (2014)
13. Marron, D., Bifet, A., Morales, G.D.F.: Random forests of very fast decision trees on GPU for mining evolving big data streams. In: Proceedings of the Twenty-First European Conference on Artificial Intelligence, ECAI 2014, pp. 615–620 (2014)
14. Michalewicz, Z.: Genetic Algorithms + Data Structures = Evolution Programs, 3rd edn. Springer, Heidelberg (1996). https://doi.org/10.1007/978-3-662-03315-9
15. Reska, D., Jurczuk, K., Kretowski, M.: Evolutionary induction of classification trees on spark. In: Rutkowski, L., Scherer, R., Korytkowski, M., Pedrycz, W., Tadeusiewicz, R., Zurada, J.M. (eds.) ICAISC 2018, Part I. LNCS (LNAI), vol. 10841, pp. 514–523. Springer, Cham (2018). https://doi.org/10.1007/978-3-319-91253-0_48
16. Rokach, L., Maimon, O.: Top-down induction of decision trees classifiers - a survey. IEEE Trans. Syst. Man Cybern. Part C (Appl. Rev.) **35**(4), 476–487 (2005)
17. Storti, D., Yurtoglu, M.: CUDA for Engineers : An Introduction to High-Performance Parallel Computing. Addison-Wesley, New York (2016)
18. Tsutsui, S., Collet, P. (eds.): Massively Parallel Evolutionary Computation on GPGPUs. Natural Computing Series. Springer, Heidelberg (2013). https://doi.org/10.1007/978-3-642-37959-8

# A Distributed Modular Scalable and Generic Framework for Parallelizing Population-Based Metaheuristics

Hatem Khalloof[✉], Phil Ostheimer, Wilfried Jakob, Shadi Shahoud,
Clemens Duepmeier, and Veit Hagenmeyer

Institute of Automation and Applied Informatics (IAI), Karlsruhe Institute
of Technology (KIT), Karlsruhe, Germany
{hatem.khalloof,wilfried.jakob,shadi.shahoud,clemens.duepmeier,
veit.hagenmeyer}@kit.edu, phil.sidney@gmx.de

**Abstract.** In the present paper, microservices, container virtualization
and the publish/subscribe messaging paradigm are exploited to develop
a distributed, modular, scalable and generic framework for parallelizing population-based metaheuristics. The proposed approach paves the
way for an easy deployment of existing metaheuristic algorithms such as
Evolutionary Algorithms (EAs) on a scalable runtime environment with
full runtime automation. Furthermore, it introduces simple mechanisms
to work efficiently with other components like forecasting frameworks
and simulators. In order to analyze the feasibility of the design, the EA
GLEAM (General Learning Evolutionary Algorithm and Method) is integrated and deployed on a cluster with 4 nodes and 128 cores for benchmarking. The overhead of the framework is measured and the obtained
results show not only low values but also a small increase with growing
number of computing nodes.

**Keywords:** Parallel EAs · Microservices · Virtualization · Container ·
Cluster · Parallel computing · Scalability · Coarse-Grained Model

## 1 Introduction

Metaheuristics have been widely used to solve several real world optimization
problems e.g. [4,5]. However, finding good local optima or the global one for a
highly complex optimization problem usually requires computational expensive
fitness function evaluations. For reducing the computational effort of population-
based metaheuristics such as EAs, parallelization attracted the attention of
researchers as a potential solution.

Cantú-Paz [1] presented and investigated three parallelization approaches of
EAs, namely the Global Model, the Fine-Grained Model and the Coarse-Grained
Model (Island Model). While the first model only parallelizes the evaluation of
the individuals and performs the genetic operators centrally, the second and third
models divide the population into subpopulations in order to apply the genetic

© Springer Nature Switzerland AG 2020
R. Wyrzykowski et al. (Eds.): PPAM 2019, LNCS 12043, pp. 432–444, 2020.
https://doi.org/10.1007/978-3-030-43229-4_37

operators and the evaluation locally. In the Coarse-Grained Model, each sub-population forms an *island* where a sequential EA is locally applied to evolve it. The islands periodically communicate with each other according to a predefined topology for exchanging some promising individuals called migrants.

In recent years, many approaches for parallelizing EAs exploiting modern hardware and software technologies such as web, cloud, Graphics Processing Unit (GPU) and Big Data technologies have been introduced e.g. [3,7–14]. However, most of these approaches e.g. [3,10,11] have a monolithic architecture which limits the scalability, modularity and the maintainability of the system, to name a few. Furthermore, they suffer from the lack of abilities to plug in problem specific functionalities (e.g. simulators) e.g. [3,7,9–11] and the restriction to one technology stack (cf. [2] for a detailed discussion).

The present work extends the framework introduced in [2], which enables the execution of EAs in a distributed and scalable runtime environment based on container and microservices technologies. In [2], the implementation of the Global Model was introduced. This work focuses on the Coarse-Grained Model not only to reduce the execution runtime but also to utilize the beneficial properties of structured populations such as preserving a high diversity and avoiding the problem of premature convergence [17]. The implementation utilizes a microservice architecture where the Coarse-Grained Model is implemented as a set of several independent services running in parallel. Each service performs a special task [6] and can use its own technology stack. To build highly parallel and scalable framework, a microservice architecture represents a good choose, since each service can use the most suitable technologies and can be scaled horizontally. Data is exchanged between the services by calling REST-APIs and using the publish/subscribe messaging paradigm to reduce the coupling among the services. The services are encapsulated by containers to maintain runtime automation and platform independence. Hence, the proposed microservice and container based architecture provides a good foundation to reach the following six aims defined in [2]:

1. Highly parallel and scalable runtime environment.
2. Maximum flexibility in mapping optimization tasks to different external applications (e.g. simulators).
3. Easy integration (pluggability) of different types of population-based meta-heuristic algorithms.
4. Full runtime automation on big cluster and cloud computing environments.
5. Easy to use web based management and execution environment.
6. Reliability of a proven and established software environment, which is used in many other applications to ensure its further development and maintenance.

The rest of the present paper is structured as follows. The next section reviews related work on state-of-the-art frameworks for parallelizing EAs. Section 3 describes the system architecture of the proposed framework. Section 4 reports on experiments with the first deployment of the framework. It describes the integration of the EA GLEAM and the deployment on a cluster. Then, the first benchmarking results measuring the overhead introduced by the framework are presented and discussed. Section 5 concludes with a summary and future work.

## 2  Related Work

Population-based metaheuristics need a lot of computational power for obtaining the optimal (or near optimal) solution for many real life optimization problems. Therefore, the parallelization of metaheuristics – especially EAs – has received a lot of attention by researchers. The execution of EAs in cluster and cloud environments is a promising step exploiting their enormous computational capabilities. Moreover, the contribution of Big Data and cloud technologies offers a great potential for an efficient parallelization of EAs. One of the first implementations using Big Data technologies, namely container technology, is the one by Salza et al. [7]. They proposed an approach to distribute Genetic Algorithms (GAs) according to the Global Model using the Advanced Message Queuing Protocol (AMQP), software containers, RabbitMQ and CoreOS. RabbitMQ provides the implementation of AMQP and acts as a message broker to accept and forward messages between the master and the slaves. The underlying communication paradigm of AMQP is the publish/subscribe pattern. The master and the slaves are executed in Docker[1] containers. CoreOS allocates resources and orchestrates the containers. They reported linear speedup for large evaluation times [8]. García-Valdez et al. [11] implemented evospace-js based on an event-driven architecture and an asynchronous I/O model to parallelize EAs. Several web technologies such as Node.js, JavaScript, RESTful web services and an in-memory database, namely Redis[2] are used. The EvoSpace Model [10] is the baseline platform of evospace-js. The two main components of the framework are the evospace-js population repository and the remote clients called EvoWorkers. While the evospace-js population repository manages the population in the pool, the EvoWorkers take a part of the population from the pool and execute a local evolutionary search. The KafkEO framework by Guervós et al. [9] introduces a cloud technologies based architecture to parallelize EAs. They describe their underlying parallelization model as an asynchronous Coarse-Grained Model. The framework is functionally equivalent to the EvoSpace Model [10] and there is no strict topology between the islands. Therefore, it can rather be described as a pool model based implementation. The serverless framework, namely Open-Whisk is used to deploy the services. Kafka is used as a communication channel that provides the message queues. Jurczuk et al. [12–14] developed and implemented a GPU-based application to distribute evolutionary induction of decision tree for large-scale data applying the Global Model.

In Khalloof et al. [2], a framework for the parallelization of EAs using microservices and container technologies was introduced. In a first implementation, the Global Model is integrated and evaluated. The proposed approach in [2] is strictly based on a microservice architecture in which each service can be

---

[1]  www.docker.com.

[2]  www.redis.io.

scaled and developed independently. The functionalities are distributed across six different microservices where each one can be deployed and executed in one or more containers. The solution in [2] shows a linear increase of the overhead with growing population sizes and an almost linear increase of the overhead with the number of slaves. In contrast to the other proposed approaches, the underlying microservice architecture combined with container technologies allows building highly modular, flexible and scalable systems. In addition to these beneficial traits, external simulators can also be easily incorporated.

Despite the clear advantages of so far implemented of parallelization models based on microservices and container technologies, there is no implementation of the Coarse-Grained Model using microservices and container technologies, the authors are aware of. Therefore, the present framework is designed according to the microservice architecture which splits the different parts of the Coarse-Grained Model into distinct services for allowing an efficient development as well as an easy deployment and integration of existing EAs. Moreover, utilizing container technology to deploy the framework on a cluster enables full runtime automation.

## 3    The New Framework for the Deployment of EAs in a Scalable Runtime Environment

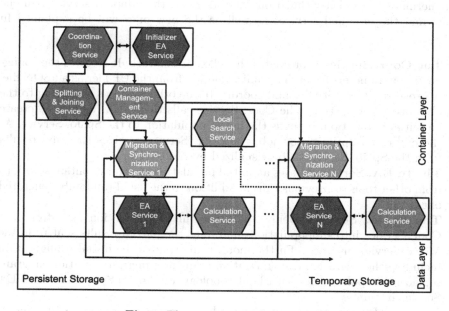

**Fig. 1.** The container and data layer

The conceptual architecture of the proposed framework extends the one introduced in [2]. The framework has two main tiers, namely the User Interface (UI)

Tier and the Cluster Tier. The Container Layer and the Data Layer shown in Fig. 1 are the two sub-layers of the Cluster Tier. In the following sections, we will describe the Container Layer with the implemented microservices and the Data Layer in detail, since they represent the core of the framework.

### 3.1 Container Layer

Executing parallel population-based metaheuristics in a highly distributed environment requires several tasks. For instance, coordinating the parallel execution, distributing the data, executing the parallel EAs, managing the computation resources, starting external simulators or extending EAs to Memetic Algorithms (MAs) by adding local search. The proposed framework carries out these tasks based on eight decoupled and cohesive microservices, see Fig. 1. Each microservice is deployed into one or more containers. Most of the services i.e. the Coordination Service (Coor. Service), the Splitting & Joining Service (Sp.Jo. Service), the Container Management Service (Co.Ma. Service), the Local Search Service (L.S. Service), the Calculation Service (Ca. Service) and the Evolutionary Algorithm Service (EA Service) are adapted from [2].

However, to support the Coarse-Grained Model, two additional services, namely the Migration & Synchronization Service (Mi.Sy. Service) and the Initializer EA Service (In.EA. Service) are introduced as new architecture elements. Furthermore, several new functions are added to the adapted services. In the following, the modified services as well as the new ones will be explained in detail.

1. **The Coor. Service** coordinates the whole framework. It receives the configuration parameters for each optimization job from the UI Tier as a JSON file. It processes it to initialize and coordinate the other services according to the obtained configuration. The Coor. Service calls the In.EA. Service to create an initial population. It sends the initial population to the Sp.Jo. Service. At the end of an optimization job, the Coor. Service receives the final results from the Sp.Jo. Service to be visualized by the UI Tier.
2. **The In.EA. Service** creates an initial population that may contain solutions from other tools or previous runs in addition to the usual randomly generated ones.
3. **The Sp.Jo. Service** evenly splits the initial population sent from the Coor. Service into subpopulations and joins the partial results sent from the Mi.Sy. Service instances. Furthermore, it distributes the initial configuration including the migration rate, migration frequency, migrant selection, migrant replacement policy and the island topology to the Mi.Sy. Service and EA Service instances.
4. **The Co.Ma. Service** creates the number of requested containers (instances) for the Mi.Sy. Service as well as for EA Service.
5. **The Mi.Sy. Service** performs all tasks related to the migration policy, synchronizes the execution and checks if the global termination criterion is met.

6. **The EA Service** is called by the Mi.Sy. Service to evolve the population. This includes parent selection, application of genetic operators and the acceptance of the offspring. Moreover, the EA Service evaluates the offspring either internally or externally by sending them to the Ca. Service.
7. **The Ca. Service** is called by the EA Service to evaluate the offspring. This enables the integration of the Global Model into the framework in order to accelerate the evaluation process.
8. **The L.S. Service** is an extension of a deployed EA to support MAs. This can be especially useful for complex tasks or other heuristics if appropriate local search methods are available [16].

Since the focus of this work is not on MAs and hierarchical parallelization models like the combination of the Global Model with the Coarse-Grained Model, the last two services are not presented in the following. For simplicity's sake, the term *island* in the following sections refers to the Mi.Sy. Service and EA Service.

## 3.2 Mapping of the Coarse-Grained Model Pseudo-code to the Proposed Microservices

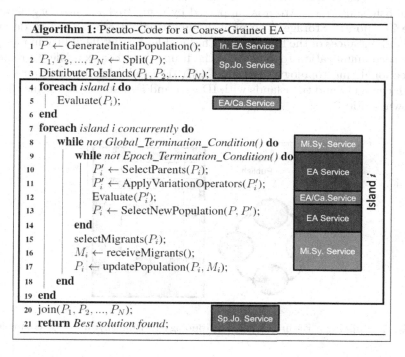

**Algorithm 1:** Pseudo-Code for a Coarse-Grained EA

1  $P \leftarrow$ GenerateInitialPopulation();   In. EA Service
2  $P_1, P_2, ..., P_N \leftarrow$ Split($P$);   Sp.Jo. Service
3  DistributeToIslands($P_1, P_2, ..., P_N$);
4  **foreach** *island i* **do**
5  | Evaluate($P_i$);   EA/Ca.Service
6  **end**
7  **foreach** *island i concurrently* **do**
8  | **while** *not Global_Termination_Condition()* **do**   Mi.Sy. Service
9  | | **while** *not Epoch_Termination_Condition()* **do**
10 | | | $P_i' \leftarrow$ SelectParents($P_i$);   EA Service
11 | | | $P_i' \leftarrow$ ApplyVariationOperators($P_i'$);
12 | | | Evaluate($P_i'$);   EA/Ca.Service
13 | | | $P_i \leftarrow$ SelectNewPopulation($P, P'$);   EA Service
14 | | **end**
15 | | selectMigrants($P_i$);
16 | | $M_i \leftarrow$ receiveMigrants();   Mi.Sy. Service
17 | | $P_i \leftarrow$ updatePopulation($P_i, M_i$);
18 | **end**
19 **end**
20 join($P_1, P_2, ..., P_N$);   Sp.Jo. Service
21 **return** *Best solution found*;

**Fig. 2.** Mapping of the coarse-grained EA pseudo-code to the proposed microservice architecture

The introduced microservices can be divided into two groups. While the first group contains the services that directly perform the functions of the Coarse-Grained Model, the other one includes the services that coordinate the whole framework. The pseudocode for Coarse-Grained EAs is mapped to the services in the first group, see Fig. 2. The In.EA. Service executes line 1, the Sp.Jo. Service executes the lines 2, 3, 20 and 21, the Mi.Sy. Service implements the lines 8 and 15–17, the EA Service corresponds to the lines 9–11 and 13, and the EA Service (by internal evaluation) or the Ca. Service (by external evaluation) executes the lines 5 and 12.

### 3.3  Data Layer

The Data Layer stores the subpopulations and the configuration. Additionally, the partial results are stored by each Mi.Sy Service in the Data Layer. At the end of each optimization task, the Sp.Jo. Service merges the partial results to form the final results i.e. the best solutions found and persistently stores them in the Data Layer for further use. To accomplish these two storages, the Data Layer provides a Temporary Storage and a Persistent Storage. The Temporary Storage is responsible for storing the subpopulations, migrants, configuration and partial results. Moreover, the message exchange functionality in the form of the publish/subscribe pattern is supported by using Redis in-memory database as the Temporary Storage. The publish/subscribe pattern is used to decouple the island instances of the framework. This enables further flexibility in mapping different communication topologies to the framework. For example, to realize a bi-directional ring topology the island with ID $i$ publishes the migrants to its publish channel $i$ and the islands with ID $i+1$ and $i-1$ subscribe to this channel as shown in Fig. 3.

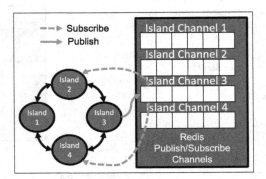

**Fig. 3.** Mapping a Bi-directional ring topology to Publish/Subscribe channels

# 4   Evaluation of the Proposed Approach

In the following, a short description of the EA used as a test case will be introduced. Afterwards, we describe the deployment steps on the cluster in greater detail and discuss the obtained results from the evaluation regarding the overhead of the framework.

## 4.1   Integration of an Evolutionary Algorithm

The EA GLEAM [15,16] is used as a test case for the framework. However, any other EA can be used to validate the functionality of the proposed framework. GLEAM is integrated into two services, namely the In.EA. Service and the EA Service. Since GLEAM is written in C, the interaction between the implemented services and GLEAM is restricted to command-line calls. Data exchange is accomplished by file I/O. The communication interfaces of the two EA-related services provide a GLEAM-specific layer that allows to integrate GLEAM as it is. The GLEAM functions for reading and saving a complete population are used, so that the populations between the epochs are accessible for the Mi.Sy. Service. The same mechanism is used for passing the start population formed by the In. EA Service to the Sp.Jo. Service. Each epoch includes the application of the genetic operations and the internal evaluation of the individuals. As a test case, the non-convex, non-linear multimodal Rastrigin function [18] is chosen. Since the focus of the work is on the estimation of the overhead of the framework and not on the solution quality, the only relevant information thereby is the size of one individual which is approximately 2 KB containing 20 genes. In other words, the other EA settings e.g. hamming distance for the crossover and recombination or initialization strategy for the start population are irrelevant for this evaluation.

## 4.2   Deployment and Execution on a Cluster

Figure 4 gives an overview of the involved technological layers of the framework. On the lowest level, the cluster hardware runs a certain Operating System (OS). All microservices run in a containerized environment that is independent from the underlying OS. On top of the OS, we have the container orchestration system, namely Kubernetes[3]. It dynamically allocates the resources to Pods that execute the microservices. Inside the pods, the implemented services are run by the Docker Engine. In addition to the implemented services, Redis runs in a Pod acting as a Temporary Storage and an implementation for the publish/subscribe pattern. We use a cluster with four nodes with 32 Intel cores (2,4 GHz), 128 GB RAM and a SSD disk each. The nodes are connected to each other by a LAN with 10 GBit/s bandwidth.

---

[3] www.kubernetes.io.

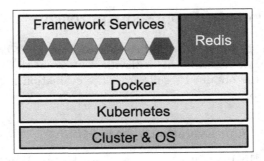

**Fig. 4.** The technological layers of the framework

### 4.3  Framework Migration Overhead (FMO)

In order to assess the performance of the framework, we analyze the effects on the overhead by changing the following parameters:

1. Topology: We choose a simple ring, a bi-directional ring, a ladder and a completely connected graph as communication topologies among the islands.
2. Migration Rate: We choose migration rates of 1, 2, 4, 8, 16, 32, 64, 128, 256, 512, 1024, 2048, 4096, 8192 and 16384 migrants. This results in an amount of exchanged data per epoch between 2 KB and 32 MB.
3. Number of Islands: The number of islands is chosen of 8, 16, 32, 64 and 120, so that at minimum two cores are left on each node for the OS.

For each combination of parameters we execute 100 epochs so that the overhead related functions are executed 100 times to minimize the effect of single delays from the OS. The execution time of each epoch is subdivided into the amount of time needed to run the deployed EA called EA Execution Time and the Framework Overhead which includes the communication overhead, FMO and the amount of time that the framework needs to carry out the other tasks e.g. initialization, splitting the initial population and joining the partial results. The results of the experiments shows that the governing factor for the Framework Overhead is the FMO which arises from the change of the migration rate. The rest amount of the Framework Overhead (i.e. by excluding FMO) remains small and almost constant for the considered migration rates as shown in Fig. 5. Figure 6a shows the results for varying the migration rates between 1 and 16384 for the four topologies with a fixed number of 8 islands. As shown in Fig. 6b, the framework shows an almost constant FMO for the migration rates between 1 to 64 and from 128 migrants the FMO starts to increase and the difference between the four topologies is more and more significant. In other words, with an increased number of connections between the islands such as the complete topology, the FMO rises proportionally with increasing migration rate. In order to show the influence of the number of islands on the FMO, their number is varied between 16 and 120 and the migration rate between 1 and 128, see Figs. 6c, d, e and f. Since the ring, bi-directional ring and the ladder always have a constant number of neighbors of 1, 2 and 3 respectively, the increasing number of islands has hardly any influence on the FMO. However, the number of neighbors for a complete graph increases exponentially with a growing number of islands. Hence, the FMO also increases exponentially. For 16 and 32 islands Figs. 6c and d, exhibit an increase of the FMO for the complete graph topology starting with a migration rate of 32. However for 64 and 120 islands, the complete graph topology shows peaks for 16 migrants and for 8 migrants respectively, see Figs. 6e and f. The reason for these peaks for lower migration rates is the effect of the exponential growth of connections between the neighbors by increasing number of islands.

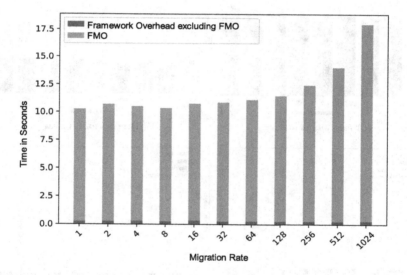

**Fig. 5.** The rest of framework Overhead vs. FMO for 100 Epochs and a ring topology with 8 Islands

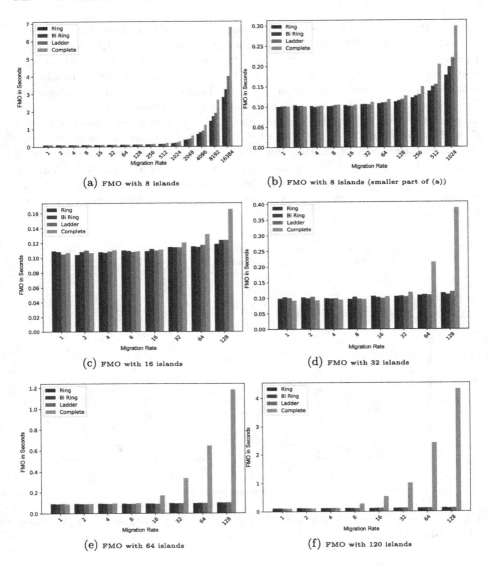

**Fig. 6.** The framework migration overhead (FMO) per Epoch for a Ring, Bi-Directional Ring, Ladder and Complete graph topology

## 5   Conclusion and Future Work

The Coarse-Grained parallelization Model of EAs is evaluated by designing and developing a new framework based on microservices, container virtualization and the publish/subscribe messaging paradigm. The combination of microservices and container virtualization allows for an easy integration of existing EAs written in different programming languages. Furthermore, it guarantees a seamless deployment in a scalable runtime environment, such as a cluster or a cloud. The use of the publish/subscribe messaging paradigm ensures a full decoupling between the services. This facilitates the implementation of several possible communication topologies among the services. Several setup configurations with four topologies are investigated. The observed results shows that the proposed approach provides almost constant migration overhead increasing the migration rate till a fixed threshold, namely 128 migrants, which is an unrealistic high number for real applications. This threshold could not be eliminated but it could be shifted. Hence, the new proposed framework is a promising approach to parallelize existing EAs. As part of future work, the Coor. Service has to be extended to change the topologies dynamically during one optimization task. In addition, the performance of the framework will be analyzed in comparison to other approaches regarding the quality of the solutions e.g. which migration rate would be enough to provide satisfied solution diversity and avoid premature convergence and achievable speed-up rates. Shifting the above threshold by using for example faster computer-networking communications should be also considered.

# References

1. Cantú-Paz, E.: A survey of parallel genetic algorithms. Calculateurs paralleles, reseaux et systems repartis **10**(2), 141–171 (1998)
2. Khalloof, H., et al.: A generic distributed microservices and container based framework for metaheuristic optimization. In: Proceedings of the Genetic and Evolutionary Computation Conference Companion, pp. 1363–1370, ACM (2018) https://doi.org/10.1145/3205651.3208253
3. Salza, P., Ferrucci, F., Sarro, F.: elephant56: design and implementation of a parallel genetic algorithms framework on Hadoop MapReduce. In: Proceedings of the 2016 on Genetic and Evolutionary Computation Conference Companion, pp. 1315–1322, ACM (2016) https://doi.org/10.1145/2908961.2931722
4. Biethahn, J., Nissen, V.: Evolutionary Algorithms in Management Applications. Springer, Heidelberg (2012). https://doi.org/10.1007/978-3-642-61217-6
5. Dasgupta, D., Michalewicz, Z.: Evolutionary Algorithms in Engineering Applications. Springer, Heidelberg (2013). https://doi.org/10.1007/978-3-662-03423-1
6. Newman, S.: Building Microservices: Designing Fine-Grained Systems. O'Reilly Media, Inc., Newton (2015)
7. Salza, P., Ferrucci, F.: An Approach for Parallel Genetic Algorithms in the Cloud using Software Containers. Computing Research Repository (CoRR), pp. 1–7 (2016)
8. Salza, P., Ferrucci, F.: Speed up genetic algorithms in the cloud using software containers. Future Gener. Comput. Syst. **92**, 276–289 (2019). https://doi.org/10.1016/j.future.2018.09.066
9. Merelo Guervós, J.J., García-Valdez, J.M.: Introducing an event-based architecture for concurrent and distributed evolutionary algorithms. In: Auger, A., Fonseca, C.M., Lourenço, N., Machado, P., Paquete, L., Whitley, D. (eds.) PPSN 2018. LNCS, vol. 11101, pp. 399–410. Springer, Cham (2018). https://doi.org/10.1007/978-3-319-99253-2_32

10. García-Valdez, M., Trujillo, L., Merelo, J.J., De Vega, F.F., Olague, G.: The EvoSpace model for pool-based evolutionary algorithms. J. Grid Comput. **13**(3), 329–349 (2015). https://doi.org/10.1007/s10723-014-9319-2

11. García-Valdez, M., Merelo, J.J.: evospace-js: asynchronous pool-based execution of heterogeneous metaheuristics. Proceedings of the Genetic and Evolutionary Computation Conference Companion, pp. 1202–1208, ACM (2017). https://doi.org/10.1145/3067695.3082473

12. Jurczuk, K., Czajkowski, M., Kretowski, M.: Multi-GPU approach for big data mining: global induction of decision trees. In: Proceedings of the Genetic and Evolutionary Computation Conference Companion, pp. 175–176, ACM, July 2019. https://doi.org/10.1145/3319619.3322045

13. Jurczuk, K., Reska, D., Kretowski, M.: What are the limits of evolutionary induction of decision trees? In: Auger, A., Fonseca, C.M., Lourenço, N., Machado, P., Paquete, L., Whitley, D. (eds.) PPSN 2018. LNCS, vol. 11102, pp. 461–473. Springer, Cham (2018). https://doi.org/10.1007/978-3-319-99259-4_37

14. Jurczuk, K., Czajkowski, M., Kretowski, M.: Evolutionary induction of a decision tree for large-scale data: a GPU-based approach. Soft Comput. **21**(24), 7363–7379 (2017). https://doi.org/10.1007/s00500-016-2280-1

15. Blume, C., Jakob, W.: GLEAM - an evolutionary algorithm for planning and control based on evolution strategy. In: GECCO Late Breaking Papers, pp. 31–38 (2002)

16. Jakob, W.: A general cost-benefit-based adaptation framework for multimeme algorithms. Memetic Comput. **2**(3), 20–218 (2010). https://doi.org/10.1007/s12293-010-0040-9

17. Herrera, F., Lozano, M., Moraga, C.: Hierarchical distributed genetic algorithms. Int. J. Intell. Syst. **14**(11), 1099–1121 (1999). https://doi.org/10.1002/(sici)1098-111x(199911)14:11⟨1099::aid-int3⟩3.0.co;2-o

18. Rastrigin, L.A.: Systems of Extremal Control. Mir, Moscow (1974)

# Parallel Processing of Images Represented by Linguistic Description in Databases

Danuta Rutkowska[1]([✉])(iD) and Krzysztof Wiaderek[2]

[1] Information Technology Institute, University of Social Sciences,
90-113 Lodz, Poland
drutkowska@san.edu.pl
[2] Institute of Computer and Information Sciences,
Czestochowa University of Technology, 42-201 Czestochowa, Poland
krzysztof.wiaderek@icis.pcz.pl

**Abstract.** This paper concerns an application of parallel processing to color digital images characterized by linguistic description. Attributes of the images are considered with regard to fuzzy and rough set theories. Inference is based on the CIE chromaticity color model and granulation approach. By use of the linguistic description represented in databases, and the rough granulation, the problem of image retrieval and classification is presented.

**Keywords:** Parallel processing · Image processing · Linguistic description · Databases · Fuzzy and rough sets · Information granulation · CIE chromaticity color model

## 1 Introduction

The paper presents a parallel processing approach in application to several problems concerning color digital images characterized by linguistic description. This subject refers to author's previous articles [19–26], especially [25,26], and [24]. In all these papers, the CIE chromaticity color model (see e.g. [9]) is applied to digital images, and fuzzy sets [28] are employed to distinguish color areas on the CIE chromaticity diagram (see Fig. 1).

In [25] a parallel processing intelligent system that generates a linguistic description of an input image or a collection of color digital images is presented. This system, by use of the inference based on fuzzy IF-THEN rules [28,29], can produce a natural language description of the following form, e.g.: *The image contains "yellowish green" color in the right side of the picture, and "red" color and very small clusters of "pink" color pixels in the left down corner, and a small cluster of "red" and "pink" pixels in the central upper area, approximately.*

In [26] the linguistic description of image attributes are considered as represented in databases, with regard to image retrieval. Fuzzy sets [28] are used as color granules, and fuzzy numbers determine color participation in an image. To illustrate the color participation, a fuzzy histogram is introduced in [24].

R. Wyrzykowski et al. (Eds.): PPAM 2019, LNCS 12043, pp. 445–456, 2020.
https://doi.org/10.1007/978-3-030-43229-4_38

This paper is focused on another artificial intelligence approach, that is the rough set theory [11], and rough granulation (see e.g. [13, 15, 17]. The linguistic description of image attributes represented in databases is now considered with regard to inference based on the rough sets. In this way, problems of image granulation, clustering and classification are studied. In addition, different examples, including face images, are analyzed. Of course, the presented approach applied to processing large number of color digital images requires parallelism.

## 2   Granular Color Segmentation of Images

In [23] a granular system for generating linguistic description of color images is introduced, and then developed in [24–26]. In those papers, first of all, fuzzy granules [14, 30] – defined by membership functions – are employed as color and location granules. Only with regard to the shape attribute, the rough set theory [11] is considered in [19–23]. Now, such a system that can be used in order to generate the linguistic description of image attributes of the form represented in databases, is extended within the framework of the rough set theory.

Of course, we still use the fuzzy color regions (fuzzy granules) in the CIE chromaticity diagram, shown in Fig. 1, where the color granules 1–23 have the following names (labels): 1 - white, 2 - yellowish green, 3 - yellow green, 4 - greenish yellow, 5 - yellow, 6 - yellowish orange, 7 - orange, 8 - orange pink, 9 - reddish orange, 10 - red, 11 - purplish red, 12 - pink, 13 - purplish pink, 14 - red purple, 15 - reddish purple, 16 - purple, 17 - bluish purple, 18 - purplish blue, 19 - blue, 20 - greenish blue, 21 - bluegreen, 22 - bluish green, 23 - green; see [9].

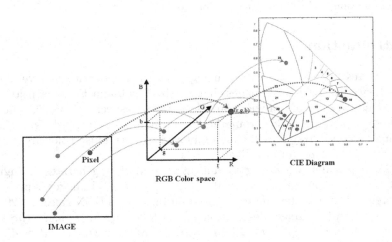

**Fig. 1.** Color segmentation of an image (Color figure online)

In our approach, these labels are linguistic values associated with the corresponding fuzzy sets – fuzzy color granules, denoted as $C_1$, $C_2$, ..., $C_{23}$, respectively. As a matter of fact, a color segmentation of an image (fuzzy segmentation) is realized when pixels of an image are grouped into the color granules (fuzzy regions of the CIE triangle). This procedure is illustrated in Fig. 1. Pixels that constitute an image are usually represented by (r, g, b) values in the RGB color model. In our approach, every (r, g, b) pixel is transformed into the CIE model, by use of formulas well known in the literature (see e.g. [9,18]).

For every pixel of an image, its RGB values are transformed into the appropriate point in the CIE triangle, and then its membership value is determined. Hence, for every image we obtain a matrix of the membership values. As a matter of fact, this is a multidimensional matrix – composed of the membership matrices for every color granule, from 1 to 23.

The 23-dimensional matrix can be processed in parallel, producing particular color segmentation of an image. Of course, not every color granule occurs in an image, so we may obtain a matrix (or matrices) containing all values equal zero. This is visualized in Fig. 2 where color granules 13–23 do not exist in the image. However, the membership values are calculated for every pixel, and for each of 23 color granules.

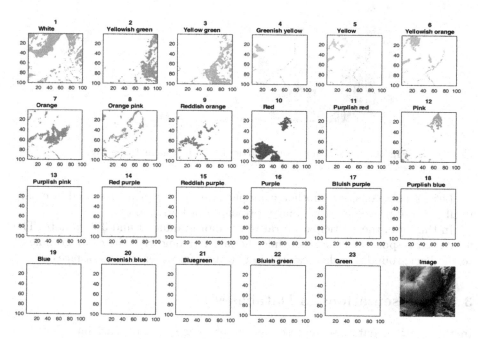

**Fig. 2.** Color granules in the input image (Image 1) (Color figure online)

Figure 3 portrays the histogram that illustrates participation rates of particular colors (fuzzy color granules) in Image 1. We see zero participation rates for color granules 13–23, very low participation rate for color granule 11, and highest participation rate for color granule 1. This histogram corresponds to Fig. 2. Value $p$ denotes the participation rate of every color granule (from 1 to

23), assuming that each of them participates in the image with the same rate. This value is applied as the unit of participation of particular colors in an image, and considered as a fuzzy number (see [24] and [26]); details concerning fuzzy numbers one can find e.g. in [8].

It is worth explaining that color saturation decreases towards the center of the CIE diagram (see e.g. [9]), Let us notice that color granule 1, labeled as "white", is located in the center (see Fig. 1). The CIE diagram is the two-dimensional model to describe hue and saturation. In this model, chromaticity is an objective specification of the quality of a color regardless of its luminance. It separates the three dimensions of color into one luminance dimension and a pair of chromaticity dimension (x and y coordinates). Taking into account the luminance (brightness), the "white" color domain (color granule 1) as a matter of fact contains points from "white" to "black" color.

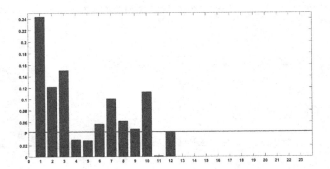

**Fig. 3.** Histogram of color participation in Image 1; see Fig. 2

Let us now consider another example – an image of a face, and compare results of the processing this image, presented in Figs. 4 and 5.

In this case, the participation rates for color granules 1 and 3 dominate. The former corresponds to the face while the latter (along with the color granule 2) to the background. Similar results are obtained for other images of faces.

## 3   Representations in Databases

In this section, database representations of image descriptions, introduced in [26], are presented and considered with regard to image retrieval, clustering, and classification.

Table 1 is a database table that includes data concerning participation rates of particular color granules, from 1 to 23, denoted as $C_1$, $C_2$, ..., $C_{23}$, in input images. The values in the table correspond to the histograms, e.g. Fig. 3 – for Image 1 presented in Fig. 2; Image $k$ is one of the images from the image collection under consideration.

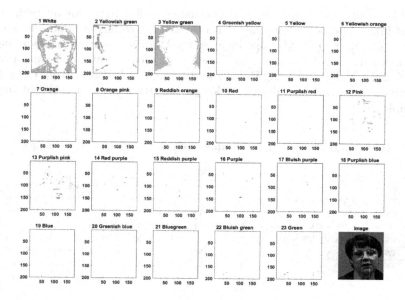

**Fig. 4.** Color granules in the input image of a face (Color figure online)

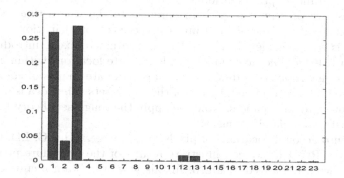

**Fig. 5.** Histogram of color participation in the image of the face from Fig. 4

**Table 1.** Participation rates of particular color granules, from 1 to 23, in input images

| File of image | Participation of $C_1$ | Participation of $C_2$ | ... | Participation of $C_{23}$ |
|---|---|---|---|---|
| Image 1 | 0.24 | 0.12 | ... | 0.00 |
| ... | ... | ... | ... | ... |
| Image $k$ | 0.27 | 0.01 | ... | 0.00 |
| ... | ... | ... | ... | ... |

Table 2 contains linguistic values obtained according to the membership functions of fuzzy sets, defined in [26], that describe the color participation rates in the images. First row of this table includes the linguistic values corresponding to the numerical values for Image 1 in Table 1. The membership functions of fuzzy sets VS, S, M, B, VB, denoting *Very Small, Small, Medium, Big, Very Big,* respectively, as linguistic values of color participation rates in an input image, are employed.

**Table 2.** Table with linguistic values of color participation in images

| Im. | $C_1$ | $C_2$ | $C_3$ | $C_4$ | $C_5$ | $C_6$ | $C_7$ | $C_8$ | $C_9$ | $C_{10}$ | $C_{11}$ | $C_{12}$ | $C_{13}$ | $C_{14}$ | $C_{15}$ | $C_{16}$ | $C_{17}$ | $C_{18}$ | $C_{19}$ | $C_{20}$ | $C_{21}$ | $C_{22}$ | $C_{23}$ |
|---|---|---|---|---|---|---|---|---|---|---|---|---|---|---|---|---|---|---|---|---|---|---|---|
| 1 | S | S | S | VS | VS | VS | S | VS | VS | S | VS | VS | VS | VS | VS | VS | VS | VS | VS | VS | VS | VS | VS |
| 2 | S | VS | S | VS | VS | VS | S | VS | VS | S | VS | VS | VS | VS | VS | VS | VS | VS | VS | VS | VS | VS | VS |
| 3 | M | VS | S | VS | VS | VS | VS | VS | VS | VS | VS | VS | VS | VS | VS | VS | VS | VS | VS | VS | VS | VS | VS |
| 4 | VS | M | S | VS | VS | VS | VS | VS | VS | S | VS | VS | VS | VS | VS | VS | VS | VS | VS | VS | VS | VS | VS |
| 5 | S | VS | VS | B | VS | VS | VS | VS | VS | VS | VS | VS | VS | VS | VS | VS | VS | VS | VS | VS | VS | VS | VS |
| 6 | S | VS | VS | VS | VS | VS | VS | S | VS | S | VS | B | VS | VS | VS | VS | VS | VS | VS | VS | VS | VS | VS |
| 7 | M | VS | S | VS | VS | VS | VS | VS | VS | VS | VS | S | VS | VS | VS | VS | VS | VS | VS | VS | VS | VS | VS |
| ... | ... | ... | ... | ... | ... | ... | ... | ... | ... | ... | ... | ... | ... | ... | ... | ... | ... | ... | ... | ... | ... | ... | ... |

Fuzzy databases are studied in the literature, e.g. [3, 6] More details concerning the color granules participation in images, with regard to representations in databases, are presented in [26]. In addition to the values that denote the participation rates of colors in the images, the color participation in parts of the images is also considered. This refers to the macropixels, introduced and employed in [19–25]. The fuzzy macropixels indicate locations within an image. Based on the maxropixels of different size, approximate shapes of color objects on a picture can be determined, by use of the rough sets. This idea is introduced in the authors' previous papers. Now, we apply the rough set theory in order to retrieve, cluster and classify images.

Taking into account locations of pixels in an image, LU, LC, LD, CU, MU, MC, MD, RU, RC, RD, we can illustrate results of the procedure presented in Fig. 1 in the form portrayed in Fig. 6. In this way, we obtain a cube composed of three-dimensional elements that contain values of participation rates of particular color granules, $C_1, C_2, ..., C_{23}$, in locations LU, ..., RD, where LU - left upper, LC - left center, LD - left down, MU - medium upper, MC - medium center, MD - medium down, RU - right upper, RC - right center, RD - right down.

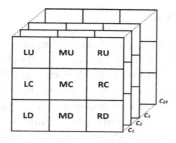

**Fig. 6.** Three-dimensional cube of color participation rates

Table 3 corresponds to the cube portrayed in Fig. 6 but presented in two-dimensional form of matrix with elements including participation values of color granules, $C_1$, $C_2$, ..., $C_{23}$, in locations LU, ..., RD. The exemplary values concern the particular locations of an image, analogously as the values in Table 1 express the color granules participation in the whole image.

**Table 3.** Table of image representation

|          | LU   | MU   | RU   | LC   | MC   | RC   | LD   | MD   | RD   |
|----------|------|------|------|------|------|------|------|------|------|
| $C_1$    | 0.18 | 0.20 | 0.30 | 0.27 | 0.16 | 0.25 | 0.36 | 0.30 | 0.35 |
| $C_2$    | 0.13 | 0.02 | 0.00 | 0.13 | 0.00 | 0.00 | 0.06 | 0.00 | 0.01 |
| $C_3$    | 0.52 | 0.14 | 0.52 | 0.44 | 0.00 | 0.18 | 0.52 | 0.00 | 0.16 |
| ...      |      |      |      |      |      |      |      |      |      |
| $C_{23}$ | 0.00 | 0.00 | 0.00 | 0.00 | 0.00 | 0.00 | 0.01 | 0.00 | 0.00 |

In this approach, every image is characterized by use of their cube or table representation, as shown in Fig. 6 or Table 3, respectively. Their elements, instead of numerical values, can include linguistic labels of membership functions of fuzzy sets – similarly as Table 2 is generated based on Table 1.

A big advantage of the cube representation is its similarity to the OLAP cube uses in warehouses; see e.g. [1]; OLAP stands for OnLine Analytical Processing. In this way, we can analyze color participations in bigger or smaller locations (macropixels) of an image; see also [26]. This means a hierarchical way, similarly like the time dimension in a typical OLAP cube can be analyzed with regard to e.g. year, month, day, hour.

In the case of 9 macropixels, like in Fig. 6, where locations LU ..., RD are considered, and 23 color granules of the CIE diagram, $C_1$, $C_2$, ..., $C_{23}$, we have 807 elements in the cube representation of an image. Of course, the more smaller macropixels are considered the more elements of the cube representation (as well as the table representation) we have. Therefore, a parallel approach should be applied.

## 4 Rough Granulation

In the rough set theory, the problem of classifying objects is based on accessible information about the objects. In particular, objects characterized by the same accessible information are considered as indiscernible [27]. In our case, objects are images from an image collection. Every image is characterized by the linguistic description generated by the parallel processing intelligent system [25].

Let $S = (U, A)$ where $U$ is the universe (a finite set of objects), and $A$ is a set of attributes (features, variables), with every attribute $a \in A$; meaning that attribute $a$ belongs to the considered set of attributes $A$. The domain of $a$ is a set of associated values, denoted as $V_a$. Each attribute $a$ defines an information function $f_a : U \rightarrow V_a$. According to [12], using the terminology of the rough set theory, $S = (U, A)$ is an information system.

Thus, knowledge representation, in the rough set theory, is realized by use of the information system of a tabular form that is similar to databases. Objects are characterized by attributes, and the attribute values are included in the tables.

In our case, $U$ is the set of images, and attributes are e.g. participation rates of particular color granules $C_1$, $C_2$, ..., $C_{23}$, in locations (macropixels) LU, ..., RD. The information function, $f_a$, transforms images into their table or cube representation, with elements expressing values of the attributes (see Fig. 6 or Table 3).

According to the rough set philosophy [12], objects in a universe of discourse – characterized by the same information – are indiscernible (similar) in view of the available information about them. Thus, the indiscernibility relation is defined. Any set of all indiscernible (similar) objects is called an elementary set, and forms a basic granule (atom) of knowledge about the universe.

For every set of attributes $B \subset A$, an indiscernibility relation $IND(B)$ on $U$ is defined in the following way: two objects are indiscernible by the set of attributes $B$ in $A$ if, for every attribute $b \in B$, values of these attributes are the same for both objects.

The tables, used in the rough set analysis, are called decision tables. Their columns are labeled by attributes, and rows by objects of interest. Attributes of the decision table are divided into two disjoint groups called condition and decision attributes, respectively. The decision tables include values of the attributes. Any decision table in S induces a set of IF-THEN decision rules.

With any rough set a pair of precise sets, called the lower and upper approximation of the rough set, is associated. The lower approximation consists of all objects which surely belong to the set and the upper approximation contains all objects which possibly belong to the set. The difference between the upper and the lower approximation constitutes the boundary region of the rough set. For details, see [12].

## 5 Image Retrieval and Classification

This section presents an application of the linguistic description of images in databases, as well as the rough granulation approach, to image retrieval, clustering and classification. From the rough set theory point of view, we want to cluster (or classify) or retrieve color digital pictures (images) that are indiscernible with regard to a specified equivalence (indiscernibility) relation.

In databases described in Sect. 3, every table representation of an image (see Table 3) has a name, the same as the associated image that it represents. For the image portrayed in Fig. 2, the name is "Image 1". Thus, we can compare images based on their table representations. Moreover, analyzing cube representations of images, we can compare their table representations. Each single table (and the corresponding cube) includes data that characterize only one image.

As mentioned earlier, every image is viewed as its cube or table representation with their values of linguistic labels of fuzzy numbers that express participation of color granules in particular locations (macropixels). By use of these image representations, we can easily retrieve (select) an image (or images) that e.g. contain many pixels of a particular color in a specific location. For the examples illustrated in Figs. 2 and 4, we see that the table (or cube) representation of the former image include more empty (zero value) elements than the latter one. In addition, there are bigger values of participation of the dominating colors in the image of the face.

In the rough set theory, the so-called information tables (information systems) are two-dimensional tables where rows correspond to objects and columns to their attributes. In our approach, objects (color images) are represented by two-dimensional tables (Table 3), equivalent to three-dimensional cubes (Fig. 6).

As a result of the problem of image retrieval, as well as image clustering or classification, we can obtain a rough set of images. This means that as the answer to such a query formulated to the database, we can get images that match the query, and those that may or not fulfill requirements for that query. The decision is inferred by use of IF-THEN rules within the framework of the rough set theory. Hence, to answer the query, images that are similar with regard to the indiscernibility relation are inferred.

For example, consider a query to select an image, from a collection of images, that contains a big number of pixels of orange color located in the right upper part of the picture. As the answer to this query, we get images that are indiscernible (similar) in view of color and location attributes: color $C_7$ and location RU. The linguistic descriptions in the query: "big number" of pixels, as well as "orange" color and "right upper" location are labels of fuzzy sets. With regard to the linguistic variable "number of pixels", the linguistic values, VS - very small, S - small, M - medium, B - big, VB -very big, are defined as membership functions of fuzzy sets, in the same way as the fuzzy numbers expressing the color participation in Table 2. For details concerning fuzzy sets and their membership functions, see e.g. [16].

# 6  Conclusions and Final Remarks

The rough set theory [11], similar as the fuzzy sets [28] and fuzzy logic [29], are powerful mathematical tools for processing imprecise and incomplete data. These artificial intelligence methods are widely applied in data mining and pattern recognition; see e.g. [2,10,14]. In our approach, the rough set theory is applied to the databases that include fuzzy values. Thus, the equivalence relation that we employ in the image retrieval, clustering or classification, is rather viewed as the similarity relation than strictly the indiscernibility relation.

When data mining tools are implemented on high-performance parallel computers, they can analyze massive databases in a reasonable time [5]. The parallel computing approach to data mining algorithms is known as *Parallel Data Mining*. The so-called *SPMD Parallelism (Single Program Multiple Data)* – that is exploited when a set of processes execute in the same algorithm on different partitions of a data set, and processes cooperate to exchange partial results – is suitable for the problems presented in this paper. This concerns the analysis of particular macropixels, and a large number of images; both issues are described and illustrated in [25].

The so-called *Independent Parallelism* [5] – exploited when processes are executed in parallel in independent way – can also be applied in order to obtain the color image segmentation (see examples in Figs. 2 and 4). In this case, each process has access to the whole image, and is realized for different parameter that is the color granule in the CIE diagram.

The *Distributed Data Mining* should also be considered to analyze images from multi-site repositories, not only stored in centralized collections like warehouses.

This paper presents basic concepts of the parallel approach to the problem of processing color images represented by linguistic description in databases. The main idea is to realize in parallel both generating the linguistic description of images and analysis of the databases in order to retrieve, cluster or classify color digital images.

It is obvious that processing big data, that means large and streaming data, requires parallelization that significantly decreases computational time; see e.g. [7]. It is also known that the MATLAB software – which is suitable for our applications – realizes the *SPMD* approach [4] in the *Matlab Parallel Toolbox*.

# References

1. Alain, K.M., Nathanael, K.M., Rostin, M.M.: Integrating fuzzy concepts to design a fuzzy data warehouse. Int. J. Comput. **27**(1), 112–132 (2017)
2. Bello, R., Falcon, R., Pedrycz, W., Kacprzyk, J. (eds.): Granular Computing: At the Junction of Rough Sets and Fuzzy Sets. Springer, Heidelberg (2008). https://doi.org/10.1007/978-3-540-76973-6
3. Buckles, B.P., Petry, F.E.: Extending the fuzzy database with fuzzy numbers. Inform. Sci. **34**(2), 145–155 (1984)
4. Cao, J.-J., Fan, S.-S., Yang, X.: SPMD performance analysis with parallel computing of MATLAB. In: Proceedings of the 2012 Fifth International Conference on Intelligent Networks and Intelligent Systems, pp. 80–83. IEEE (2012)

5. Cesario, E., Talia, D.: From parallel data mining to grid-enabled distributed knowledge discovery. In: An, A., Stefanowski, J., Ramanna, S., Butz, C.J., Pedrycz, W., Wang, G. (eds.) RSFDGrC 2007. LNCS (LNAI), vol. 4482, pp. 25–36. Springer, Heidelberg (2007). https://doi.org/10.1007/978-3-540-72530-5_3

6. De, S.K., Biswas, R., Roy, A.R.: On extended fuzzy relational database model with proximity relations. Fuzzy Sets Syst. **117**(2), 195–201 (2001)

7. Devi, V.S., Meena, L.: Parallel MCNN (PMCNN) with application to prototype selection on large and streaming data. J. Artif. Intell. Soft Comput. Res. **7**(3), 155–169 (2017)

8. Dubois, D., Prade, H.: Fuzzy Sets and Systems: Theory and Applications. Academic Press, New York (1980)

9. Fortner, B.: Number by color. Part 5. SciTech J. **6**, 30–33 (1996)

10. Pal, S.K., Meher, S.K., Dutta, S.: Class-dependent rough-fuzzy granular space, dispersion index and classification. Pattern Recognit. **45**, 2690–2707 (2012)

11. Pawlak, Z.: Rough Sets. Theoretical Aspects of Reasoning about Data. Kluwer Academic Publishers, Dordrecht (1991)

12. Pawlak, Z.: Rough set theory and its applications. J. Telecommun. Inf. Tech. **3**, 7–10 (2002)

13. Pawlak, Z.: Granularity of knowledge, indiscernibility and rough sets. In: Fuzzy Systems Proceedings. IEEE World Congress on Computational Intelligence, vol. 1, pp. 106–110 (1998)

14. Pedrycz, W., Park, B.J., Oh, S.K.: The design of granular classifiers: a study in the synergy of interval calculus and fuzzy sets in pattern recognition. Pattern Recognit. **41**, 3720–3735 (2008)

15. Rakus-Andersson, E.: Approximation and rough classification of letter-like polygon shapes. In: Skowron, A., Suraj, Z. (eds.) Rough Sets and Intelligent Systems, pp. 455–474. Springer, Heidelberg (2013). https://doi.org/10.1007/978-3-642-30341-8_24

16. Rutkowska, D.: Neuro-Fuzzy Architectures and Hybrid Learning. Springer, Heidelberg (2002). https://doi.org/10.1007/978-3-7908-1802-4

17. Skowron, A., Stepaniuk, J.: Information granules: towards foundations of granular computing. Int. J. Intell. Syst. **16**(1), 57–85 (2001)

18. Wiaderek, K.: Fuzzy sets in colour image processing based on the CIE chromaticity triangle. In: Rutkowska, D., Cader, A., Przybyszewski, K. (eds.) Selected Topics in Computer Science Applications, pp. 3–26. Academic Publishing House EXIT, Warsaw (2011)

19. Wiaderek, K., Rutkowska, D.: Fuzzy granulation approach to color digital picture recognition. In: Rutkowski, L., Korytkowski, M., Scherer, R., Tadeusiewicz, R., Zadeh, L.A., Zurada, J.M. (eds.) ICAISC 2013, Part I. LNCS (LNAI), vol. 7894, pp. 412–425. Springer, Heidelberg (2013). https://doi.org/10.1007/978-3-642-38658-9_37

20. Wiaderek, K., Rutkowska, D., Rakus-Andersson, E.: Color digital picture recognition based on fuzzy granulation approach. In: Rutkowski, L., Korytkowski, M., Scherer, R., Tadeusiewicz, R., Zadeh, L.A., Zurada, J.M. (eds.) ICAISC 2014, Part I. LNCS (LNAI), vol. 8467, pp. 319–332. Springer, Cham (2014). https://doi.org/10.1007/978-3-319-07173-2_28

21. Wiaderek, K., Rutkowska, D., Rakus-Andersson, E.: Information granules in application to image recognition. In: Rutkowski, L., Korytkowski, M., Scherer, R., Tadeusiewicz, R., Zadeh, L.A., Zurada, J.M. (eds.) ICAISC 2015, Part I. LNCS (LNAI), vol. 9119, pp. 649–659. Springer, Cham (2015). https://doi.org/10.1007/978-3-319-19324-3_58

22. Wiaderek, K., Rutkowska, D., Rakus-Andersson, E.: New algorithms for a granular image recognition system. In: Rutkowski, L., Korytkowski, M., Scherer, R., Tadeusiewicz, R., Zadeh, L.A., Zurada, J.M. (eds.) ICAISC 2016, Part II. LNCS (LNAI), vol. 9693, pp. 755–766. Springer, Cham (2016). https://doi.org/10.1007/978-3-319-39384-1_67

23. Wiaderek, K., Rutkowska, D., Rakus-Andersson, E.: Linguistic description of color images generated by a granular recognition system. In: Rutkowski, L., Korytkowski, M., Scherer, R., Tadeusiewicz, R., Zadeh, L.A., Zurada, J.M. (eds.) ICAISC 2017. LNCS (LNAI), vol. 10245, Part I, pp. 603–615. Springer, Cham (2017). https://doi.org/10.1007/978-3-319-59063-9_54

24. Wiaderek, K., Rutkowska, D.: Linguistic description of images based on fuzzy histograms. In: Choraś, M., Choraś, R.S. (eds.) IP&C 2017. AISC, vol. 681, pp. 27–34. Springer, Cham (2018). https://doi.org/10.1007/978-3-319-68720-9_4

25. Wiaderek, K., Rutkowska, D., Rakus-Andersson, E.: Parallel processing of color digital images for linguistic description of their content. In: Wyrzykowski, R., Dongarra, J., Deelman, E., Karczewski, K. (eds.) PPAM 2017, Part I. LNCS, vol. 10777, pp. 544–554. Springer, Cham (2018). https://doi.org/10.1007/978-3-319-78024-5_47

26. Wiaderek, K., Rutkowska, D., Rakus-Andersson, E.: Image retrieval by use of linguistic description in databases. In: Rutkowski, L., Scherer, R., Korytkowski, M., Pedrycz, W., Tadeusiewicz, R., Zurada, J.M. (eds.) ICAISC 2018, Part II. LNCS (LNAI), vol. 10842, pp. 92–103. Springer, Cham (2018). https://doi.org/10.1007/978-3-319-91262-2_9

27. Wolkenhauer, O.: Possibility Theory with Applications to Data Analysis. UMIST Control Systems Centre Series. Wiley, New York (1998)

28. Zadeh, L.A.: Fuzzy sets. Inf. Control **8**, 338–353 (1965)

29. Zadeh, L.A.: Fuzzy logic = computing with words. IEEE Trans. Fuzzy Syst. **4**, 103–111 (1996)

30. Zadeh, L.A.: Toward a theory of fuzzy information granulation and its centrality in human reasoning and fuzzy logic. Fuzzy Sets Syst. **90**, 111–127 (1997)

# An OpenMP Parallelization of the $K$-means Algorithm Accelerated Using KD-trees

Wojciech Kwedlo[(✉)] and Michał Łubowicz

Faculty of Computer Science, Białystok University of Technology,
Wiejska 45a, 15-351 Białystok, Poland
w.kwedlo@pb.edu.pl

**Abstract.** In the paper a KD-tree based filtering algorithm for $K$-means clustering is considered. A parallel version of the algorithm for shared memory systems, which uses OpenMP tasks both for KD-tree construction and filtering in the assignment step of $K$-means, is proposed. In our approach, an OpenMP task is created for a recursive call performed by tree construction and filtering procedures. A data partitioning step during the tree construction is also parallelized by OpenMP tasks. In computational experiments we measured runtimes of the parallel and serial version of the filtering algorithm and a parallel version of classical Lloyd's algorithm for six datasets sampled from two distributions. The results of experiments, performed on a 24-core system indicate that our version filtering algorithm has very good parallel efficiency. Its runtime is up to four orders of magnitude shorter than the runtime of parallel Lloyd's algorithm.

**Keywords:** $K$-means clustering · OpenMP tasks · KD-trees

## 1 Introduction

Clustering [8] is an important problem in machine learning and data mining. The aim of clustering can be defined as dividing a set of objects into $K$ disjoint groups, called clusters, in such a way that the objects within the same group are very similar, whereas the objects in different groups are very distinct. In this work, we assume that the number of clusters $K$ is known a priori. Clustered objects are represented by a learning set $X$ consisting of $N$ feature vectors $x(1), \ldots, x(n)$. It is assumed that $x(i) \in \mathbb{R}^M$, where $M$ is the dimension of the feature space. Clustering problem can be then defined as the problem of searching for partition of $X$ minimizing a criterion function. A sum of squared error (SSE) is a commonly used criterion. It is defined as the sum of squared distances between feature vectors and nearest cluster centroids. The clustering problem with the $SSE$ criterion and Euclidean distance is known to be NP-hard for $K > 1$ [1].

© Springer Nature Switzerland AG 2020
R. Wyrzykowski et al. (Eds.): PPAM 2019, LNCS 12043, pp. 457–466, 2020.
https://doi.org/10.1007/978-3-030-43229-4_39

The $K$-means algorithm [12] is a popular heuristic for minimization of the SSE. It is an iterative refinement method, which given an initial solution (a set of cluster centroids) produces a sequence of solutions with a decreasing values of SSE. Its most popular formulation is also called Lloyd's method [11]. An iteration of the algorithm consists of an assignment step and an update step. In the assignment step each feature vector is allocated to the cluster represented by the nearest centroid. In the update step the cluster centroids are recalculated as sample means of allocated feature vectors. The iterations consisting of the assignment and the update steps are performed until no feature vector changes its assigned cluster.

The Lloyd's algorithm is a brute force method, which in each assignment step computes distances between all the feature vectors and all the centroids. In recent years a lot of research work has been devoted to faster alternatives, which are exact, i.e, give the same output as the Lloyd's method. These accelerated approaches can be roughly divided into two groups. The first group consists of the algorithms [7], which use the bounds on distances and the triangle inequality to skip some distance calculations. While they work very well in high-dimensional feature spaces (for large $M$), their speed in low-dimensional feature spaces is considerably lower. The second group (e.g., [9,15]) consists of methods, which use a KD-tree structure to organize the clustered data. These algorithms are very fast only in low dimensional spaces [9]. Clustering low-dimensional data is used in many applications, e.g., in seismology (grouping earthquakes along seismic faults) or astronomy (grouping stars in galaxies).

The main contribution of this work is development, implementation and experimental evaluation of a parallel version of a KD-tree based, *filtering* algorithm [9,15] for shared memory systems. This version uses OpenMP tasks to balance the load. While parallel versions of the filtering algorithm for systems with distributed memory, parallelized using MPI bindings for Java, do exist (e.g., [16]) their efficiency is greatly reduced because of poor load balancing. Other parallel version of $K$-means algorithms are limited to Lloyd's method or the triangle inequality-based approaches [7,10].

The remainder of the paper is organized as follows. In the next section Lloyd's algorithm is presented. Section 3 contains the description of the filtering algorithm. Section 4 describes the proposed parallelization of this algorithm. The results of computational experiments performed on a 24-core system are presented in Sect. 5. The last section concludes the paper.

## 2   Lloyd's Algorithm

Denote by $a(i) \in \{1, \ldots, K\}$, where $i = 1, \ldots, N$, an index of a cluster to which a feature vector $x(i)$ is assigned, and by $c(j) \in \mathbb{R}^M$ the centroid of $j$-th cluster, where $j = 1, \ldots, K$. The pseudo-code of Lloyd's method is shown on Algorithm 1.

In the assignment step a feature vector $x(i)$ is allocated to the cluster represented by the nearest centroid. $d(x, y)$ denotes a distance between two vectors in $\mathbb{R}^M$. In the update step the cluster centroids are computed as sample

**Data**: Initial cluster centroids $c(1), c(2), \ldots, c(K)$
**repeat**
> {assignment step}
> **for** $i \leftarrow 1$ **to** $N$ **do**   $a(i) \leftarrow \underset{j=1\ldots K}{\arg\min}\, \mathrm{d}(x(i), c(j))$
>
> {update step}
> **for** $j \leftarrow 1$ **to** $K$ **do**   $y(j) \leftarrow \mathbf{0}, z(j) \leftarrow 0$
> **for** $i \leftarrow 1$ **to** $N$ **do**
>> $y(a(i)) \leftarrow y(a(i)) + x(i)$
>> $z(a(i)) \leftarrow z(a(i)) + 1$
>
> **for** $j \leftarrow 1$ **to** $K$ **do**   $c(j) \leftarrow y(j)/z(j)$

**until** assignment $a$ does not change

**Algorithm 1.** Lloyd's algorithm [7, 10]

means of allocated feature vectors. The most time-consuming part of the step is the loop iterating over all the feature vectors. This loop accumulates the so-called sufficient statistics: sums of feature vectors assigned to each cluster in $y(1), y(2), \ldots, y(K)$, and counts of feature vectors assigned to each cluster in $z(1), z(2), \ldots, z(K)$. Afterwards, the vector sums are divided by the corresponding counts, giving new centroid coordinates.

The parallelization of Lloyd's algorithm using OpenMP threads can be easily done using data decomposition. The learning set $X$ and the corresponding assignment array $a$ are partitioned into equally-sized portions. Each OpenMP thread is responsible for a single portion. In the assignment step each thread independently computes the assignments for its portion of feature vectors. In the update step each thread first computes private sufficient statistics for its portion of feature vectors. Next the private sufficient statistics are summed giving the global sufficient statistics. We implemented [10] this phase using a minimum-spanning tree reduction algorithm [4]. Then, the cluster centroids are computed from the sufficient statistics by the master thread.

## 3   The Filtering Algorithm

The filtering algorithm [9,15] organizes the feature vectors in a variant of a KD-tree [3], which is a binary search tree partitioning the feature space by axis-orthogonal hyperplanes. Each node of the tree stores information about the feature vectors contained in a hyperrectangle (called a *cell*) represented by two $M$-length boundary vectors. The root's cell is the bounding box of the dataset.

The tree is constructed using a recursive partitioning strategy. If a cell contains a number of vectors smaller than a small constant $S$ it is declared as a *leaf*. Otherwise it is split into two subcells by an axis-orthogonal hyperplane. The vectors stored in the cell are partitioned into one or the other side of the hyperplane. The tree nodes representing the two subcells are the *left child* and *right child* of the original node. There are several methods for selecting a splitting hyperplane. Following [9] we employed the *sliding midpoint* rule, which tries to divide the cell through its middle point using a hyperplane orthogonal to its the longest side. However, in case of a *trivial split* with all the feature vectors of the original cell lying on one side of the splitting hyperplane the rule "slides" the hyperplane toward the vectors until it encounters the first vector.

A version of KD-tree used by the filtering algorithm has to store some additional information in a node $u$: the sum (denoted as $u.y$) and the count (denoted as $u.z$) of all the feature vectors contained in the node. It must be also able to quickly identify feature vectors contained in a node. In our implementation, we store the feature vectors as rows of a two-dimensional array. During the construction phase, when partitioning of feature vectors contained in a cell, we physically rearrange rows of the array, using the method similar to the popular quicksort algorithm [5]. This approach ensures that each node of the tree is associated with a continuous block of rows. It is thus sufficient, for a node $u$ to store in $u.first$ and $u.last$ the indices of the first and the last feature vector in the block.

The pseudo-code of the filtering algorithm, in a form of a recursive function **Filter**, which combines the assignment step with computation of sufficient statistics $y$ and $z$, is shown on Algorithm 2.

> **Filter**(KDTreeNode $u$, CentroidIndices $C$)
>> **if** $u$ is a leaf **then**
>>> **for** $i \leftarrow u.first$ **to** $u.last$ **do**
>>>> $a(i) \leftarrow \arg\min_{j \in C} d(x(i), c(j))$
>>>> $y(a(i)) \leftarrow y(a(i)) + x(i)$
>>>> $z(a(i)) \leftarrow z(a(i)) + 1$
>>
>> **else**
>>> $j^* \leftarrow \arg\min_{j \in C} d(c(j), u.midpoint)$
>>> **foreach** $j \in C \setminus j^*$ **do**
>>>> **if IsFarther** $(c(j), c(j^*), u.cell)$ **then** $C \leftarrow C \setminus j$
>>>
>>> **if** $|C = 1|$ **then**
>>>> **for** $i \leftarrow u.first$ **to** $u.last$ **do** $a(i) \leftarrow j^*$
>>>> $y(j^*) \leftarrow y(j^*) + u.y$
>>>> $z(j^*) \leftarrow z(j^*) + u.z$
>>>
>>> **else**
>>>> **Filter** $(u.left, C)$
>>>> **Filter** $(u.right, C)$

**Algorithm 2.** The filtering algorithm.

The arguments of the function are a node of the tree $u$ and a set of *candidate centroid* indices $C$. A candidate centroid is a centroid which can be the nearest one for any point contained in a node's cell. The candidate centroids for the tree root $r$ are all the centroids, i.e., the initial call is **Filter**$(r, \{1, 2, \ldots, K\})$.

If a node $u$ is a leaf, then an iteration of Lloyd's algorithm, for all feature vectors contained in $u$ using candidate centroids is performed. Otherwise, the candidate centroids indices are filtered out and propagated down the tree. First a candidate centroid with index $c(j^*)$, which is closest to cell midpoint (denoted as $u.midpoint$) is found. Then, for every remaining candidate centroid $c(j)$ the algorithm checks, using **IsFarther** function if no part of $u.cell$ is closer to $c(j)$ than it is to $c(j^*)$. If so, $c(j)$ cannot be the nearest centroid to any feature vector in $u.cell$ and thus its index can be removed from the candidates. The details of **IsFarther** function [9] are omitted in this paper due to space limitations.

If, after the filtering only one candidate index (it must be $j^*$) remains in $C$ then all the feature vectors are allocated to the cluster $j^*$ and the vector sums and counts of all the feature vectors contained in $u$ are added to $y(j^*)$ and $z(j^*)$, respectively. Otherwise we apply a recursion on the children of $u$.

## 4   Parallelization with OpenMP Tasks

When an OpenMP thread encounters a `task` construct [14], the related block of code is made ready for execution. The execution may be immediate (e.g., by an encountering or idle thread) or deferred to some time in future. Tasks may be executed in any order. The `taskwait` directive is provided for waiting on all child tasks generated by the current task. OpenMP tasks are a useful mechanism for parallelization in situations, where amount of parallelism is unknown, for instance when traversing a pointer-based tree structure.

In our implementation both the KD-tree construction and the **Filter** function are parallelized using a similar approach. When the serial tree construction procedure or the **Filter** function makes a recursive call its parallel counterpart creates a new OpenMP task. To avoid overwhelming of an OpenMP runtime with a huge number of small tasks, a new task is not created if the number of feature vectors in a tree node cell is less than $N_{\min}$ parameter. In this case our code reverts to sequential execution, and no further tasks on lower recursion levels are created.

Similarly to parallel Lloyd's algorithm our approach maintains private per-thread sufficient statistics $y$ and $z$. When a task has to update the statistics it uses the private copy of its executing thread. After the completion of the **Filter** function the private statistics are reduced, in the same way as in parallel Lloyd's method, using a minimum-spanning tree reduction [4].

A costly step in a tree construction procedure is the partition of feature vectors contained in a tree node by a splitting hyperplane. For the root of the tree, when only one OpenMP task is active the cost of this step is $O(MN)$. Sequential execution of partition would result in a high load imbalance at higher levels of the tree (e.g., only one core active for the tree root). For this reason we parallelized the partition [6] using OpenMP tasks.

Initial profiling of our parallel code indicated that memory allocation took a large share of runtime of the tree construction. To alleviate this bottleneck, we implement a private memory pool for each thread. Each thread first preallocates a pool of $N_{\text{th}}$ objects (i.e., tree nodes). The requests for new nodes by tasks are served from the pools of executing threads, without use of the C++ heap memory manager. Only when a pool is exhausted a new heap allocation for a new pool of $N_{th}$ objects is performed.

## 5   Experimental Results

In this section the results of the computational experiments with Lloyd's and filtering algorithms are reported. The experiments had two objectives. The first of them was the assessment of strong scaling efficiency of our parallel version of the filtering algorithm. The second objective was the comparison of runtime of parallel filtering algorithm with the runtime of parallel Lloyd's method.

When measuring the runtime of the filtering approach a question arises whether to include the time needed to build a KD-tree into runtime of the K-means algorithm. On the one hand, doing so would be more conservative. On the other hand, a KD-tree need to be build only once for a given dataset and can be used in all subsequent clustering experiments. In our experiments we have decided to use both options.

All the experiments were carried out on single nodes of Tryton supercomputer installed at the Academic Computer Centre in Gdańsk, Poland. A node of Tryton is a system with two 12-core Intel Xeon E5-2670 v3 (2.3 GHz) CPUs with 128 or 256 GiB DDR4 DRAM. The algorithms were implemented in C++ and compiled by Intel C++ compiler (icpc version 19.0.2) using the following set of optimization options: `-Ofast -ipo -xCORE-AVX2`.

The experiments were performed on datasets sampled from two three-dimensional distributions. The first of them, denoted as *mixture3*, is a Gaussian Mixture Model, which consists of 1024 very well separated Gaussian clusters. The parameters of the model were obtained using the simulator proposed in [13]. The second distribution, denoted as *uniform3* is a continuous uniform distribution on $[0, 1)$.

In all the experiments we used the same set of parameters. $S$, the minimal number of feature vectors in KD-tree leaf was set to 64. $N_{min}$, the minimal number of feature vectors in a node which resulted the two child nodes processed by two OpenMP tasks was set to 10000. The memory pool size $N_{th}$ was set to 4096.

For each of two distributions we sampled three datasets using $N \in \{10^7, 10^8, 10^9\}$. For each of six datasets we used six values of $K \in \{16, 64, 256, 1024, 4096, 16384\}$. For each combination of a dataset and $K$ we generated a initial single solution using the K-means|| initialization method [2]. This solution was used to initialize the serial version of the filtering algorithm, the parallel version of the filtering algorithm, and the parallel version of Lloyd's method. Because the same solution was used to initialize three versions of $K$-means algorithm, they always achieved the convergence after the same number of iterations. The average iteration time of the algorithms is shown on Table 1 To reduce effect of the system noise we repeated each experiment three times; the median of three

**Table 1.** Average iteration times (in seconds) for different $K$-means algorithms. The fourth column shows the number of iterations required for the convergence.

| Dataset family | $N$ | $K$ | #iter | Filtering sequential | Filtering parallel | Lloyd's parallel |
|---|---|---|---|---|---|---|
| mixture3 | $10^7$ | 16 | 61 | 0.0556(0.00152) | 0.00339(0.000269) | 0.0551 |
| | | 64 | 35 | 0.0964(0.00212) | 0.0057(0.000273) | 0.185 |
| | | 256 | 15 | 0.223(0.00333) | 0.0133(0.000382) | 0.702 |
| | | 1024 | 39 | 0.0939(0.00928) | 0.00576(0.00081) | 2.77 |
| | | 4096 | 263 | 0.233(0.22) | 0.0121(0.0113) | 11.7 |
| | | 16384 | 366 | 0.6(0.591) | 0.0296(0.0291) | 45.1 |
| | $10^8$ | 16 | 41 | 0.904(0.011) | 0.0431(0.00134) | 0.553 |
| | | 64 | 37 | 0.998(0.00891) | 0.0479(0.00111) | 1.85 |
| | | 256 | 22 | 1.67(0.00991) | 0.079(0.00123) | 7.04 |
| | | 1024 | 70 | 0.55(0.0266) | 0.0264(0.00202) | 27.7 |
| | | 4096 | 715 | 1.01(0.956) | 0.0483(0.0459) | 117 |
| | | 16384 | 1098 | 2.34(2.3) | 0.106(0.104) | 455 |
| | $10^9$ | 16 | 80 | 5.28(0.0462) | 0.245(0.00515) | 5.51 |
| | | 64 | 93 | 4.54(0.0386) | 0.21(0.00395) | 18.5 |
| | | 256 | 26 | 16.2(0.06) | 0.744(0.00682) | 70.3 |
| | | 1024 | 72 | 5.92(0.1) | 0.274(0.00624) | 277 |
| | | 4096 | 1754 | 4.61(4.37) | 0.212(0.201) | 1.16e3 |
| | | 16384 | 2159 | 10.6(10.4) | 0.489(0.48) | 4.49e3 |
| uniform3 | $10^7$ | 16 | 189 | 0.0564(0.0374) | 0.00319(0.00211) | 0.0552 |
| | | 64 | 581 | 0.0728(0.0666) | 0.00383(0.00349) | 0.185 |
| | | 256 | 898 | 0.126(0.122) | 0.00625(0.00602) | 0.702 |
| | | 1024 | 2415 | 0.213(0.211) | 0.00997(0.00989) | 2.78 |
| | | 4096 | 1281 | 0.386(0.383) | 0.0181(0.0179) | 11.7 |
| | | 16384 | 640 | 0.801(0.795) | 0.0382(0.0377) | 45.7 |
| | $10^8$ | 16 | 970 | 0.217(0.174) | 0.0106(0.00864) | 0.549 |
| | | 64 | 931 | 0.364(0.319) | 0.0174(0.0154) | 1.85 |
| | | 256 | 3264 | 0.592(0.579) | 0.0282(0.0276) | 7.01 |
| | | 1024 | 2357 | 1.01(0.995) | 0.0474(0.0466) | 27.7 |
| | | 4096 | 4291 | 1.68(1.67) | 0.0762(0.0757) | 117 |
| | | 16384 | 3588 | 2.89(2.88) | 0.129(0.128) | 449 |
| | $10^9$ | 16 | 1234 | 1.2(0.812) | 0.055(0.0381) | 5.5 |
| | | 64 | 1039 | 1.96(1.49) | 0.089(0.0691) | 18.5 |
| | | 256 | 3715 | 2.85(2.71) | 0.131(0.126) | 70.1 |
| | | 1024 | 4391 | 4.71(4.6) | 0.217(0.213) | 277 |
| | | 4096 | 13590 | 7.67(7.63) | 0.355(0.353) | 1.16e3 |
| | | 16384 | 9353 | 12.7(12.6) | 0.581(0.578) | 4.49e3 |

runtimes is reported. For two versions of the filtering algorithm we report two average iteration times. The greater value includes the time needed to build a KD-tree. The smaller value (in parentheses) excludes the KD-tree building time.

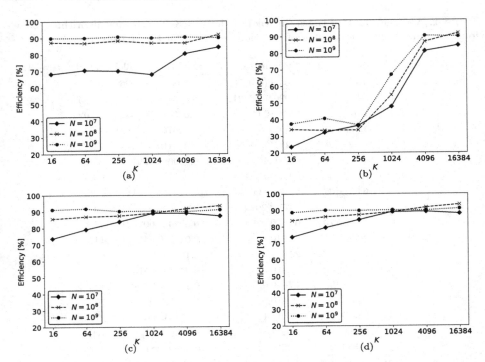

**Fig. 1.** Parallel efficiency depending on the number of clusters $K$ and learning set size $N$ for (a) *mixture3* datasets, (b) for *mixture3* datasets without KD-tree building, (c) *uniform3* datasets, (d) for *uniform3* datasets without KD-tree building

Based on the average iteration times we calculated the strong scaling efficiency of the filtering algorithm as the ratio (in percents) of parallel speedup (equal to the runtime of serial version divided by the runtime of parallel version) to the ideal linear speedup equal 24 (a node of Tryton has 24 cores total). The efficiency plots in are shown on Fig. 1.

The plots demonstrate that in most cases algorithm scales very well, with the efficiency near or greater than 70%. The only exception to this rule are mixture3 datasets for $K \leq 1024$, when the tree building time is omitted from the clustering runtime. The Table 1 shows, that in this case the number of iterations is small. We conjecture, that the parallel efficiency of the **Filter** function is lower at the initial iterations of the $K$-means algorithm.

To compare the runtimes of parallel algorithms we have calculated so called algorithmic speedup [7] of the filtering algorithm as a ratio of the average iteration time of Lloyd's method to the average iteration time of the filtering algorithm. The algorithmic speedup plots are shown on Fig. 2.

The plots indicate that, even if we include the time needed to build a KD-tree into the runtime of the $K$-means, the filtering algorithm is much faster. The speedup ranges from tens (for small $K$) to thousands (for large $K$).

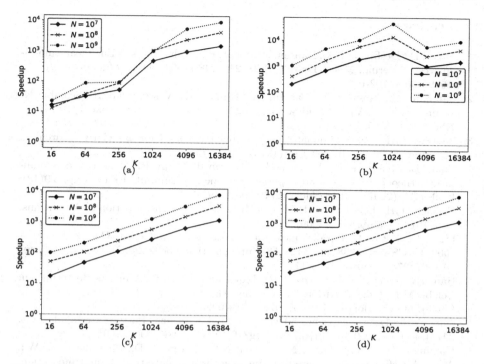

**Fig. 2.** Algorithmic speedup of the parallel filtering algorithm over parallel Lloyd's algorithm depending on the number of clusters $K$ and learning set size $N$ for (a) *mixture3* datasets, (b) for *mixture3* datasets without KD-tree building, (c) *uniform3* datasets, (d) for *uniform3* datasets without KD-tree building

## 6 Conclusions

In this paper we proposed a shared memory parallel version of the filtering algorithm for the $K$-means clustering. To parallelize recursive and irregular problems of tree construction and traversal we used OpenMP tasks. The strong scalability of our method was evaluated experimentally on six datasets. The experiments demonstrated that, excluding several cases of small $K$, our method scales very well. Moreover, it runs much faster than the parallel Lloyd's algorithm.

The results of this study allow us to recommend our parallel version of the filtering algorithm over Lloyd's method for clustering in low dimensional feature spaces. An open question remains, at which dimension $M$ the increased costs of the algorithm (which rise exponentially with $M$ [9]) make it slower than Lloyd's method. We are going to investigate this problem in further works.

**Acknowledgments.** This work was supported by Białystok University of Technology grant S/WI/2/2018 funded by Polish Ministry of Science and Higher Education. The calculations were carried out at the Academic Computer Centre in Gdańsk, Poland.

# References

1. Aloise, D., Deshpande, A., Hansen, P., Popat, P.: NP-hardness of Euclidean sum-of-squares clustering. Mach. Learn. **75**(2), 245–248 (2009). https://doi.org/10.1007/s10994-009-5103-0
2. Bahmani, B., Moseley, B., Vattani, A., Kumar, R., Vassilvitskii, S.: Scalable K-means++. Proc. VLDB Endow. **5**(7), 622–633 (2012). https://doi.org/10.14778/2180912.2180915
3. Bentley, J.: Multidimensional binary search trees used for associative searching. Commun. ACM **18**(9), 509–517 (1975)
4. Chan, E., Heimlich, M., Purkayastha, A., van de Geijn, R.: Collective communication: theory, practice, and experience. Concurr. Comput. Pract. Exp. **19**(13), 1749–1783 (2007). https://doi.org/10.1002/cpe.1206
5. Cormen, T.H., Leiserson, C.E., Rivest, R.L., Stein, C.: Introduction to Algorithms. MIT Press, Cambridge (2009)
6. Frias, L., Petit, J.: Parallel partition revisited. In: McGeoch, C.C. (ed.) WEA 2008. LNCS, vol. 5038, pp. 142–153. Springer, Heidelberg (2008). https://doi.org/10.1007/978-3-540-68552-4_11
7. Hamerly, G., Drake, J.: Accelerating Lloyd's algorithm for $k$-means clustering. In: Celebi, M.E. (ed.) Partitional Clustering Algorithms, pp. 41–78. Springer, Cham (2015). https://doi.org/10.1007/978-3-319-09259-1_2
8. Jain, A.K.: Data clustering: 50 years beyond K-means. Pattern Recognit. Lett. **31**(8), 651–666 (2010). https://doi.org/10.1016/j.patrec.2009.09.011
9. Kanungo, T., Mount, D.M., Netanyahu, N.S., Piatko, C.D., Silverman, R., Wu, A.Y.: An efficient K-means clustering algorithm: analysis and implementation. IEEE Trans. Pattern Anal. Mach. Intell. **24**(7), 881–892 (2002). https://doi.org/10.1109/TPAMI.2002.1017616
10. Kwedlo, W., Czochański, P.J.: A hybrid MPI/OpenMP parallelization of K-means algorithms accelerated using the triangle inequality. IEEE Access **7**, 42280–42297 (2019). https://doi.org/10.1109/ACCESS.2019.2907885
11. Lloyd, S.: Least squares quantization in PCM. IEEE Trans. Inf. Theory **28**(2), 129–137 (1982). https://doi.org/10.1109/TIT.1982.1056489
12. MacQueen, J.: Some methods for classification and analysis of multivariate observations. In: Proceedings of the Fifth Berkeley Symposium on Mathematical Statistics and Probability, vol. 1, pp. 281–297 (1967)
13. Maitra, R., Melnykov, V.: Simulating data to study performance of finite mixture modeling and clustering algorithms. J. Comput. Graph. Stat. **19**(2), 354–376 (2010)
14. OpenMP Architecture Review Board: OpenMP application program interface version 4.5 (2015). http://www.openmp.org/wp-content/uploads/openmp-4.5.pdf
15. Pelleg, D., Moore, A.: Accelerating exact K-means algorithms with geometric reasoning. In: Proceedings of the 5th ACM SIGKDD International Conference on Knowledge Discovery and Data Mining, pp. 277–281 (1999). https://doi.org/10.1145/312129.312248
16. Pettinger, D., Di Fatta, G.: Scalability of efficient parallel K-means. In: Proceedings of the 5th IEEE International Conference on e-Science, Workshop on Computational e-Science, pp. 96–101 (2009). https://doi.org/10.1109/ESCIW.2009.5407991

# Evaluating the Use of Policy Gradient Optimization Approach for Automatic Cloud Resource Provisioning

Włodzimierz Funika[(✉)][iD] and Paweł Koperek[iD]

Faculty of Computer Science, Electronics and Telecommunication,
Department of Computer Science, AGH,
al. Mickiewicza 30, 30-059 Kraków, Poland
funika@agh.edu.pl, pkoperek@gmail.com

**Abstract.** Reinforcement learning is a very active field of research with many practical applications. Success in many cases is driven by combining it with Deep Learning. In this paper we present results of our attempt to use modern advancements in this area for automated management of resources used to host distributed software. We describe the use of three policy training algorithms from the policy gradient optimization family, to create a policy used to control the behavior of an autonomous management agent. The agent is interacting with a simulated cloud computing environment, which is processing a stream of computing jobs. We discuss and compare the policy performance aspects and the feasibility to use them in real-world scenarios.

**Keywords:** Reinforcement learning · Deep neural networks · Autonomous cloud management · OpenAI Gym · OpenAI Spinning Up · CloudSim Plus

## 1 Introduction

Reinforcement learning techniques have been known for a long time [9, 20, 21, 24]. Until recently they were mostly applicable to relatively simple problems, where observing the whole environment was easy and the space of different states did not have many dimensions. The new advancements in the area allow to tackle domains which are much more complicated, like computer games [12], control of robots [6, 10] or the game of Go [19]. The state-of-the-art results can be obtained thanks to application of neural networks, e.g. in form of Deep Q Learning [11], Double Q-Learning [7] or Asynchronous Actor-Critic Agents (A3C) [13]. This enables learning complex behaviors, by directly observing an environment and a handful of associated metrics. In some cases this approach allowed to achieve results surpassing human performance.

Such successes encourage further experimenting with applying Deep Learning to Reinforcement Learning in other domains. One particular area, where

© Springer Nature Switzerland AG 2020
R. Wyrzykowski et al. (Eds.): PPAM 2019, LNCS 12043, pp. 467–478, 2020.
https://doi.org/10.1007/978-3-030-43229-4_40

this technique might deliver a lot of benefits, is the management of resources of distributed software applications, especially in cases where there are many additional constraints, e.g. SLA agreements [14]. Managing such an infrastructure in a way, which minimizes the monetary cost while maintaining the business requirements, is a very complex subject. Researchers experimented with various approaches, including the classic Q-learning algorithm [25] or rule-based policies [15]. There are also first attempts to exploit Deep Reinforcement Learning, e.g. the Deep Q Learning [23].

In Deep-Q-learning, the training process attempts to discover an approximation of the $Q$ function denoting the sum of future reward in case the agent follows an optimal policy. This means that the training optimizes the agent's performance *indirectly*. The advantage of this approach is that it can use all the information collected about agent's interactions with the environment by storing them in a buffer. It is more *sample efficient* - needs less interactions with environment to conduct the learning [16]. Unfortunately this also makes the optimization process less stable and the training process can even diverge from the optimum in some cases [22].

The area of Reinforcement Learning is rapidly expanding with new algorithms and approaches being published often. One of the main areas of development are the policy gradient optimization methods. Comparing to e.g. Q-learning, they allow to avoid problems with convergence - are more stable and reliable. They are also usually simpler to implement and optimize the policy directly.

In this paper we attempt to use the recent advancements in the policy optimization area (Trust Region Policy Optimization [17] and the Proximal Policy Optimization [18] algorithms) to the cloud resources provisioning problem. To our best knowledge this is the first attempt to apply those techniques in the mentioned context.

The paper is structured as follows: in Sect. 2 we present background and related work, Sect. 3 describes the architecture of the training environment. Section 4 explains the experiment setup and Sect. 5 provides results. Section 6 summarizes our research and outlines next steps.

## 2    Background and Related Work

Below we present the related work setting the foundation for our research.

### 2.1    Reinforcement Learning

Reinforcement Learning (RL) [21] is a machine learning paradigm which focuses on discovering a policy for autonomous agents, which take actions within a specific environment. The goal of the policy is to maximize a specific reward, whose value can be presented to the agent with a delay. The RL approach differs from *supervised learning*: training the agent is based on the fact that knowledge is gathered in a trial and error process, where the agent interacts with the environment and observes results of its actions. There is no *supervising entity*, which would be capable of providing feedback on whether certain actions are better than others. RL is also different from unsupervised learning: it focuses on maximizing the reward signal instead of finding a structure hidden in collections of unlabeled data.

In this paper we focus on *model-free* methods, which do not learn or have access to a model of the environment. There are two main approaches in this branch of RL:

- *Q-learning* - which focuses on creating an approximator $Q_\theta(s, a)$ for the $Q^*(s, a)$ optimal action-value function. $Q^\pi(s, a)$ denotes a function, which provides the expected return, given that the agent starts in state $s$ and takes an action $a$ and forever after acts according to some policy $\pi$. This means that the learning process influences the actual policy in an indirect way, the policy can be defined prior to the learning process and remains unchanged.
- *Policy optimization* - which directly changes the behavior of the agent by optimizing parameters $\theta$ of a policy denoted as $\pi_\theta$. Optimization is performed usually *on-policy* meaning that each update uses the data collected while acting according to the latest version of the policy. The policy itself does not have to consult a value function (a function $V^\pi(s)$ which provides the expected return if agent starts in state $s$ and follows the policy $\pi$ forever after).

## 2.2 Policy Gradient Optimization Methods

Policy gradient methods improve the policy parameters $\Theta$ based on the gradient of some estimated scalar performance objective $J(\pi_\theta)$ in respect to the policy parameters. These methods seek to maximize performance (measured as reward obtained from interactions with the environment) and as such they change the parameters according to an iterative process in which the changes to parameters approximate a *gradient ascent* in $J$:

$$\Theta_{k+1} = \Theta_k + \alpha \nabla_\Theta J(\pi_{\Theta_k}) \tag{1}$$

where $\Theta_k$ denotes policy's parameters in the $k$-th iteration of the training process.

Methods, which apart from learning an approximation of the optimal policy, learn also an approximate value function, are often called *actor-critic* methods [5]. The *actor* refers to the learned policy $\pi$ and *critic* refers to the learned value function.

There are many variants of policy gradient optimization methods, however in this paper we focus on three of them: *"Vanilla" Policy Gradient* [13], *Trust Region Policy Optimization* [17], *Proximal Policy Optimization* [18].

The first one, *Vanilla* Policy Gradient (VPG), defines the gradient of the performance objective as:

$$\nabla_\theta J(\pi_\theta) = \mathbb{E}_{\tau \sim \pi} \left[ \sum_{t=0}^{T} \nabla_\theta log \pi_\theta(a_t|s_t) A^{\pi_\theta}(s_t, a_t) \right] \tag{2}$$

where:

- $\tau$ is a trajectory (a sequence of states and actions)
- $t$ denotes the number of the step in a trajectory
- $\pi_\theta$ is the policy with parameters $\theta$ and $\pi_\theta(a_t|s_t)$ is the probability of taking action $a_t$ in state $s_t$
- $A^{\pi_\theta}$ is the advantage function for the current policy, which describes to what extent taking an action $a_t$ in a state $s_t$ is more profitable than selecting a random action, following by acting according to $\pi$ forever after

This algorithm is usually considered a poor choice because of poor data efficiency (a sample interaction with environment is used once and then discarded) and poor robustness. Often the policy prematurely converges to a local maximum which leads to suboptimal behavior of the agent. It is hard to find the right step size of gradient ascent. Even small differences in the parameter space might lead to drastic changes in performance, a single wrong step can have catastrophic effects and recovering from them might be very hard.

The Trust Region Policy Optimization (TRPO) approach aims to improve on this problem. It tries to take the largest step possible, however it adds a special constraint on how close the policies (before and after an update of parameters) are allowed to be. The gradient update is defined as:

$$\Theta_{k+1} = \underset{\theta}{argmax}\ L(\Theta_k, \Theta) \tag{3}$$

$$s.t.\ \overline{D}_{KL}(\Theta||\Theta_k) \leq \delta \tag{4}$$

where $L$ is called the *surrogate advantage* and provides a measure how policy $\pi_\theta$ performs comparing to the old policy $\pi_{\theta_k}$.

$$L(\Theta_k, \Theta) = \underset{s,a \sim \pi_{\theta_k}}{\mathbb{E}} \left[ \frac{\pi_\theta(a|s)}{\pi_{\theta_k}(a|s)} A^{\pi_{\theta_k}}(s_t, a_t) \right] \tag{5}$$

The $\overline{D}_{KL}$ is an average KL-divergence between policies across the states which were visited by the old policy. Unfortunately this update formula needs to be approximated in order to speed up the calculations, e.g. by using Taylor expansions. The whole algorithm is relatively complex and hard to tune.

Proximal Policy Optimization (PPO) aims to solve the same problem as TRPO, however using an approach that is simpler to implement. It tries to strike a balance between sample efficiency, ease of tuning and ease of implementation. The algorithm aims to compute a parameter update at each step, that on the one hand minimizes the cost function, while at the same time ensures the difference to the previous policy is relatively small. In PPO, instead of introducing additional conditions, the objective is modified in such a way that it ensures the updates to parameters are not too big. The objective is defined by the following function:

$$L^{CLIP}(\Theta) = \mathbb{E}_t \left[ min(r_t(\Theta)A_t, clip(r_t(\Theta), 1 - \epsilon, 1 + \epsilon)A_t) \right] \qquad (6)$$

where $r_t$ denotes probability ratio $r_t(\Theta) = \frac{\pi_\Theta(a_t|s_t)}{\pi_{\Theta_{old}}(a_t|s_t)}$ ($\Theta_{old}$ is the vector of policy parameters before the update), $clip(r_t(\Theta), 1 - \epsilon, 1 + \epsilon)$ function keeps the $r_t(\Theta)$ value within specified limits (*clips* it at the end of the range) and $\epsilon$ is a hyperparameter (its typical value is between 0.1 and 0.3).

Each of the algorithms (VPG, TRPO, PPO) requires setting a slightly different set of hyperparameters. Depending on the problem to be solved, certain algorithm might be easier to fine tune and produce policies which would perform more efficiently. In our exploration we attempt to experimentally verify which one is most suitable to automated management of cloud resources.

## 3   Architecture of the Training Environment

The simulation environment has been implemented following the results of our prior research [4]. The main simulation process is implemented with the use of the CloudSim Plus simulation framework [3] which is being used in a wide range of studies [8].

It is wrapped with the interface provided by the Open AI Gym framework [2]. This allows for decoupling the simulation from other elements of the system, which in turn allows to easily reuse the same environment in experiments with different algorithms. This also helps to parallelize the execution in situations where multiple simulations need to be run simultaneously.

The architecture of our training environment is presented in Fig. 1.

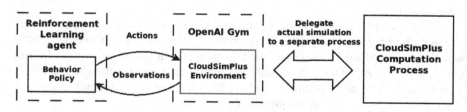

**Fig. 1.** Components of the discussed system; arrows denote interactions between them.

During the experiment (in both training and evaluation mode) the agent becomes a part of the simulation. Agent's behavior is defined by a policy, which is an approximation of a function, which given a state of the environment returns the action to be executed (in a form of probability of whether an action is executed for stochastic policies or as the action itself in case of deterministic policies). A state is represented by a set of metrics which are calculated in each step of the simulation. These metrics include: the number of running virtual machines, 90-th percentile of task queue wait time, 99-th percentile of task queue wait time, average CPU utilization, 90-th percentile of CPU utilization, total task queue wait time, wait queue size.

The approximation is a neural network, which means that the policy is parameterized by the values of weights of network's layers. This enables using an iterative process which improves the parameters based on the results of interactions of the policy with the environment.

As a basis for implementation of the policy training algorithms, we used the OpenAI SpinningUp project [1].

## 4    Experiment Setup

The primary objective of the experiment is to create a dynamic resource allocation policy using the policy gradient optimization methods (Vanilla Policy Gradient, PPO, TRPO) and compare their performance.

To speed up the training process and avoid to disruption of actual applications, we chose to evaluate the training process in a simulated environment. The simulated environment consists of a single datacenter which could host up to a 1000 virtual machines. Each virtual machine provides a 4 core processor with 16 GB of RAM, similar as *xlarge* Amazon EC2 instances. The initial number of virtual machines is pre-configured to provide a uniform starting point for all experiment runs. Each algorithm uses its own training schedule, however was evaluated by executing over the whole simulation, until all tasks are processed.

The workload simulation is based on logs of actual jobs executed on IBM SP2 cluster working in the Swedish Royal Institute of Technology. It consists of 28490 batch jobs executed between October 1996 thru August 1997. Jobs have varying time of execution and use different amounts of resources. The configuration of simulated CPU cores is adjusted to the configuration of the original execution environment.

The reward function is set up as the negative cost of running the infrastructure including the SLA penalties. This enables us to formulate the optimization task as maximization of the objective function (minimizing the cost of the running infrastructure). The cost of running virtual machines is set to $0.2 per hour of their work. The SLA penalty is set to $0.00001 for every second of delay in task execution (e.g. waiting in queue for execution).

To carry out collecting the measurements in a reasonable amount of time, simulation time is speeded up 1000 times and we limited the number of steps to 10000 per episode.

We ran the three mentioned algorithms independently. Each of them trained a policy, which was driving the decisions to add or remove resources from the processing infrastructure. We ran all of them in training sessions of 100 episodes, where a single episode consisted of 10000 steps of simulation. The agent could potentially have 1M interactions with the environment over the course of training.

# 5   Experiment Results

Below we present the results of training a policy using VPG, TRPO and PPO methods. First we compare results of training a policy using different algorithms, then we discuss whether the best policy can be applied also to other workloads. Finally we analyze whether the policy can be used in an environment in which the time has been speeded up by a different factor.

## 5.1   Training with Different Algorithms

Figure 2 presents the average reward per learning episode for each method. In all three cases we experimented with multiple combinations of values of learning process parameters. For comparison purposes we selected the best combination for each algorithm.

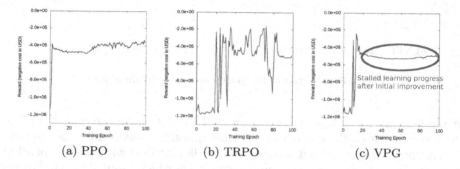

|                | (a) PPO | (b) TRPO | (c) VPG |

**Fig. 2.** Average reward (USD) per training episode.

All the algorithms started with a similarly low reward level. PPO after a few episodes reduced the cost per episode by ca. 57% and started to slowly improve it over time by ca. 22%. In VPG training looks similar, however the initial jump by 56% took more time. The cost stabilized between 475K and 525K USD, the policy seemed to stall after the initial rapid improvement. The training in TRPO was less stable: the policy was changing between very low and high costs. Apparently the agent's behavior was stabilized with the cost reduced by 54%.

To analyze the resulting policies in more details, we have observed the behavior of the trained policies over the complete simulation. Figure 3 presents the behavior of the agent (changes of the virtual machines count) in the context of the reward it was receiving and the number of jobs in queue. The agent's actions resulted in accruing the following costs over the course of the whole simulation: PPO - 1586K USD, TRPO - 1650K USD, VPG - 1768K USD.

All the policies increase the number of virtual machines to the maximum and then maintain it, even if the utilization is dropping. The policy trained with the use of the VPG algorithm simply increased the number of virtual machines

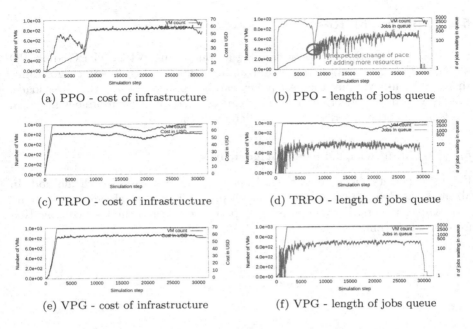

(a) PPO - cost of infrastructure

(b) PPO - length of jobs queue

(c) TRPO - cost of infrastructure

(d) TRPO - length of jobs queue

(e) VPG - cost of infrastructure

(f) VPG - length of jobs queue

**Fig. 3.** Sample run of the policy over the full simulation.

at the beginning and kept it until the end of simulation. In the case of TRPO, the agent increased the number of machines quickly, however from time to time it attempted to reduce their number. The policy trained with the use of PPO was initially adding the resource in a slower pace, which resulted in building up a bigger initial queue of jobs to calculate. The backlog was quickly processed once the number of machines was increased to a certain level. That approach rendered the best results, thus we chose to further evaluate the policy trained with the use of PPO.

## 5.2   Managing Different Workload

To verify whether the selected policy memorized information about the training workload or found generalized rules how to behave, we attempted to apply it to another workload recorded at the Atlas cluster from Lawrence Livermore National Laboratory (LLNL). The simulated environment consisted of a single datacenter which could host up to a 1152 virtual machines with AMD Opteron CPUs with total of 9216 cores. Other parameters of the simulation were the same as in the training workload. The progress of the simulation is presented in Fig. 4.

The policy actively managed the resources (added and removed VMs). The cost of resources was kept on a relatively low level and the number of jobs held in the waiting queue was stable. After investigating the spikes, we discovered that they were caused by scheduling huge jobs requiring a lot of resources.

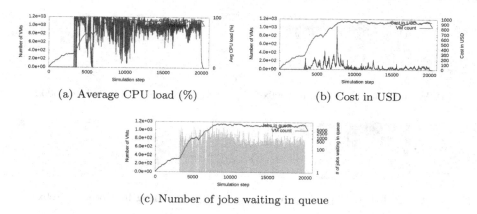

(a) Average CPU load (%)

(b) Cost in USD

(c) Number of jobs waiting in queue

**Fig. 4.** Executing the policy on a different workload (LLNL Atlas).

Overall the policy was able to work in a scenario, in which it was not trained. This suggests that the rules governing how the policy makes decisions, are more general and applicable to the situations which were not presented to the agent in the past.

## 5.3  Managing Resources on Different Time Scales

Finally, we tested how the best policy behaved when applied to the same workload on different time scales. Figure 5 presents the course of the simulation with the use of the workload from Atlas cluster (LLNL) with a 1000x speed up (the same as training), 100x speed up and without changing time. For brevity, in the last example we present only a small fraction of the overall simulation.

Although the behavior of the policy is a bit different when the time scale changes, there are some similarities. The cluster was being used to a full extent in all cases, however for smaller speed ups it seemed that the number of used VMs was reduced more aggressively. The number of jobs in the queue was maintained on a similar, stable level and was lower for smaller speed ups. It seemed that the smaller the speedup was, the more accurate the behavior of the policy was. This effect can be explained by the fact that the environment was not changing so rapidly allowing more time for response.

This result suggests that the policy can be trained with a speeded up simulation, however there are limits to that. The bigger the speed up is the less accurate the simulation is, what might lead to training a policy which ignores signals indicating a change is required.

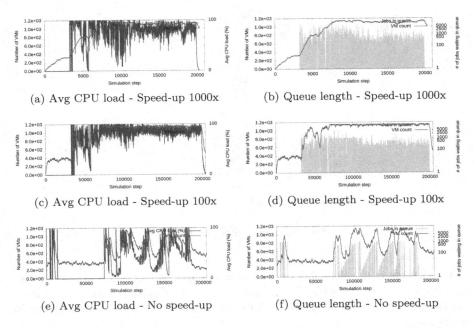

(a) Avg CPU load - Speed-up 1000x

(b) Queue length - Speed-up 1000x

(c) Avg CPU load - Speed-up 100x

(d) Queue length - Speed-up 100x

(e) Avg CPU load - No speed-up

(f) Queue length - No speed-up

**Fig. 5.** Executing the policy on different time scales (LLNL Atlas workload).

## 6    Conclusions and Further Work

In this paper we presented the results of evaluation of three algorithms from the policy gradient optimization family in the context of automatic management of software infrastructures. We showed the results of training policies and using them to drive decisions of autonomous agents which control the infrastructure processing a stream of computation jobs. The agent's environment was a simulation of a cloud-based distributed application. Resources could be scaled horizontally. The agents observed the environment through some monitoring metrics, interacted with it by adding or removing virtual machines and as a reward received the current negative cost of running the infrastructure. As a foundation for our implementation we used OpenAI Gym, OpenAI Spinning Up and CloudSim Plus projects.

The results of this experiment are promising. The algorithms we have chosen proved to be quite flexible, e.g. with minor modifications we were able to reuse the PPO used previously to create autonomous bots playing computer games at a professional human level. The policies learned to reduce the overall cost of application by avoiding penalties coming from an unnecessarily long waiting queue, instead of drastically limiting the amount of resources. The agent's behavior over the course of the full simulation was stable, i.e. they did not increase or reduce the number of virtual machines in a chaotic way. The policy which achieved the lowest cost was created using the PPO algorithm. Further experiments revealed that this policy was generalizing knowledge, i.e. was able to manage the infrastructure processing different workloads. Speeding up the flow of time during the simulation enabled reducing the time of training without significantly impacting the quality of decisions made by the policy.

Unfortunately, there are a few disadvantages of the presented approach. Most importantly the behavior of the agent is difficult to explain. After training of the policy, it does not provide information why it is making particular decisions. The policy needs to be re-trained to be up-to-date with the application and changes to the distributed environment. The application needs to be prepared for frequent resource changes, otherwise agent's actions might lead to unexpected failures. Finally, the algorithms require tuning to a specific application. The knowledge accumulated in one policy most likely will not be transferrable to a different application.

At the current stage of development, we believe that a fully autonomous RL management agent is premature for a real-world application. On the other hand the policy could provide a great value as a part of a supportive system: guide the human operators by providing suggestions on what actions should be taken.

We plan to continue this area of research. The next research steps include extending the simulator in order to simulate more complex situations. We also want to explore the possibility to create a continuous training loop, where the agent could be changed and improved, e.g. using an evolutionary approach.

**Acknowledgements.** The paper was partially financed by AGH University of Science and Technology Statutory Fund. Computational experiments were carried out on the PL-Grid infrastructure.

# References

1. Achiam, J.: OpenAI spinning up (2018). https://github.com/openai/spinningup. Accessed 30 Apr 2019
2. Brockman, G., et al.: OpenAI gym. CoRR abs/1606.01540 (2016). http://arxiv.org/abs/1606.01540
3. Filho, M.C.S., Oliveira, R.L., Monteiro, C.C., Inácio, P.R.M., Freire, M.M.: CloudSim plus: a cloud computing simulation framework pursuing software engineering principles for improved modularity, extensibility and correctness. In: 2017 IFIP/IEEE Symposium on Integrated Network and Service Management (IM), pp. 400–406, May 2017
4. Funika, W., Koperek, P., Kitowski, J.: Repeatable experiments in the cloud resources management domain with use of reinforcement learning. In: Cracow Grid Workshop 2018, pp. 31–32. ACC Cyfronet AGH, Kraków (2018)
5. Grondman, I., Busoniu, L., Lopes, G., Babuska, R.: A survey of actor-critic reinforcement learning: standard and natural policy gradients. IEEE Trans. Syst. Man Cybern. Part C Appl. Rev. **42**(6), 1291–1307 (2012)
6. Gu, S., Holly, E., Lillicrap, T., Levine, S.: Deep reinforcement learning for robotic manipulation with asynchronous off-policy updates. In: Proceedings 2017 IEEE International Conference on Robotics and Automation (ICRA). IEEE, Piscataway, May 2017

7. Van Hasselt, H., Guez, A., Silver, D.: Deep reinforcement learning with double q-learning. In: Proceedings of the Thirtieth AAAI Conference on Artificial Intelligence, AAAI 2016, pp. 2094–2100. AAAI Press (2016)
8. Hussain, A., Aleem, M., Azhar, M., Muhammad, I., Islam, A.: Investigation of cloud scheduling algorithms for resource utilization using CloudSim. Comput. Inf. **38**, 525–554 (2019)
9. Kaelbling, L.P., Littman, M.L., Moore, A.W.: Reinforcement learning: a survey. CoRR cs.AI/9605103 (1996). http://arxiv.org/abs/cs.AI/9605103
10. Kalashnikov, D., et al.: QT-Opt: scalable deep reinforcement learning for vision-based robotic manipulation. CoRR abs/1806.10293 (2018). http://arxiv.org/abs/1806.10293
11. Mnih, V., et al.: Playing atari with deep reinforcement learning (2013). http://arxiv.org/abs/1312.5602
12. Mnih, V., et al.: Human-level control through deep reinforcement learning. Nature **518**(7540), 529–533 (2015)
13. Mnih, V., et al.: Asynchronous methods for deep reinforcement learning. In: Balcan, M.F., Weinberger, K.Q. (eds.) Proceedings of the 33rd International Conference on Machine Learning, vol. 48, pp. 1928–1937. PMLR, June 2016
14. Nikolow, D., Slota, R., Polak, S., Pogoda, M., Kitowski, J.: Policy-based SLA storage management model for distributed data storage services. Comput. Sci. **19**, 405 (2018)
15. Rufus, R., Nick, W., Shelton, J., Esterline, A.C.: An autonomic computing system based on a rule-based policy engine and artificial immune systems. In: MAICS. CEUR Workshop Proceedings, vol. 1584, pp. 105–108. CEUR-WS.org (2016)
16. Schaul, T., Quan, J., Antonoglou, I., Silver, D.: Prioritized experience replay. In: ICLR (2016)
17. Schulman, J., Levine, S., Moritz, P., Jordan, M.I., Abbeel, P.: Trust region policy optimization. CoRR abs/1502.05477 (2015)
18. Schulman, J., Wolski, F., Dhariwal, P., Radford, A., Klimov, O.: Proximal policy optimization algorithms. CoRR abs/1707.06347 (2017). http://arxiv.org/abs/1707.06347
19. Silver, D., et al.: Mastering the game of go without human knowledge. Nature **550**, 354–359 (2017)
20. Sutton, R.S., Barto, A.G.: Reinforcement Learning: An Introduction. MIT Press, Cambridge (1998). http://www.cs.ualberta.ca/ sutton/book/the-book.html
21. Sutton, R.S.: Temporal credit assignment in reinforcement learning. Ph.D. thesis (1984)
22. Szepesvari, C.: Algorithms for Reinforcement Learning. Morgan and Claypool Publishers, San Rafael (2010)
23. Wang, Z., Gwon, C., Oates, T., Iezzi, A.: Automated cloud provisioning on AWS using deep reinforcement learning. CoRR abs/1709.04305 (2017). http://arxiv.org/abs/1709.04305
24. Witten, I.H.: An adaptive optimal controller for discrete-time Markov environments. Inform. Control **34**, 286–295 (1977)
25. Xu, C.Z., Rao, J., Bu, X.: URL: a unified reinforcement learning approach for autonomic cloud management. J. Parallel Distrib. Comput. **72**(2), 95–105 (2012)

# Improving Efficiency of Automatic Labeling by Image Transformations on CPU and GPU

Łukasz Karbowiak$^{(\boxtimes)}$ (ID)

Czestochowa University of Technology, Czestochowa, Poland
lkarbowiak@icis.pcz.pl

**Abstract.** Automatic image labeling is a process which detects and creates labels on various objects of an image. This process is widely used for autonomous vehicles, when original images collected by cameras placed in the car are sent to the artificial neural network (ANN). Improving the efficiency of labeling is vital for further development of such vehicles. In this study, three main aspects of image labeling are analyzed and tested, namely, (i) time required for labeling, (ii) accuracy, and (iii) power consumption.

In this paper, an approach is proposed to improve the efficiency of the serial and parallel implementation of labeling, performed respectively on CPU and GPU. One of transformations used in this approach is converting an original image to its monochrome equivalent. Other transformations, such as sharpening and change of colors intensity, are based on using the image histogram. The testing and validation of image transformations are performed on a test dataset containing frames of proprietary videos collected by an onboard camera. For both serial and parallel labeling, the same ANN is used. Preliminary results of these tests show promising improvements when considering the above-mentioned aspects of automatic labeling.

**Keywords:** Automatic labeling · Image transformations on CPU and GPU · Energy consumption

## 1 Introduction

Automatic image labeling [1] is a process which detects and creates labels on various objects of an image. Labeling can mark animals, plants, food, vehicles, people, etc. In this article, the main focus is on recognizing vehicles by an artificial neural network. This type of labeling is used for autonomous vehicles, when original images collected by cameras placed in the car are sent to the artificial neural network. If we can reduce the labeling time, we will increase the efficiency of the entire vehicles, and finally achieve increasing the control precision of that vehicles and/or saving the energy. In fact, decreasing the required time will allow

© Springer Nature Switzerland AG 2020
R. Wyrzykowski et al. (Eds.): PPAM 2019, LNCS 12043, pp. 479–490, 2020.
https://doi.org/10.1007/978-3-030-43229-4_41

increasing the number of processed images, what in consequence will improve the quality of vehicle control. The lower power consumption will permit extending the usage of a vehicle.

The main idea in this study is to transform the images before they are delivered to the neural network. The paper presents the way of modifying original images provided by the camera. To achieve the desired effect, all image transformations should be very fast, because the total time of modification and labeling must be shorter than the time of labeling the original image. Reducing the time of labeling may ensure a lower energy/power consumption. Thus, the power consumption will be tested for each proposed image transformation as an additional parameter determining improvements over the basic scenario without transformations.

## 2   Control Parameters

In this study, the following parameters are used to determine the efficiency of considered algorithms: time, accuracy and energy (power) consumption.

The time required to get a result from the neural network is measured from the moment when the image is loaded for labeling. The final value is obtained as the median of 100 measurements for a test program in which 1000 photos with various vehicles are labeled. The accuracy is measured based on the number of labels correctly determined by the artificial neural network. The power consumed during labeling [2] is calculated as the average of all runs of a test program.

The goal of this study is to develop transformations that will allow the greatest possible reduction of time needed for labeling, while increasing or preserving the level of accuracy as well as reducing or preserving the power consumption. All these parameters will be tested on both CPU and GPU.

## 3   Image Transformations

To reduce the time complexity of labeling, all of considered transformations can not be too complicated. The first of them is converting the original RGB images to their grayscale (monochrome) equivalents (see Fig. 1). It is performed by calculating intensity values $L$ for pixels according to the following formula [3]:

$$L = 0.2989 * R + 0.5870 * G + 0.1140 * B. \tag{1}$$

The second transformation is the histogram equalization [4] applied for grayscale images. This transformation is performed in few steps. Let $h$ denotes the normalized histogram of an image. The first step is calculating the histogram of the image:

$$h_k = n_k/(MxN), k \in [0, ..L-1]. \tag{2}$$

Here: $n_k$ is the number of pixels with the intensity value $L$ equal to $k$, where $L$ is in the range $<0, 255>$, and $M \times N$ is the number of pixels in an image.

**Fig. 1.** Image before and after RGB-to-monochrome conversion

In the second step, the histogram equalization is performed using the following expression:

$$I_{i,j} = floor((L-1) \sum_{n=0}^{k} h_k).$$ (3)

Here: floor rounds down to the nearest integer, and $I_{i,j}$ is the resulting image obtained after the histogram equalization. This transformation rescales $I_{i,j}$ from the range $<0,1>$ to $<0, L-1>$. An example of effect of using the histogram equalization is presented in Fig. 2, while Fig. 3 illustrates the result of applying Eq. (3) to the monochrome image.

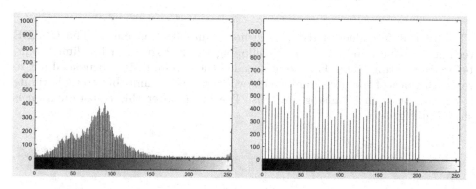

**Fig. 2.** Histogram before and after transformation defined by Eq. (3).

In the third step, instead of a single histogram, the adaptive histogram equalization (AHE) [5] uses several of them. They correspond to different sections of the image. The transformation is proportional to the cumulative distribution function calculated for pixel intensity values in a neighbourhood region. An example of effect of using the AHE technique to a histogram is shown in Fig. 4.

The contrast limited adaptive histogram equalization (CLAHE) [6] is an improved version of AHE with the contrast enhancement having a limit. For all neighbourhood regions, the algorithm performs a contrast limiting procedure.

**Fig. 3.** Result of histogram equalization.

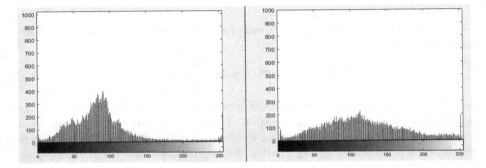

**Fig. 4.** Original histogram versus histogram obtained using AHE.

This limit is intended to reduce the noise amplification effect. The CLAHE technique reduces amplification by clipping the histogram in the limit point, before calculating CDF. This point is called the clip limit. It is beneficial not to reject the part of the histogram which overstep the clip limit but to redistribute it equally between all histogram bins. The image after this transformation is shown in Fig. 5.

**Fig. 5.** Effect of using CLAHE.

Unsharp masking [7] is an image sharpening technique which uses an original image and its blurred version. The resulting sharpened image is a result of subtracting the blurred version from the original image (see Fig. 6).

**Fig. 6.** Effect of unsharp masking.

The last tested transformation is increasing contrast [8] of the image. It is performed by saturating the bottom and top 1% of all pixel values (see Fig. 7).

**Fig. 7.** Effect of increasing contrast.

## 4   Benchmarking CPU Implementation

### 4.1   CPU Implementation and Tested Scenarios

The test program for CPU is implemented in a serial fashion using the Matlab software. The hardware used in these tests is the quad-core Intel Core i5-4460 CPU clocked at 3.20 GHz and equipped with 16 GB RAM. In the first (or basic) scenario, an artificial neural network (ANN) is used for processing original RGB images, without the proposed transformation. This network corresponds to the convolutional neural network GoogLeNet [10] that is 22 layers deep. The origin of the test dataset (see Fig. 8) are frames of proprietary videos collected by an onboard camera. For a package of 1000 images, the basic scenario allows us to obtain the labeling time $T_L = 760.43$ s. The network we use permits processing correctly $R = 52.8\%$ of labels. The power consumption $P_L = 28.88$ W. It is measured using the RAPL infrastructure [2].

In the second scenario, the conversion of the original RGB image to its monochrome (grayscale) equivalent is applied before feeding images into the ANN. In this case, no improvement in the processing time is achieved since now

**Fig. 8.** An example of image from the test dataset and results of its transformations.

we have $T_L = 760.90\,\text{s}$, but the labeling accuracy is improved to the level of $R = 70.1\%$, as well the power consumption as now $P_L = 29.01\,\text{W}$.

The third scenario assumes that the combination of the RGB-grayscale conversion and histogram equalization is used in the preprocessing phase. This approach results in increasing the labeling time slightly, but permits improving the accuracy of results achieved by the ANN significantly. In fact, the average labeling time is now equal to $767.57\,\text{s}$. A the same time, this scenario allows us to improve the accuracy by $23.5\%$ compared to the first scenario, to the level of $R = 76.3\%$, with a minimum decrease in the power consumption by $0.03\,\text{W}$.

The fourth scenario is based on using the CLAHE technique for monochrome images. This leads to increasing labeling time by approximately $2.79\%$ as compared to the first scenario, but allows us to achieve $100\%$ of accuracy. The power consumption is equal to $28.76\,\text{W}$ in this case.

In the fifth scenario, the technique of unsharp masking is directly applied for RGB images. This scenario results in increasing the processing time to $T_L = 888,93\,\text{s}$, but improving the image sharpness permits us to increase the accuracy to $R = 78.2\%$, with the average power consumption equal to $28.78\,\text{W}$. At the same time, the sixth scenario which utilize unsharped masking for monochrome images allows decreasing the labeling time to $T_L = 774.88\,\text{s}$ with the improved accuracy of $R = 88.5\%$, and the power consumption $P_L = 28.78\,\text{W}$.

The last (seventh) scenario includes converting an RGB image to a monochrome one and then increasing its contrast. As a result, the following set of control parameters is achieved: $T_L = 758.40\,\text{s}$, $R = 71.6\%$, $P_L = 28.93\,\text{W}$.

### 4.2  Conclusions for CPU Implementation

Concerning the processing time of the serial CPU implementation (Fig. 9), the best effect is obtained by for the scenario which applies the conversion of an RGB image to the monochrome one, followed by increasing its contrast. This conclusion is confirmed (Fig. 10) for different numbers of images in analyzed packages (1000, 500 and 100 images). What is important, this scenario also allows increasing the accuracy of labeling by $18.8\%$ compared to the basic scenario. At the same time, it has been found that using the CLAHE technique for monochrome images gives $100\%$ accuracy of labeling (see Fig. 11).

Regarding the power consumed by the CPU (Fig. 12), it can be concluded that additional transformations do not have a clear impact on this parameter. In fact, differences in the power consumption for all tested scenarios are very small; they can be caused, e.g., by a difference in the CPU temperature.

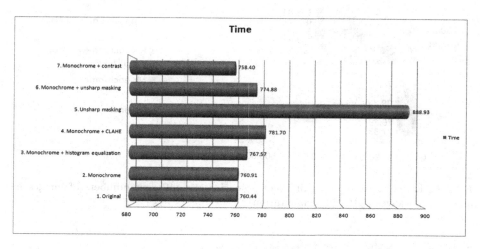

**Fig. 9.** Labeling time [s] for all tested scenarios of CPU implementation (packages with 1000 images).

## 5   Benchmarking GPU Implementation

### 5.1   GPU Implementation and Tested Scenarios

The advantages of parallel processing are used to accelerate computations on a GPU card, which is NVIDIA GeForce GTX 960 with 1024 CUDA cores and 2 GB GDDR5 memory (released in 2015 and based on the Maxwell architecture). This card is connected to the same CPU as described in the previous section. The test program has been modified [9] to offload transformations and labeling to the GPU card, based on using *gpuArray* object from the Matlab environment. The normal power consumption of the tested GPU card is about 14.15 W. It is measured using the GPU-Z software tool [11].

For packages with 1000 images, the labeling time of the basic scenario is $T_L = 500.39$ s (Fig. 13). It is definitely less than for the serial processing by the CPU. The accuracy of this labeling corresponds to 528 correctly returned results, which gives the accuracy of 52.8%. The GPU power consumption is 29.28 W.

The monochromatic transformation in the second scenario does not results in improving the labeling time since now $T_L = 501.83$ s, but the efficiency is improved by 17.3% at the cost of a slight increase in the power consumption ($P_L = 29.43$ W).

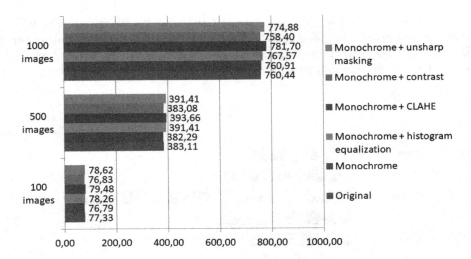

**Fig. 10.** Labeling time [s] for different scenarios and different numbers of images in analyzed packages (CPU implementation).

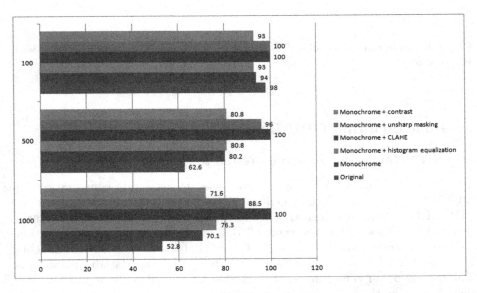

**Fig. 11.** Labeling accuracy [%] for different scenarios and different numbers of images in analyzed packages (CPU implementation).

Combining the monochromatic and histogram equalization transformations in the third scenario leads to a bit worse labeling time ($T_L = 501.49\,\text{s}$), but allows increasing the accuracy to the level of $R = 76.3\%$ together with some deterioration in the power efficiency ($P_L = 29.56\,\text{W}$).

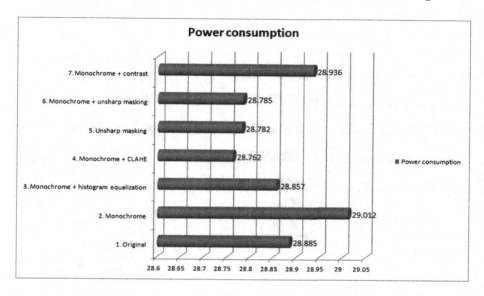

**Fig. 12.** Power consumption [W] for all tested scenarios of CPU implementation (packages with 1000 images).

As before, the use of the CLAHE technique for monochrome images in the fourth scenario is the best solution when considering only the accuracy parameter, since this scenario permits us to achieve 100% of accuracy. At the same time, the labeling time is increased to $T_L = 505.83$ s, as well as the power consumption ($P_L = 29.61$ W).

Applying the technique of unsharp masking for RGB images (5th scenario) allows 78.2% accuracy of labeling. The disadvantage of this scenario is a significant increase in the labeling time because now $T_L = 580.63$ s. At the same time, considerable better results gives the usage of unsharp masking for monochrome images that ensures 88.5% of labeling accuracy with the labeling time of $T_L = 520.20$ s, and the power consumption of 29.64 W.

The last scenario, which assumes increasing contrast of monochrome images, results in reducing the labeling time ($T_L = 498.10$ s), with the accuracy of $R = 71.6\%$ and power consumption equal to 29.44 W.

## 5.2 Conclusions for GPU Implementation

The main conclusion is that the parallel processing on the GPU provides a definitely less labeling time than the serial implementation by CPU (see Fig. 14 as well). It can be also concluded that among the considered scenarios the largest improvement in the processing time is achieved for the combination of the RGB-to-monochrome transformation with increasing the contrast.

Concerning the accuracy parameter, their values are exactly the same for the CPU and GPU implementations. This is the expected effect because both

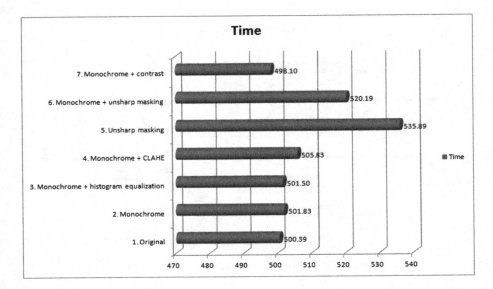

**Fig. 13.** Labeling time [s] for all tested scenarios of GPU implementation (packages with 1000 images).

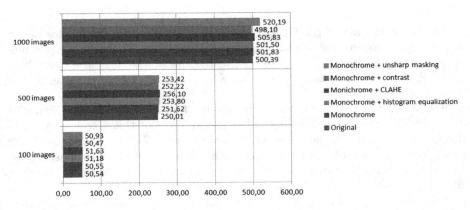

**Fig. 14.** Labeling time [s] for different scenarios and different numbers of images in analyzed packages (GPU implementation).

of them use the same algorithms for the image processing. Regarding the power consumption, the GPU implementation is only slightly worse than in the case of the CPU (see Fig. 15). Like the CPU implementation, differences in the power consumption for various scenarios are very small, and can be result of a difference in the GPU temperature.

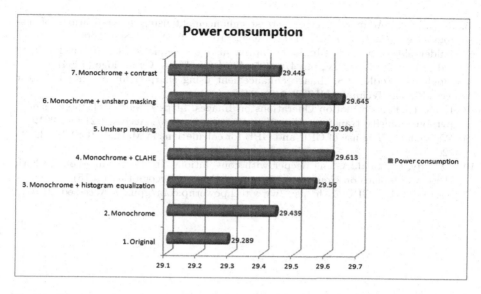

**Fig. 15.** Power consumption [W] for all tested scenarios of GPU implementation (packages with 1000 images).

## 6  Overall Summary

This paper studies the usage of the set of simple transformations of original images in the process of image labeling before delivering them to the artificial neural network. Some transformations (or filters) result in decreasing the labeling time, others allow increasing the accuracy of labeling. A combination of filters has been found that simultaneously reduces the labeling time and improves the accuracy. The third considered parameter, which is the power consumption for the CPU and GPU, can be skipped while determining the best choice for transformations. This is due to relatively small power overheads introduced by these transformations as compared to image processing by the artificial neural network. In consequence, differences in the power consumption for various transformations are very small, and can be caused by a difference in the temperature of the CPU and GPU.

## References

1. Szeliski, R.: Computer Vision: Algorithms and Applications. Springer, London (2010). https://doi.org/10.1007/978-1-84882-935-0
2. Intel RAPL. https://01.org/rapl-power-meter/. Accessed 21 Oct 2019
3. MATLAB: RGB image to grayscale image conversion. https://www.geeksforgeeks.org/matlab-rgb-image-to-grayscale-image-conversion/. Accessed 21 Oct 2019
4. Gonzalez, R.C., Woods, R.E.: Digital Image Processing, 3rd edn. Pearson, New York (2008)

5. Stark, J.A.: Adaptive image contrast enhancement using generalizations of histogram equalization. IEEE Trans. Image Process. **9**(5), 889–896 (2000)
6. Zuiderveld, K.: Contrast limited adaptive histogram equalization. In: Heckbert, P. (ed.) Graphics Gems IV, pp. 474–485. Academic Press, Cambridge (1994)
7. Singh, H., Sodhi, J.S.: Image enhancement using sharpen filters. Int. J. Latest Trends Eng. Technol. (IJLTE) **2**(2), 84–94 (2013)
8. Intensity Transformation Operations on Images. https://www.geeksforgeeks.org/python-intensity-transformation-operations-on-images/. Accessed 21 Oct 2019
9. Woźniak, J.: The use of CPU and GPU for calculations in Matlab. JCSI **10**, 32–35 (2019)
10. Szegedy, C., et al.: Going deeper with convolutions. In: Proceedings of the 28th IEEE Conference on Computer Vision and Pattern Recognition (2015)
11. TechPowerUp GPU-Z. https://www.techpowerup.com/gpuz/. Accessed 21 Oct 2019

# Special Session on GPU Computing

# Efficient Triangular Matrix Vector Multiplication on the GPU

Takahiro Inoue, Hiroki Tokura, Koji Nakano$^{(\boxtimes)}$, and Yasuaki Ito

Department of Information Engineering, Hiroshima University,
1-4-1 Kagamiyama, Higashi-Hiroshima 739-8527, Japan
nakano@cs.hiroshima-u.ac.jp

**Abstract.** The main purpose of this paper is to present a very efficient GPU implementation to compute the trmv, the product of a triangular matrix and a vector. Usually, developers use cuBLAS, a linear algebra library optimized for each of various generations of GPUs, to compute the trmv. To attain better performance than cuBLAS, our GPU implementation of the trmv uses various acceleration technique for latest GPUs. More specifically, our GPU implementation has the following features: (1) only one kernel is called; (2) maximum number of threads are invoked; (3) all memory access to the global memory is coalesced; (4) all memory access to the shared memory has no bank conflict; and (5) shared memory access is minimized by a warp shuffle function. Experimental results for five generations of NVIDIA GPUs for matrices of sizes from $32 \times 32$ to $16K \times 16K$ for fp32 show that our GPU implementation is faster than cuBLAS and muBLAS for almost all matrix sizes and GPU generations.

**Keywords:** Matrix multiplication · Trmv · Parallel algorithm · GPGPU

## 1 Introduction

*The GPU* (Graphics Processing Unit), is a specialized circuit designed to accelerate computation for building and manipulating images [4,5,7,12,13]. Latest GPUs are designed for general purpose computing and can perform computation in applications traditionally handled by the CPU. Hence, GPUs have recently attracted the attention of many application developers. NVIDIA provides a parallel computing architecture called *CUDA* (Compute Unified Device Architecture) [10], the computing engine for NVIDIA GPUs. CUDA gives developers access to the virtual instruction set and memory of the parallel computational elements in NVIDIA GPUs. NVIDIA also provides many libraries of primitives running on the GPU. For example, cuBLAS [11] is a basic linear algebra library which includes primitives such as matrix multiplication.

Let $X$ be an $n \times n$ matrix and $Y$ be an $n$-element vector. The product of them is an $n$-element vector $Z$ such that $Z(i) = \sum_{j=0}^{n-1} X(i,j)Y(j)$ for all $i$ $(0 \leq i \leq n-1)$. The computation of matrix-vector multiplication is called *gemv*.

© Springer Nature Switzerland AG 2020
R. Wyrzykowski et al. (Eds.): PPAM 2019, LNCS 12043, pp. 493–504, 2020.
https://doi.org/10.1007/978-3-030-43229-4_42

If $X$ is a lower/upper triangular matrix such that $X(i,j) = 0$ for all $i$ and $j$ ($i < j/i > j$), then the computation of lower/upper triangular matrix is called *trmv*. For simplicity, we assume that the trmv is the multiplication of a lower triangle matrix and a vector. We follow the address mapping of cuBLAS [11] to store matrices and vectors. More specifically, elements in an matrix are stored in the address space of a memory in column major order as illustrated in Fig. 1 (left) for the compatibility with cuBLAS. For example, element $X(i,j)$ ($0 \leq i, j \leq n-1$) is stored in offset $i + jn$ of the address space for $X$. Thus, a segment in a column of $X$ can be accessed very efficiently by coalesced memory access. Conversely, access to a segment in a row takes a lot of time due to stride memory access.

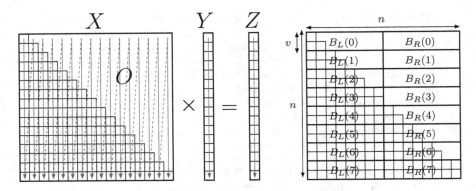

**Fig. 1.** Illustrating the trmv and half bands $B_L(i)$ and $B_R(i)$ for the trmv.

In the area of scientific computing, linear algebra computations such as scalar-vector, matrix-vector, and matrix-matrix vector multiplication are frequently used [14]. Developers can accelerate such multiplications by calling cuBLAS APIs [11]. Note that cuBLAS is not an open source software, it is very hard to guess the details of the implementations, which are optimized for individual NVIDIA GPU architectures such as Kepler, Maxwell, Pascal, Volta, and Turing. Furthermore, many works have been devoted to accelerate the gemv and the trmv [1–3,6,8]. For example, Muramatsu *et al.* [9] uses the trmv to accelerate the computation of the eigenvalue by Hessenberg reduction. Recently, Mukunoki *et al.* [8] showed an efficient GPU implementation for the trmv called *muBLAS*, which attains better performance than cuBLAS. In their implementation, an $n \times n$ matrix $X$ is partitioned into $\frac{n}{v}$ bands $B(0), B(1), \ldots, B(\frac{n}{v} - 1)$ of size $v \times n$ and each product $B(i) \cdot Y$ is computed using one CUDA block. Since elements in $B$ are arranged in column major order, $v$ must be large enough for coalesced memory access to the global memory. Clearly, the computational cost for $B(\frac{n}{v} - 1)$ is much larger than that for $B(0)$. To equalize the computational costs for all CUDA blocks, they partition each $B(i)$ into the left and the right halves $B_L(i)$ and $B_R(i)$ as illustrated in Fig. 1 (right). In the figure, each CUDA block $i$ ($0 \leq i \leq 7$) computes the product for each $B_L(i)$ and writes it in the global memory. After that, CUDA blocks 0, 1, 2, and 3 computes the that for each $B_R(4), B_R(5), B_R(6)$, and $B_R(7)$ and write them in the global memory. By summing the product appropriately, the trmv can be computed. Their implementation uses only $\frac{n}{v}$ CUDA blocks, hence memory access throughput can not be maximized. Also, they use three kernel calls with large overheads. Furthermore, muBLAS uses auto-tuning technique to find best parameters of the value of $v$, the number of threads per CUDA block, the dimension of threads, etc. It measures the time for the trmv for all combinations of parameters to find best parameter in advance to attain the best performance for the trmv.

The main contribution of this paper is to present a more efficient GPU implementation for the trmv than cuBLAS and muBLAS. In our GPU implementation, a matrix $X$ is partitioned into blocks of fixed size $32 \times 64$ and the product for each block is computed using a CUDA block with 256 threads. Thus, our GPU implementations use much more threads than muBLAS. Also, the product for each block is computed very efficiently using a warp shuffle function __shfl_sync(), which directly copies the value stored in the register of another thread. In addition, our implementation invokes only one kernel call. We have evaluated the performance of cuBLAS, muBLAS, and our GPU implementation for matrices of sizes from $32 \times 32$ to $16K \times 16K$ using five generations of GPUs, Kepler, Maxwell, Pascal, Volta, and Turing, Our implementation always attains better performance than cuBLAS and muBLAS on latest GPUs with Volta and Turing architectures. Also, it runs faster for matrices of size from $128 \times 128$ to $2K \times 2K$ for obsolete GPU architectures Kepler, Maxwell, and Pascal. Although our implementation uses the same CUDA program with fixed parameters for all GPU generations, it runs faster than cuBLAS and muBLAS almost all cases.

This paper is organized as follows. In Sect. 2, we presents three GPU algorithms, COLUMN-WISE, ROW-WISE, and THREAD-WISE, which compute some group-wise sums and will be used in our GPU implementation for the trmv. We then go on to show the details of our GPU implementation in Sect. 3. Section 4 shows the experimental results using various generations of GPU architectures. Section 5 concludes our work.

## 2   Row-Wise, Column-Wise, and Thread-Wise Sums Computation on the GPU

This section shows GPU implementations to compute row-wise, column-wise, and thread-wise sums of a matrix using a warp shuffle function of the GPU. These algorithms are key ingredients of our matrix vector multiplication algorithms. Due to the stringent limitation, we omit the explanation of GPU architecture necessary to understand this section. Please see [10] for the details of GPU architecture and CUDA software architecture.

Suppose that each thread $k$ ($0 \le k \le 31$) in a warp of 32 threads has 32-bit register $a$ and let $a[k]$ denote register $a$ of thread $k$. We assume that these 32 registers are arranged in a $2^p \times \frac{32}{2^p}$ matrix in column-major order as illustrated in Fig. 2, where $p$ ($0 \le p \le 5$) is an integer parameter. *The column-wise sums*

|   | 0 | 1 | 2 | 3 | 4 | 5 | 6 | 7 |
|---|---|---|---|---|---|---|---|---|
| 0 | 0 | 4 | 8 | 12 | 16 | 20 | 24 | 28 |
| 1 | 1 | 5 | 9 | 13 | 17 | 21 | 25 | 29 |
| 2 | 2 | 6 | 10 | 14 | 18 | 22 | 26 | 30 |
| 3 | 3 | 7 | 11 | 15 | 19 | 23 | 27 | 31 |

$2^2$

$\frac{32}{2^2}$

**Fig. 2.** Arrangement of 32 registers of 32 threads to a $2^p \times \frac{32}{2^p}$ matrix in column-major order when $2^p = 4$

are $a[j2^p] + a[j2^p + 1] + a[j2^p + 2] + \cdots + a[j2^p + 2^p - 1]$ for all $j$ $(0 \le j \le \frac{32}{2^p} - 1)$ and *the row-wise sums* are $a[i] + a[i + 1 \cdot 2^p] + a[i + 2 \cdot 2^p] + \cdots + a[i + (\frac{32}{2^p} - 1) \cdot 2^p]$ for all $i$ $(0 \le i \le 2^p - 1)$.

We will show that the column-wise sums and the row-wise sums can be computed efficiently by warp shuffle function `__shfl_sync(mask,var,src,width)`, which directly returns the value of register `var` of another thread [10]. More specifically, 32 threads in a warp are partitioned into groups of `width` threads, and the value of register `var` of a thread in the same group specified by `src` is returned. Further, `mask` specifies active threads and it is `0xffffffff = -1` if all 32 threads are active. For example, if every thread $k$ $(0 \le k \le 31)$ executes `__shfl_sync(-1,a,k+2,4)`. then it returns the value of $a[k + 2]$ if thread $k + 2$ is in the same group of thread $k$. Otherwise, it returns the value of $a[k + 2 - 4]$, which is always in the same group.

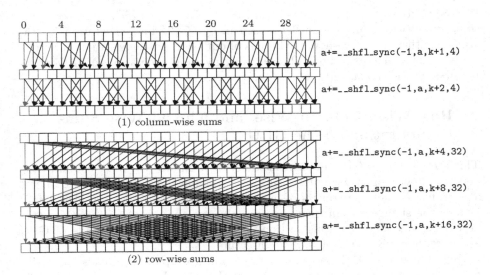

**Fig. 3.** GPU algorithms for the column-wise and row-wise sums for $2^p = 4$ using a warp shuffle function

Figure 3 illustrates GPU implementations for computing the column-wise sums and the row-wise sums for $2^p = 4$, respectively. We can see that the column-wise sums are computed by executing the warp shuffle function twice. Note that, all threads in the same group stores the sum after executing this algorithm. For example, all of $a[0], a[1], a[2], a[3]$ store the column-wise sum $a[0] + a[1] + a[2] + a[3]$. The row-wise sums are computed by executing the warp shuffle function three times. For example, we can see that $a[29]$ stores the row-wise sum $a[1] + a[5] + a[9] + \cdots + a[29]$. Further, all of $a[1], a[5], a[9], \ldots, a[29]$ store this row-wise sum. For later reference, let $COLUMN\text{-}WISE(2^p)$ and $ROW\text{-}WISE(\frac{32}{2^p})$ denote the GPU algorithm for computing the column-wise sums and the row-wise sums for a $2^p \times \frac{32}{2^p}$ matrix, respectively, Clearly, both COLUMN-WISE($2^p$) and ROW-WISE($2^p$) execute warp shuffle function $p$ times and no memory access to the shared memory/global memory is necessary, and they runs very fast.

Next, we will show $THREAD\text{-}WISE(2^q)$, a GPU algorithm to compute the sum of registers in threads with the same ID of multiple warps, where $q$ ($1 \leq q \leq 5$) is an integer parameter. Suppose that $32 \cdot 2^q$ threads of $2^q$ warps ($2 \leq 2^q \leq 32$) has register $a$. Figure 4 (1) illustrates registers when $2^q = 8$. Let $a[w][k]$ ($0 \leq w \leq 2^q - 1; 0 \leq k \leq 31$) denote register $a$ of thread $k$ in warp $w$. The thread-wise sum is $t[k] = a[0][k] + a[1][k] + \cdots + a[2^q - 1][k]$ for each $k$, that is, the sum of registers of threads $k$ over all warps. In Fig. 4 (1), the thread-wise sums correspond to the column-wise sums. We will show that all thread-wise sums can be computed efficiently using a $2^q \times 32$ matrix $s$ in the shared memory illustrated in Fig. 4 (2). Note that $s[0][k], s[1][k], \ldots, s[2^q - 1][k]$ are in the same bank and if two or more threads in the same warp access to these elements at the same time then bank conflicts occur. To avoid bank conflicts, each thread $k$ of warp $w$ writes $a[w][k]$ in $s[w][(k + w) \bmod 32]$ as illustrated in Fig. 4 (2), in which the values of $k$ are shown. After that, each thread $k$ of warp $w$ reads $s[k \bmod 2^q][(k + w) \bmod 32]$, which stores the value of $a[k \bmod 2^q][(k + w - (k \bmod 2^q)) \bmod 32]$. From $k - (k \bmod 2^q) = \lfloor \frac{k}{2^q} \rfloor \cdot 2^q$, it reads $a[k \bmod 2^q][(\lfloor \frac{k}{2^q} \rfloor \cdot 2^q + w) \bmod 32]$. Hence, the values of $a[0][k], a[1][k], \ldots, a[2^q - 1][k]$ are read by consecutive $2^q$ threads in the same warp. For example, in Fig. 4 (3), the first 8 threads of warp 0 read $a[0][0], a[1][0], \ldots, a[7][0]$. Thus, each warp $w$ ($0 \leq w \leq 2^q - 1$) can compute the thread-wise sums $t[w], t[w + 2^q], t[w + 2 \cdot 2^q], \ldots, t[w + (\frac{32}{2^q} - 1)2^q]$ by COLUMN-WISE($2^q$) as shown in Fig. 4 (4).

## 3 GPU Implementation for the Trmv

In this section, we show our GPU implementation for the trmv. We assume that a triangular matrix $X$ with $n \times n$ 32-bit float numbers and a vector $Y$ with $n$ 32-bit float numbers are stored in the global memory of the GPU. Our goal is to compute $Z = X \cdot Y$ and to store the resulting values in the global memory.

Suppose that an $n \times n$ matrix $X$ into blocks of size $32 \times 64$ each as illustrated in Fig. 5. In the figure, a $256 \times 256$ matrix $X$ is partitioned into blocks of size $32 \times 64$ each. Since $X$ is a triangular matrix, we ignore all-zero blocks and assign unique IDs to non-zero blocks in column-major order as illustrated in Fig. 5. Since our

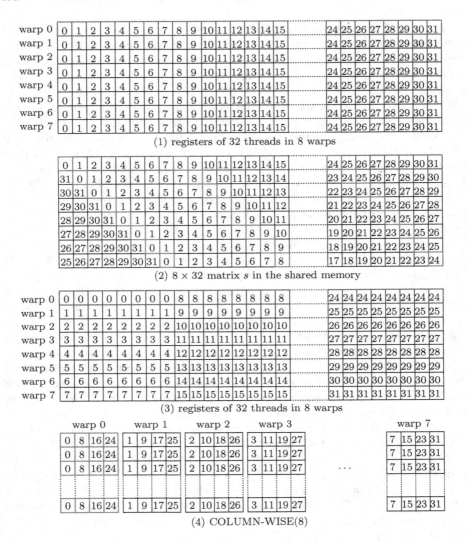

(1) registers of 32 threads in 8 warps

(2) 8 × 32 matrix $s$ in the shared memory

(3) registers of 32 threads in 8 warps

(4) COLUMN-WISE(8)

**Fig. 4.** Matrix $s$ in the shared memory when $2^q = 8$.

goal is to compute the matrix vector multiplication, it makes sense to partition vector $Y$ into $\frac{n}{64}$ sub-vectors of 64 elements each, and resulting vector $Z$ into $\frac{n}{32}$ sub-vectors of 32 elements each. In our GPU implementation, the product of each block and the corresponding sub-vector is computed, and the resulting values are added to a resulting sub-vector of $Z$ appropriately. For example, in Fig. 5, the product of block 15 and sub-vector 2 is computed, and it is added to the resulting sub-vector 5. Clearly, by executing the same operation for all non-zero blocks, the resulting vector $Z$ stores the product of $X$ and $Y$. Further, to reduce the computing overhead, we can omit memory access and computation

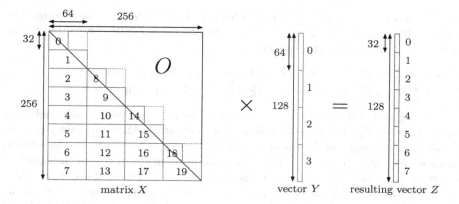

**Fig. 5.** Partitioning of matrix $X$ into blocks, vector $Y$ into sub-vectors, and resulting vector $Z$ into resulting sub-vectors

for the right half of a block, if all elements in it are zero. For example, in Fig. 5, we do not have to access the right half of blocks 0, 8, 14, and 18.

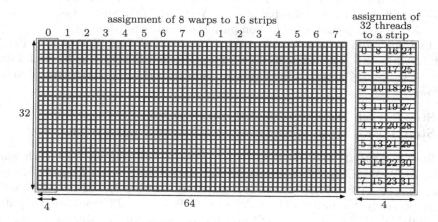

**Fig. 6.** A $32 \times 64$ block $X'$ of a matrix $X$ partitioned into 16 strips of size $32 \times 4$

For a fixed block and a sub-vector, we will show how the product of them is computed. Let $X'(i,j)$ $(0 \le i \le 31; 0 \le j \le 63)$ denote elements in the block and $Y'(j)$ $(0 \le j \le 63)$ denote elements in the corresponding sub-vector. We will show how each $Z'(i) = \sum_{j=0}^{63} X'(i,j)Y'(j)$ is computed. We use one CUDA block with 8 warps of $256 = 8 \cdot 32$ threads for this task. The block $X'$ is partitioned into 16 strips of size $32 \times 4$ as illustrated in Fig. 6(left). In other words, each strip $w$ $(0 \le w \le 15)$ consists of all elements $X'(i,j)$ $(0 \le i \le 31; 4w \le j \le 4w + 3)$. Each warp $w$ $(0 \le w \le 7)$ is assigned to strip $w$ and $w + 8$. Further, each of 32 threads in a warp is assigned to four adjacent elements in the same column of a strip as shown in Fig. 6(right). More specifically, thread $k$ $(0 \le k \le 31)$ of warp $w$ $(0 \le w \le 7)$ is assigned 8 elements

- $X'((4k \bmod 32) + p, 4w + \lfloor \frac{k}{8} \rfloor)$ $(0 \le p \le 3)$ in strip $w$, and
- $X'((4k \bmod 32) + p, 32 + 4w + \lfloor \frac{k}{8} \rfloor)$ $(0 \le p \le 3)$ in strip $w + 8$.

The algorithm has three steps. In Step 1, each thread reads these eight elements in two strips from the global memory and stores them in eight registers. Recall that elements in $X$ are arranged in column-major order in the address space as we have shown in Fig. 1. Hence, four elements in a strip assigned to a thread are consecutive in the memory space, and float4 data type, which combines four 32-bit float numbers, can be used to access these four elements efficiently. Thus, all elements in block $X'$ can be copied to registers of 256 threads by coalesced access very efficiently. Next, each thread $k$ $(0 \le k \le 31)$ of warp $w$ $(0 \le w \le 7)$ reads $Y'(4w + \lfloor \frac{k}{8} \rfloor)$, and $Y'(32 + 4w + \lfloor \frac{k}{8} \rfloor)$ in sub-vector $Y'$ necessary to compute the product. After that, it computes the sum of the products $S((4k \bmod 32) + p, 4w + \lfloor \frac{k}{8} \rfloor) = X'((4k \bmod 32) + p, 4w + \lfloor \frac{k}{8} \rfloor) \cdot Y'(4w + \lfloor \frac{k}{8} \rfloor) + X'((4k \bmod 32) + p, 32 + 4w + \lfloor \frac{k}{8} \rfloor) \cdot Y'(32 + 4w + \lfloor \frac{k}{8} \rfloor)$ for each $p$ $(0 \le p \le 3)$ by local computation. Figure 7(above) illustrates how each element of $32 \times 32$ matrix $S$ is stored in a register of one of 32 threads in 8 warps. The reader should have no difficulty to confirm that $Z'(i) = \sum_{j=0}^{31} S(i, j)$ holds, and so sub-vector $Z' = X' \cdot Y'$ can be obtained by computing the row-wise sums of $S$.

In Steps 2 and 3, the row-wise sums of $S$ are computed as illustrated in Fig. 7. Let $T(i, w) = S(i, 4w) + S(i, 4w + 1) + S(i, 4w + 2) + S(i, 4w + 3)$ for each $i$ and $w$ $(0 \le i \le 3; 0 \le w \le 7)$. The row-wise sums of $S$ is computed in two steps as follows:

**Step 2.** Each $T(i, w)$ is computed in each warp $w$.

**Step 3.** Each $T(i, 0) + T(i, 1) + \cdots + T(i, 7)$ $(= Z'(i))$ is computed.

In Step 2, $T(0, w)$, $T(4, w)$, $T(8, w)$, ..., $T(28, w)$ are computed by each warp $w$ by ROW-WISE(4). We can assume that the resulting sums $T(0, w)$, $T(4, w)$, $T(8, w)$, ..., $T(28, w)$ are stored in registers of threads 0, 1, 2, ..., 7 of warp $w$, respectively. Similarly, $T(1, w)$, $T(5, w)$, $T(9, w)$, ..., $T(29, w)$ are computed by ROW-WISE(4), and the resulting sums are stored in registers of threads 8, 9, 10, ..., 15 of warp $w$ respectively. The same procedure is repeated twice to compute all $T(i, w)$'s. After Step 2, $T(0, w)$, $T(1, w)$, $T(2, w)$, $T(3, w)$, $T(4, w)$, $T(5, w)$, $T(6, w)$, $T(7, w)$, ..., $T(31, w)$ are stored in a register of thread 0, 8, 16, 24, 1, 9, 17, 25, ..., 31 of warp $w$. Step 3 executes THREAD-WISE(8) for these registers in 8 warps of 32 threads each, we can obtain the value of each $Z'(i)$. The resulting row-wise sums are added to the corresponding sub-vector of $Z$ in the global memory. Note that, this addition operation must be performed using atomic read-add-write function atomicAdd [10], because the other CUDA blocks may perform addition operation to the same resulting sub-vector of $Z$ at the same time. The large overhead of atomicAdd is negligible, because atomicAdd function is executed only once for each $32 \times 64$ block.

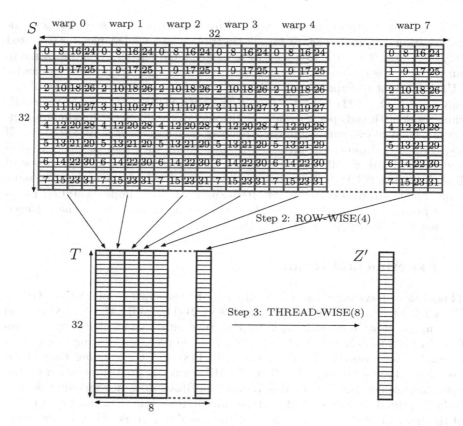

**Fig. 7.** Illustrating elements of $32 \times 32$ matrix $S$ and $32 \times 8$ matrix $T$ computed by each of 32 threads of 8 warps

For the reader's benefits, we summarize our GPU implementation for the trmv in 3 steps as follows:

**Step 1.** Each warp $w$ reads a block of $X$ and the corresponding sub-vector in $Y$ and computes the sum $S(i,j)$ of the products.

**Step 2.** It computes every $T(i,j)$ by executing ROW-WISE(4) four times.

**Step 3.** Compute all row-wise sums of $T$ by executing THREAD-WISE(8) and the resulting values are added to the corresponding resulting sub-vector of $Z$ by atomicAdd.

Step 1 performs the global memory access to read $X$ and $Y$. Since Step 2 simply performs ROW-WISE(4), it performs no memory access to the global memory or the shared memory. Step 3 performs shared memory access to execute THREAD-WISE(8) and global memory access for writing $Z'$ in the corresponding sub-vector of $Z$ by atomicAdd.

This implementation can be executed in only one kernel. Also, barrier synchronization __syncthreads() for all 256 threads in a CUDA block is executed only after Step 2. So, the overhead by kernel call and barrier synchronization in our algorithm is quite small. Each thread uses only eight 32-bit registers and a CUDA block of 8 warps with 256 threads uses the shared memory space of size only $4 \times 8 \times 32 = 1024$ bytes for executing THREAD-WISE(8). Since the maximum resident threads per streaming multiprocessor is 2048, the maximum number of CUDA blocks running in the streaming multiprocessor is $2048/256 = 8$. If a streaming multiprocessor invokes 8 CUDA blocks, they uses the shared memory space of size $8 \cdot 1024 = 8192$ bytes. Since a streaming multiprocessor of Tesla V100 has 96KB, it is possible to allocate 8 CUDA blocks to it at the same time. The theoretical occupancy is 100%, and our GPU implementation maximizes memory access throughput. As we are going to show later, the achieved occupancy is very close to 100% for large matrices.

## 4 Experimental Results

This section shows experimental results using five generations of NVIDIA GPUs. Table 1 shows the running time of cuBLAS(v9.2) [11], muBLAS(v1.5.38) [8], and our implementation on each GPU for the trmv with a triangular matrix of size from $32 \times 32$ to $16K \times 16K$ of fp32 (32-bit single precision floating point numbers). We have also used CUDA nvcc(v9.2.148) and the running time is the average of 1000 iterations. Recall that muBLAS needs pre-computation to find optimal parameters. The running time of muBLAS does not include the time to find optimal parameters, which takes much more time than the computation of the trmv. The table also shows that the speedup factors of our implementation over cuBLAS and muBLAS, respectively. From the table, we can see that our implementation run faster than cuBLAS and muBLAS for all matrices on Volta and Turing GPUs, which are latest architectures. On obsolete GPUs with Kepler, Maxwell, and Pascal architectures, muBLAS is a bit faster than our implementation for $16K \times 16K$ matrices, because muBLAS can invoke enough threads if the matrix is large and optimal parameter selection of muBLAS can attain better performance than our fixed parameter implementation. We can see that our implementation run faster than cuBLAS and muBLAS for almost all sizes of matrices with a very few exceptions.

Table 2 shows the achieved occupancy on Tesla V100 measured on each warp scheduler using hardware performance counters to count the number of active warps on that scheduler every clock cycle. To sufficiently hide latency between dependent instructions, the latency must be large enough. As we can see in the table, our implementation is very close to 100% for large matrices, while the that of cuBLAS is quite small. Thus our implementation runs much faster than cuBLAS for large $n$.

**Table 1.** The running time in $\mu$s to compute the product of a $n \times n$ triangular matrix and $n$-element vector

| $n$ | 32 | 64 | 128 | 256 | 512 | 1K | 2K | 4K | 8K | 16K |
|---|---|---|---|---|---|---|---|---|---|---|
| Tesla K40 (Kepler architecture) | | | | | | | | | | |
| cuBLAS | 22.9 | 25.2 | 30.3 | 43.8 | 72.0 | 248 | 317 | 623 | 1332 | 4880 |
| muBLAS | 40.6 | 42.5 | 42.7 | 51.0 | 57.2 | 69.3 | 115 | 281 | 825 | 3015 |
| Our implementation | 23.9 | 25.3 | 25.8 | 26.1 | 28.9 | 40.5 | 77.2 | 226 | 840 | 3370 |
| Speedup:cuBLAS | 0.958 | 0.996 | 1.17 | 1.68 | 2.49 | 6.12 | 4.11 | 2.76 | 1.59 | 1.45 |
| Speedup:muBLAS | 1.70 | 1.68 | 1.66 | 1.95 | 1.98 | 1.71 | 1.49 | 1.24 | 0.982 | 0.895 |
| GeForce GTX 980 (Maxwell architecture) | | | | | | | | | | |
| cuBLAS | 8.79 | 9.84 | 14.2 | 18.2 | 28.8 | 126 | 233 | 477 | 1109 | 4024 |
| muBLAS | 18.0 | 18.7 | 19.3 | 19.7 | 21.3 | 27.8 | 73.4 | 217 | 794 | 3040 |
| Our implementation | 8.43 | 9.30 | 9.71 | 10.1 | 10.5 | 18.1 | 57.9 | 200 | 771 | 3067 |
| Speedup:cuBLAS | 1.04 | 1.06 | 1.46 | 1.80 | 2.74 | 6.96 | 4.02 | 2.39 | 1.44 | 1.31 |
| Speedup:muBLAS | 2.14 | 2.01 | 1.99 | 1.95 | 2.03 | 1.54 | 1.27 | 1.09 | 1.03 | 0.991 |
| GeForce GTX 1080 (Pascal architecture) | | | | | | | | | | |
| cuBLAS | 7.53 | 8.66 | 11.2 | 14.8 | 23.4 | 90.9 | 187 | 369 | 855 | 3045 |
| muBLAS | 14.2 | 14.9 | 14.9 | 16.6 | 17.0 | 22.0 | 53.7 | 160 | 558 | 2296 |
| Our implementation | 7.51 | 8.08 | 8.11 | 8.12 | 9.37 | 14.7 | 42.0 | 143 | 555 | 2338 |
| Speedup:cuBLAS | 1.00 | 1.07 | 1.38 | 1.82 | 2.50 | 6.18 | 4.45 | 2.58 | 1.54 | 1.3.0 |
| Speedup:muBLAS | 1.89 | 1.84 | 1.84 | 2.04 | 1.81 | 1.5.0 | 1.28 | 1.12 | 1.01 | 0.982 |
| Tesla V100 (Volta architecture) | | | | | | | | | | |
| cuBLAS | 9.46 | 11.8 | 13.9 | 19.6 | 36.5 | 103 | 216 | 414 | 844 | 2665 |
| muBLAS | 17.8 | 18.3 | 19.2 | 19.6 | 20.5 | 22.1 | 31.7 | 60.6 | 174 | 629 |
| Our implementation | 8.78 | 9.22 | 9.26 | 9.34 | 9.54 | 10.7 | 19.2 | 47.9 | 160 | 615 |
| Speedup:cuBLAS | 1.08 | 1.28 | 1.50 | 2.10 | 3.83 | 9.63 | 11.3 | 8.64 | 5.28 | 4.33 |
| Speedup:muBLAS | 2.03 | 1.98 | 2.07 | 2.10 | 2.15 | 2.07 | 1.65 | 1.27 | 1.09 | 1.02 |
| GeForce RTX 2080Ti (Turing architecture) | | | | | | | | | | |
| cuBLAS | 9.72 | 10.7 | 12.7 | 20.6 | 39.2 | 78.8 | 84.2 | 170 | 364 | 1129 |
| muBLAS | 20.6 | 22.2 | 23.7 | 23.8 | 24.7 | 25.3 | 37.9 | 86.9 | 272 | 1135 |
| Our implementation | 9.68 | 10.2 | 10.5 | 10.8 | 12.3 | 12.5 | 25.8 | 71.2 | 261 | 996 |
| Speedup:cuBLAS | 1.00 | 1.05 | 1.21 | 1.91 | 3.19 | 6.30 | 3.26 | 2.39 | 1.39 | 1.13 |
| Speedup:muBLAS | 2.13 | 2.18 | 2.26 | 2.20 | 2.01 | 2.02 | 1.47 | 1.22 | 1.04 | 1.14 |

**Table 2.** The achieved occupancy in percent of three implementations on Tesla V100

| $n$ | 32 | 64 | 128 | 256 | 512 | 1K | 2K | 4K | 8K | 16K |
|---|---|---|---|---|---|---|---|---|---|---|
| cuBLAS | 10.7 | 8.63 | 8.41 | 12.1 | 23.6 | 24.4 | 6.20 | 6.23 | 6.24 | 6.24 |
| muBLAS | 5.42 | 8.32 | 9.06 | 11.0 | 13.8 | 20.0 | 29.1 | 43.0 | 59.2 | 69.4 |
| Our implementation | 12.0 | 12.1 | 12.2 | 12.2 | 12.2 | 36.3 | 79.4 | 93.0 | 95.4 | 96.5 |

## 5  Conclusion

We have shown very efficient GPU implementation for the trmv. Experimental results for five generations of NVIDIA GPUs show that for matrices of sizes from $32 \times 32$ to $16K \times 16K$ show that our GPU implementation is faster than cuBLAS and muBLAS for almost all sizes and GPU generations. Our trmv algorithm requires that the matrix size is a multiple of 32. However, if this is not the case, we can use zero-padding technique.

## References

1. Charara, A., Ltaief, H., Keyes, D.: Redesigning triangular dense matrix computations on GPUs. In: Dutot, P.-F., Trystram, D. (eds.) Euro-Par 2016. LNCS, vol. 9833, pp. 477–489. Springer, Cham (2016). https://doi.org/10.1007/978-3-319-43659-3_35
2. Fujimoto, N.: Faster matrix-vector multiplication on GeForce 8800GTX. In: Proceedings of International Symposium on Parallel and Distributed Processing, April 2008
3. He, G., Gao, J., Wang, J.: Efficient dense matrix-vector multiplication on GPU. Concurr. Comput. Pract. Exp. **30**(19), e4705 (2018)
4. Honda, T., Yamamoto, S., Honda, H., Nakano, K., Ito, Y.: Simple and fast parallel algorithms for the Voronoi map and the Euclidean distance map, with GPU implementations. In: Proceedings of International Conference on Parallel Processing, pp. 362–371, August 2017
5. Hwu, W.W.: GPU Computing Gems Emerald Edition. Morgan Kaufmann, Burlington (2011)
6. Karwacki, M., Stpiczynski, P.: Improving performance of triangular matrix-vector BLAS routines on GPUs. Adv. Parallel Comput. **22**, 405–412 (2012)
7. Matsumura, N., Tokura, H., Kuroda, Y., Ito, Y., Nakano, K.: Tile art image generation using conditional generative adversarial networks. In: Proceedings of International Symposium on Computing and Networking Workshops, pp. 209–215 (2018)
8. Mukunoki, D., Imamura, T., Takahashi, D.: Automatic thread-block size adjustment for memory-bound BLAS kernels on GPUs. In: Proceedings of International Symposium on Embedded Multicore/Many-Core Systems-on-Chip, June 2016
9. Muramatsu, J., Fukaya, T., Zhang, S.L., Kimura, K., Yamamoto, Y.: Acceleration of Hessenberg reduction for nonsymmetric eigenvalue problems in a hybrid CPU-GPU computing environment. Int. J. Netw. Comput. **1**(2), 132–143 (2011)
10. NVIDIA Corporation: NVIDIA CUDA C programming guide version 4.0 (2011)
11. NVIDIA Corporation: CUBLAS LIBRARY user guide, February 2019. https://docs.nvidia.com/cuda/cublas/index.html
12. Ogawa, K., Ito, Y., Nakano, K.: Efficient Canny edge detection using a GPU. In: Proceedings of International Conference on Networking and Computing, pp. 279–280. IEEE CS Press, November 2010
13. Takeuchi, Y., Takafuji, D., Ito, Y., Nakano, K.: ASCII art generation using the local exhaustive search on the GPU. In: Proceedings of International Symposium on Computing and Networking, pp. 194–200, December 2013
14. Tokura, H., et al.: An efficient GPU implementation of bulk computation of the eigenvalue problem for many small real non-symmetric matrices. Int. J. Netw. Comput. **7**(2), 227–247 (2017)

# Performance Engineering for a Tall & Skinny Matrix Multiplication Kernels on GPUs

Dominik Ernst[1]([✉]), Georg Hager[1], Jonas Thies[2], and Gerhard Wellein[1]

[1] Erlangen Regional Computing Center (RRZE), 91058 Erlangen, Germany
dominik.ernst@fau.de
[2] Simulation and Software Technology, German Aerospace Center (DLR),
Cologne, Germany

**Abstract.** General matrix-matrix multiplications (GEMM) in vendor-supplied BLAS libraries are best optimized for square matrices but often show bad performance for tall & skinny matrices, which are much taller than wide. Nvidia's current CUBLAS implementation delivers only a fraction of the potential performance (as given by the roofline model) in this case. We describe the challenges and key properties of an implementation that can achieve perfect performance. We further evaluate different approaches of parallelization and thread distribution, and devise a flexible, configurable mapping scheme. A code generation approach enables a simultaneously flexible and specialized implementation with autotuning. This results in perfect performance for a large range of matrix sizes in the domain of interest, and at least 2/3 of maximum performance for the rest on an Nvidia Volta GPGPU.

**Keywords:** Tall & skinny · Matrix multiplication · GPU

## 1 Introduction

### 1.1 Tall & Skinny Matrix Multiplications (TSMM)

The general matrix-matrix multiplication (GEMM) is such an essential linear algebra operation used in many numerical algorithms that hardware vendors usually supply an implementation that is perfectly optimized for their hardware. In case of Nvidia, this is part of CUBLAS [7]. However, since these implementations are focused on mostly square matrices, they often perform poorly for matrices with unusual shapes.

In this paper, we cover the operation $\mathbf{C} = \mathbf{A}^T \mathbf{B}$, with matrices $\mathbf{A}$ and $\mathbf{B}$ being tall & skinny, i.e., much taller than they are wide. If $\mathbf{A}$ and $\mathbf{B}$ are of size $K \times M$ and $K \times N$, the shared dimension $K$ is long (of the order of $10^6$), whereas the dimensions $M$ and $N$ are short, which we define here as the range $[1, 64]$. $\mathbf{A}$ and $\mathbf{B}$ are both stored in row-major order. We are interested in a highly efficient implementation of this operation on the Nvidia Volta GPGPU (Fig. 1).

© Springer Nature Switzerland AG 2020
R. Wyrzykowski et al. (Eds.): PPAM 2019, LNCS 12043, pp. 505–515, 2020.
https://doi.org/10.1007/978-3-030-43229-4_43

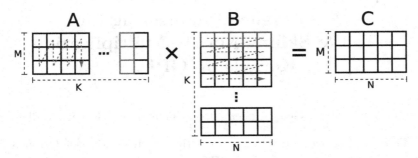

**Fig. 1.** Illustration of $\mathbf{A}^T\mathbf{B} = \mathbf{C}$ with $\mathbf{A}$ and $\mathbf{B}$ being tall & skinny matrices. Note that $\mathbf{A}$ is transposed in the illustration.

## 1.2   Application

Row-major tall & skinny matrices are the result of combining several vectors to block vectors. *Block Vector Algorithms* are linear algebra algorithms that compute on multiple vectors simultaneously for improved performance. For instance, by combining multiple, consecutive *Sparse Matrix Vector* (SpMV) multiplications to a *Sparse Matrix Multiple Vector* (SpMMV) multiplication, the matrix entries are loaded only once and used for the multiple vectors, which reduces the overall memory traffic and consequently increases performance of this memory bound operation. This has first been analytically shown in [3] and is used in many applications such as described in [10].

The simultaneous computation on multiple vectors can also be used to gain numerical advantages. This has been shown for block vector versions of the Lanczos algorithm [1], of the biconjugate gradient algorithm [9], and of the Jacobi-Davidson Method [10], each of which use block vectors to compute multiple eigenvectors simultaneously. Many such algorithms require multiplications of block vectors. For example, the tall & skinny matrix matrix multiplication $\mathbf{A}^T\mathbf{B}$ occurs in classical Gram-Schmidt orthogonalization of a number of vectors represented by $\mathbf{B}$ against an orthogonal basis $\mathbf{A}$.

## 1.3   Roofline Model

We use the roofline model [12] to obtain an upper limit for the performance of this operation. Under the assumption that the three matrices are transferred just once from memory and that $2MNK$ floating point operations are performed, the arithmetic intensity assuming $K \gg M, N$ and $M = N$ is

$$I_C = \frac{2MNK}{(MK + NK + MN) \times 8} \frac{\text{flop}}{\text{byte}} \stackrel{K \gg M,N}{\approx} \frac{2MN}{(M + N) \times 8} \frac{\text{flop}}{\text{byte}} \stackrel{M=N}{=} \frac{M}{8} \frac{\text{flop}}{\text{byte}}. \tag{1}$$

In this symmetric case, the arithmetic intensity grows linearly with $M$. This paper will show measurements only for the symmetric case, although the non-symmetric case is not fundamentally different, with the intensity being proportional to the harmonic mean of both dimensions and consequently dominated by the smaller number. With the derived intensity, the model predicts $P = \max\left(M/8 \times B_s, P_{peak}\right)$ as the "perfect" performance yardstick.

Usually the GEMM is considered a classic example for a compute-bound problem with high arithmetic intensity. However, at $M, N = 1$, the arithmetic intensity is just $1/8$ flop/byte, which is far below the roofline knee of modern compute devices and therefore strongly memory bound. This is not surprising given that a matrix multiplication with $M, N = 1$ is the same as a scalar product. At the other endpoint of the considered spectrum, at $M, N = 64$, the arithmetic intensity is $8$ flop/byte, which is close to the inverse machine balance of a V100 GPU (see below). Therefore the performance character of the operation changes from extremely memory bound at $M, N = 1$ to simultaneously memory and compute bound at $M, N = 64$. An implementation with perfect performance thus needs to fully utilize the memory bandwidth at all sizes and reach peak floating point performance for the large sizes. The very different performance characteristics make it hard to write an implementation that fits well for both ends of the spectrum. Different optimizations are required for both cases.

With the performance as given by the roofline model, it is possible to judge the quality of an implementation's performance as the percentage of the roofline limit. This is plotted in Fig. 2 for CUBLAS. The graph shows a potential performance headroom of $5\times$ to $100\times$.

## 1.4   Contribution

This paper shows the necessary implementation techniques to achieve near-perfect performance for double precision tall & skinny matrix-matrix multiplications on an Nvidia V100 GPGPU.

Two different parallel reduction schemes are implemented and analyzed as to their suitability for matrices with lower row counts.

A code generator is implemented that generates code for specific matrix sizes and tunes many configuration options specifically to that size. This allows to exploit regularity, where size parameters allow it, while still generating the least possible overhead where they do not.

## 1.5   Related Work

*CUBLAS* is NVIDIA's BLAS implementation. The GEMM function interface in BLAS only expects column-major matrices. Treating the matrices as transposed column major matrices and executing $\mathbf{AB}^T$ is an equivalent operation.

*CUTLASS* [8] is a collection of primitives for multiplications especially of small matrices, which can be composed in different ways to form products of larger matrices. One of these is the splitK kernel, which additionally parallelizes the inner summation of the matrix multiplication. Square matrix multiplications usually do not do this, but it is what is required for sufficient parallelism. An adapted version of the "06_splitK_gemm" example is used for benchmarking.

**Table 1.** Measured memory bandwidth on a Tesla V100-PCIe-16 GB of a read only kernel with different amount of load parallelism (ILP) and occupancies

| % of occupancy | ILP, Gbyte/s | | |
|---|---|---|---|
| | 1 | 4 | 16 |
| 1 block, 4 warps | 3.0 | 10.1 | 16.3 |
| 6.25% | 228 | 629 | 815 |
| 12.5% | 419 | 824 | 877 |
| 25% | 681 | 872 | 884 |
| 50% | 834 | 884 | 887 |
| 100% | 879 | 891 | 877 |

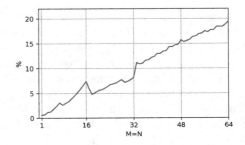

**Fig. 2.** Percentage of roofline predicted performance achieved by CUBLAS for the range $M = N \in [1, 64]$ on a Tesla V100-PCIe-16 GB

## 1.6  Hardware

In this work we use Nvidia's V100-PCIe-16 GB GPGPU (Volta architecture) with CUDA 10.0. The hardware data was collected with our own CUDA microbenchmarks, available at [2] together with more detailed data.

*Memory Bandwidth.* The STREAM benchmarks [6] all contain a write stream, while the TSMM does not. We thus use a thread-local sum reduction to estimate the achievable memory bandwith (see Table 1). The maximum is above 880 Gbyte/s; but this is only possible with sufficient parallelism, either through high occupancy or instruction level parallelism (ILP) in the form of multiple read streams, achieved here through unrolling.

*Floating Point Throughput.* The V100 can execute one 32-wide double precision (DP) floating point multiply add (FMA) per cycle on each of its 80 streaming multiprocessors (SMs) and runs at a clock speed of 1.38 GHz for a DP peak of $80 \times 32 \times 2 \times 1.38$ Gflop/s $= 7066$ Gflop/s. One SM quadrant can process a 32 warp lanes-wide instruction every four cycles at a latency of eight cycles. Full throughput can already be achieved with a single warp per quadrant, if instructions are independent.

*L1 Cache.* The L1 cache plays an instrumental role in achieving the theoretically possible arithmetic intensity. The per-SM private L1 cache can transfer a full 128-byte cache line per cycle. Therefore a 32-wide, unit-stride DP load takes two cycles but this number increases by a cycle for each 128-byte cache line that is affected. This amounts to a rate of one load for two FMA instructions.

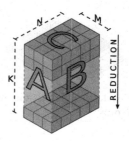

```
for m = 0...M:
  for n = 0...N:
    for k = 0...K:
      C[m][n] += A[k][m] * B [k][n]
```

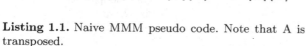

**Listing 1.1.** Naive MMM pseudo code. Note that A is transposed.

**Fig. 3.** Illustration of the iteration space

## 2  Implementation Strategies

The arithmetic intensity (1) was derived assuming perfect caching, i.e. that **A** and **B** are transferred from memory just once. The fastest way to reuse values is to use a register and have the thread, the register belongs to, perform all required operations on this data. Data used by multiple threads can preferably be shared in the L1 cache for threads in the same thread block or in the L2 cache otherwise. This works best only with some spatial and temporal access locality in place.

### 2.1  Thread Mapping Options

The parallelization scheme, i.e., the way in which work is mapped to GPU threads, plays an important role for data flow in the memory hierarchy. The canonical formulation of an MMM is the three-level loop nest shown in Listing 1.1.

The two outer loops, which are completely independent and therefore well parallelizable, are usually the target of an implementation focused on square matrices. For skinny matrices, these loops are much too short to yield enough parallelism for a GPU. In consequence, the loop over the long K dimension has to be parallelized as well, which also involves parallelizing the sum inside the loop. There are many more terms in the parallel reduction than threads, so that each thread can first serially compute a partial sum, which is afterwards reduced to a total sum.

The iteration space of an MMM can be visualized as the cuboid spanned by the outer product of **A** and **B** (see Fig. 3), where each cell contains a multiplication. A sum reduction over the $K$ dimension yields the result matrix **C**. Each horizontal slice requires to load a row of **A** and **B** of length $M$ and $N$, respectively, and computes a rectangle with $M \times N$ FMAs in it.

For data locality, the two small loops have to be moved into the $K$ loop. This creates $MN$ intermediate sums that have to be stored. Depending on whether and how the two small loops are parallelized, each thread computes only some of these intermediates. Figures 4, 5 and 6 visualize this by showing a slice of

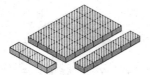

**Fig. 4.** Parallelization over $K$ only

**Fig. 5.** Parallelization over $K$ loop and an inner loop

**Fig. 6.** Parallelization over $K$ and tiling of the two inner loops, here with tile size $2 \times 3$

the multiplication cube and which values a single thread would compute. The number of loads that each thread has to do are the affected values in the row of **A** and **B**, also visible in the illustrations, while the number of FMA operations is the number of highlighted cells in the slice. It is favorable to have a high FMA/loads ratio, as the L1 cache is not as fast as the FP units. This can be achieved by maximizing the area and the squareness of the area that is computed by a single thread. At the same time, more intermediate results per thread increase the register count, which decreases occupancy and eventually leads to spilling.

The easiest approach to achieve this goal would be to only parallelize the $K$ loop, as visualized in Fig. 4. While this maximizes the arithmetic intensity already in the L1 cache, the $MN$ intermediate results occupy $2MN$ registers, so the maximum of 256 registers per thread is already exceeded at $M, N > 11$, causing spilling and poor performance.

One of the inner loops could be parallelized as well, leading to the pattern in Fig. 5. The amount of registers required is only $M$ or $N$, so there is no spilling even at $M, N = 64$. However, the narrow shape results in a FMA/loads ratio below 1 and therefore a low arithmetic intensity in the L1 cache.

A better approach that combines managable register requirements with a more square form is to subdivide the two smaller loops into tiles like in Fig. 6. This mapping also allows for much more flexibility, as the tile sizes can be chosen small enough to avoid spilling or reach a certain occupancy goal but also large enough to create a high FMA/loads ratio.

## 2.2  Global Reduction

After each thread has serially computed its partial, thread-local result, a global reduction is required, which is considered overhead. It depends only on the thread count, though, whereas the time spent in the serial summation grows linearly with the row count and therefore becomes marginal for large enough row counts. However, the authors of [11] argue that the performance at small row counts is still relevant, as the available GPU memory is shared by more data structures than just the two tall & skinny matrices, which limits the data set size.

Since the *Pascal* architecture, atomic add operations are available for global memory, making global reductions more efficient than on older systems. Each thread can just use an atomicAdd of its partial value to the final results. The throughput of `atomicAdd` operations is limited by the amount of contention, which grows for smaller matrix sizes. We improve on this global atomic reduction variant with a local atomic variant that reduces the amount of global `atomicAdd` operations by first computing thread block local partial results using shared memory atomics. This is followed by a global reduction of the local results. Additionally, we opportunistically reduce the amount of launched threads for small row counts.

## 2.3  Leapfrogging and Unrolling

On Nvidia's GPU architectures, load operations can overlap with each other. The execution will only stall at an instruction that requires an operand from an outstanding load. The compiler maximizes this overlap by moving all loads to the beginning of the loop body, followed by the FP instructions that consume the loaded values. At least one or two of the loads come from memory, which take longer to complete than queueing all load operations, so that execution stalls at the first FP instruction. A way to circumvent this stall is to load the inputs one loop iteration ahead into a separate set of *next* registers, while the computations still happen on the *current* values. At the end of the loop, the *next* values become the *current* values of the next loop iteration by assignment. These assignments are the first instructions that depend on the loads and thus the computations can happen while the loads are still in flight. A similar effect can be achieved by unrolling the $K$ loop. The compiler can move the loads of multiple iterations to the front, therefore creating more overlap.

## 2.4  Code Generation

A single implementation cannot be suitable for all matrix sizes. In order to engineer the best code for each size, some form of metaprogramming is required. C++ templates allow some degree of metaprogramming but are limited in their expressivity or require convoluted constructs. Usually the compiler unrolls and eliminates short loops with known iteration count in order to reduce loop overhead, combine address calculations, avoid indexed loads from arrays for the thread-local results, deduplicate and batch loads, and much more. A direct manual generation of the code offers more control, however. For example, in order to enable tile sizes that are not divisors of the matrix dimensions, guarding statements have to be added around computations that could exceed the matrix size. These should be omitted wherever it is safe in order to not compromise performance for dividing tile sizes. We therefore use a code generating script in python, which allows to prototype new techniques much quicker and with more control. Many different parameters can be configured easily, for example whether leapfrogging and unrolling is used, how the reduction is performed, and what tile sizes to set. The same reasoning for code generation is made by [4], where it is used to generate small matrix multiplication kernels for CPUs.

## 3   Results

### 3.1   Impact of Reductions

Figure 7 shows the relative performance of our TSMM implementation versus row count with respect to a baseline without any reduction for a selection of inner matrix sizes and tile sizes, choosing either of the two reduction methods described in Sect. 2.2. As expected, the impact of the reduction generally decreases with increasing row count. The method with only global atomics is especially slow for the narrower matrices $(M, N = 4)$. Many threads writing to a small amount of result values leads to contention and causes a noticeable impact even for a device memory filling matrix $(K = 10^8)$. The local atomic variant drastically reduces the number of writing threads, resulting in less than 10% overhead even for the smallest sizes and near perfect performance for $K > 10^6$. For the wider matrices, the difference is smaller. The global atomic version is not as slow because writes spread out over more result values and the local atomic variant is not as fast because the larger tile size requires more work in the local reduction. Both variants incur less than 10% overhead just above $10^4$, a point where only about 0.2% of the GPU memory is used.

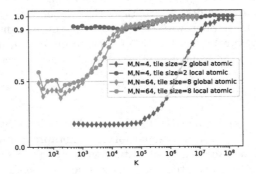

**Fig. 7.** Ratio of achieved performance of four kernels with two matrix sizes and two different final reduction methods vs. the long dimension $K$ compared to the performance of a kernel that lacks the final reduction

### 3.2   Tile Sizes

Figure 8 shows the dependence of performance on tile sizes $T_M$ and $T_N$ for the case $M, N = 32$ with leapfrogging. Performance drops off sharply if the tile sizes become too large and too many registers are used. The number of registers can be approximated by $2 \times (T_M T_N + 2(T_M + T_N))$, which accounts for the thread-local sums $(T_M T_N)$ and the loaded values $(T_M + T_N)$. Leapfrogging requires two registers for each of the latter and double precision doubles the overall number of needed registers. The graph shows the isolines of 112 and 240 registers, which, accounting for some registers used otherwise, represent the occupancy drop from 25% to 12.5% at 128 registers and the start of spilling at 256 registers.

The best-performing tile sizes generally sit on these lines, maximizing the area of the tile without requiring too many registers. The dimensions are largely symmetric but not perfectly, as threads are mapped to tiles in $M$ direction first. While there are some patterns, where some tile sizes would generally be faster than others, there is no clear trend that either dividers of the matrix width or powers of two are favored in any way.

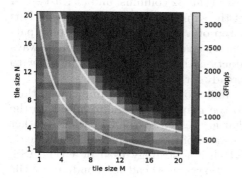

**Fig. 8.** Performance for different tile sizes, $M, N = 32$, $K = 2^{24}$, leapfrogging enabled. The two white lines are $2 \times (T_M T_N + 2(T_M + T_N)) = R$, with $R = 112,240$ to mark approximate boundaries of register usage.

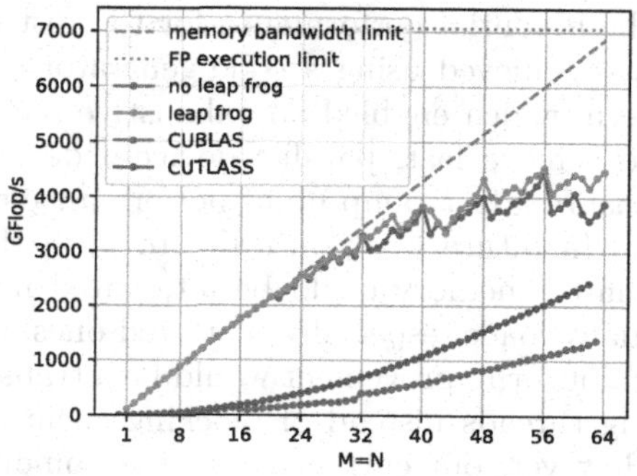

**Fig. 9.** Best achieved performance for each matrix size with $M = N$ in comparison with the roofline limit, CUBLAS and CUTLASS, with $K = 2^{23}$

## 3.3  Best Performance

An exhaustive search was used to find the best tile size for each matrix size. Figure 9 shows the results with and without leapfrogging. The performance meets the roofline prediction (gray dashed line) perfectly until $M, N = 20$. Until $M, N = 36$, the best performance stays within 95% of the limit. Beyond that, the growing arithmetic intensity does not translate into a proportional speedup anymore, although we are still about a factor of two away from peak. The best performance appears to plateau at about 4500 Gflop/s, or 2/3 of peak. Leapfrogging gives about 10–15% advantage for the large sizes. Note that the best tile size changes when leapfrogging is used as it requires more registers.

The cause of the observed performance limit is insufficient memory parallelism. In a large tile, like the $10 \times 8$ which performs best for $M, N = 64$, only 1–2 of the 18 loads go to memory and are not backed by a cache. At 12.5% occupancy and 1–2 read streams, Table 1 shows an achieved memory bandwidth between 400 and 700 Gbyte/s, far below the 880 Gbyte/s needed to meet the roofline expectation.

Both CUBLAS' and CUTLASS' performance is far below the potential performance, especially for the small sizes. Their peformance increases with wider matrices and it can be expected that CUTLASS would overtake the presented implementation at around $M, N = 100$, and CUBLAS somewhat later.

## 4    Conclusion and Outlook

We have shown how to get perfect performance on a V100 GPU for the multiplication of tall & skinny matrices narrower than 32 columns, and at least $2/3$ of the potential performance for the rest of the skinny range until width 64. This was achieved using a code generator on top of a suitable thread mapping pattern, which enabled an exhaustive parameter space search. Two different ways to achieve fast, parallel device-wide reductions have been devised in order to ensure a fast ramp up of performance already for shorter matrices.

In future work, in order to push the limits of the current implementation, shared memory could be integrated into the mapping scheme to speed up the many loads, especially scattered ones, that are served by the L1 cache. Another point for improvement would be a transposition of tiles and threads by interleaving threads instead of blocking them, which is a strategy used by CUTLASS. However, our current setup has somewhat lower overhead for nondividing tile sizes and configurations, where a warp would work on multiple slices.

The presented performance figures were obtained by parameter search. An advanced performance model, currently under development, could be fed with code characteristics such as load addresses and instruction counts generated with the actual code and then used to eliminate bad candidates much faster. It will also support a better understanding of performance limiters.

Prior work by us in this area is already part of the sparse matrix toolkit *GHOST* [5] and we plan to integrate the presented work as well.

**Acknowledgements.** This work was supported by the ESSEX project in the DFG Priority Programme SPPEXA.

## References

1. Cullum, J., Donath, W.E.: A block Lanczos algorithm for computing the $q$ algebraically largest eigenvalues and a corresponding eigenspace of large, sparse, real symmetric matrices. In: 1974 IEEE Conference on Decision and Control Including the 13th Symposium on Adaptive Processes, pp. 505–509, November 1974
2. Ernst, D.: CUDA Microbenchmarks. http://tiny.cc/cudabench
3. Gropp, W.D., Kaushik, D.K., Keyes, D.E., Smith, B.F.: Towards realistic performance bounds for implicit CFD codes. In: Proceedings of Parallel CFD 1999, pp. 233–240. Elsevier (1999)
4. Herrero, J.R., Navarro, J.J.: Compiler-optimized kernels: an efficient alternative to hand-coded inner kernels. In: Gavrilova, M.L., Gervasi, O., Kumar, V., Tan, C.J.K., Taniar, D., Laganá, A., Mun, Y., Choo, H. (eds.) ICCSA 2006. LNCS, vol. 3984, pp. 762–771. Springer, Heidelberg (2006). https://doi.org/10.1007/11751649_84

5. Kreutzer, M., et al.: GHOST: building blocks for high performance sparse linear algebra on heterogeneous systems. Int. J. Parallel Program., 1–27 (2016). https://doi.org/10.1007/s10766-016-0464-z
6. McCalpin, J.D.: Memory bandwidth and machine balance in current high performance computers. In: IEEE Computer Society Technical Committee on Computer Architecture (TCCA) Newsletter, pp. 19–25, December 1995
7. NVIDIA: CUBLAS reference (2019). https://docs.nvidia.com/cuda/cublas. Accessed 05 May 2019
8. NVIDIA: CUTLASS (2019). https://github.com/NVIDIA/cutlass. Accessed 05 May 2019
9. O'Leary, D.P.: The block conjugate gradient algorithm and related methods. Linear Algebra Appl. **29**, 293–322 (1980). http://www.sciencedirect.com/science/article/pii/0024379580902475. Special Volume Dedicated to Alson S. Householder
10. Röhrig-Zöllner, M., et al.: Increasing the performance of the Jacobi-Davidson method by blocking. SIAM J. Sci. Comput. **37**(6), C697–C722 (2015). https://doi.org/10.1137/140976017
11. Thies, J., Röhrig-Zöllner, M., Overmars, N., Basermann, A., Ernst, D., Wellein, G.: PHIST: a pipelined, hybrid-parallel iterative solver toolkit. ACM Trans. Math. Softw. (2019)
12. Williams, S., Waterman, A., Patterson, D.: Roofline: an insightful visual performance model for multicore architectures. Commun. ACM **52**(4), 65–76 (2009). https://doi.org/10.1145/1498765.1498785

# Reproducible BLAS Routines with Tunable Accuracy Using Ozaki Scheme for Many-Core Architectures

Daichi Mukunoki[1(✉)], Takeshi Ogita[2], and Katsuhisa Ozaki[3]

[1] RIKEN Center for Computational Science, Hyogo, Japan
daichi.mukunoki@riken.jp
[2] Tokyo Woman's Christian University, Tokyo, Japan
ogita@lab.twcu.ac.jp
[3] Shibaura Institute of Technology, Saitama, Japan
ozaki@sic.shibaura-it.ac.jp

**Abstract.** Generally, floating-point computations comprise rounding errors; the result may be inaccurate and not identical (non-reproducible). Particularly, heterogeneous computing has many factors that affect reproducibility. The loss of accuracy and reproducibility could be a crucial issue in debugging complex codes and the reliability of computations. In this paper, we propose high-performance implementations of reproducible basic linear algebra subprograms (BLAS) routines with tunable accuracy for many-core architectures. Our approach is based on an accurate matrix-multiplication method, Ozaki scheme, which can be constructed on level-3 BLAS that performs standard floating-point operations. We demonstrate the performance of three routines: inner product (DOT), matrix-vector multiplication (GEMV), and matrix-multiplication (GEMM) on NVIDIA's Volta GPU by comparing these with the standard routines provided by the vendor. Furthermore, we demonstrate the reproducibility between CPU and GPU and its accuracy.

**Keywords:** Accurate · Reproducible · BLAS

## 1 Introduction

Floating-point operations suffer from loss of accuracy due to round-off errors, as well as their accumulations. It is observed not only for ill-conditioned but also for regular problems. Hence there is a constant demand for accurate computations. Moreover, reproducibility is another issue caused by round-off errors. In this paper, "reproducibility" means an ability to obtain a bit-level identical result every time on the same input data regardless of the computational environment. But as floating-point computation is non-associative, the result may not always be the same depending on the order of the arithmetic operation.

© Springer Nature Switzerland AG 2020
R. Wyrzykowski et al. (Eds.): PPAM 2019, LNCS 12043, pp. 516–527, 2020.
https://doi.org/10.1007/978-3-030-43229-4_44

The loss of accuracy and reproducibility may become a crucial issue in the development and maintenance (including porting and debugging) of complex codes, as well as the reliability of the result of computations. Both the issues of accuracy and reproducibility are becoming more significant as the number of operations in a computation increases and the code becomes complex toward the exa-scale computing era. In particular, with respect to reproducibility, recent many-core architectures have many factors affecting the issue: the order of the arithmetic operations can be changed depending on the number of processes/threads and whether atomic operations have been used. Besides, the use or non-use of fused multiply-add may affect the rounding error. Also, in heterogeneous computations, dynamic load balancing may change the target hardware of the computation being carried out.

Therefore, as the first step toward accurate and reproducible numerical computations, several research works [1,2,6] have proposed accurate and reproducible basic linear algebra subprograms (BLAS). In general, BLAS is an essential building block for numerical computations. Although several BLAS implementations supporting accurate and reproducible computations have been proposed, the cost of the software development and the performance overhead against the highly optimized standard implementations, in particular on level-3 routines [3], are still high. Besides, the implementation for the full set of the BLAS routines requires extensive time and labor, even if it does not include the performance optimization for a particular platform.

To address these problems, we propose high-performance implementations of reproducible BLAS routines for many-core architectures using the Ozaki scheme [10]. This method achieves reproducibility and tunable accuracy including correct rounding on computations based on inner product such as matrix-matrix multiplication. The greatest advantage of the Ozaki scheme is that it is constructed based on level-3 BLAS. The highly optimized implementation is usually provided by the system vendor on most platforms. Thus, we have inferred that this scheme is useful in providing accurate and reproducible BLAS routines with high performance at low development cost for many platforms. Furthermore, we present our implementation of the three routines from level-1 to level-3 BLAS: inner product (DOT), matrix-vector multiplication (GEMV), and matrix-matrix multiplication (GEMM) for both x86 CPUs and NVIDIA GPUs. Our evaluations demonstrate its reproducibility, accuracy, and performance compared to the standard BLAS implementations provided by the vendors.

## 2   Related Work

Although reproducibility is not guaranteed, using high-precision arithmetic is the most simple and easy approach to achieve accurate results. XBLAS[1] [8] and MPLAPACK[2] [9] are BLAS and LAPACK implementations with the high-precision arithmetic approach. It can achieve reproducibility only if the precision

---

[1]  http://www.netlib.org/xblas.

[2]  http://mplapack.sourceforge.net.

is enough to achieve correct-rounding, but the required precision depends on the input, and the existing implementations do not provide any mechanisms to determine the required precision automatically. For reproducibility only (i.e., without accuracy), Intel provides the conditional numerical reproducible (CNR) mode [12] for their linear algebra library known as Intel math kernel library (MKL)[3], which guarantees reproducibility for the same version of MKL on any processor. In NVIDIA cuBLAS[4], symmetric matrix-vector multiplication routine (SYMV) has two modes: a mode guaranteeing reproducibility in each execution on the same device, and another one achieving better speed with atomic add instructions but does not guarantee reproducibility. These examples suggest the need for reproducibility, but the results still rely on the library implementation (and its target hardware); therefore, they cannot address the reproducibility issue between CPU and GPU.

On the other hand, several BLAS have been being developed to address both accurate computations and full reproducibility regardless of the computational environment. For example, ReproBLAS[5] [2] is a reproducible BLAS developed based on a summation algorithm using a pre-rounding technique. The accuracy is tunable through the use of multiple bins to store the value. It targets on CPUs as well as distributed parallel environments, but the current version does not support thread-level parallelism. Moreover, ExBLAS[6] [6] is another reproducible BLAS implementation, which guarantees reproducibility through correct-rounding based on a long accumulator with error-free transformations. It supports thread-level parallelism not only on CPUs but also on GPUs through OpenCL. Besides, there is an experimental implementation on CPUs called RARE-BLAS [1], which also guarantees reproducibility with correct-rounding computation based on an accurate summation algorithm.

Our approach, the Ozaki scheme, supports full reproducibility and tunable accuracy like ReproBLAS. Ozaki et al. [10] and Ichimura et al. [7] have reported the performance of matrix-multiplication using the Ozaki scheme on many-core CPUs. On the other hand, the contributions of this paper include (1) the first adaptation of the Ozaki scheme to reproducible BLAS routines from level-1 to 3 routines and (2) the first high-performance implementations of reproducible BLAS routines with tunable accuracy for both CPUs and GPUs. Also, ExBLAS supports GPUs, but it does not support tunable accuracy, and the performance has not been discussed compared to the theoretical peak. Moreover, despite our attempts, the version as of April 2019 did not work on the latest GPU architecture. Furthermore, compared with ReproBLAS, ExBLAS, and RARE-BLAS, the greatest advantage of the Ozaki scheme is that it can be constructed on existing standard BLAS implementations; it does not require full-scratch implementation, unlike the existing approaches. This is a great benefit in terms of development and maintenance costs.

---

[3] https://software.intel.com/en-us/mkl.

[4] https://developer.nvidia.com/cublas.

[5] https://bebop.cs.berkeley.edu/reproblas.

[6] https://exblas.lip6.fr.

## 3   Ozaki Scheme

In this section, we briefly describe the overview of the Ozaki scheme as full details of this scheme are found in [10]. Hereinafter, we explain the case of an inner product of two vectors $x \in \mathbb{F}^n$ and $y \in \mathbb{F}^n$, where $\mathbb{F}$ is the set of floating-point numbers. Let $\mathtt{fl}(\cdots)$ denote a computation performed with IEEE 754 floating-point arithmetic. First, using Algorithm 1, we split the input vectors into several split vectors as follows:

$$x = x^{(1)} + x^{(2)} + \cdots + x^{(s_x)}, s_x \in \mathbb{N}, x^{(p)} \in \mathbb{F}^n$$
$$y = y^{(1)} + y^{(2)} + \cdots + y^{(s_y)}, s_y \in \mathbb{N}, y^{(q)} \in \mathbb{F}^n$$

Then, as shown in Algorithm 2, the Ozaki scheme computes the inner product $x^T y$ as follows:

$$\begin{aligned}
x^T y &= (x^{(1)} + x^{(2)} + \cdots + x^{(s_x)})^T (y^{(1)} + y^{(2)} + \cdots + y^{(s_y)}) \\
&= \mathtt{fl}((x^{(1)})^T y^{(1)}) + \mathtt{fl}((x^{(1)})^T y^{(2)}) + \cdots + \mathtt{fl}((x^{(1)})^T y^{(s_y)}) \\
&\quad + \cdots \\
&\quad + \mathtt{fl}((x^{(s_x)})^T y^{(1)}) + \mathtt{fl}((x^{(s_x)})^T y^{(2)}) + \cdots + \mathtt{fl}((x^{(s_x)})^T y^{(s_y)})
\end{aligned}$$

Then, the splitting by Algorithm 1 for each is performed to meet the following two properties:

1. if $x^{(p)}(i)$ and $y^{(q)}(j)$ are non-zero elements,

$$|x^{(p)}(i)| \geq |x^{(p+1)}(i)|, |y^{(q)}(j)| \geq |y^{(q+1)}(j)|$$

2. $(x^{(p)})^T y^{(q)}$ to be error-free:

$$(x^{(p)})^T y^{(q)} = \mathtt{fl}((x^{(p)})^T y^{(q)}), 1 \leq p \leq s_x, 1 \leq q \leq s_y$$

The former means that the accuracy of the final result can be controlled by omitting some lower split vectors. The latter means that the inner products of the split vectors can be computed using the standard BLAS routine (DOT) performed by the standard floating-point operations. Besides, as $\mathtt{fl}((x^{(p)})^T y^{(q)})$ has no round-off error (error-free), even if it is computed with some non-reproducible method, the result can be reproducible. After that, the computation results of the split vectors are summed by using some accurate and reproducible algorithm like NearSum [11].

This method can be applied to matrix operations based on inner product such as matrix-matrix multiplication. For example, in matrix-multiplication, the splitting can be performed at a matrix level, and then, the computations for those split matrices can be performed using the standard GEMM routine[7].

In this paper, we introduce $d \in \mathbb{N}, d \leq \max(s_x, s_y)$, as the computation degree of inner products of the split vectors to control the accuracy of the final result.

---

[7] It must be implemented based on the standard floating-point inner product without the use of the divide-and-conquer approach such as Strassen's algorithm.

**Algorithm 1.** Splitting of a vector $x \in \mathbb{F}^n$ in the Ozaki scheme (u denotes the unit round-off of IEEE 754: $\mathbf{u} = 2^{-53}$ for double-precision. $x$ and $x_{\mathrm{split}}[j]$ are vectors, and the others are scalar values. Lines 9 and 10 are computations of $x_i$ and $x_{\mathrm{split}}[j]_i$ for $0 \leq i \leq n - 1$)

---

1: **function** $(x_{\mathrm{split}}[s_x] \leftarrow \mathtt{Split}(x, n))$
2:     $\rho = \mathtt{ceil}((\mathtt{log2}(\mathbf{u}^{-1}) + \mathtt{log2}(n + 1))/2)$
3:     $\mu = \max_{0 \leq i \leq n-1}(|x_i|)$
4:     $j = 0$
5:     **while** $(\mu \neq 0)$ **do**
6:         $j = j + 1$
7:         $\tau = \mathtt{ceil}(\mathtt{log2}(\mu))$
8:         $\sigma = 2^{(\rho + \tau)}$
9:         $x_{\mathrm{split}}[j] = \mathtt{fl}((x + \sigma) - \sigma)$
10:         $x = \mathtt{fl}(x - x_{\mathrm{split}}[j])$
11:         $\mu = \max_{0 \leq i \leq n-1}(|x_i|)$
12:     **end while**
13:     $s_x = j$
14: **end function**

---

**Algorithm 2.** An inner product: $r = x^T y$ $(x, y \in \mathbb{F}^n)$ with the Ozaki scheme

---

1: **function** $(r \leftarrow \mathtt{Ozaki\_DOT}(n, x, y))$
2:     $x_{\mathrm{split}}[s_x] = \mathtt{Split}(x, n)$
3:     $y_{\mathrm{split}}[s_y] = \mathtt{Split}(y, n)$
4:     $r = 0$
5:     **for** $(q = 1, q \leq s_y, q = q + 1)$ **do**
6:         **for** $(p = 1, p \leq s_x, p = p + 1)$ **do**
7:             $r = r + \mathtt{fl}((x_{\mathrm{split}}[p])^T y_{\mathrm{split}}[q])$
8:         **end for**
9:     **end for**
10: **end function**

---

The constant $d$ corresponds to the number of split vectors used during the computation. In this case, at most $d^2$ inner products are computed. However, when $d$ is specified, we can omit the computations for $p + q > d + 1$ in $(x^{(p)})^T y^{(q)}$ with the trade-off of the computation cost and a certain degree of accuracy loss (fast-mode). In this case, the number of inner products can be reduced to $d(d + 1)/2$ at most. The accuracy that can be obtained with a certain $d$ depends on the length of the input vectors and the range of the absolute values in the input vectors – the degree of $d$ required to achieve the best accuracy depends on the input vectors.

## 4    Implementation

In this section, we present the implementation of our BLAS routines named "OzBLAS"[8]. Our implementation only supports the following operations: DOT

---

[8] Available at http://www.math.twcu.ac.jp/ogita/post-k/results.html.

$(r = x^T y)$, GEMV $(y = Ax)$, and GEMM $(C = AB)$ on x86 CPUs and NVIDIA GPUs. We designed our routines to be interface-compatible with the standard double-precision (IEEE 754 binary64) BLAS routines except for a handler (a structure), which holds some parameters required for the Ozaki scheme such as the $d$. However, for simplicity, the current version does not support the full functionality of the BLAS interface: our implementation is intended only to perform the above operations. For example, scalar options and transposed operations are not supported. OzBLAS provides an initialization function that allocates working memory for storing the split vectors/matrices in advance as the dynamic allocation consumes a non-negligible execution time when the problem size increases.

Subsequently, we mainly describe our GPU version, but the proposed implementation techniques are also adaptive to x86 CPUs as well. We developed our GPU implementation using CUDA 10.0. Each routine consists of the following three parts: splitting (Algorithm 1), computation (with BLAS), and summation (NearSum). We implemented the splitting and summation as independent CUDA kernels. We implemented a splitting kernel for a vector and two splitting kernels for a matrix: row and column directions since the two input matrices on matrix-multiplication are split using the maximum values of the row and column directions, respectively. In the splitting algorithm, the max at line 11 should be obtained while the line 10 processes to reduce memory access. We compute $2^{\rho+\tau}$ using scalbn (double x, int n) function, which is a function computes scaling $x$ by two to the power of $n$.

The computation part can be performed using the corresponding double-precision cuBLAS routine (i.e., cublasDdot for DOT and cublasDgemv for GEMV), but we can easily increase the performance as follows. On DOT and GEMV, the computations of the split data can be performed using GEMM (cublasDgemm) by storing the split vectors as a matrix. Since the computation of those is memory-bound, we can improve the performance by increasing the data reuse. On DOT, however, the computation may become a skinny matrix-multiplication that outputs a small matrix. Usually, multi-threaded GEMM routines are implemented by utilizing the two-dimensional data parallelism of the matrix. Consequently, on a skinny matrix-multiplication, the data parallelism can be insufficient against the degree of parallel processing on many-core processors. On GPUs, as the cublasDgemm is not optimized for such a skinny matrix-multiplication, we used an original GEMM implementation that was optimized by us. We implemented it by modifying a typical inner product kernel so that it computes multiple vectors. However, we note that this work is not an essential point of our approach because such an optimization is an issue to be dealt with in the BLAS developer. We note that, on DOT, if the computation is performed using GEMM, the fast-mode, which omits the computations for $p + q > d + 1$ in $(x^{(p)})^T y^{(q)}$, cannot contribute to the performance improvement. This is because the computation of the split vectors is finished at one time using GEMM either with or without fast-mode. If fast-mode is specified, we need to ignore the results of those computations that should be omitted in fast-mode to keep the consistency of the result on every routine. For instance, the result of an inner product computed by DOT with fast-mode must be the same as the one computed by GEMM (with $n = m = 1$) with fast-mode. Because of that, the DOT with fast-mode fills zero into those results after the computations using GEMM.

On GEMM, we used a batched GEMM, which is a new BLAS interface to compute multiple independent BLAS operations simultaneously [4]. In general, the batched BLAS is intended to improve the use of the computation resource on many-core processors by computing multiple small problems when the problem size of each is not large enough compared to the number of cores. As the Ozaki scheme computes multiple products of split matrices or vectors independently, it is expected to accelerate the performance by utilizing batched GEMM when the problem size and/or the block size is relatively small. Then, we used cublasDgemmBatched, which is the batched DGEMM in cuBLAS, to compute the matrix-multiplications of the split matrices simultaneously.

In addition to the above, we deployed a blocking technique in our implementation to reduce memory consumption. The Ozaki scheme consumes huge memory spaces for storing the split matrices. For instance, on matrix-multiplication, the naive implementation requires at least $d^2 + 2d + 1$ times (or $d(d+1)/2 + 2d + 1$ times on fast-mode) extra memory space[9]. Therefore, on GEMV and GEMM, we performed in the block manner of the whole procedure (splitting, computation, and summation) by dividing a matrix into a rectangle shape. When the block size is one, the memory consumption is minimized, but the performance is also degraded since the matrix-multiplication is performed with inner products. Currently, our implementation uses some fixed block size the user-defined, but the determination of the optimal block size can be automated.

Furthermore, the original paper [10] initially proposed to utilize a sparse matrix representation when the split matrix has enough sparsity, though we did not implement it in this paper. Split matrices may contain many zero elements when the range of values in the input matrix is relatively wide. If a high-performance sparse matrix-matrix multiplication routine is provided in the environment, we may be able to enhance the performance by switching dense operation to sparse operation depending on the sparsity of the matrix.

Similarly, we implemented the CPU version like the above GPU version except for the following two points: (1) the computation part is performed using Intel MKL (but other BLAS implementations such as OpenBLAS can be used) instead of cuBLAS. The MKL also provides the batched GEMM routine. (2) the other parts are multi-threaded through OpenMP instead of CUDA to achieve parallelism.

## 5    Evaluation

The performance evaluation was conducted on NVIDIA Titan V, a Volta architecture GPU with compute-capability 7.0, with CUDA 10.0 and the driver

---

[9] "+1" corresponds the working space for storing $x$ at line 10 in Algorithm 1, which can be shared between the two input matrices.

**Table 1.** Maximum relative error to MPRF (2048-bit) on GEMM ($n = m = k = 1000$). The MPFR result is also rounded to double-precision. "fast" represents fast-mode.

| Input (min/max exp) | $\phi = 0\,(-7/-1)$ | $\phi = 4\,(-10/+7)$ | $\phi = 8\,(-19/+15)$ | $\phi = 12\,(-28/+23)$ |
|---|---|---|---|---|
| cuBLAS (double) | 2.63E−10 | 2.95E−11 | 4.16E−11 | 7.78E−11 |
| OzBLAS ($d = 2$, fast) | 6.32E−08 | 3.32E−02 | 3.65E+02 | 5.14E+05 |
| OzBLAS ($d = 2$) | 0 | 4.61E−03 | 4.26E+02 | 4.89E+05 |
| OzBLAS ($d = 3$, fast) | 0 | 1.80E−08 | 1.25E−02 | 9.84E+01 |
| OzBLAS ($d = 3$) | 0 | 1.17E−09 | 2.08E−02 | 1.46E+02 |
| OzBLAS ($d = 4$, fast) | 0 | 7.42E−15 | 1.99E−08 | 1.91E+00 |
| OzBLAS ($d = 4$) | 0 | 8.25E−16 | 9.62E−09 | 1.91E+00 |
| OzBLAS ($d = 6$, fast) | 0 | 0 | 0 | 5.50E−12 |
| OzBLAS ($d = 6$) | 0 | 0 | 0 | 2.36E−12 |

410.73. It has 7449.6 GFlops theoretical peak performance on double-precision with 2560 cores and 12 GB HBM2 of 652.8 GB/s theoretical peak bandwidth. The host machine has an Intel Xeon W-2123 CPU and the operating system is CentOS Linux release 7.4.1708 (3.10.0-693.2.2.el7.x86_64). The GPU codes were compiled by NVIDIA CUDA compiler (nvcc) release 10.0 with "-O3 -gencode arch=compute_60, code=sm_70". For comparison, we also evaluated the performance of cuBLAS routines. On DOT, the result is returned to the global memory. Only on GEMM, the blocking was applied due to the memory shortage (the size is 2560).

First, we demonstrate the accuracy. Table 1 shows the maximum relative error against the result by MPFR[10] [5] with 2048-bit on GEMM ($n = m = k = 1000$). '$d$' represents the degree of the products of the split vectors or matrices to control the accuracy. The input matrices were initialized as (rand − 0.5) × exp($\phi$ × randn), where rand is an uniform random number $[0, 1)$ and randn is a random number from the standard normal distribution. In this way, the range of the absolute value of the input can be controlled by $\phi$: the larger $\phi$ is, the wider the range becomes. Note that the Ozaki scheme with a certain $d$ may be less accurate than the standard double-precision operations when the input matrices have a relatively wide range of absolute values. However, our approach returns the reproducible result if the same $d$ and the same mode (fast-mode or not) are specified.

Second, Table 2 shows examples that demonstrate the reproducibility between CPU and GPU on DOT. The CPU implementation of OzBLAS is developed on Intel MKL version 2016.2.181. The OzBLAS may be less accurate than Intel MKL and NVIDIA cuBLAS depending on $d$ (as well as the absolute range of the input and the dimension) but always guarantees reproducibility between CPU and GPU.

Furthermore, we conducted performance evaluation to see the performance with different $d$. We initialized inputs with $\phi = 12$, which is large enough not to stop the splitting before the defined $d$. Thus, Fig. 1(a) shows the execution

---

[10] http://www.mpfr.org.

**Table 2.** Demonstration of reproducibility: the results with various implementations on DOT ($n = 1000$, shown in hexadecimal format)

| | CPU (Intel Xeon W-2123) | GPU (NVIDIA Titan V) |
|---|---|---|
| (a) $\phi = 0$ | | |
| MPFR (2048-bit) | 0x1.1d87b7f379caap-2 | N/A |
| cuBLAS (double) | N/A | 0x1.1d87b7f379ca**8**p-2 |
| MKL (double) | 0x1.1d87b7f379c**94**p-2 | N/A |
| OzBLAS ($d = 1$) | 0x1.1d8**703d2468**p-2 | 0x1.1d8**703d2468**p-2 |
| OzBLAS ($d = 2$) | 0x1.1d87b7f379caap-2 | 0x1.1d87b7f379caap-2 |
| (b) $\phi = 4$ | | |
| MPFR (2048-bit) | 0x1.9f0b6f909ac28p+25 | N/A |
| cuBLAS (double) | N/A | 0x1.9f0b6f909ac28p+25 |
| MKL (double) | 0x1.9f0b6f909ac28p+25 | N/A |
| OzBLAS ($d = 1$) | 0x1.9**efaf072**p+25 | 0x1.9**efaf072**p+25 |
| OzBLAS ($d = 2$) | 0x1.9f0b6f**8fb2af6**p+25 | 0x1.9f0b6f**8fb2af6**p+25 |
| OzBLAS ($d = 3$) | 0x1.9f0b6f909ac28p+25 | 0x1.9f0b6f909ac28p+25 |

time overhead in times compared to the corresponding standard double-precision BLAS routine in cuBLAS (i.e., the relative execution time normalized to the cuBLAS routine). Our DOT, GEMV, and GEMM compute the split vectors/matrices using our original GEMM implementation, cublasDgemm, and cublasDgemmBatched, respectively. Our implementation achieves the corresponding performance to the theoretical overheads (shown with the straight lines without markers in the figures). The theoretical overheads were obtained as follows. On DOT and GEMV, they can be estimated based on the number of vector and matrix references, respectively. On DOT, as the split and computation part is $O(dn)$ which the summation part is $O(d)$, we only take into account of the cost of the split and computation parts. The number of memory references, in terms of the number of elements, is $6dn$ (see footnote[11]) on the split part and $2dn$ on the computation part. Since the standard DOT routine requires $2n$, the estimated overhead is $4d$ times compared to it. On GEMV, similarly, if we ignore the cost of lower order terms in terms of $d$ and $n$, the split and computation parts reference $3dn^2$ and $dn^2$ elements, respectively. Since the standard GEMV accesses $n^2$ elements, the theoretical overhead becomes $4d$ times as well as the case of DOT. On GEMM, the performance of the computation part is compute-bound and dominant in the total execution time. Accordingly, the performance can be explained based on the number of GEMMs in the computation, and the cost of the other part can be negligible. Thus, the overhead compared to the

---

[11] In Algorithm 1, line 9 loads $x$ and stores $x_{\text{split}}$, and line 10 stores $x$. Line 11 can be performed through line 10 and does not need to be performed at the end. Instead, line 3 is needed to be performed at the first time. Thus, three vector accesses occur per $d$ on each vector.

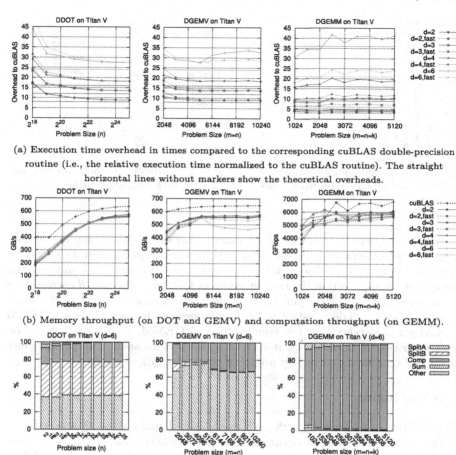

(a) Execution time overhead in times compared to the corresponding cuBLAS double-precision routine (i.e., the relative execution time normalized to the cuBLAS routine). The straight horizontal lines without markers show the theoretical overheads.

(b) Memory throughput (on DOT and GEMV) and computation throughput (on GEMM).

(c) Performance breakdown when $d = 6$ (SplitA/B correspond to left/right operand).

**Fig. 1.** Performance evaluation results on Titan V GPU. "fast" represents fast-mode.

standard GEMM is the number of GEMMs in the computation part: $d(d + 1)/2$ on the fast-mode or $d^2$. Figure 1(b) shows the throughput in terms of GB/s on DOT and GEMV (memory-bound) and GFlops/s on GEMM (compute-bound). At the largest size, our implementations achieved more than 83% of the performance which the corresponding cuBLAS routine achieved, except for the case of GEMV with $d = 6$. Figure 1(c) shows the breakdown of the execution time of each routine when $d = 6$ with fast-mode. The computation part dominates the execution time on GEMM, but the split cost dominates it on DOT and GEMV. The cost for the summation with NearSum is negligibly small on all operations. As you can see in the breakdown, it is inferred that the performance degradation on GEMV with $d = 6$ is caused by the computation part performed with cublasDgemm, though the reason is unknown.

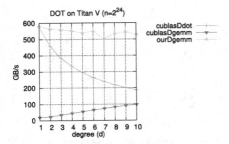

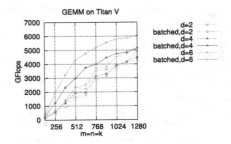

**Fig. 2.** Throughput improvement of DOT with special GEMM routine (ourDgemm)

**Fig. 3.** Throughput improvement of GEMM with batched routine

Finally, we demonstrate the effectiveness of optimizations. Figure 2 demonstrates the effectiveness of the use of GEMM for the computation of DOT. As we mentioned in Sect. 4, cublasDgemm is not optimized for such a skinny matrix-multiplication appeared in the computation on DOT. In this case, $d$ corresponds to the dimension of the output matrix, and $n$ corresponds to that of the inner product. Also, Fig. 3 demonstrates the effectiveness of the use of batched GEMM (cublasDgemmBatched) for the computation of GEMM. By computing several small matrices simultaneously, it increases the performance. The performance shown in Fig. 1 was observed by applying the same optimizations.

## 6    Conclusion

In this paper, we proposed high-performance implementations of reproducible BLAS routines with tunable accuracy using the Ozaki scheme and demonstrated that our implementations achieved nearly the ideal performance on a GPU. In our approach, the performance overhead compared to the standard double-precision BLAS routines depends on the range of the absolute values of the input data, the dimension, and the parameter $d$ corresponding to the number of split vectors/matrices used in the computation: the accuracy can be controlled through $d$. If $d$ is not sufficient, the accuracy can be less accurate than the standard DGEMM, but the result is still reproducible regardless of the computational environment. The most advantage of the Ozaki scheme, when compared to the other existing methods, is the low development cost. Although the method is adaptive only to inner product based operations, it enables us to relatively easily provide accurate and reproducible BLAS routines on any platform if some high-performance BLAS implementation is provided on it. Even though the other existing approaches may be able to outperform our implementations with performance optimization, our approach will still be significant in terms of the development cost.

**Acknowledgment.** This research was partially supported by MEXT as "Exploratory Issue on Post-K computer" (Development of verified numerical computations and super high-performance computing environment for extreme researches) and the Japan Society for the Promotion of Science (JSPS) KAKENHI Grant Number 19K20286.

# References

1. Chohra, C., Langlois, P., Parello, D.: Reproducible, accurately rounded and efficient BLAS. In: Desprez, F., et al. (eds.) Euro-Par 2016. LNCS, vol. 10104, pp. 609–620. Springer, Cham (2017). https://doi.org/10.1007/978-3-319-58943-5_49
2. Demmel, J., Ahrens, P., Nguyen, H.D.: Efficient reproducible floating point summation and BLAS. Technical report, UCB/EECS-2016-121, EECS Department, University of California, Berkeley (2016)
3. Dongarra, J.J., Du Croz, J., Hammarling, S., Duff, I.S.: A set of level 3 basic linear algebra subprograms. ACM Trans. Math. Softw. **16**(1), 1–17 (1990). https://doi.org/10.1145/77626.79170
4. Dongarra, J., Hammarling, S., Higham, N.J., Relton, S.D., Valero-Lara, P., Zounon, M.: The design and performance of batched BLAS on modern high-performance computing systems. In: International Conference on Computational Science (ICCS 2017), vol. 108, pp. 495–504 (2017). https://doi.org/10.1016/j.procs.2017.05.138
5. Fousse, L., Hanrot, G., Lefèvre, V., Pélissier, P., Zimmermann, P.: MPFR: a multiple-precision binary floating-point library with correct rounding. ACM Trans. Math. Softw. **33**(2), 131–1315 (2007). https://doi.org/10.1145/1236463.1236468
6. Iakymchuk, R., Collange, S., Defour, D., Graillat, S.: ExBLAS: reproducible and accurate BLAS library. In: Proceedings of the Numerical Reproducibility at Exascale (NRE2015) at SC 2015 (2015)
7. Ichimura, S., Katagiri, T., Ozaki, K., Ogita, T., Nagai, T.: Threaded accurate matrix-matrix multiplications with sparse matrix-vector multiplications. In: 2018 IEEE International Parallel and Distributed Processing Symposium Workshops (IPDPSW), pp. 1093–1102 (2018). https://doi.org/10.1109/IPDPSW.2018.00168
8. Li, X.S., et al.: Design, implementation and testing of extended and mixed precision BLAS. ACM Trans. Math. Softw. **28**(2), 152–205 (2000). https://doi.org/10.1145/567806.567808
9. Nakata, M.: The MPACK; multiple precision arithmetic BLAS (MBLAS) and LAPACK (MLAPACK). http://mplapack.sourceforge.net
10. Ozaki, K., Ogita, T., Oishi, S., Rump, S.M.: Error-free transformations of matrix multiplication by using fast routines of matrix multiplication and its applications. Numer. Algorithms **59**(1), 95–118 (2012). https://doi.org/10.1007/s11075-011-9478-1
11. Rump, S., Ogita, T., Oishi, S.: Accurate floating-point summation part II: sign, K-fold faithful and rounding to nearest. SIAM J. Sci. Comput. **31**(2), 1269–1302 (2009). https://doi.org/10.1137/07068816X
12. Todd, R.: Introduction to Conditional Numerical Reproducibility (CNR) (2012). https://software.intel.com/en-us/articles/introduction-to-the-conditional-numerical-reproducibility-cnr

# Portable Monte Carlo Transport Performance Evaluation in the PATMOS Prototype

Tao Chang[1]($\boxtimes$), Emeric Brun[1], and Christophe Calvin[2]

[1] DEN-Service d'Etudes des Réacteurs et de Mathématiques Appliquées (SERMA), CEA, Univ. Paris-Saclay, Gif-sur-Yvette, France
{tao.chang,emeric.brun}@cea.fr
[2] CEA DRF, Maison de la Simulation CEA, Université Paris-Saclay, 91191 Gif-sur-Yvette, France
christophe.calvin@cea.fr

**Abstract.** A heterogeneous offload version of Monte Carlo neutron transport has been developed in the framework of PATMOS prototype via several programming models (OpenMP thread, OpenMP offload, OpenACC and CUDA). Two algorithms are implemented, including both history-based method and pseudo event-based method. A performance evaluation has been carried out with a representative benchmark, slabAllNuclides. Numerical results illustrate the promising gain in performance for our heterogeneous offload MC code. These results demonstrate that pseudo event-based approach outperforms history-based approach significantly. Furthermore, by using pseudo event-based method, the OpenACC version is competitive enough, obtaining at least 71% performance comparing to the CUDA version, wherein the OpenMP offload version renders low performance for both approaches.

**Keywords:** Monte Carlo transport · History-based method · Pseudo event-based method · Programming model · GPU

## 1 Introduction

Monte Carlo (MC) neutron transport simulation is a stochastic method that is widely used in the nuclear field to perform reference calculations. Instead of solving the neutron transport equation by introducing discretizations and physical approximations, the MC method simulates the life of a large number of particles from their birth to their death. Their life consists in a succession of random flights and collisions. From the caracheristics of these events, physical quantities can be computed such as the density of particles, the reaction rates, the heat power. The very few approximations introduced make the MC method a precise approach of neutron transport simulation under complex conditions. However it incurs a much higher computational cost comparing to the deterministic method routinely used in the industry.

© Springer Nature Switzerland AG 2020
R. Wyrzykowski et al. (Eds.): PPAM 2019, LNCS 12043, pp. 528–539, 2020.
https://doi.org/10.1007/978-3-030-43229-4_45

## 1.1  Cross Section Calculation

The most consuming part of MC neutron transport simulation is caused by the total cross section calculations, as illustrated in Eq. (1). $\Sigma(E, T)$ and $\sigma_{t,i}(E, T)$ are respectively the macroscopic total cross section $(cm^{-1})$ and the microscopic total cross section of nuclide $i$ $(cm^2)$ at energy $E$ and temperature $T$, $N_i$ is the atomic density of nuclide $i$ $(at/cm^2)$.

$$\Sigma(E, T) = \sum_i N_i * \sigma_{t,i}(E, T) \tag{1}$$

Basically, cross section is a physical quantity that represents the interaction probability of the particle with the nuclides composing the crossed material. Figure 1 shows the MC simulation workflow, including mainly three steps (initialization, particle tracking and tally computation). We can see that the total and partial cross sections calculations are required to sample distance and interaction type for each iteration in particle tracking process.

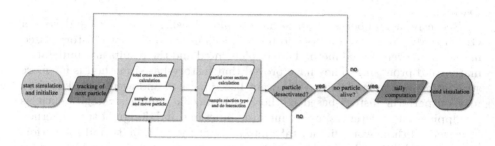

**Fig. 1.** Monte Carlo neutron transport simulation workflow.

## 1.2  Related Work

In order to alleviate this high computational cost, researchers and developers of many MC transport solvers have turned to solutions by porting MC codes to modern architectures and the vectorization of MC codes comes to be a major aspect to dig into. However, the conventional MC algorithm (Fig. 1), also known as the history-based method is considered as an embarassingly parallel algorithm wherein particles are being simulated independently during their lifetime. Each thread tracks one particle at one time, feeded with different instructions, leading to great thread divergence. This main feature makes it an unsatisfactory candidate for data-level parallelism.

As an alternative solution, the event-based method was proposed by regrouping particles according to their different event types (absorption, collision, migration, scattering) and undertaking the simulation of banked particles in parallel [1]. This approach is better suited for vectorization on modern computers, while it introduces extra workload of consolidating surviving particles during the simulation. Furthermore, event-based method requires restructuring the control flow and redesigning data structure, making it difficult to implement with full-physics capabilities.

Except for different particle tracking strategies expressed above, there are mainly two approaches for cross section calculations, pre-tabulated method and on-the-fly Doppler Broadening method. With respect to pre-tabulated method, a binary search is employed to find energy lower-upper bound in cross section tables of given temperatures and the required cross section is retrieved by performing a linear-linear interpolation based on this bound. Previous work done by Wang [2] has proved that there are few opportunities to exploit vectorization for Monte Carlo algorithms based on pre-tabulated cross section. With respect to on-the-fly Doppler Broadening method, instead of storing all cross section tables in memory, temperature dependent cross section data are calculated on-the-fly whenever they are requested by performing an integral computation [3]. The compute-intensive FLOP work between frequent memory loads introduced by on-the-fly Doppler Broadening may mitigate the latency-bound bottleneck mainly induced by the binary search in the pre-tabulated cross section approach [4].

Recently, a number of studies have explored using Intel MIC and Nvidia GPU for MC transport solvers and micro-benchmarks [5–8]. Both history-based method and event-based method are implemented and the results are quite informative and promising. They have proved that history-based method is far more straightforward to implement than event-based method and event-based method may outperform history-based method with specific tuning strategies such as remapping data references, use of intrinsic functions and design of trivial kernel. However, all these work did not take into account the portability and the performance portability of MC codes, which are two significant factors to consider for software development. To our knowledge, the only work related to the implementation of MC transport solver with concern of portability is carried on by Bleile and his group [9] where they developed a monoenergetic event-based MC neutron transport solver relying on the Nvidia Thrust library and discovered that Thrust version can only obtain a maximum of 36% performance comparing to CUDA version and this percentage keeps decreasing while increasing the number of particles. The investigation highlights the trade-off between portability and performance for MC transport solvers and shows the lack of optimizations for portable codes.

### 1.3 Our Approach

Generally speaking, all work mentioned above only focused on pre-tabulated cross section approach, offloading either entire simulation process or several micro-benchmarks (cross section lookup, sampling collision distances) to accelerators. A CPU-based profiling of Monte Carlo simulation implemented in PAT-MOS (more details in Sect. 2) in terms of run time percentage is illustrated in Table 1. All results were retrieved by C++ native clock and perf. It has shown that the total cross section calculations account for up to 95% of total run time.

Thus, we adopt a heterogeneous offloading strategy, which performs all the particle tracking and scoring on host, and offloads microscopic total cross section calculations with on-the-fly Doppler Broadening method to accelerators.

**Table 1.** Typical Monte Carlo neutron transport run time percentage in PATMOS.

| Processing Step | Run Time Percentage (%) |
|---|---|
| Total Cross Section | 95.4 |
| *exp* | 17.6 |
| *erfc* | 49.4 |
| binary_search | 2.4 |
| compute_integral | 79.2 |
| Partial Cross Section | 1.7 |
| *exp* | 0.2 |
| *erfc* | 0.6 |
| binary_search | 0.1 |
| compute_integral | 1.4 |
| Initialization | 1.8 |
| buildMedium | 1.5 |
| Others | 1.1 |

where `binary_search`, `compute_integral` and `buildMedium` are user-defined functions mainly used for the calculation of cross section and the initialization of simulation. *erfc* and *exp* are the most consuming mathematical functions to calculate the complementary error and the base $e$ exponential which needed for on-the-fly Doppler Broadening and used in `compute_integral` function.

The objective of this paper is indeed to address the challenge above by performing a performance evaluation of existing programming models in the framework of MC neutron transport codes with our offloading strategy and on-the-fly Dopper Broadening approach on heterogeneous architectures (CPU + GPU). The rest of the paper is organized as follows. Section 2 gives a brief introduction of MC neutron transport application, programming models, algorithms, as well as implementations for experiments. Section 3 covers the comparison of performance of a benchmark called slabAllNuclides based on different algorithms and architectures. Several concluding remarks are drawn in Sect. 4, as well as certain plans for future development.

## 2    Background and Development

### 2.1    The PATMOS Monte Carlo Prototype

PATMOS is a prototype of Monte Carlo neutron transport under development at CEA dedicated to the testing of algorithms for high-performance computations on modern architectures. It relies on a hybrid parallelism based on MPI for distributed memory parallelism and OpenMP or C++ native threads for shared memory parallelism. One of the goals is to perform pin-by-pin full core depletion calculations for large nuclear power reactors with realistic temperature fields. PATMOS is entirely written in C++, with a heavy use of polymorphism in order to always allow the choice between competing algorithms such as the mix of nuclides with pre-computed Doppler-broadened cross sections and on-the-fly Doppler broadening [10].

The physics of PATMOS is simplified with two types of particles (mono-kinetic pseudo-particles and neutrons). Four types of physical interactions including elastic scattering, discrete inelastic scattering, absorptions and simplified fission have been implemented. The scoring part is encapsulated into a scorer class which deals with tally computation during the simulation and gathers statistical results afterwards.

## 2.2 Programming Models

PATMOS allows two levels of parallelism via MPI + Multi-thread libraries (OpenMP, C++ threads). Each MPI rank performs history tracking for a batch of particles and the average scores of simulation are calculated via inter-node communications. In the shared memory parallellism, each CPU thread is in charge of tracking a particle. It allows all threads to share the non-mutable data and also the scores which are concurrently updated through atomic operations. Since the offloading part of our MC simulation introduces no inter-node communication, we carried out only a set of intra-node experiments using test cases mentioned above with a simple hybrid programming model OpenMP thread + $\{X\}$, where $\{X\}$ can be any languages which are capable of parallel programming on modern accelerators. The programming languages that we have used for implementation are listed below:

1. *CUDA*: A low-level programming language created by Nvidia that leverages Nvidia GPUs to solve complex computational problems [11]. It allows developers to interoperate with C, C++, Fortran, Python or other languages, which provides an efficient interface to manipulate Nvidia GPUs in parallel programming. To explore maximal computing power of Nvidia GPU architectures, one shall take advantage of CUDA thread and memory hierarchies.
2. *OpenACC*: A user-driven performance-portable accelerator programming language supporting implicit offload computing by directives [12]. The directives can be used to accomplish data transfer, kernel execution and stream synchronization.
3. *OpenMP offload*: From OpenMP 4.0, the specification starts to provide a set of directives to instruct the compiler and runtime to do offload computing targeting to devices such as GPUs, FPGAs, Intel MIC, etc.

CUDA provides three key abstractions to explore maximal computing power of GPU architectures (thread hierarchy, memory hierarchy and barrier synchronization). From the thread hierarchy perspective, CUDA makes use of three levels of work units to describe a `block-warp-thread` parallelism. From the memory hierarchy point of view, CUDA exposes a group of programmable memory types such as registers, shared memory, constant memory, and global memory. Concerning barrier synchronization, CUDA provides system-level and block-level of barrier synchronization.

As a comparision, OpenACC and OpenMP offload both offer a set of directives to express thread hierarchy (OpenACC: `gang-worker-vector`, OpenMP offload: `team-parallel-simd`) whereas they do not offer programming interface to on-chip memory and thread synchronization. The lack of access to entire CUDA's feature set may lead to a considerable penalty of performance for OpenACC and OpenMP offload [13]. The key difference between OpenACC and OpenMP offload is that OpenACC supports CUDA asynchronous multistreaming with the directive `async(stream_id)` while OpenMP offload offers `nowait` clause.

## 2.3 Algorithms and Implementations

Algorithm 1 shows the procedure of partial offloading history-based method implemented in slabAllNuclides benchmark. Miscroscopic cross section lookup is the only part offloaded to device. Once the required data for cross section lookups are transfered from host to device, calculations on device begin and the host summarizes macroscopic total cross section after the results being transfered back. It is obvious that our design brings in too many back-and-forth data transfers between host and device. The size of data movement for each calculation (a group of nuclides in one material) is quite small but the large number of memcpy calls induces many launch overheads which may degrade performance in an overwhelming way.

---

**Algorithm 1.** History-based algorithm.

```
1  foreach particle generated from source do
2      while particle is alive do
3          calculation of macroscopic cross section:
4              • do microscopic cross section lookups ⟹ offloaded;
5              • sum up total cross section;
6          sample distance, move particle, do interaction;
7      end
8  end
```

---

We managed to use zero-copy pinned memory to mitigate this bottleneck for the CUDA version, whereas the OpenACC and OpenMP offload versions have no support for such technique. As an alternative solution, we also tried to merge multiple small kernels into a big one by grouping a set of calculations for different particles together. This is achieved by banking multiple particles into one group and offloading microscopic cross section lookups for all these particles. In this way, the number of data transfers can be reduced and the amount of work for each kernel is increased.

---

**Algorithm 2.** Pseudo event-based algorithm.

---

1 **foreach** *bank of N particles generated from source* **do**
2     **while** *particles remain in bank* **do**
3         **foreach** *remaining particle in bank* **do**
4             bank required data for microscopic cross section lookups;
5         **end**
6         • do microscopic cross section lookups $\Longrightarrow$ offloaded;
7         **foreach** *remaining particle in bank* **do**
8             • sum up total cross section;
9             sample distance, move particle, do interaction;
10         **end**
11     **end**
12 **end**

---

**Algorithm 3.** Microscopic cross section lookup

---

**Input**: randomly sampled a group of $N$ tuples of materials, energies and temperatures, $\{(m_i, E_i, T_i)\}_{i \in N}$
**Result**: caculated microscopic cross sections for $N$ materials, $\{\sigma_{ik}\}_{i \in N, k \in |m_i|}$

1 **CUDA Threadblock Level**
2 #pragma acc parallel loop gang **or**
3 #pragma omp target teams distribute
4 **for** $(n_{ik}, E_i, T_i)$ *where* $n_{ik} \in m_i$ **do**
5     $\sigma_{ik} = pre\_calcul()$;
6     **CUDA Thread Level**
7     #pragma acc loop vector **or**
8     #pragma omp parallel for
9     **foreach** *thread in warp* **do**
10         $\sigma_{ik} \mathrel{+}= compute\_integral()$;
11     **end**
12 **end**

---

To fulfill this tuning strategy, our history-based method is redesigned to a new "pseudo event-based" approach. The details of this new method are described in Algorithm 2. The overall procedure of history tracking is reorganized into two **for** loops, where the first loop stores all required data for calculation of microscopic cross sections of all alive particles and the second loop takes the responsibility to do particle movement and interaction. Between two loops, the microscopic cross section lookups are executed on host or device.

Algorithm 3 shows the algorithm of microscopic cross section lookup using CUDA thread hierarchy. The implementation of CUDA version requires to rewritting the computing kernel in CUDA with some specific tuning strategies such as zero-copy and warp shuffling. On the contrary, OpenACC and OpenMP offload make use of `#pragma acc parallel loop gang` and `#pragma omp target teams distribute` directly in CPU codes to achieve CUDA block level parallelism. Each warp calculates microscopic cross section of a nuclide in parallel. At CUDA thread level, `#pragma acc loop vector` and `#pragma omp parallel for` are adopted to parallelize an inner-loop of integral calculation injected by on-the-fly Doppler Broadening computation.

# 3   Benchmarking Results

A series of intra-node tests on two architectures has been carried out. The architectures are listed as follows:

- **Ouessant:**       $2\times$ 10-core IBM Power8, SMT8       $+$ $4\times$ Nvidia P100
- **Cobalt-hybrid:** $2\times$ 14-core Intel Xeon E5-2680 v4, HT $+$ $2\times$ Nvidia P100

On Ouessant, PATMOS was compiled with GCC 7.1, CLANG 3.8.0, PGI 18.10, and XLC 16.1.1 combined with CUDA 9.2. On Cobalt-hybrid, GCC 7.1, CLANG 5.0, PGI 18.7 and Intel compiler 17.0 were used with CUDA 9.0.

## 3.1   Benchmark

To evaluate performance of different programming models in the framework of PATMOS, a benchmark named slabAllNuclides was used to perform a fixed source MC simulation with on-the-fly Doppler Broadening method using a slab geometry with an arbitrary number of heterogeneous regions (10,000 volumes). Each material contains 388 nuclides of the ENDFBVIIr0 library at 900K. The main components of the mixture are H1 and U238 so as to be representative of a Pressurized Water Reactor (PWR) calculation.

The input parameters of slabAllNuclides are fixed to $2 \times 10^4$ particles, 10 cycles for the following tests. The bank size of pseudo event-based method is set to 100, since too small bank size cannot maximize performance and too large bank size causes CPU idle threads because there are no enough tasks to be distributed to all threads (**number of banks** cannot be divisible by **number of threads**) as illustrated in Fig. 2.

We firstly performed a CPU test to obtain the baseline performance on each architecture. We find that history-based method and pseudo event-based method make little difference on CPU. It turns out that the peak performances of slabAllNuclides are $4.7 \times 10^2$ particles/s (PGI, 20 cores, SMT8) and $10.1 \times 10^2$ particles/s (Intel compiler, 28 cores, HT2) on Ouessant and Cobalt-hybrid.

The performance gap between two machines is produced by different level of vectorization achieved by compilers. Intel compiler automatically takes advantage of processing power of Intel architecture and leads to performance improvement comparing to other compilers ($2.2\times$ speedup).

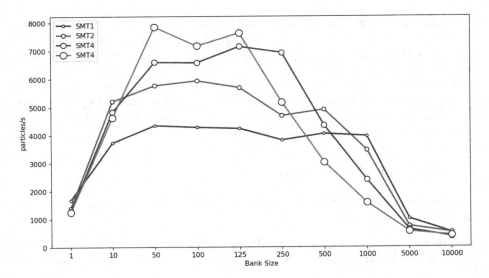

**Fig. 2.** slabAllNuclides bank size analysis on Ouessant.

### 3.2    Performance Evaluation

A series of tests were carried out either on CPU or on GPU so as to evaluate the performance of the different programming models. Numerical results of peak performance are illustrated in Table 2.

We find that among the implemented programming models, the CUDA and OpenACC versions allow to obtain better performance (at maximal $65.8 \times 10^2$ and $52.4 \times 10^2$ particles/s on Ouessant, $48.5 \times 10^4$ and $34.5 \times 10^4$ particles/s on Cobalt-hybrid). The OpenMP offload version is not competitive with the OpenACC version, merely leading to a tracking rate of $12.2 \times 10^2$ particles/s on Ouessant.

Figure 3 provides a straightforward view for comparision of performance speedup among different programming models and architectures. They reflect that (1) the performance of slabAllNuclides running on Cobalt-hybrid is better than the performance running on Ouessant. (2) Pseudo event-based method (*PEB*) outperforms history-based method (*HB*) with a factor of 3–9 speedup. (3) The OpenMP offload version cannot be compiled on Cobalt-hybrid, it renders much lower performance comparing to the CUDA and OpenACC versions.

Overall, it is obvious that on GPU, pseudo event-based method is more suitable than history-based method. The OpenACC version is competitive with CUDA version via pseudo event-based method. The OpenMP offload version provokes a huge performance degradation via either history-based method or pseudo event-based method. This performance penalty is caused by underdeveloped support of CUDA streams for our OpenMP offload implementation achieved by IBM XL compiler. As for CPU, on Ouessant the performances of OpenACC and OpenMP offload versions are close to that of OpenMP thread version because

**Table 2.** Particle tracking rate via different programming models.

| Machine | | Programming model | slabAllNuclides ($\times 10^2$ particles/s) | |
|---|---|---|---|---|
| | | | HB | PEB |
| Ouessant | CPU (20 cores, SMT8) | OMPth | 4.7 | 4.7 |
| | | OMPth+ACC | 4.6 | 4.5 |
| | | OMPth+offload | 3.7 | 3.7 |
| | 1P100 | OMPth+CUDA | 6.7 | 27.2 |
| | | OMPth+ACC | 2.7 | 22.2 |
| | | OMPth+offload | 2.5 | 3.6 |
| | 2P100 | OMPth+CUDA | 13.0 | 47.5 |
| | | OMPth+ACC | 4.5 | 40.2 |
| | | OMPth+offload | 4.3 | 6.7 |
| | 4P100 | OMPth+CUDA | 23.7 | 65.8 |
| | | OMPth+ACC | 9.4 | 52.4 |
| | | OMPth+offload | 5.0 | 12.2 |
| Cobalt-hybrid | CPU (28 cores, HT2) | OMPth | 10.1 | 8.7 |
| | | OMPth+ACC | 5.6 | 5.0 |
| | 1P100 | OMPth+CUDA | 6.8 | 25.4 |
| | | OMPth+ACC | 2.7 | 18.5 |
| | 2P100 | OMPth+CUDA | 16.4 | 48.5 |
| | | OMPth+ACC | 5.6 | 34.5 |

where `OMPth` refers to OpenMP thread, `OMPth+offload` means OpenMP host and offload functionalities, `ACC` is equal to OpenACC.

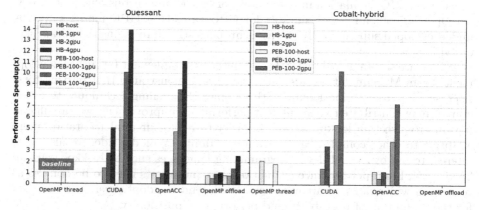

**Fig. 3.** slabAllNuclides performance speedup - baseline for the performance is obtained on Ouessant with 20 cores, SMT8.

all compilers achieve no vectorization for on-the-fly Doppler broadening algorithm. However, on x86 architecture, the best performance is obtained by Intel C++ compiler which manages to vectorize the loop with 256-bit vector register

(AVX2). The OpenACC version introduces a performance penalty because it is compiled with PGI that does not vectorize the main loop of on-the-fly Doppler broadening algorithm.

## 4   Conclusion

This paper depicts the implementation of a PATMOS benchmark slabAllNuclides in the framework of MC neutron transport via different programming languages (OpenMP threads, OpenMP offload, OpenACC). The total microscopic cross section lookup is the unique part which is offloaded to accelerators. This implementation is the first study to use OpenACC and OpenMP offloading functionality for MC simulation and to offer comparisons between programming models in terms of performance. We describe an alternative algorithm "pseudo event-based method" for the purpose of mitigating performance bottleneck incited by conventional "history-based" method.

Performance results are provided across two computing architectures (Ouessant and Cobalt-hybrid). It is clear that the GPU performance via pseudo event-based method surpasses significantly history-based method. The OpenACC version can obtain at least 71% CUDA performance with pseudo event-based method. In contrast, the one with history-based method is limited to only 40% of the CUDA version. With respect to the OpenMP offload version, both history-based method and pseudo event-based method can only exploit around 33% CUDA performance which is completely unsatisfactory. We can conclude that (1) OpenACC is a good choice for the development of portable pseudo event-based MC simulation in the context of our heterogeneous offload strategy. (2) OpenMP offload is not suitable for CPU threads + GPU model since the underdeveloped support of CUDA asynchronous stream restrains the parallelism on GPU side.

There are several capabilities that we intend to implement for future development. From MC simulation side, since we have demonstrated that porting total microscopic cross section lookup with Doppler Broadening techniques to accelerators may contribute to a significant performance improvement, we can offload partial cross section lookup as well in order to make our implementation more adaptive to other complex cases. From programming model side, we also have interest to use other high-level programming languages such as Kokkos [14] and SYCL [15]. From performance evaluation side, more tests need to be done so as to cover a wider range of architectures. We intend to adopt several metrics [16] for the evaluation of portability and performance portability.

**Acknowledgments.** This work was performed using HPC resources from GENCI-IDRIS (in the framework of Cellule de Veille Technologique) and CEA-CCRT (Research and Technology Computing Center).

# References

1. Brown, F.B., Martin, W.R.: Monte Carlo methods for radiation transport analysis on vector computers. Prog. Nucl. Energy **14**(3), 269–299 (1984). https://doi.org/10.1016/0149-1970(84)90024-6
2. Wang, Y., Brun, E., Malvagi, F., Calvin, C.: Competing energy lookup algorithms in Monte Carlo neutron transport calculations and their optimization on CPU and Intel MIC architectures. J. Comput. Sci. **20**, 94–102 (2017). https://doi.org/10.1016/j.jocs.2017.01.006
3. Cullen, D.E., Weisbin, C.R.: Extract doppler broadening of tabulated cross sections. Nucl. Sci. Eng. **60**(3), 199–229 (1976)
4. Tramm, J.R., Siegel, A.R.: Memory bottlenecks and memory contention in multi-core Monte Carlo transport codes. In SNA+ MC 2013-Joint International Conference on Supercomputing in Nuclear Applications+ Monte Carlo, p. 04208. EDP Sciences (2014). https://doi.org/10.1051/snamc/201404208
5. Ozog, D., Malony, A.D., Siegel, A.R.: A performance analysis of SIMD algorithms for Monte Carlo simulations of nuclear reactor cores. In: 29th IEEE International Parallel and Distributed Processing Symposium, pp. 733–742. IEEE Press, Hyderabad (2015). https://doi.org/10.1109/IPDPS.2015.105
6. Hamilton, S.P., Slattery, S.R., Evans, T.M.: Multigroup Monte Carlo on GPUs: Comparision of history- and event-based algorithms. Ann. Nucl. Energy **113**, 506–518 (2018). https://doi.org/10.1016/j.anucene.2017.11.032
7. Hamilton, S.R., Evans, T.M.: Continuous-energy Monte Carlo neutron transport on GPUs in the shift code. Ann. Nucl. Energy **128**, 236–247 (2019). https://doi.org/10.1016/j.anucene.2019.01.012
8. Bergmann, R.M., Vujić, J.L.: Algorithmic choices in WARP-a framework for continuous energy Monte Carlo neutron transport in general 3D geometries on GPUs. Ann. Nucl. Energy **77**, 176–193 (2015). https://doi.org/10.1016/j.anucene.2014.10.039
9. Bleile, R.C., Brantley, P.S., Dawson, S.A., O'Brien, M.J., Childs, H.: Investigation of portable event-based monte Carlo transport using the Nvidia thrust library (No. LLNL-CONF-681383). Lawrence Livermore National Lab. (LLNL), Livermore, CA (United States) (2016)
10. Brun, E., Chauveau, S., Malvagi, F.: PATMOS: a prototype Monte Carlo transport code to test high performance architectures. In: M&C 2017 - International Conference on Mathematics & Computational Methods Applied to Nuclear Science & Engineering, Jeju, Korea, 16–20 April (2017)
11. Nvidia: CUDA Programming Guide (2018)
12. The OpenACC Application Programming Interface. https://www.openacc.org
13. Hoshino, T., Maruyama, N., Matsuoka, S., Takaki, R.: CUDA vs OpenACC: performance case studies with kernel benchmarks and a memory-bound CFD application. In: 13th IEEE/ACM International Symposium on Cluster, Cloud, and Grid Computing, 136–143 (2013). https://doi.org/10.1109/CCGrid.2013.12
14. Edwards, H.C., Trott, C.R., Sunderland, D.: Kokkos: enabling manycore performance portability through polymorphic memory access patterns. J. Parallel Distrib. Comput. **74**(12), 3202–3216 (2014). https://doi.org/10.1016/j.jpdc.2014.07.003
15. SYCL Overview. https://www.khronos.org/sycl/
16. Pennycook, S.J., Sewall, J.D., Lee, V.W.: A metric for performance portability. In: 7th International Workshop in Performance Modeling, Benchmarking and Simulation of High Performance Computer Systems, arXiv:1611.07409 (2016)

# Special Session on Parallel Matrix Factorizations

# Multifrontal Non-negative Matrix Factorization

Piyush Sao$^{(\boxtimes)}$ and Ramakrishnan Kannan

Oak Ridge National Laboratory, Oak Ridge, USA
{saopk,kannanr}@ornl.gov

**Abstract.** Non-negative matrix factorization (NMF) is an important tool in high-performance large scale data analytics with applications ranging from community detection, recommender system, feature detection and linear and non-linear unmixing. While traditional NMF works well when the data set is relatively dense, however, it may not extract sufficient structure when the data is extremely sparse. Specifically, traditional NMF fails to exploit the structured sparsity of the large and sparse data sets resulting in dense factors. We propose a new algorithm for performing NMF on sparse data that we call multifrontal NMF (MF-NMF) since it borrows several ideas from the multifrontal method for unconstrained factorization (e.g. LU and QR). We also present an efficient shared memory parallel implementation of MF-NMF and discuss its performance and scalability. We conduct several experiments on synthetic and realworld datasets and demonstrate the usefulness of the algorithm by comparing it against standard baselines. We obtain a speedup of 1.2x to 19.5x on 24 cores with an average speed up of 10.3x across all the real world datasets.

**Keywords:** Sparse matrix computations · Non-negative matrix factorization · Data analysis · Multifrontal methods

## 1 Introduction

Non-negative Matrix Factorization (NMF) is the problem of determining two non-negative matrices $W \in \mathbb{R}_+^{m \times k}$ and $H \in \mathbb{R}_+^{k \times n}$ such that $A \approx WH$ for the

This manuscript has been authored by UT-Battelle, LLC under Contract No. DE-AC05-00OR22725 with the U.S. Department of Energy. This research was funded by Oak Ridge National Laboratory-Laboratory Directed Research and Development (LDRD) and used resources of the Oak Ridge Leadership Computing Facility, which is a DOE Office of Science User Facility supported under Contract DE-AC05-00OR22725. The United States Government retains and the publisher, by accepting the article for publication, acknowledges that the United States Government retains a non-exclusive, paid-up, irrevocable, world-wide license to publish or reproduce the published form of this manuscript, or allow others to do so, for United States Government purposes. The Department of Energy will provide public access to these results of federally sponsored research in accordance with the DOE Public Access Plan (http://energy.gov/downloads/doe-public-access-plan).

© Springer Nature Switzerland AG 2020
R. Wyrzykowski et al. (Eds.): PPAM 2019, LNCS 12043, pp. 543–554, 2020.
https://doi.org/10.1007/978-3-030-43229-4_46

given matrix $A \in \mathbb{R}^{m \times n}$ with $m$ samples and $n$ features. Formally,

$$\underset{W \geqslant 0, H \geqslant 0}{\arg \min} \ \|A - WH\|_F^2 \tag{1}$$

where $\|X\|_F^2 = \sum_i \sum_j x_{ij}^2$.

With the advent of large scale internet data and interest in Big Data, researchers have started studying scalability of many foundational Data Mining and Machine Learning (DM/ML) algorithms. In the typical case of sparse matrices from internet data sets such as webgraph, bag of words matrices, $k \ll \min(m, n)$; for problems today, $m$ and $n$ can be on the order of millions or more, and $k$ is on the order of few tens to thousands.

NMF is widely used in DM/ML as a dimension reduction method and for many real world problems as the non-negativity is inherent in many representations of real-world data and the resulting low rank factors are expected to have a natural interpretation. The applications of NMF range from text mining, computer vision and bioinformatics to blind source separation, unsupervised clustering and many other areas.

There is a vast literature on algorithms for NMF and their convergence properties [14]. The commonly adopted NMF algorithms are – (i) Multiplicative Update (MU) [17] (ii) Hierarchical Alternating Least Squares (HALS) [1,12] (iii) NMF based on Alternating Nonnegative Least Squares and Block Principal Pivoting (ABPP) [15], and (iv) Stochastic Gradient Descent (SGD) Updates [7]. As in Algorithm 1, most of the algorithms in NMF literature are based on alternately optimizing each of the low rank factors $W$ and $H$ while keeping the other fixed, in which case each subproblem is a constrained convex optimization problem. In this paper, we are considering HALS for explaining our proposed algorithm and experiments. But without loss of generality, it can be easily extended for other NMF algorithms as well.

It is trivial to understand that to approximate the sparse input matrix $A$, we need sparse $W$ and $H$. To promote sparsity in the factors $W$ and $H$ [16], the above Eq. (1), is extended with sparse $\ell_1$ constraints as

$$\underset{W \geqslant 0, H \geqslant 0}{\arg \min} \ \|A - WH\|_F^2 + \|W\|_1 + \|H\|_1 \tag{2}$$

where $\|X\|_1 = \sum_i \sum_j abs(x_{ij})$.

Even though the factors from Eq. (2) yields sparser factors over Eq. (1), still the approximation error will be high, i.e in the order of 0.8 and 0.9. This is because, the low rank approximation $WH$ will have non-zero entries in the place of zeros on the input matrix $A$, resulting in huge error. That is $\text{nnz}(WH) \gg \text{nnz}(A)$. As a side effect, sparse NMF algorithm takes longer to converge.

To overcome this problem, we are proposing a Multifrontal Non-negative Matrix Factorization (MF-NMF) algorithm. Specifically, traditional NMF fails to exploit the structured sparsity of the large and sparse data sets. We propose a new algorithm for performing NMF on sparse data that we call multifrontal NMF (MF-NMF) since it borrows several ideas from the multifrontal method for unconstrained factorization (e.g. LU and QR). We show from Figs. 5 and 6 that the MF-NMF algorithm achieves 1.2x to 19.5x speed up with an average speedup of 10.3x on 24 cores without compromise in accuracy.

## 2   Background

In this section, we introduce the baseline NMF algorithm based on Hierarchical Alternative Least Squares (HALS) and explain the problems of the NMF algorithm on sparse data.

### 2.1   Non-negative Matrix Factorization (NMF)

The NMF [13] algorithms alternate between updating one of $W$ and $H$ using the given input matrix $A$ and other 'fixed' factor - $H$ for updating $W$ or $W$ for updating $H$. We show the structure of any NMF algorithm in Algorithm 1 (Fig. 1).

---

**Algorithm 1.** $[W, H] = \mathrm{NMF}(A, k)$

**Require:** $A$ is an $m \times n$ matrix, low rank $k$
1: Initialize $H$ with a non-negative matrix in $\mathbb{R}_+^{n \times k}$.
2: **while** stopping criteria not satisfied **do**
3:    Update $W$ as $\arg\min_{\tilde{W} \geqslant 0} \left\| A - \tilde{W} H \right\|_F$
4:    Update $H$ as $\arg\min_{\tilde{H} \geqslant 0} \left\| A - W \tilde{H} \right\|_F$

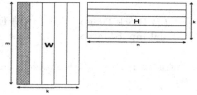

**Fig. 1.** HALS Algorithms that determines $2k$ vectors of $W$ and $H$

---

The NMF algorithms vary in the realization of the steps 3 and 4 of Algorithm 1. In this paper, we are using a specific NMF algorithm called Hierarchical Alternating Least Squares (HALS) [1] that has the following update rule for solving Eq. (2)

$$H^i \leftarrow [(H^i + W^T A)^i - (W^T W + 2\beta\mathbf{1}_k)_i H]_+$$
$$W_i \leftarrow [(HH^T + 2\beta\mathbf{1}_k)_{ii} W_i + (AH^T)_i - W(HH^T + 2\beta\mathbf{1}_k)_i]_+ \qquad (3)$$

where $\mathbf{1}_k$ is a matrix of $k \times k$ with all one's, $H^i$ is row vector, $W_i$ is column vector and $\alpha, \beta$ are some positive scalars.

The convergence properties of different NMF algorithms are discussed in detail by Kim, He and Park [14]. While we focus only on the HALS algorithm in this paper, we highlight that our algorithm is not restricted to this, and is seamlessly extensible to other NMF algorithms as well, including HALS, ABPP, Alternating Direction Method of Multipliers (**ADMM**) [22], and Nesterov-based methods [10].

It can be shown that the Eq. (3) results in a relatively dense $W$ and $H$ for low rank $k$. Also, we can observe from Eq. (3), that there is no inherent parallelism on HALS algorithm other than the BLAS level parallelism available for sparse-dense matrix multiplication.

## 2.2  NMF on Sparse Datasets

In the baseline NMF algorithm Algorithm 1, the factor matrix $W$ and $H$ are dense even when $A$ matrix is sparse. Thus, the approximant matrix $\hat{A} = WH$ is dense. This is undesirable in the case of structured sparse matrix since dense approximant $\hat{A}$ will not preserve the patterned sparsity of the matrix $A$ such as the topic-modeling on bag of words representation and sparse adjacency matrix representation of large scale social network graphs.

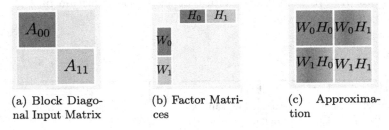

(a) Block Diago-    (b) Factor Matri-    (c)  Approxima-
nal Input Matrix    ces                 tion

**Fig. 2.** Dense factors problem on illustrative block diagonal sparse NMF

Consider the NMF of $2 \times 2$ sparse block diagonal matrix as shown in Fig. 2a. This block-diagonal structure of the matrix suggests that the data is *separable* into two disjoint sets described by matrices $A_{00}$ and $A_{11}$. When we perform NMF of this matrix, we obtain two dense factors $W = \begin{bmatrix} W_0 \\ W_1 \end{bmatrix}$ and $H = \begin{bmatrix} H_0 & H_1 \end{bmatrix}$ (Fig. 2b). Hence the non-negative approximation of the matrix $A$ is dense (Fig. 2c). Thus this approximation loses the separability property of the input matrix.

Furthermore, the error of non-negative approximation for the $2 \times 2$ sparse block diagonal matrix is given by

$$\|A - WH\|_F^2 \approx \|A_{00} - W_0 H_0\|_F^2 + \|A_{11} - W_1 H_1\|_F^2 + \|W_1 H_0\|_F^2 + \|W_0 H_1\|_F^2.$$

Note that the first two terms $\|A_{00} - W_0 H_0\|_F^2$ and $\|A_{11} - W_1 H_1\|_F^2$ are at least equal to the approximation error when we perform NMF on blocks $A_{00}$ and $A_{11}$, respectively. Additionally, $\|W_1 H_0\|_F^2 \; \|W_0 H_1\|_F^2$ are spurious and non-zero and they further worsen the approximation error.

We summarize the drawbacks of performing Algorithm 1 to compute NMF of a sparse matrix as (a) $W$ and $H$ factors are dense (b) NMF approximation does not preserve the structural properties of the data and (c) the error of approximation (and thus the approximation quality) is not optimal.

**Informal Problem Statement:** To overcome the limitations of Algorithm 1, we seek an NMF algorithm that preserve the structural sparsity of the input data in the approximation $\hat{A} = WH$. Since the factor matrices are non-negative, no numerical cancellation can occur. Hence if the approximation $\hat{A}$ is sparse then $W$ and $H$ will also be sparse.

# 3    Multifrontal Non-negative Matrix Factorization

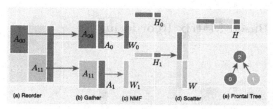 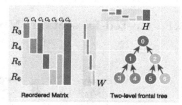

(a)  MF-NMF on $2 \times 3$ block sparse matrix

(b)  MF-NMF factors with two level ND

**Fig. 3.** On (a), we show the four steps of MF-NMF and the frontal tree for one level-ND. (a) Shows reordered matrix, $W$ and $H$ factors and frontal tree for two level-ND.

First, we describe the working of MF-NMF on a $2 \times 3$ block-sparse matrix shown in Fig. 3a-a. In MF-NMF, we reorder the input matrix $A$ to expose the sparse block structure as shown in Fig. 3a-a and instead of performing rank-$k$ NMF on the whole matrix $A$, we divide it into two smaller sub-problems $A_0 = \begin{bmatrix} A_{00} & A_{02} \end{bmatrix}$ and $A_1 = \begin{bmatrix} A_{10} & A_{12} \end{bmatrix}$, and perform a $k/2$ rank NMF on $A_0 \approx_+ W_0 H_0$ and $A_1 \approx_+ W_1 H_1$. Finally, $W_0$ and $W_1$, and $H_0$ and $H_1$, are scattered to form the final $W$ and $H$ factors. We perform this process in the following four steps (Fig. 3a).

1. Reorder: we compute a column permutation $P$, and reorder the matrix to obtain a block sparse matrix structure as shown in Fig. 3a.
2. Gather: We gather the sparse blocks into multiple smaller subproblems $A_i$.
3. NMF: We perform independent NMF on each $A_i \approx_+ W_i H_i$ using Algorithm 1.
4. Scatter: We scatter all $W_i$ and $H_i$ to construct the final factor matrices $W$ and $H$.

---

**Algorithm 2.** $[W, H] = \text{MF-NMF}(A, k, h)$

---

**Require:** $A$ is an $m \times n$ matrix, low rank $k$, height $h$ of frontal tree
1: $F, P \leftarrow$ Reorder $(A, h)$ % $F$ is frontal tree and $P$ is permutation (Section 3.1)
2: $\ell \leftarrow \#\text{leaf}(F)$
3: $A_0, A_1, \cdots, A_\ell \leftarrow$ Gather $(A, F, P)$
4: **for** $i = 1$ to $\ell$ **do**
5:     $[W_i, H_i] \leftarrow \text{NMF} \left(A_i, \frac{k}{\ell}\right)$
6: $W, H \leftarrow$ Scatter $\left([W_1, H_1], [W_2, H_2], \cdots [W_\ell, H_\ell]\right)$

---

The only non-trivial step in MF-NMF is reordering. In this paper, we use the so nested-dissection (ND) based graph-partition to obtain desired ordering.

## 3.1 Nested-Dissection (ND) Based Matrix Reordering

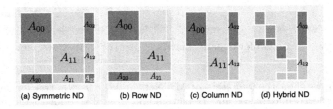

(a) Symmetric ND    (b) Row ND    (c) Column ND    (d) Hybrid ND

**Fig. 4.** Different ND reorderings. In the first order shows a symmetric reordering, and the second and third case show row and column ND respectively. In the last case, we do a column-ND at the top level; and perform a row-ND for $A_{00}$ and $A_{11}$.

Without loss of generality, we assume that the input sparse matrix $A$ is a tall rectangular matrix. Typical ND is performed on a symmetric sparse matrix. So we first describe ND for the symmetric case and then we show how do we use symmetric ND for ordering rectangular sparse matrices.

**Symmetric ND:** For a symmetric sparse matrix $A \in \mathbb{R}^{n \times n}$, there is a corresponding undirected graph $G = (V, E)$ with $|V| = n$ vertices corresponding to each column in $A$; and $|E| = nnz(A)$ edges, where any edge $e_{ij}$ corresponds to a non-zero entry $A_{ij}$.

In ND of the matrix $A$, we find a vertex separator $S \subset V$, that partitions the $V = C_1 \cup S \cup C_2$, such that there are no edges between any vertex in $C_1$ to any vertex in $C_2$. Using this partition, we reorder the matrix so that (a) vertices in each set $C_1$, $C_2$ and $S$ numbered consecutively; and (b) vertices in $S$ are numbered at the end. In the symmetric case, the matrix is reordered symmetrically, i.e., if $P$ is the permutation matrix, then reordered matrix $\hat{A} = P^T A P$, so $\hat{A}$ is also symmetric.

We can obtain such a permutation using graph partitioning tools such as Metis and Scotch. The re-ordered matrix has an *arrow-head* structure as shown in Fig. 4a. In Fig. 4a, $A_{00}$, $A_{11}$ and $A_{22}$ correspond to $C_1$, $C_2$ and $S$, respectively.

**Unsymmetric ND:** When $A$ is a rectangular matrix, we can perform ND on $B_c = A^T A$ or $B_r = AA^T$. First consider $B_c = A^T A$. Let $P_c$ be the permutation matrix obtained by performing ND on $B_c = A^T A$. Using $P_c$, we permute the columns of the matrix $A$ to obtain $\hat{A}_c = AP_c$. We call this reordering scheme as column-ND. $\hat{A}_c$ has a rotated-staircase structure as shown in Fig. 4c. Similarly, we can perform ND on $B_r$ to obtain a permutation matrix $P_r$ and permute the matrix $A$ to obtain $\hat{A}_r = P_r A$, which has a reverse-staircase structure as shown in Fig. 4b. We call this reordering scheme as row-ND.

In general, the approximation quality $\frac{\|A - WH\|_F}{\|A\|_F}$ of factors can vary significantly with the choice of reordering scheme, and it depends on the nature of data itself. Informally, the reordering scheme that best captures the natural structure of the data, will result in best approximation quality. One may use more complex reordering scheme such as alternating between column-ND and row-ND. For example, in Fig. 4d we use column-ND in the first step and us row-ND for partitioning $A_{00}$ and $A_{11}$. In this paper, we keep our discussion limited to column-ND.

### 3.2 Gather Substep

In the Gather step, we gather smaller block sparse matrices to form disjoint submatrices $A_i$ that we call frontal matrix. To do so, we need to know the number of frontal matrices $A_i$, and its row and column set. The number of submatrices depends on the *depth* of ND performed in the reordered step. If we perform ND only once as in the case of Fig. 3a, we get two submatrices and for ND of depth two in Fig. 3b we get four submatrices.

**Frontal Tree:** The column structure of frontal can be expressed conveniently using a tree, that we call *frontal-tree*, which is analogous to the so called-elimination tree of the sparse Cholesky factorization. The frontal tree for ND of depth one and two are shown in Fig. 3a and b, respectively. Each node $i$ in the frontal tree corresponds to a subset of column $C_i$. The number of frontal matrices is the number of leaf nodes in the frontal tree.

The row set $R_i^F$ of a frontal matrix $A_i$ corresponding to a leaf node $i$ is the set of non-zero rows in $A_{:,C_i}$ where $C_i$ is the set of columns corresponding to $i$-th node in the frontal tree. The column set $C_i^F$ of a frontal matrix $A_i$ is the union of columns in all the ancestors of the leaf node $i$ in the frontal tree and itself. For example, in Fig. 3b, the leaf node 3 has ancestors 1 and 0 so the frontal matrix corresponding to will have the $C_3^F = C_3 \cup C_1 \cup C_0$ as the column set. Using $R_i^F$ and $C_i^F$, the $i$-th frontal matrix can be obtained as $A_i = A_{R_i^F, C_i^F}$.

After the Gather step, we perform independent NMF on all the frontal matrix and, if required, in Scatter the factors to form the final $W$ and $H$ matrices.

The complete algorithm appears in Algorithm 2. We use the desired height of the frontal tree, $h$ as a user input, which controls the number of frontal matrices.

Note that the number of frontal matrices can be at most $k$, where $k$ is the rank of the final non-negative factors.

## 4   Experiments and Evaluation

In this section, we present results from a series of numerical experiment to understand the scalability of 3D sparse triangular solver algorithm.

### 4.1   Experimental Set-Up

The entire experiment was conducted on a node in Rhea commodity-type Linux cluster at Oak Leadership Computing Facility. Every node has two Intel® Xeon® E5-2650 at 2.0 GHz with each having 8 physical cores and 16 Hyperthreads with a total of 16 cores and 32 hyperthreads per node. We compile the code with gcc 6.2.0 compiler with "-fopenmp" flag for openmp threads.

### 4.2   Datasets

For the synthetic experiments, we generated standard normal random matrices and the details of the real world datasets are presented in Table 1.

We use a mix of matrices from real world and scientific applications. The matrices *psse0 and psse1* are related to powergrid and the rest of the matrices belong to scientific applications such as combinatorial, linear programming and quantum chemistry.

Both the real world and the synthetic matrices are shifted with the minimum value such that there is no negative elements in the input matrix.

**Table 1.** Test sparse matrices used in experiments

| Name | Source | $m$ | $n$ | $\frac{nnz}{\max(m,n)}$ | Name | Source | $m$ | $n$ | $\frac{nnz}{\max(m,n)}$ |
|------|--------|-----|-----|-------------------------|------|--------|-----|-----|-------------------------|
| Franz11 | Comb | 4.7e4 | 3.0e4 | 7 | Franz5 | Comb | 7.3e3 | 2.8e3 | 5.9 |
| Franz7 | Comb | 1.0e4 | 1.7e3 | 4.1 | Catears24 | Comb | 1.0e3 | 2.6e3 | 2.9 |
| kneser831 | Comb | 1.5e4 | 1.5e4 | 2.9 | mk10-b2 | Comb | 3.1e3 | 6.3e2 | 3 |
| rosen2 | Linear prog | 1.0e3 | 3.0e3 | 15.4 | SiNa | Quantum chemistry | 5.7e3 | 5.7e3 | 34.61 |
| psse0 | Power | 2.6e4 | 1.1e4 | 3.83 | psse1 | Power | 1.4e4 | 1.1e4 | 4.0 |

### 4.3   Metrics

We are reporting relative error as a quality metric and per iteration time as performance metric.

*Relative Error:* Relative error defines the approximation of the obtained factor matrices $W$ and $H$ against the input matrix as defined below.

$$\text{Relative Error} = \frac{\|A - WH\|_F^2}{\|A\|_F^2} \tag{4}$$

*Average Per Iteration Time in Seconds:* NMF algorithms are iterative in nature. There are many stopping criterion and the most common among them are (a) difference of relative error between two successive iterations (b) tolerance on the relative error and (c) number of iterations. In this paper, we are specifically using HALS NMF algorithm and stopping after 20 iterations. All the baselines and the proposed algorithm were run for 20 iterations. For most of the datasets, it is common in the community to run for 20–30 iterations, some datasets might need longer. Hence, we are reporting the average per iteration time in seconds as the performance metric.

For both the time and the error, we also report the relative performance by normalizing with the minimum value on every dataset. That is., the in both Figs. 5 and 6 for *franz11*, we normalized every accuracy and the time with the minimum from that experiment.

### 4.4 Real World Data

**Parallel Performance.** As in Fig. 5, we obtained a speedup of 1.2x to 19.5x on *mk10b2* with an average speed up of 10.3x across all the real world datasets. The level of parallelism and the speedup is directly proportional to the height of the tree that is dependent on the sparsity pattern. For realworld data with good sparsity pattern like *psse0* and *Franz*, we are able to obtain trees with height up to level 7. Where as, *cat-ears* and *rosen2*, cannot obtain more than two levels. In the case of *Franz7*, *Franz11*, at height 7, the speedup is 16x and 10x for *psse*. Sometimes it is difficult to obtain balanced partition for a given level – that is., equal number of nnz's across partition; hurting the speedup. For example, in *mk10b2*, at level 1, with two leaf nodes, we obtain a speedup of only 1.2x which is alleviated in higher levels achieving a peak speedup of 19.5x. Overall, the proposed Algorithm 2, gave good speedup for sufficiently sparse matrices with good sparsity pattern.

**Accuracy.** We are reporting the relative accuracy with the global matrix. That is., we compute the relative error of the proposed Algorithm 2 and normalize with the relative error of global matrix. Every red bar in Fig. 6 is for the global matrix with relative accuracy always 1.0 and all the rest are normalized with this absolute value of the relative error. The relative accuracy of Algorithm 2 is in between 1.0 to 1.04 – that is., in a worst case there was only 4% deviation from the global relative error. The absolute relative error for global matrices for *Franz11*, *Franz5*, *Franz7*, *cat-ears*, *knesser*, *mk10b2*, *psse0*, *psse1*, *rosen2* and *sina* are 0.998246, 0.971278, 0.97998, 0.97805, 0.996353, 0.988146, 0.934138, 0.893648, 0.904814 and 0.998638 respectively.

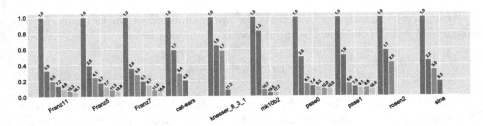

**Fig. 5.** Parallel performance of MF-NMF on real world datasets by varying the height of frontal tree. Every bar shows the relative speedup with respect to the baseline Algorithm 1

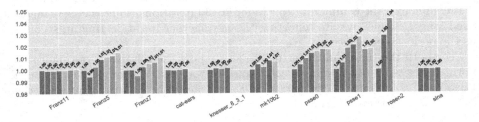

**Fig. 6.** Relative accuracy of MF-NMF with respect to the baseline Algorithm 1 by varying the height of the frontal tree (Color figure online)

## 5    Related Work

Our proposed MF-NMF borrows heavily from direct methods for sparse linear system of equations which goes back to seminal work George [8]. Comprehensive discussion of sparse LU and Cholesky factorization, including matrix reordering and the elimination tree can be found in [3,11,20]. Matrix reordering used for MF-NMF is more closely related to sparse QR factorization [2] and unsymmetric LU factorization [9].

The popular algorithm for NMF is Multiplicative Update (MU) and there are literature that focuses on distributed implementation of MU on Hadoop [18,19,23] and Spark. These explored Matrix multiplication, element-wise multiplication, and element-wise division are the building blocks of the MU algorithm.

While we do not discuss parallelism aspect of MF-NMF, design of MF-NMF is also motivated by improving the parallelism in NMF. Sparse NMF algorithms on shared memory environment are heavily reliant on parallelism available on BLAS/LAPACK libraries for Sparse-Dense matrix operations [4,21] and by using the alternate formulations of NMF [5,6]. For the first time in the literature, the proposed MF-NMF exposes more parallelism in shared memory environment by exploiting the sparsity of the data set.

# 6   Conclusion

The traditional NMF fails to exploit the structured sparsity of the large and sparse data sets. In this paper we proposed a MF-NMF algorithm that uses concepts from sparse direct methods and graph partitioning to exploit the structured sparsity. Our initial assessment suggests MF-NMF can be much faster than traditional NMF while incurring negligible penalty in terms of accuracy. Our reordering scheme based on ND, is only a first step towards exploiting sparsity for NMF and we believe exploring more sophisticated reordering schemes based on domain knowledge is warranted. In the future, we would like to extend the work with (a) the distributed communication avoiding variant and (b) quantify the usefulness of the factors obtained from the proposed algorithm on internet scale data.

# References

1. Cichocki, A., Zdunek, R., Phan, A.H., Amari, S.: Nonnegative Matrix and Tensor Factorizations: Applications to Exploratory Multi-way Data Analysis and Blind Source Separation. Wiley, Chichester (2009)
2. Davis, T.: Multifrontral multithreaded rank-revealing sparse QR factorization. In: Dagstuhl Seminar Proceedings. Schloss Dagstuhl-Leibniz-Zentrum für Informatik (2009)
3. Duff, I.S., Erisman, A.M., Reid, J.K.: Direct Methods for Sparse Matrices. Clarendon Press, Oxford (1986)
4. Fairbanks, J.P., Kannan, R., Park, H., Bader, D.A.: Behavioral clusters in dynamic graphs. Parallel Comput. **47**, 38–50 (2015)
5. Flatz, M., Kutil, R., Vajteršic, M.: Parallelization of the hierarchical alternating least squares algorithm for nonnegative matrix factorization. In: 2018 IEEE 4th International Forum on Research and Technology for Society and Industry (RTSI), pp. 1–5. IEEE (2018)
6. Flatz, M., Vajteršic, M.: A parallel algorithm for nonnegative matrix factorization based on Newton iteration. In: Proceedings of the IASTED International Conference Parallel and Distributed Computing and Networks (PDCN 2013), pp. 600–607. ACTA Press (2013)
7. Gemulla, R., Nijkamp, E., Haas, P.J., Sismanis, Y.: Large-scale matrix factorization with distributed stochastic gradient descent. In: Proceedings of the KDD, pp. 69–77. ACM (2011)
8. George, A.: Nested dissection of a regular finite element mesh. SIAM J. Numer. Anal. **10**(2), 345–363 (1973)
9. Grigori, L., Boman, E.G., Donfack, S., Davis, T.A.: Hypergraph-based unsymmetric nested dissection ordering for sparse LU factorization. SIAM J. Sci. Comput. **32**(6), 3426–3446 (2010)
10. Guan, N., Tao, D., Luo, Z., Yuan, B.: NeNMF: an optimal gradient method for nonnegative matrix factorization. IEEE Trans. Signal Process. **60**(6), 2882–2898 (2012)
11. Heath, M.T., Ng, E., Peyton, B.W.: Parallel algorithms for sparse linear systems. SIAM Rev. **33**(3), 420–460 (1991)

12. Ho, N.-D., Van Dooren, P., Blondel, V.D.: Descent methods for nonnegative matrix factorization. CoRR, abs/0801.3199 (2008)
13. Kannan, R., Ballard, G., Park, H.: MPI-FAUN: an MPI-based framework for alternating-updating nonnegative matrix factorization. IEEE Trans. Knowl. Data Eng. **30**(3), 544–558 (2018)
14. Kim, J., He, Y., Park, H.: Algorithms for nonnegative matrix and tensor factorizations: a unified view based on block coordinate descent framework. J. Global Optim. **58**(2), 285–319 (2014)
15. Kim, J., Park, H.: Fast nonnegative matrix factorization: an active-set-like method and comparisons. SIAM J. Sci. Comput. **33**(6), 3261–3281 (2011)
16. Kim, W., Chen, B., Kim, J., Pan, Y., Park, H.: Sparse nonnegative matrix factorization for protein sequence motif discovery. Expert Syst. Appl. **38**(10), 13198–13207 (2011)
17. Lee, D.D., Sebastian Seung, H.: Algorithms for non-negative matrix factorization. In: NIPS, vol. 13, pp. 556–562 (2001)
18. Liao, R., Zhang, Y., Guan, J., Zhou, S.: CloudNMF: a MapReduce implementation of nonnegative matrix factorization for large-scale biological datasets. Genomics Proteomics Bioinform. **12**(1), 48–51 (2014)
19. Liu, C., Yang, H.C., Fan, J., He, L.-W., Wang, Y.-M.: Distributed nonnegative matrix factorization for web-scale dyadic data analysis on MapReduce. In: Proceedings of the WWW, pp. 681–690. ACM (2010)
20. Sao, P., Li, X.S., Vuduc, R.: A communication-avoiding 3D LU factorization algorithm for sparse matrices. In: Proceedings of the IEEE International Parallel and Distributed Processing Symposium (IPDPS), Vancouver, BC, Canada, May 2018
21. Sao, P., Vuduc, R., Li, X.S.: A distributed CPU-GPU sparse direct solver. In: Silva, F., Dutra, I., Santos Costa, V. (eds.) Euro-Par 2014. LNCS, vol. 8632, pp. 487–498. Springer, Cham (2014). https://doi.org/10.1007/978-3-319-09873-9_41
22. Sun, D.L., Févotte, C.: Alternating direction method of multipliers for non-negative matrix factorization with the beta-divergence. In: 2014 IEEE International Conference on Acoustics, Speech and Signal Processing (ICASSP), pp. 6201–6205, May 2014
23. Yin, J., Gao, L., Zhang, Z.M.: Scalable nonnegative matrix factorization with block-wise updates. In: Calders, T., Esposito, F., Hüllermeier, E., Meo, R. (eds.) ECML PKDD 2014. LNCS (LNAI), vol. 8726, pp. 337–352. Springer, Heidelberg (2014). https://doi.org/10.1007/978-3-662-44845-8_22

# Preconditioned Jacobi SVD Algorithm Outperforms PDGESVD

Martin Bečka and Gabriel Okša[✉]

Institute of Mathematics, Slovak Academy of Sciences, Bratislava, Slovak Republic
{Martin.Becka,Gabriel.Oksa}@savba.sk

**Abstract.** Recently, we have introduced a new preconditioner for the one-sided block-Jacobi SVD algorithm. In the serial case it outperformed the simple driver routine DGESVD from LAPACK. In this contribution, we provide the numerical analysis of applying the preconditioner in finite arithmetic and compare the performance of our parallel preconditioned algorithm with the procedure PDGESVD, the ScaLAPACK counterpart of DGESVD. Our Jacobi based routine remains faster also in the parallel case, especially for well-conditioned matrices.

**Keywords:** Singular value decomposition · Parallel computation · Dynamic ordering · One-sided block-Jacobi algorithm · Preconditioning

## 1 Introduction

The singular value decomposition (SVD) is an important matrix factorization problem with many practical applications. There are two main branches of SVD algorithms: bidiagonalization methods and the Jacobi methods. The Jacobi methods are considered to be slow, but they are often highly accurate and easily parallelizable. Recently, it has been realized that these methods can be improved using the blocking, suitable orderings and preconditioners [6]. It was shown in [2] that the serial one-sided block-Jacobi algorithm (OSBJA) can be accelerated by applying the matrix $W$ of eigenvectors of $A^T A$ as a preconditioner, i.e., by using $AW$ as an input for the Jacobi algorithm. The performance of this approach was comparable to the LAPACK divide-and-conquer procedure DGESDD, but it clearly outperformed the simple driver routine DGESVD. Since the only procedure for the SVD in ScaLAPACK is PDGESVD (the parallel implementation of DGESVD) and our previous parallel OSBJA was comparable with that [4,6], we can also expect a successful application of the preconditioner in the parallel case.

## 2 One-Sided Block-Jacobi Algorithm with New Preconditioning

The task is to compute efficiently the SVD of a real $m \times n$ $(m \geq n)$ matrix $A$,

$$A = U \begin{pmatrix} \Sigma \\ 0 \end{pmatrix} V^T,$$

© Springer Nature Switzerland AG 2020
R. Wyrzykowski et al. (Eds.): PPAM 2019, LNCS 12043, pp. 555–566, 2020.
https://doi.org/10.1007/978-3-030-43229-4_47

where $U(m \times m)$ and $V(n \times n)$ are orthonormal matrices and $\Sigma = \mathrm{diag}(\sigma_i)$, where $\sigma_1 \geq \sigma_2 \geq \cdots \geq \sigma_n \geq 0$ are singular values of $A$. When $m \gg n$, it can be more efficient to compute the QR decomposition of $A$ first and then the SVD of R-factor, which is of size $n \times n$. Afterwards, the SVD of $A$ can be easily restored. In this paper, the SVD is applied to a matrix $A$ directly.

Our previous research was devoted to (mostly parallel) Jacobi SVD algorithms, where we have combined three approaches: (i) the Jacobi algorithm as the method for SVD computation, (ii) the classical Jacobi choice of a pivot, and (iii) the eigenvalue decomposition (EVD) of $A^T A$ as a way to compute the SVD.

When analyzing the serial OSBJA, each Jacobi rotation corresponds to an orthonormal $n \times n$ matrix $J_i$, so that

$$(J_{\mathrm{it}}^T \ldots (J_2^T (J_1^T (A^T A) J_1) J_2) \ldots J_{\mathrm{it}})$$
$$= V^T (A^T A) V = (AV)^T AV = \mathrm{diag}(\sigma_i^2)$$

for $V = J_1 J_2 \ldots J_{\mathrm{it}}$. If the iterative process starts with $A^{(0)} = A$, and $A^{(k)} = A^{(k-1)} J_k$ at iteration $k$, then $J_k$ is chosen so that the level of orthogonality between the columns of $A^{(k-1)}$ is increased. If $A^{(\mathrm{it})}$ is orthogonal with respect to the prescribed precision, then the iterative process finishes and $A^{(\mathrm{it})} = U\Sigma$.

It is obvious that one way how to speed up the algorithm is to keep the number of iterations as small as possible. Originally, when computing the EVD of a symmetric matrix, Jacobi proposed to eliminate two off-diagonal elements $a_{ij} = a_{ji}$, $i \neq j$, of the largest magnitude at each step, giving the largest possible reduction of the off-diagonal Frobenius norm of $A$. This was considered inefficient as it introduces an $O(n^2)$ search for each rotation, but its application requires just $O(n)$ flops [6,7]. However, this is true for the first index pair $(i,j)$ only, and the following index pairs can be found in $O(n)$ steps, since each rotation changes just two rows and columns of $A$. This idea was generalized to the block case and the notion of *dynamic ordering* of tasks has been introduced [3,4].

The application of dynamic ordering was successful both in serial and parallel cases. Consider the serial OSBJA where a matrix $A$ of size $m \times n$ is divided into $\ell$ block columns, $A = (A_1, A_2, \ldots, A_\ell)$, each of size $m \times n/\ell$. It is necessary to define a weight $w_{rs}$, $1 \leq r < s \leq \ell$, which measures a departure from orthogonality between subspaces spanned by columns of $A_r$ and $A_s$. The exact weight $\overline{w}_{rs} = \|A_r^T A_s\|_F$ costs $O(m\, n^2/\ell^2)$ flops, which is equal to the cost of the whole matrix update; i.e., this approach is too expensive.

Let $\tilde{A}_r$ be the normalized block column $A_r$, so that each column of $\tilde{A}_r$ has the unit Euclidean norm. Define the vector $c_s$,

$$c_s \equiv \frac{\tilde{A}_s\, e}{\|e\|_2}, \quad e = (1,1,\ldots,1)^T$$

as the representative vector of subspace $\mathrm{span}(A_s)$. Then the *approximate* weight $w_{rs}$ was defined in [4] as

$$w_{rs} \equiv \|\tilde{A}_r^T c_s\|_2 = \frac{\|\tilde{A}_r^T \tilde{A}_s\, e\|_2}{\|e\|_2}.$$

In the serial case, the update of weights after the application of one rotation costs $O(mn)$ flops [2]. It does not dominate in the computation of one iteration step for $\ell \leq \sqrt{n}$, which is a reasonable condition. Moreover, it has been shown in [10] that the expectation value of the square of approximate weight is proportional to the square of exact weight. This fact enables to use the approximate weights in the stopping criterion of the OSBJA.

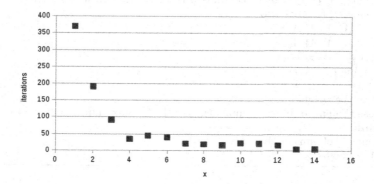

**Fig. 1.** Number of iterations for the maximum weight of order $10^{-x}$

But still, the number of iterations may remain too high as shown in Fig. 1, which depicts a typical behavior of the OSBJA. The horizontal axis is calibrated using the negative logarithm of the maximum weight, while the vertical axis is calibrated in the number of executed iterations spent for a given maximum weight. At the beginning of computation, the number of iterations is large and decreases with a decreasing maximum weight. However, when the value of maximum weight is of order $10^{-7}$ and less, the number of iterations remains almost constant and much less as compared to that at the beginning of computation. This is the consequence of the asymptotic quadratic convergence of the OSBJA [12]. So it would be desirable to find such an orthogonal transformation, which could "jump" over the slow phase of convergence.

The SVD computation by using the EVD of $A^T A$ can be numerically inaccurate because of roundoff errors in the computation of $A^T A$ (recall that $\kappa(A^T A) = \kappa^2(A)$) [6]. But for a well-conditioned matrix $A$, the eigenvector matrix $W$ of $A^T A$ can be a good estimate of the right singular vector matrix $V$.

In exact arithmetic, $AW$ has orthogonal columns and $AW = U\Sigma$. In finite floating point arithmetic, the columns of $AW$ may be far from orthogonal, but it can be more efficient to start the Jacobi process with $AW$ instead of the original $A$ [2]. A necessary condition of the usability of this approach is the availability of a fast EVD procedure giving a highly orthogonal $W$ together with a fast procedure for matrix multiplication. The serial preconditioned one-sided block-Jacobi algorithm (P_OSBJA) is listed below as Algorithm 2. Details of

P_OSBJA with Dynamic Ordering
  Input: $A = (A_1, A_2, \ldots A_\ell)$, each block column is $m \times n/\ell$
  Compute the Gram matrix: $B = A^T A$
  $[W, \Lambda] = \text{EVD}(B)$
  $A = A * W$
  Set: $V = W$
  Compute the weights $w_{rs}$
  Choose the pair $(i, j)$ of block columns with the maximum weight $maxw$
  **while** $maxw \leq (n/\ell)\, \varepsilon_M$ **do**
$$G_{ij} = \begin{pmatrix} A_i^T A_i & A_i^T A_j \\ A_j^T A_i & A_j^T A_j \end{pmatrix}$$
     $[X_{ij}, \Lambda_{ij}] = \text{EVD}(G_{ij})$
     $(A_i, A_j) = (A_i, A_j) * X_{ij}$
     $(V_i, V_j) = (V_i, V_j) * X_{ij}$
     Update the weights $w_{rs}$
     Choose the pair of block columns $(i, j)$ with the maximum weight $maxw$
  **end while**
  $\sigma_r$: norms of columns of $A_r$
  $U_r = A_r * \text{diag}(\sigma_r^{-1})$

its implementation can be found in [2]. In the next section, numerical properties of the preconditioning in Algorithm 2 are analyzed in finite floating point arithmetic.

# 3   Quality of Preconditioning in Finite Arithmetic

In this section, the arithmetic operations are denoted with the prefix fl($\cdot$) and the computed matrices and vectors are denoted by a hat (like $\hat{B}$, $\hat{v}$). We assume that neither overflow nor underflow occurs during the computations.

## 3.1   Eigenvalue Decomposition of fl($A^T A$)

Consider a matrix $A$ of order $m \times n$, $m \geq n$, with the *exact* SVD $A = U\Sigma V^T$, where $U$ and $V$ is the orthogonal matrix of left and right singular vectors, respectively, and the singular values (diagonal elements of the diagonal matrix $\Sigma$) are ordered non-increasingly, $\sigma_1 \geq \sigma_2 \geq \ldots \geq \sigma_n \geq 0$. Using [9, p. 32], let us define the *gap* $g_i$ for each singular value $\sigma_i$ by

$$g_i \equiv \min_{\sigma_j \neq \sigma_i} |\sigma_i^2 - \sigma_j^2|. \tag{1}$$

Then, in exact arithmetic, the Gram matrix $A^T A$ has the eigenvalues $\sigma_i^2$, $1 \leq i \leq n$, and its eigenvectors are corresponding columns of $V$, which are exact right singular vectors of $A$.

  Assume that in the finite floating point arithmetic with the round-off unit $\varepsilon$ the computed Gram matrix is symmetric (e.g., compute only its upper triangle including the diagonal and then copy the matrix elements into the lower triangle). In finite arithmetic, the computed symmetric Gram matrix $\hat{B} = \mathrm{fl}(A^T A)$ differs from the exact one $B = A^T A$. Using the result of [8, p. 78], $\hat{B} = B + \Delta B$ with a symmetric perturbation $\Delta B$, and

$$\|\Delta B\|_2 = \|\hat{B} - B\|_2 \leq \gamma_n \|A^T\|_2 \|A\|_2 = \sigma_1^2 \gamma_n. \tag{2}$$

Here $\gamma_n \equiv n\varepsilon/(1 - n\varepsilon)$ and $\|A\|_2$ is the spectral norm of $A$.

Now assume that the EVD $\hat{W}\hat{\Sigma}^2\hat{W}^T$ of $\hat{B}$ is computed using some backward stable algorithm, and the computed eigenvalues are $\hat{\sigma}_i^2$, $1 \leq i \leq n$. Then, according to [1, p. 83], there exists a modestly growing function $\alpha(n)$ such that

$$|\hat{\sigma}_i^2 - \sigma_i^2| \leq \alpha(n)\sigma_1^2\varepsilon, \; 1 \leq i \leq n, \tag{3}$$

so that the large eigenvalues are computed with a high relative accuracy whereas the smallest ones may loose it. However, it should be stressed that the EVD of $\mathrm{fl}(A^T A)$ is *not* used for the computation of singular values of $A$ in our algorithm. Additionally, $\hat{W}\hat{\Sigma}^2\hat{W}^T$ is nearly the exact EVD of a perturbed symmetric matrix $\mathrm{fl}(A^T A) + E$, i.e.,

$$\mathrm{fl}(A^T A) + E = (\hat{W} + \Delta\hat{W})\hat{\Sigma}^2(\hat{W} + \Delta\hat{W})^T, \tag{4}$$

where $\hat{W} + \Delta\hat{W}$ is orthonormal with $\|\hat{W} + \Delta\hat{W}\|_2 = 1$, and

$$\|\Delta\hat{W}\|_2 \leq \alpha(n)\varepsilon. \tag{5}$$

Using [1, p. 83] and $\hat{\sigma}_1^2 \leq \sigma_1^2(1 + \alpha(n)\varepsilon)$ from (3), the perturbation $E$ is bounded by

$$\|E\|_2 \leq \|\mathrm{fl}(A^T A)\|_2 \, \alpha(n) \, \varepsilon = \hat{\sigma}_1^2 \, \alpha(n) \, \varepsilon = \sigma_1^2\alpha(n)\varepsilon + O(\varepsilon^2).$$

However, one has $\mathrm{fl}(A^T A) = A^T A + \Delta B$, and (4) transforms to

$$A^T A + (\Delta B + E) = (\hat{W} + \Delta\hat{W})\hat{\Sigma}^2(\hat{W} + \Delta\hat{W})^T, \tag{6}$$

where, using (2),

$$\|\Delta B + E\|_2 \leq \|\Delta B\|_2 + \|E\|_2 \leq \sigma_1^2 \gamma_n + \sigma_1^2\alpha(n)\varepsilon$$

$$= \left(\frac{n}{1 - n\varepsilon} + \alpha(n)\right)\sigma_1^2\varepsilon \equiv \beta(n)\sigma_1^2\varepsilon \tag{7}$$

with some another modestly growing function $\beta(n)$ depending essentially on $n$ (if $1 \gg n\varepsilon$), whereby the terms of order $O(\varepsilon^2)$ were omitted.

Equations (6) and (7) enable to use Theorem 1.1 from [9] that describes the relation between the exact eigenvector $w_i$, the computed eigenvector $\hat{w}_i$ and the gap $g_i$:

$$\text{if } g_i > 4\beta(n)\sigma_1^2\varepsilon, \text{ then } \sin\angle(\hat{w}_i, w_i) \leq \frac{2\beta(n)\sigma_1^2\varepsilon}{g_i - 4\beta(n)\sigma_1^2\varepsilon}. \tag{8}$$

According to (8), when $g_i \gg 4\beta(n)\sigma_1^2\varepsilon$ the computed eigenvector $\hat{w}_i$ is close to the exact eigenvector $w_i$. However, if $w_i$ corresponds to the eigenvalue $\sigma_i^2$ in a tight cluster with $g_i \approx 4\beta(n)\sigma_1^2\varepsilon$, the eigenvector $w_i$ is ill-conditioned and the computed $\hat{w}_i$ can be far from it.

Next, we estimate the loss of orthogonality of columns in the computed matrix $\hat{W}$. Recall that the matrix $\hat{W} + \Delta\hat{W}$ is orthonormal so that

$$I - (\hat{W} + \Delta\hat{W})^T(\hat{W} + \Delta\hat{W}) = 0.$$

Using this fact together with (5), one obtains

$$\begin{aligned}
\|I - \hat{W}^T\hat{W}\|_2 &= \|\hat{W}^T\Delta\hat{W} + (\Delta\hat{W})^T\hat{W} + (\Delta\hat{W})^T\Delta\hat{W}\|_2 \\
&= \|(\hat{W} + \Delta\hat{W})^T\Delta\hat{W} - (\Delta\hat{W})^T\Delta\hat{W} + (\Delta\hat{W})^T(\hat{W} + \Delta\hat{W}) - \\
&\quad - (\Delta\hat{W})^T\Delta\hat{W} + (\Delta\hat{W})^T\Delta\hat{W}\|_2 \\
&\leq \|(\hat{W} + \Delta\hat{W})^T\|_2\|\Delta\hat{W}\|_2 + \|\hat{W} + \Delta\hat{W}\|_2\|(\Delta\hat{W})^T\|_2 + \|\Delta\hat{W}\|_2^2 \\
&= 2\|\Delta\hat{W}\|_2 + \|\Delta\hat{W}\|_2^2 \\
&\leq 2\alpha(n)\varepsilon + O(\varepsilon^2).
\end{aligned} \tag{9}$$

Although the computed matrix $\hat{W}$ is not exactly orthogonal, we assume in the following that $\|\hat{w}_i\|_2 = 1$, $1 \leq i \leq n$, *exactly*, i.e., we ignore the floating point error of order $O(\varepsilon)$ in the normalization of $\hat{W}$'s columns at the output of the EVD procedure.

The loss of orthogonality of $\hat{W}$ can also be seen by estimating its spectral norm. Indeed,

$$\|\hat{W}\|_2 = \max_{\|y\|_2 = 1} \|\hat{W}y\|_2 \geq \|\hat{W}e_1\|_2 = \|\hat{v}_1\|_2 = 1,$$

where $e_1$ is the first column of identity. On the other hand,

$$\|\hat{W}\|_2 = \|\hat{W} + \Delta\hat{W} - \Delta\hat{W}\|_2 \leq \|\hat{W} + \Delta\hat{W}\|_2 + \|\Delta\hat{W}\|_2 \leq 1 + \alpha(n)\varepsilon,$$

so that

$$1 \leq \|\hat{W}\|_2 \leq 1 + \alpha(n)\varepsilon. \tag{10}$$

The above analysis shows that the computed matrix $\hat{W}$ is orthogonal up to the order $O(\alpha(n)\varepsilon)$. This is the crucial property for a subsequent application of the OSBJA to the product $\mathrm{fl}(A\hat{W})$. Recall that during the Jacobi iterations, the matrix $\hat{W}$ will be multiplied from the right by a series of orthogonal updates in the form of matrix-matrix multiplications. If the computed matrix $\hat{W}$ were *not* orthogonal up to the order $O(\alpha(n)\varepsilon)$ at the beginning, there would be no way how to ensure its orthogonality to this order at the end of iteration process. On the other hand, since $\hat{W}$ *is* orthogonal to the order $O(\alpha(n)\varepsilon)$, subsequent updates by orthogonal matrices (which are themselves orthogonal up to the order $O(\varepsilon)$ in the finite floating point arithmetic) lead to the final matrix of right singular vectors that is orthogonal up to the order $O(\alpha(n)\varepsilon)$ again.

## 3.2   Quality of $\mathrm{fl}(A\hat{W})$

Exact eigenvectors in $W$ are, at the same time, exact right singular vectors of $A$, so that $AW = U\Sigma$ exactly. If the computed $\hat{W}$ is to be used as the SVD preconditioner, one has to analyze how far is $\mathrm{fl}(A\hat{W})$ from $U\Sigma$ columnwise.

Let $\hat{x}_i \equiv \text{fl}(A\hat{w}_i)$, and $x_i \equiv A\hat{w}_i$ exactly. Recall that in exact arithmetic $Aw_i = \sigma_i u_i$.

To analyze the quality of preconditioning, we are interested in the upper bound for $\sigma_i^{-1}\|\hat{x}_i - Aw_i\|_2 = \|\sigma_i^{-1}\hat{x}_i - u_i\|_2$. Observe that (cf. [8, p. 78])

$$\|x_i - \hat{x}_i\|_2 \le \|A\|_2 \|\hat{w}_i\|_2 \gamma_n = \sigma_1 \gamma_n.$$

Now compare $\hat{x}_i$ with exact $Aw_i$:

$$\|\hat{x}_i - Aw_i\|_2 = \|\hat{x}_i - x_i + x_i - Aw_i\|_2 \le \|\hat{x}_i - x_i\|_2 + \|A\hat{w}_i - Aw_i\|_2$$
$$\le \sigma_1 \gamma_n + \|A\|_2 \|\hat{w}_i - w_i\|_2 = \sigma_1 \gamma_n + \sigma_1 \|\hat{w}_i - w_i\|_2.$$

The upper bound for $\|\hat{w}_i - w_i\|_2$ can be found using the isosceles triangle $(\hat{w}_i,\, w_i,\, \hat{w}_i - w_i)$ with $\|\hat{w}_i\|_2 = \|w_i\|_2 = 1$:

$$\sin \frac{\angle(\hat{w}_i, w_i)}{2} = \frac{\|\hat{w}_i - w_i\|_2}{2},$$

so that

$$\|\hat{w}_i - w_i\|_2 = 2\sin\frac{\angle(\hat{w}_i, w_i)}{2} \le 2\sin\angle(\hat{w}_i, w_i) \le \frac{4\beta(n)\sigma_1^2\varepsilon}{g_i - 4\beta(n)\sigma_1^2\varepsilon}.$$

Finally, the combination of above estimates gives

$$\|\sigma_i^{-1}\hat{x}_i - u_i\|_2 \le \frac{\sigma_1}{\sigma_i}\left(\gamma_n + \frac{4\beta(n)\sigma_1^2\varepsilon}{g_i - 4\beta(n)\sigma_1^2\varepsilon}\right). \tag{11}$$

Using the sine theorem for the triangle $(\sigma_i^{-1}\hat{x}_i,\, u_i,\, \sigma_i^{-1}\hat{x}_i - u_i)$, one can estimate the angle $\angle(\hat{x}_i, u_i) = \angle(\sigma_i^{-1}\hat{x}_i, u_i)$:

$$\sin\angle(\hat{x}_i, u_i) = \sin\angle(\sigma_i^{-1}\hat{x}_i, u_i) = \frac{\sin\omega}{\|u_i\|_2}\|\sigma_i^{-1}\hat{x}_i - u_i\|_2$$
$$\le \frac{\sigma_1}{\sigma_i}\left(\gamma_n + \frac{4\beta(n)\sigma_1^2\varepsilon}{g_i - 4\beta(n)\sigma_1^2\varepsilon}\right), \tag{12}$$

where $\|u_i\|_2 = 1$, $\omega$ is the angle opposite to $u_i$ and $\sin\omega \le 1$. Notice that this bound is the same as that in (11).

It is instructive to observe the connection between upper bounds in Eqs. (8) and (12). If the gap $g_i$ is small, then, according to (8), the computed $\hat{w}_i$ can be far from the exact $w_i$ so that one can not expect that $\sigma_i^{-1}\text{fl}(A\hat{w}_i)$ will be close to $u_i$. Indeed, (12) shows that the original error of $\hat{w}_i$ with respect to $w_i$ is essentially amplified by the factor $2\sigma_1/\sigma_i$.

The upper bound in (12) provides a suitable background for the discussion about the quality of the new SVD preconditioner. One can think of four cases:

1. Well-conditioned matrix $A$ with well-separated singular values.

   All spectral gaps $g_i$ are relatively large and all values $\sigma_1/\sigma_i$ are relatively modest. Hence, each vector $\text{fl}(A\hat{w}_i)$ will be close to the corresponding left singular vector $u_i$ so that all columns of $\hat{X} \equiv \text{fl}(A\hat{W})$ will be nearly orthogonal. Consequently, only a small number of iterations (serial or parallel) will be needed in the OSBJA applied to $\hat{X}$.

2. Well-conditioned matrix $A$ with a tight cluster of singular values.

   Assume that $A$ has $n - t$ well-separated largest singular values and a tight cluster of $t$ smallest singular values $\{\sigma_i\}_{i=n-t+1}^{n}$ (this distribution of singular values is encountered quite often in various applications). Although all values $\sigma_1/\sigma_i$ are modest, the gaps $g_i$, $n - t + 1 \leq i \leq n$, can be very small so that the last $t$ eigenvectors of the Gram matrix can be ill-conditioned and the vectors $\hat{x}_i$, $n - t + 1 \leq i \leq n$, can be far from the corresponding left singular vectors $u_i$. Let us partition the matrix $\hat{W}$ in the form $\hat{W} = (\hat{W}_1, \hat{W}_2)$ where $\hat{W}_2$ contains last $t$ columns of $\hat{W}$. Then $\hat{X} = (\mathrm{fl}(A\hat{W}_1), \mathrm{fl}(A\hat{W}_2)) = (\hat{X}_1, \hat{X}_2)$. Here, $\hat{X}_1$ will have nearly orthogonal columns whereas the columns of $\hat{X}_2$ will be orthogonal neither to the columns of $\hat{X}_1$ nor mutually inside $\hat{X}_2$. On the other hand, when the last $t$ gaps $g_i$ are not too small, the loss of orthogonality can be relatively mild so that one can apply the OSBJA to the whole matrix $\hat{X}$.

3. Ill-conditioned matrix $A$ with well-separated singular values.

   At least $\sigma_1/\sigma_n = \kappa(A)$ will be large so that at least $\hat{x}_n$ can be far from $u_n$ and loose its orthogonality to all other columns of $\hat{X}$. For other computed vectors $\hat{x}_i$, $i \neq n$, their proximity to the corresponding $\mathrm{span}(u_i)$ depends on the value $\sigma_1/\sigma_i$. One can expect that only few last columns of $\hat{X}$ will loose their orthogonality so that it may be convenient to apply the OSBJA to the whole matrix $\hat{X}$.

4. Ill-conditioned matrix $A$ with a tight cluster of singular values.

   Consider again the cluster from the case 2 above. Now, all values $\sigma_1/\sigma_i \approx \kappa(A)$, $n - t + 1 \leq i \leq n$, will be very large and all gaps $g_i$, $n - t + 1 \leq i \leq n$, will be very small. Consequently, the loss of orthogonality of columns in $\hat{X}_2$ can be much more pronounced than in the case 2, and the OSBJA applied to the whole $\hat{X}$ will need much more iterations for convergence. Hence, assuming that one has some information about the value of $t$, it might be more advantageous to apply the OSBJA first to the part $\hat{X}_2$ and afterwards, when the columns of $\hat{X}_2$ are mutually orthogonal to high level, to run it on the whole matrix $\hat{X}$.

### 3.3    Parallelization

The parallelization of Algorithm 2 for a given blocking factor $\ell$ is straightforward, since $\ell/2$ rotations can be computed independently. The pairs $(i, j)$ are selected by our dynamic ordering based on the maximum weight perfect matching greedy algorithm [4]. There are two variants of the parallel OSBJA. In the first variant, the number of processors $p$ is related to $\ell$ as $p = \ell/2$. In the second variant, the tight relation between $\ell$ and $p$ is broken and it is possible to use small blocking factors $\ell = p/k$ for some integer $k$ [5]. The second variant is faster, but it requires that the local EVD and the matrix multiplication are performed in parallel. In this paper, the fixed blocking factor $\ell = 2p$ is used exclusively. More details w.r.t. the parallelization can be found in [4,5].

# 4   Numerical Experiments

Numerical experiments were performed on the Doppler cluster at the University of Salzburg. Currently, its peak performance exceeds 20 TF/s. The used nodes were of type AMD Opteron 6274, 2.2 GHz, with 64 cores and 252 GB RAM. The new parallel preconditioned algorithm PP_OSBJA was written in FORTRAN 90 for distributed memory machines using the PBLAS, ScaLAPACK and MPI libraries. The aim was to compare its performance with the procedure PDGESVD from ScaLAPACK, which is the only parallel SVD procedure widely available today. The OpenBLAS [13] has been chosen as an implementation of BLAS and LAPACK. The procedure PDSYRK was used to generate Gram matrices, PDSYEVD was applied for the symmetric EVD (this is the parallel implementation of the divide-and-conquer algorithm), and PDGEMM computed matrix multiplications.

Inside the algorithm PP_OSBJA, two different data distributions were needed. Firstly, for the ScaLAPACK procedures used in the preconditioning, the block cyclic distribution was required. Afterwards, for the Jacobi algorithm with the blocking factor $\ell = 2p$, the block column distribution was used. For simplicity, the BLACS routine PDGEMR2D has been applied to change the data distribution, but the load balancing in the Jacobi algorithm was not always guaranteed. In the dynamic ordering part, the weights $w_{rs}$ were computed in parallel. Then, the maximum weight perfect matching on a complete graph with $\ell$ vertices and edge weights $w_{rs}$ was computed and applied (cf. [3]).

In our previous paper [2] dealing with the serial preconditioned block-Jacobi SVD algorithm with dynamic ordering, the numerical results were presented for matrices generated by the LAPACK routine DLATMS, which enables to define six modes of the singular value distribution. In that paper, four modes and three values of condition number were used. Here, for the parallel algorithm, the mode 3 was chosen, which defines (for a given $\kappa(A)$) the geometric sequence of singular values with $\sigma_i = (\kappa(A))^{[-(i-1)/n-1)]}$, $1 \leq i \leq n$. In the case of ill-conditioned matrices, the mode 3 provides tight clusters of the smallest singular values, which has a great impact on the convergence of PP_OSBJA (see the tables below).

During experiments with the serial algorithm, a different performance of OpenBLAS on Intel and AMD processors has been noticed. Serial experiments were run on the Intel based notebook [2] and also on one AMD core of the Doppler cluster. The tests were performed for three SVD procedures from LAPACK, then for DGESVJ with our preconditioner (the algorithm P_DGESVJ), and finally for our preconditioned serial algorithm P_OSBJA. Computational times in seconds are summarized in Table 1. Timings for Intel were always lower as compared to AMD, except for DGESVD. Since the parallel experiments were performed on the AMD based cluster, one can expect the approximate ratio 5 between computational times of PDGESVD and the new parallel preconditioned algorithm PP_OSBJA.

For the input matrix with parameters $\kappa(A) = 10^2$ and mode = 3, the comparison of performance between PDGESVD and PP_OSBJA is depicted in Table 2. For both algorithms, the first line contains the computational time in seconds. There is an additional information for PP_OSBJA in the form of profiling with

**Table 1.** Performance (in seconds) for $n = 4000$, $\ell = 8$, $\kappa(A) = 10^2$, mode = 3

| CPU | DGESVD | DGESDD | DGESVJ | P_DGESVJ | P_OSBJA |
|---|---|---|---|---|---|
| Intel | 1194 | 58 | 681 | 65 | 37 |
| AMD | 876 | 151 | 1244 | 201 | 155 |

**Table 2.** Performance for $n = 8000$, $\kappa(A) = 10^2$, mode = 3

| Algorithm | | Number of cores | | | | | |
|---|---|---|---|---|---|---|---|
| | | 10 | 20 | 30 | 40 | 50 | 60 |
| PDGESVD | $T$ [s] | 934 | 645 | 524 | 354 | 419 | 467 |
| PP_OSBJA | $T$ [s] | 178 | 98 | 76 | 91 | 94 | 89 |
| | G:EVD:MM [s] | 14:67:23 | 7:39:12 | 5:29:8 | 6:36:11 | 5:32:9 | 5:29:8 |
| | Jacobi [s] | 74 | 40 | 34 | 38 | 48 | 47 |
| | # it | 10 | 22 | 38 | 55 | 79 | 94 |

respect to the important computations. The line with the header G:EVD:MM (the Gram matrix, the EVD, the matrix multiplication) gives the time spent in preconditioning. The next line stands for the Jacobi algorithm and the last line depicts the number of parallel iterations in the Jacobi part. The number of processors (cores) varies from $p = 10$ to $p = 60$. As can be seen, the new algorithm PP_OSBJA was always faster than PDGESVD and the ratio of their performance was between 3.9 and 6.9. As expected, the dominant part of the preconditioner was the EVD computation. (Notice, however, that the used divide-and-conquer algorithm is the fastest one in the ScaLAPACK library for the EVD of symmetric matrices.) The EVD and the Jacobi part exhibited an increase of the portion of parallel execution time when using more than 30 cores. As was already shown in [3], the number of parallel iterations in the Jacobi algorithm increases superlinearly with the increase of blocking factor $\ell = 2p$. Notice that the parallel execution time of the procedure PDGESVD started to increase a bit later, for more than 40 cores. Obviously, for a matrix of a given size, both algorithms have the optimal performance for a certain number of processors, i.e., for a certain size of matrix blocks. This is connected with the hierarchical structure and size of the core's memory.

Table 3 depicts the results for an ill-conditioned matrix with $\kappa(A) = 10^8$ and mode = 3. Now the performance ratio was lower, between 1.8 and 3.0. The procedure PDGESVD was even a bit faster (as compared to itself) for $p \leq 40$ than in the case of a well-conditioned matrix. The time values of G:EVD:MM in preconditioning were very similar to those in Table 2 and they seem to be independent of $\kappa(A)$. But the quality of preconditioner for an ill-conditioned matrix was a problem. In accordance with the analysis from Sect. 3, the matrix $\mathrm{fl}(A\hat{W})$ was far from orthogonal and its column space was far from range($U$). Consequently, substantially more parallel iteration steps in the Jacobi part of

**Table 3.** Performance for $n = 8000, \kappa(A) = 10^8$, mode = 3

| Algorithm | | Number of processors | | | | | |
|---|---|---|---|---|---|---|---|
| | | 10 | 20 | 30 | 40 | 50 | 60 |
| PDGESVD | $T$ [s] | 859 | 499 | 461 | 345 | 555 | 610 |
| PP_OSBJA | $T$ [s] | 408 | 202 | 190 | 196 | 183 | 316 |
| | G:EVD:MM [s] | 14:61:23 | 7:36:12 | 8:36:13 | 7:34:11 | 5:31:9 | 5:29:8 |
| | Jacobi [s] | 310 | 147 | 134 | 144 | 138 | 274 |
| | # it | 33 | 72 | 119 | 158 | 199 | 491 |

PP_OSBJA were needed for the convergence than in the case of a well-conditioned matrix (compare the corresponding values of # it in Tables 2 and 3).

The accuracy of computation was checked by computing relative errors of the loss of orthogonality of computed right and left singular vectors, and of the computed minimal singular value. The accuracy results were of the same order of magnitude as those for the serial algorithm (see Tables 1 and 3 for mode = 3 in [2]).

## 5   Conclusions

The new parallel preconditioned one-sided block-Jacobi SVD algorithm with dynamic ordering has been developed and tested. It achieves a substantial speed-up when compared to the ScaLAPACK procedure PDGESVD. In accordance with the performed analysis, its performance is better for well-conditioned matrices than for the ill-conditioned ones. The difference is caused by the numerical properties of the matrix multiplication $\mathrm{fl}(A\hat{W})$ where $\hat{W}$ is the computed matrix of eigenvectors of the Gram matrix $\mathrm{fl}(A^T A)$. In the case of an ill-conditioned matrix $A$, the columns of the preconditioned matrix $\mathrm{fl}(A\hat{W})$ may not be orthogonal enough and can be far from range($U$). Consequently, substantially more parallel iteration steps in the Jacobi algorithm are needed for the convergence than in the case of a well-conditioned matrix. Note that the numerical analysis of preconditioning in Sect. 3 also explains the results of numerical experiments for the serial preconditioned OSBJA in [2].

A possible way, how to avoid the computation of the Gram matrix $A^T A$ and its EVD, is to compute the polar decomposition of $A$ by fast Halley iterations and then to compute the EVD of the symmetric, positive semi-definite polar factor [11]. This EVD can be computed by the PP_OSBJA without computing the Gram matrix of Hermitian polar factor, and the SVD of an original matrix $A$ can be easily restored by an additional matrix multiplication.

**Acknowledgment.** Authors were supported by the VEGA grant no. 2/0004/17.

# References

1. Anderson, A., et al.: LAPACK Users' Guide, 3rd edn. SIAM, Philadelphia (1999)
2. Bečka, M., Okša, G., Vidličková, E.: New preconditioning for the one-sided block-Jacobi SVD algorithm. In: Wyrzykowski, R., Dongarra, J., Deelman, E., Karczewski, K. (eds.) PPAM 2017. LNCS, vol. 10777, pp. 590–599. Springer, Cham (2018). https://doi.org/10.1007/978-3-319-78024-5_51
3. Bečka, M., Okša, G., Vajteršic, M.: Dynamic ordering for a parallel block Jacobi SVD algorithm. Parallel Comput. **28**, 243–262 (2002). https://doi.org/10.1016/S0167-8191(01)00138-7
4. Bečka, M., Okša, G., Vajteršic, M.: New dynamic orderings for the parallel one-sided block-Jacobi SVD algorithm. Parallel Proc. Lett. **25**, 1–19 (2015). https://doi.org/10.1142/S0129626415500036
5. Bečka, M., Okša, G.: Parallel one-sided Jacobi SVD algorithm with variable blocking factor. In: Wyrzykowski, R., Dongarra, J., Karczewski, K., Waśniewski, J. (eds.) PPAM 2013. LNCS, vol. 8384, pp. 57–66. Springer, Heidelberg (2014). https://doi.org/10.1007/978-3-642-55224-3_6
6. Dongarra, J., et al.: The singular value decomposition: anatomy of optimizing an algorithm for extreme scale. SIAM Rev. **60**, 808–865 (2018). https://doi.org/10.1137/17M1117732
7. Golub, G.H., van Loan, C.F.: Matrix Computations, 4th edn. The John Hopkins University Press, Baltimore (2013)
8. Higham, N.J.: Accuracy and Stability of Numerical Algorithms, 2nd edn. SIAM, Philadelphia (2002)
9. Jia, Z.: Using cross-product matrices to compute the SVD. Numer. Algorithms **42**, 31–61 (2006). https://doi.org/10.1007/s11075-006-9022-x
10. Kudo, S., Yamamoto, Y., Bečka, M., Vajteršic, M.: Performance analysis and optimization of the parallel one-sided block-Jacobi algorithm with dynamic ordering and variable blocking. Concurr. Comput. Pract. Exp. **29**, 1–24 (2017). https://doi.org/10.1002/cpe.4059
11. Nakatsukasa, Y., Higham, N.J.: Stable and efficient spectral divide and conquer algorithms for the symmetric eigenvalue decomposition and the SVD. SIAM J. Sci. Comput. **35**, 1325–1349 (2013). https://doi.org/10.1137/120876605
12. Okša, G., Yamamoto, Y., Vajteršic, M.: Asymptotic quadratic convergence of the block-Jacobi EVD algorithm for Hermitian matrices. Numer. Math. **136**, 1071–1095 (2017). https://doi.org/10.1007/s00211-016-0863-5
13. http://www.openblas.net

# A Parallel Factorization for Generating Orthogonal Matrices

Marek Parfieniuk[(✉)]

University of Bialystok, K. Ciolkowskiego 1M, 15-245 Bialystok, Poland
marekpk@wp.pl
http://marekparfieniuk.pl

**Abstract.** A new factorization of orthogonal matrices is proposed that is based on Givens-Jacobi rotations but not on the QR decomposition. Rotations are arranged more uniformly than in the known factorizations that use them, so that more rotations can be computed in parallel, and fewer layers of concurrent rotations are necessary to model a matrix. Therefore, throughput can be increased, and latency can be reduced, compared to the known solutions, even though the obtainable gains highly depend on application specificity, software-hardware architecture and matrix size. The proposed approach allows for developing more efficient algorithms and hardware for generating random matrices, for optimizing matrices, and for processing data with linear transformations. We have verified this by implementing and evaluating a multi-threaded Java application for generating random orthogonal matrices.

**Keywords:** Matrix · Orthogonal · Rotation · Factorization · Parallel

## 1 Introduction

Orthogonal matrices, as well as unitary matrices, the generalization of the former, have found many applications in diverse areas of science and engineering [6,9,11,15]. This is because such matrices can be used to model various phenomena and to extract information from data, having nice properties. They describe linear transforms that preserve data geometry [17] and are surely and easily invertible, as the inverse of an orthogonal matrix $\mathbf{A}$ is simply given by the transpose, $\mathbf{A}^{-1} = \mathbf{A}^{\mathrm{T}}$.

Three main issues are related to computations with orthogonal matrices. One is to generate a random matrix, the second is to optimize a matrix so as to satisfy some requirements, or to match data, and the last is to efficiently and accurately compute a transform described by an orthogonal or unitary matrix.

These problems are commonly solved by factorizing matrices into products of simpler ones, especially of rotations. Givens-Jacobi rotations can be considered to be the essential orthogonal matrices, as they have the form

$$\mathbf{R}(\alpha) = \begin{bmatrix} \cos\alpha & -\sin\alpha \\ \sin\alpha & \cos\alpha \end{bmatrix} \tag{1}$$

which describes the data flow graph in Fig. 1.

© Springer Nature Switzerland AG 2020
R. Wyrzykowski et al. (Eds.): PPAM 2019, LNCS 12043, pp. 567–578, 2020.
https://doi.org/10.1007/978-3-030-43229-4_48

Alternatively, an orthogonal matrix can be factorized into Householder reflectors [10]. On the other hand, it is possible to extract orthogonal matrices from general ones with random elements [6,16]. But it has been verified that these representations are inferior to rotation-based ones in terms of performance [1].

A plethora of rotation-based factorizations have been developed. The earliest, of the XIX century, were aimed at theoretical considerations and analytical derivations [4]. Then, solutions have appeared that are aimed at numerical computations on a single computer [1]. Finally, various solutions have been devised to organize rotations in ways that facilitate parallel computations [7]. Many factorizations are aimed at particular structures of matrices or particular software-hardware architectures, see e.g. [3] and [5].

Known rotation-based factorizations are based on the QR decomposition of an orthogonal matrix [2,4]. This algorithm is used to prove that a factorization is correct and to determine rotation angles. In this paper, we show that it is possible to factorize orthogonal matrices without using the QR algorithm. This allows for organizing rotations in a uniform way, well suited to parallel processing. Our new factorization allows for speeding up computations, which has been shown by evaluating its single- and multi-threaded Java-based implementations. However, it is difficult to determine angles, so the solution is useful mainly when the task is to generate random orthogonal matrices or to search for an optimal matrix.

## 2    Known Rotation-Based Factorizations of Orthogonal Matrices

The widely-known method for modelling an orthogonal matrix is to multiply matrices like

$$
\mathbf{G}(2,4,\alpha) =
\begin{bmatrix}
1 & 0 & 0 & 0 & 0 & 0 & 0 & 0 \\
0 & 1 & 0 & 0 & 0 & 0 & 0 & 0 \\
0 & 0 & \cos\alpha & 0 & -\sin\alpha & 0 & 0 & 0 \\
0 & 0 & 0 & 1 & 0 & 0 & 0 & 0 \\
0 & 0 & \sin\alpha & 0 & \cos\alpha & 0 & 0 & 0 \\
0 & 0 & 0 & 0 & 0 & 1 & 0 & 0 \\
0 & 0 & 0 & 0 & 0 & 0 & 1 & 0 \\
0 & 0 & 0 & 0 & 0 & 0 & 0 & 1
\end{bmatrix}
\tag{2}
$$

which are obtained by embedding a $2 \times 2$ Givens-Jacobi rotation matrix into an identity matrix. Geometrically, a plane (2D) rotation is converted into a rotation in one of planes of a many-dimensional space. We will denote such a matrix by $\mathbf{G}(m,n,\alpha)$, where $m$ and $n$ are the indexes of rows and columns whose crossings determine the rotation submatrix, and $\alpha$ is the rotation angle.

An $N \times N$ orthogonal matrix $\mathbf{A}$ can be represented as a product of $N(N-1)/2$ rotation matrices, whose angles and planes must be appropriately selected. Two

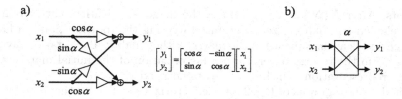

**Fig. 1.** Rotation: (a) data flow graph and (b) symbol used in higher-level graphs

rotation arrangements, or factorizations, are widely known, one of which is

$$\mathbf{A} = \prod_{m=1}^{N-1} \prod_{n=1}^{N-m} \mathbf{G}(n, n+1, \alpha_{\kappa(m,n)}) \tag{3}$$

while the second is

$$\mathbf{A} = \prod_{m=1}^{N-1} \prod_{n=1}^{N-m} \mathbf{G}(1, n+1, \alpha_{\kappa(m,n)}) \tag{4}$$

where $\kappa(m, n)$ varies from 0 to $N(N-1)/2$, with $m$ and $n$.

For $N = 6$, the arrangements take the forms

$$\begin{aligned}
\mathbf{A} = {} & \mathbf{G}(1,2,\alpha_0)\mathbf{G}(2,3,\alpha_1)\mathbf{G}(3,4,\alpha_2)\mathbf{G}(4,5,\alpha_3)\mathbf{G}(5,6,\alpha_4) \\
& \cdot \mathbf{G}(1,2,\alpha_5)\mathbf{G}(2,3,\alpha_6)\mathbf{G}(3,4,\alpha_7)\mathbf{G}(4,5,\alpha_8) \\
& \cdot \mathbf{G}(1,2,\alpha_9)\mathbf{G}(2,3,\alpha_{10})\mathbf{G}(3,4,\alpha_{11}) \\
& \cdot \mathbf{G}(1,2,\alpha_{12})\mathbf{G}(2,3,\alpha_{13}) \\
& \cdot \mathbf{G}(1,2,\alpha_{14})
\end{aligned} \tag{5}$$

and

$$\begin{aligned}
\mathbf{A} = {} & \mathbf{G}(1,2,\alpha_0)\mathbf{G}(1,3,\alpha_1)\mathbf{G}(1,4,\alpha_2)\mathbf{G}(1,5,\alpha_3)\mathbf{G}(1,6,\alpha_4) \\
& \cdot \mathbf{G}(1,2,\alpha_5)\mathbf{G}(1,3,\alpha_6)\mathbf{G}(1,4,\alpha_7)\mathbf{G}(1,5,\alpha_8) \\
& \cdot \mathbf{G}(1,2,\alpha_9)\mathbf{G}(1,3,\alpha_{10})\mathbf{G}(1,4,\alpha_{11}) \\
& \cdot \mathbf{G}(1,2,\alpha_{12})\mathbf{G}(1,3,\alpha_{13}) \\
& \cdot \mathbf{G}(1,2,\alpha_{14})
\end{aligned} \tag{6}$$

respectively, and describe the data flow graphs in Fig. 2. Such a graph represents **A** indirectly, consisting of blocks that realize operations necessary to compute $\mathbf{y} = \mathbf{A}\mathbf{x}$, or to multiply an input vector by the matrix. In order to recover **A** from angles, the identity matrix must be column-by-column passed through a graph.

Such factorizations can be derived by considering the QR decomposition of an orthonormal matrix [4,14]. The matrix is multiplied by a matrix like (2), so as to zero one of its elements outside the diagonal. The resulting matrix is multiplied by a subsequent rotation that zeros another element, preserving existing zero elements. After $N(N-1)/2$ rotations, the diagonal, identity matrix remains, whose elements in half are zeros because targeting the rotations, and in half as a side effect, as a consequence of the matrix orthogonality. The rotation inverses, multiplied in the reverse order, form a factorization of the initial matrix, or its indirect representation, which can be parametrized.

The data flow graphs of Fig. 2, or the factorizations behind them, are related to QR decompositions like that which has been demonstrated in Fig. 3, for an exemplary orthogonal matrix. Please notice that rotations of the QR decomposition are inverses (use negated angles) of rotations of the corresponding factorization or data flow graph. Moreover, they appear in reversed order.

Please also notice effects of subsequent rotations, or order in which matrix elements are zeroed. By modifying QR decomposition, so that elements are zeroed in another order, many other factorizations can be derived. In particular, subsequent rotations can operate in disjoint planes, and thus can be computed in parallel. Nevertheless, the parallelism of QR-related arrangements is limited by the key property of this approach: that each rotation zeroes one element.

A straightforward way to implement such data flows is to develop a hardware systolic array [8]. It should consists of $N(N-1)/2$ processing elements, one per rotation, which are connected exactly as in a graph and cooperate so as to produce results. However, for larger matrices, a systolic array would comprise hundreds or thousands of elements, so that its cost and resource consumption are hardly acceptable. Another problem is that such architectures are aimed at continuously processing massive streams of input vectors, not at generating a single matrix from time to time.

Another extreme is to use only one processing element, which is switched among data lines and among angles, so as to iteratively compute all rotations one-after-another. This approach suffers from poor throughput, as $N(N-1)/2$ iterations are necessary to produce output.

If a parallel computer is accessible, then it is reasonable to trade-off performance for the number of processors in use. This is possible for the known factorizations, as one can identify layers of rotations that can be computed in parallel, as they depend on results of the same preceding rotations. For example, in (5) and in Fig. 2a, these are the rotations by $\alpha_2$ and $\alpha_5$; by $\alpha_3$ and $\alpha_6$, by $\alpha_4$, $\alpha_7$, and $\alpha_9$; by $\alpha_8$ and $\alpha_{10}$; and by $\alpha_{11}$ and $\alpha_{12}$. They are preceded and followed by rotations that must be computed one at a time. So, limited parallelism can be achieved, but it is difficult to fully utilize hardware, as the number of processors in use varies from stage to stage of the algorithm.

The problem affects all factorizations that are based on the QR decomposition, including those that use Householder reflectors. Parallelism opportunities can be enhanced by reordering rotations or reflectors, but in limited extent. So it is reasonable to look for novel factorizations that are better suited to parallel computations.

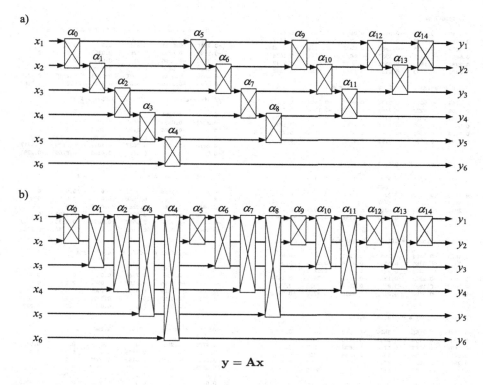

**Fig. 2.** Data flow graph for computing matrix-vector products in accordance with conventional factorizations of orthogonal matrix

## 3   New Parallel Rotation-Based Factorization

Thinking about the conventional factorizations, we have drawn three conclusions. First, in order to be able to represent an arbitrary $N \times N$ orthogonal matrix, rotations must cover all planes of the $N$-dimensional space. Second, a plane can be covered implicitly, by a combination of rotations that act in other planes. Third, it is not important how rotations are ordered and combined, provided they cover all planes.

These conclusions and our long-standing experience in factorization-based design and implementation of transforms and filter banks, please see e.g. [12] and [13], have led us to considering the following factorization

$$\mathbf{A} = \prod_{m=2}^{N} \prod_{n=1}^{N/2} \mathbf{G}((2n - 2 + (m \bmod 2)(N - 1)) \bmod N + 1,$$

$$(2n - 1 + (m \bmod 2)(N - 1)) \bmod N + 1, \alpha_{\kappa(m,n)})$$

(7)

```
// Original matrix: A

  0.0695    0.5297    0.3592    0.0803    0.1673    0.7424
  0.5471   -0.5176    0.3534    0.4925   -0.2127    0.1417
  0.5349    0.6015    0.1963   -0.0038   -0.1328   -0.5439
  0.5664   -0.2323   -0.1337   -0.6679    0.3651    0.1674
  0.2982    0.1881   -0.8303    0.3587   -0.0923    0.2217
 -0.0040   -0.0190    0.0152    0.4198    0.8760   -0.2362

// 1th rotation: A = G(1, 2, -1.759) * A

  0.5244   -0.6077    0.2798    0.4687   -0.2402    0.0
 -0.1708   -0.4232   -0.4191   -0.1712   -0.1244   -0.7558
  0.5349    0.6015    0.1963   -0.0038   -0.1328   -0.5439
  0.5664   -0.2323   -0.1337   -0.6679    0.3651    0.1674
  0.2982    0.1881   -0.8303    0.3587   -0.0923    0.2217
 -0.0040   -0.0190    0.0152    0.4198    0.8760   -0.2362

// 2th rotation: A = G(2, 3, -2.195) * A

  0.5244   -0.6077    0.2798    0.4687   -0.2402    0.0
  0.5340    0.7355    0.4042    0.0969   -0.0351   -0.0
 -0.1738   -0.0078    0.2255    0.1412    0.1786    0.9312
  0.5664   -0.2323   -0.1337   -0.6679    0.3651    0.1674
  0.2982    0.1881   -0.8303    0.3587   -0.0923    0.2217
 -0.0040   -0.0190    0.0152    0.4198    0.8760   -0.2362

// 3th rotation: A = G(1, 2, -4.858) * A

 -0.4525   -0.8156   -0.3594   -0.0281    0.0       0.0
  0.5961   -0.4950    0.3353    0.4778   -0.2428    0.0
 -0.1738   -0.0078    0.2255    0.1412    0.1786    0.9312
  0.5664   -0.2323   -0.1337   -0.6679    0.3651    0.1674
  0.2982    0.1881   -0.8303    0.3587   -0.0923    0.2217
 -0.0040   -0.0190    0.0152    0.4198    0.8760   -0.2362

// 4th rotation: A = G(3, 4, -4.890) * A

 -0.4525   -0.8156   -0.3594   -0.0281    0.0       0.0
  0.5961   -0.4950    0.3353    0.4778   -0.2428    0.0
 -0.5882    0.2273    0.1715    0.6823   -0.3277    0.0
 -0.0709   -0.0488    0.1983    0.0208    0.2403    0.9461
  0.2982    0.1881   -0.8303    0.3587   -0.0923    0.2217
 -0.0040   -0.0190    0.0152    0.4198    0.8760   -0.2362

// 5th rotation: A = G(2, 3, -2.504) * A

 -0.4525   -0.8156   -0.3594   -0.0281    0.0       0.0
 -0.8291    0.5330   -0.1674    0.0222    0.0       0.0
  0.1178    0.1120   -0.3374   -0.8327    0.4079    0.0
 -0.0709   -0.0488    0.1983    0.0208    0.2403    0.9461
  0.2982    0.1881   -0.8303    0.3587   -0.0923    0.2217
 -0.0040   -0.0190    0.0152    0.4198    0.8760   -0.2362

// 6th rotation: A = G(1, 2, -0.901) * A

 -0.9310   -0.0881   -0.3543    0.0       0.0       0.0
 -0.1596    0.9703    0.1780    0.0359    0.0       0.0
  0.1178    0.1120   -0.3374   -0.8327    0.4079    0.0
 -0.0709   -0.0488    0.1983    0.0208    0.2403    0.9461
  0.2982    0.1881   -0.8303    0.3587   -0.0923    0.2217
 -0.0040   -0.0190    0.0152    0.4198    0.8760   -0.2362

// 7th rotation: A = G(4, 5, -4.943) * A

 -0.9310   -0.0881   -0.3543    0.0       0.0       0.0
 -0.1596    0.9703    0.1780    0.0359    0.0       0.0
  0.1178    0.1120   -0.3374   -0.8327    0.4079    0.0
 -0.3065   -0.1943    0.8536   -0.3445    0.1447    0.0
 -0.0010   -0.0046    0.0037    0.1021    0.2130    0.9717
 -0.0040   -0.0190    0.0152    0.4198    0.8760   -0.2362

...
```

```
// 8th rotation: A = G(3, 4, -1.912) * A

 -0.9310   -0.0881   -0.3543    0.0       0.0       0.0
 -0.1596    0.9703    0.1780    0.0359    0.0       0.0
 -0.3282   -0.2206    0.9173   -0.0463    0.0       0.0
 -0.0086   -0.0406    0.0326    0.9000   -0.4328    0.0
 -0.0010   -0.0046    0.0037    0.1021    0.2130    0.9717
 -0.0040   -0.0190    0.0152    0.4198    0.8760   -0.2362

// 9th rotation: A = G(2, 3, -0.659) * A

 -0.9310   -0.0881   -0.3543    0.0       0.0       0.0
 -0.3272    0.6320    0.7025    0.0       0.0       0.0
 -0.1617   -0.7686    0.6162   -0.0585    0.0       0.0
 -0.0086   -0.0406    0.0326    0.9000   -0.4328    0.0
 -0.0010   -0.0046    0.0037    0.1021    0.2130    0.9717
 -0.0040   -0.0190    0.0152    0.4198    0.8760   -0.2362

// 10th rotation: A = G(1, 2, -0.467) * A

 -0.9786    0.2059    0.0       0.0       0.0       0.0
  0.1271    0.6040    0.7868    0.0       0.0       0.0
 -0.1617   -0.7686    0.6162   -0.0585    0.0       0.0
 -0.0086   -0.0406    0.0326    0.9000   -0.4328    0.0
 -0.0010   -0.0046    0.0037    0.1021    0.2130    0.9717
 -0.0040   -0.0190    0.0152    0.4198    0.8760   -0.2362

// 11th rotation: A = G(5, 6, -4.474) * A

 -0.9786    0.2059    0.0       0.0       0.0       0.0
  0.1271    0.6040    0.7868    0.0       0.0       0.0
 -0.1617   -0.7686    0.6162   -0.0585    0.0       0.0
 -0.0086   -0.0406    0.0326    0.9000   -0.4328    0.0
  0.0041    0.0195   -0.0156   -0.4320   -0.9015    0.0
  0.0       0.0       0.0       0.0       0.0       1.000

// 12th rotation: A = G(4, 5, -2.694) * A

 -0.9786    0.2059    0.0       0.0       0.0       0.0
  0.1271    0.6040    0.7868    0.0       0.0       0.0
 -0.1617   -0.7686    0.6162   -0.0585    0.0       0.0
  0.0095    0.0451   -0.0361   -0.9983    0.0       0.0
  0.0       0.0       0.0       0.0       1.0000    0.0
  0.0       0.0       0.0       0.0       0.0       1.0000

// 13th rotation: A = G(3, 4, -3.083) * A

 -0.9786    0.2059    0.0       0.0       0.0       0.0
  0.1271    0.6040    0.7868    0.0       0.0       0.0
  0.1620    0.7699   -0.6172    0.0       0.0       0.0
  0.0       0.0       0.0       1.0000    0.0       0.0
  0.0       0.0       0.0       0.0       1.0000    0.0
  0.0       0.0       0.0       0.0       0.0       1.0000

// 14th rotation: A = G(2, 3, -4.047) * A

 -0.9786    0.2059    0.0       0.0       0.0       0.0
 -0.2059   -0.9786    0.0       0.0       0.0       0.0
  0.0       0.0       1.0000    0.0       0.0       0.0
  0.0       0.0       0.0       1.0000    0.0       0.0
  0.0       0.0       0.0       0.0       1.0000    0.0
  0.0       0.0       0.0       0.0       0.0       1.0000

// 15th rotation: A = G(1, 2, -3.349) * A

  1.0000    0.0       0.0       0.0       0.0       0.0
  0.0       1.0000    0.0       0.0       0.0       0.0
  0.0       0.0       1.0000    0.0       0.0       0.0
  0.0       0.0       0.0       1.0000    0.0       0.0
  0.0       0.0       0.0       0.0       1.0000    0.0
  0.0       0.0       0.0       0.0       0.0       1.0000
```

**Fig. 3.** Rotation effects on orthogonal matrix: conventional factorization

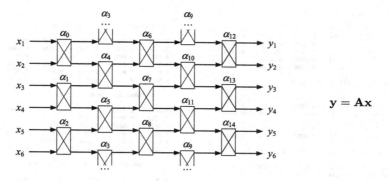

**Fig. 4.** Data flow graph for computing matrix-vector products in accordance with proposed parallel factorization of orthogonal matrix

which becomes much more clearer after expanding it for $N = 6$

$$\mathbf{A} = \mathbf{G}(1, 2, \alpha_0)\mathbf{G}(3, 4, \alpha_1)\mathbf{G}(5, 6, \alpha_2)$$
$$\cdot \mathbf{G}(6, 1, \alpha_3)\mathbf{G}(2, 3, \alpha_4)\mathbf{G}(4, 5, \alpha_5)$$
$$\cdot \mathbf{G}(1, 2, \alpha_6)\mathbf{G}(3, 4, \alpha_7)\mathbf{G}(5, 6, \alpha_8) \qquad (8)$$
$$\cdot \mathbf{G}(6, 1, \alpha_9)\mathbf{G}(2, 3, \alpha_{10})\mathbf{G}(4, 5, \alpha_{11})$$
$$\cdot \mathbf{G}(1, 2, \alpha_{12})\mathbf{G}(3, 4, \alpha_{13})\mathbf{G}(5, 6, \alpha_{14})$$

and drawing the corresponding data flow graph in Fig. 4. The idea is rather simple, but to the best of our knowledge, nothing similar has been reported in the literature.

Clearly, one advantage of our factorization is that $N/2$ rotations can be computed in parallel, in each iteration toward producing a matrix from angles. So $N/2$ processors can be utilized in a natural, direct way. If fewer processors are available, then it is easy to partition the graph, so as to assign the same number of rotations to a single processor, and to have hardware fully utilized. This will be shown by example in Sect. 4.

The second benefit is that only $N - 1$ iterations, each consisting of $N/2$ independent rotations, are necessary to produce a matrix. So the known factorizations can be outperformed in terms of both throughput and latency, especially when a solution has to be implemented in hardware.

Our factorization seems to be very similar to those resulting from the QR decomposition. But in fact, it uses rotations in a completely different way, as demonstrated in Fig. 5 for the same matrix that was used in Fig. 3 to explain the QR algorithm. Instead of zeroing elements, the first $\frac{N}{2}(\frac{N}{2} - 1)$ rotations, in $(\frac{N}{2} - 1)$ layers, preprocess the matrix so that each of the remaining rotations can zero 2 elements at once.

A separate study is necessary to explain mechanisms behind the preprocessing and behind the new factorization as a whole. In the remaining of this work, we focus our attention on showing that the factorization is general, i.e. that it

```
// Original matrix: A

 0.0695    0.5297    0.3592    0.0803    0.1673    0.7424
 0.5471   -0.5176    0.3534    0.4925   -0.2127    0.1417
 0.5349    0.6015    0.1963   -0.0038   -0.1328   -0.5439
 0.5664   -0.2323   -0.1337   -0.6679    0.3651    0.1674
 0.2982    0.1881   -0.8303    0.3587   -0.0923    0.2217
-0.0040   -0.0190    0.0152    0.4198    0.8760   -0.2362

// 1st rotation: A = G(5, 6, -7.920) * A

 0.0695    0.5297    0.3592    0.0803    0.1673    0.7424
 0.5471   -0.5176    0.3534    0.4925   -0.2127    0.1417
 0.5349    0.6015    0.1963   -0.0038   -0.1328   -0.5439
 0.5664   -0.2323   -0.1337   -0.6679    0.3651    0.1674
-0.0237   -0.0314    0.0701    0.3951    0.8802   -0.2504
-0.2973   -0.1864    0.8275   -0.3857    0.0341   -0.2055

// 2nd rotation: A = G(3, 4, -0.035) * A

 0.0695    0.5297    0.3592    0.0803    0.1673    0.7424
 0.5471   -0.5176    0.3534    0.4925   -0.2127    0.1417
 0.5546    0.5929    0.1915   -0.0274   -0.1198   -0.5376
 0.5471   -0.2534   -0.1405   -0.6673    0.3696    0.1865
-0.0237   -0.0314    0.0701    0.3951    0.8802   -0.2504
-0.2973   -0.1864    0.8275   -0.3857    0.0341   -0.2055

// 3rd rotation: A = G(1, 2, -1.319) * A

 0.5472   -0.3695    0.4317    0.4970   -0.1643    0.3220
 0.0688   -0.6418   -0.2600    0.0448   -0.2149   -0.6838
 0.5546    0.5929    0.1915   -0.0274   -0.1198   -0.5376
 0.5471   -0.2534   -0.1405   -0.6673    0.3696    0.1865
-0.0237   -0.0314    0.0701    0.3951    0.8802   -0.2504
-0.2973   -0.1864    0.8275   -0.3857    0.0341   -0.2055

// 4th rotation: A = G(4, 5, -1.557) * A

 0.5472   -0.3695    0.4317    0.4970   -0.1643    0.3220
 0.0688   -0.6418   -0.2600    0.0448   -0.2149   -0.6838
 0.5546    0.5929    0.1915   -0.0274   -0.1198   -0.5376
-0.0161   -0.0349    0.0682    0.3858    0.8852   -0.2478
-0.5474    0.2530    0.1415    0.6728   -0.3573   -0.1900
-0.2973   -0.1864    0.8275   -0.3857    0.0341   -0.2055

// 5th rotation: A = G(2, 3, -0.655) * A

 0.5472   -0.3695    0.4317    0.4970   -0.1643    0.3220
 0.3924   -0.1480   -0.0896    0.0188   -0.2434   -0.8698
 0.3980    0.8612    0.3102   -0.0490    0.0359   -0.0100
-0.0161   -0.0349    0.0682    0.3858    0.8852   -0.2478
-0.5474    0.2530    0.1415    0.6728   -0.3573   -0.1900
-0.2973   -0.1864    0.8275   -0.3857    0.0341   -0.2055

// 6th rotation: A = G(6, 1, -8.200) *

 0.0943    0.3006   -0.9248    0.1945    0.0236    0.0843
 0.3924   -0.1480   -0.0896    0.0188   -0.2434   -0.8698
 0.3980    0.8612    0.3102   -0.0490    0.0359   -0.0100
-0.0161   -0.0349    0.0682    0.3858    0.8852   -0.2478
-0.5474    0.2530    0.1415    0.6728   -0.3573   -0.1900
 0.6155   -0.2845    0.1258    0.5983   -0.1662    0.3726

// 7th rotation: A = G(5, 6, -0.727) *

 0.0943    0.3006   -0.9248    0.1945    0.0236    0.0843
 0.3924   -0.1480   -0.0896    0.0188   -0.2434   -0.8698
 0.3980    0.8612    0.3102   -0.0490    0.0359   -0.0100
-0.0161   -0.0349    0.0682    0.3858    0.8852   -0.2478
 0.0       0.0       0.1893    0.9003   -0.3774    0.1056
 0.8237   -0.3807    0.0       0.0       0.1133    0.4047
```

```
...

// 8th rotation: A = G(3, 4, 0.040) *

 0.0943    0.3006   -0.9248    0.1945    0.0236    0.0843
 0.3924   -0.1480   -0.0896    0.0188   -0.2434   -0.8698
 0.3983    0.8619    0.3072   -0.0646    0.0       0.0
 0.0       0.0       0.0807    0.3835    0.8859   -0.2480
 0.0       0.0       0.1893    0.9003   -0.3774    0.1056
 0.8237   -0.3807    0.0       0.0       0.1133    0.4047

// 9th rotation: A = G(1, 2, -7.951) * A

 0.3814   -0.1763    0.0       0.0      -0.2446   -0.8738
-0.1317   -0.2849    0.9291   -0.1954    0.0       0.0
 0.3983    0.8619    0.3072   -0.0646    0.0       0.0
 0.0       0.0       0.0807    0.3835    0.8859   -0.2480
 0.0       0.0       0.1893    0.9003   -0.3774    0.1056
 0.8237   -0.3807    0.0       0.0       0.1133    0.4047

// 10th rotation: A = G(4, 5, -1.168) * A

 0.3814   -0.1763    0.0       0.0      -0.2446   -0.8738
-0.1317   -0.2849    0.9291   -0.1954    0.0       0.0
 0.3983    0.8619    0.3072   -0.0646    0.0       0.0
 0.0       0.0       0.2058    0.9786    0.0       0.0
 0.0       0.0       0.0       0.0      -0.9630    0.2695
 0.8237   -0.3807    0.0       0.0       0.1133    0.4047

// 11th rotation: A = G(2, 3, -5.032) * A

 0.3814   -0.1763    0.0       0.0      -0.2446   -0.8738
-0.4195   -0.9077    0.0       0.0       0.0       0.0
 0.0       0.0       0.9786   -0.2058    0.0       0.0
 0.0       0.0       0.2058    0.9786    0.0       0.0
 0.0       0.0       0.0       0.0      -0.9630    0.2695
 0.8237   -0.3807    0.0       0.0       0.1133    0.4047

// 12th rotation: A = G(6, 1, -2.004) * A

-0.9077    0.4195    0.0       0.0       0.0       0.0
-0.4195   -0.9077    0.0       0.0       0.0       0.0
 0.0       0.0       0.9786   -0.2058    0.0       0.0
 0.0       0.0       0.2058    0.9786    0.0       0.0
 0.0       0.0       0.0       0.0      -0.9630    0.2695
 0.0       0.0       0.0       0.0      -0.2695   -0.9630

// 13th rotation: A = G(5, 6, -3.414) * A

-0.9077    0.4195    0.0       0.0       0.0       0.0
-0.4195   -0.9077    0.0       0.0       0.0       0.0
 0.0       0.0       0.9786   -0.2058    0.0       0.0
 0.0       0.0       0.2058    0.9786    0.0       0.0
 0.0       0.0       0.0       0.0       1.0000    0.0
 0.0       0.0       0.0       0.0       0.0       1.0000

// 14th rotation: A = G(3, 4, -0.207) * A

-0.9077    0.4195    0.0       0.0       0.0       0.0
-0.4195   -0.9077    0.0       0.0       0.0       0.0
 0.0       0.0       1.0000    0.0       0.0       0.0
 0.0       0.0       0.0       1.0000    0.0       0.0
 0.0       0.0       0.0       0.0       1.0000    0.0
 0.0       0.0       0.0       0.0       0.0       1.0000

// 15th rotation: A = G(1, 2, -3.575) *

 1.0000    0.0       0.0       0.0       0.0       0.0
 0.0       1.0000    0.0       0.0       0.0       0.0
 0.0       0.0       1.0000    0.0       0.0       0.0
 0.0       0.0       0.0       1.0000    0.0       0.0
 0.0       0.0       0.0       0.0       1.0000    0.0
 0.0       0.0       0.0       0.0       0.0       1.0000
```

...

**Fig. 5.** Rotation effects on orthogonal matrix: proposed parallel factorization

applies to an arbitrary orthonormal matrix, and allows for speeding up computations.

As our factorization has a nature different from that of the known ones, its angles cannot be determined by using the QR decomposition. We were also unsuccessful in finding a general analytic method for converting a known factorization into a parallel one.

Nevertheless, we have developed a pragmatic method for determining angles of our parallel factorization of a given matrix. It follows popular factorization-based techniques for designing filter banks, please see e.g. [17]. The idea is to use a standard optimization algorithm, like the gradient-based "fminunc" routine available in Matlab, to optimize angles of our factorization so that it produces a matrix similar to a reference one.

The algorithm should be provided with initial angles and an objective function. The angles can simply be random, but the function must be complicated. It produces a matrix from angles, in accordance with the parallel factorization, and compares the result to the matrix that has to be factorized. The difference between the matrices is returned to the optimization algorithm, so as to allow it to evaluate and choose beneficial modifications of angles.

The approach is not well suited to large matrices, as the optimization algorithm fails when it has to deal with more than a hundred or so variables. Nevertheless, for matrices up to $12 \times 12$, the method allows for empirically proving that our parallel factorization is as general as the conventional ones.

Even for smaller matrices, in many cases, it was necessary to run the optimization routine several times, so as to try various initial angles, but finally the algorithm managed to find angles for which the difference between matrices was marginal, i.e. its Frobenius norm was below $10^{-6}$. We verified this for hundreds of reference matrices that were generated from random angles by using the conventional factorizations. If so many freely-chosen matrices can be represented by both known and new factorizations, then there is a strong evidence that they both are equivalent with respect to generality, or ability to model an arbitrary orthogonal matrix.

## 4    Experimental Evaluation of Achievable Gains

The aim of our experiments was to show that our solution can be used to easily develop programs that exploit multicore processors to efficiently generate orthogonal matrices. We decided to use software and hardware platforms that are popular and should provide representative results: the Java 11 development/runtime environments, the Windows 10 operating system, and a PC computer powered by the eight-core AMD Ryzen 7 2700 Processor, clocked at 3.2 GHz.

Firstly, we have developed a single-threaded program which generates orthogonal matrices form random angles, by sequentially composing rotations in accordance with either (3) or (7). This program was necessary to gather reference results for evaluating multi-threaded implementations.

Then, a double-threaded program has been developed in accordance with Fig. 6, which shows how rotations are assigned to threads and to groups that can be computed in parallel, in periods between moments of thread synchronization. Threads need to be synchronized as some rotations assigned to one thread depend on results of rotations assigned to another thread. In particular, in Fig. 6, one can see two series of dependent rotations. One comprises rotations by $\alpha_0$, $\alpha_8$, $\alpha_{16}$ and so on. The second is formed by rotations by $\alpha_4$, $\alpha_{12}$, $\alpha_{20}$ and so on.

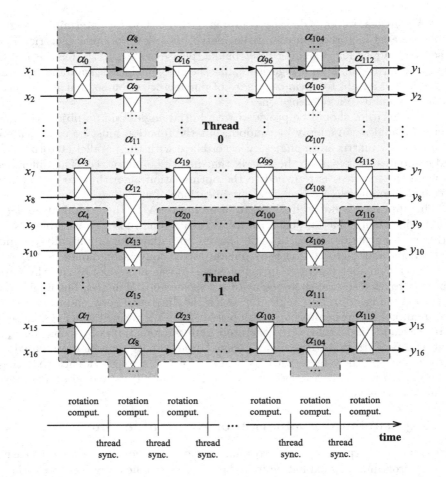

**Fig. 6.** Assignment of rotations to two threads and to computation-synchronization schedule

It is an advantage of our solution that depended rotations form uniform series that can easily be determined and handled programmatically. For known factorizations, in data flow graphs like those in Fig. 2, it is much more difficult to assign rotations to threads and to groups that can be computed without synchronization.

In Fig. 6, we show a data flow graph for a particular size of matrix, $16 \times 16$, so as to have readable values as subscripts, instead of unclear equations. It is rather easy to adapt the graph to other sizes of matrices, or to more threads.

Table 1 shows results of our performance tests, which consisted in measuring times necessary to recover an orthogonal matrix from randomly selected rotation angles. Four sizes of matrices have been considered, which differ considerably in memory consumption and computational load. For each settings, we have listed

**Table 1.** Times necessary to generate orthogonal matrices by 1- and 2-thread implementations of proposed factorization.

| Matrix size | Number of rotations | Time necessary to generate 100 matrices [ms] | | Relative change [%] |
|---|---|---|---|---|
| | | By 1 thread | By 2 threads | |
| 128 × 128 | 8128 | 84 | 127 | 151.2 |
| | | 89 | 133 | 149.4 |
| | | 97 | 148 | 152.6 |
| 256 × 256 | 32640 | 441 | 368 | 83.4 |
| | | 463 | 378 | 81.6 |
| | | 477 | 389 | 81.5 |
| 512 × 512 | 130816 | 3271 | 1947 | 59.5 |
| | | 3328 | 1988 | 59.7 |
| | | 3751 | 2401 | 64.0 |
| 1024 × 1024 | 523776 | 41646 | 20005 | 48.0 |
| | | 41914 | 20187 | 48.2 |
| | | 48238 | 21215 | 44.0 |

three times in order to prove that results are representative and reflect general trends, even though execution times are affected by activity of other processes that run on a given machine and by temporal states of cache memory.

A clear conclusion from the data in Table 1 is that the multi-threaded implementation outperforms the single-threaded one when large matrices are processed. It can be even twice as fast. For small matrices, however, the overhead related to thread management and synchronization excesses the essential computation-related load, so multi-threaded implementations can perform much worse than simple single-threaded programs.

So, our approach suits specific yet practical application area, when implemented on a general-purpose processor. The new factorization should be more useful when implemented in hardware that is aimed at computing rotations in parallel. We plan to develop a specialized digital circuit by using the FPGA technology.

## 5   Conclusion

A method for generating orthogonal matrices has been proposed that is significantly different from know ones. By giving up the QR decomposition, we were able to arrange Givens rotations uniformly, so that more rotations can be computed in parallel, and fewer layers of independent rotations are necessary to model a matrix, compared to the conventional arrangements. So throughput and latency can be reduced, even though practical gains highly depend on application specificity, software/hardware architecture and matrix size. They can be considerable, as it has been verified by evaluating a multi-threaded Java-based program for generating random orthogonal matrices.

**Acknowledgments.** This work was financially supported from the Polish Ministry of Science and Higher Education under subsidy for maintaining the research potential of the Faculty of Mathematics and Informatics, University of Bialystok.

# References

1. Anderson, T., Olkin, I., Underhill, L.: Generation of random orthogonal matrices. SIAM J. Sci. Stat. Comput. **8**(4), 625–629 (1987)
2. Arioli, M.: Tensor product of random orthogonal matrices. Technical report RAL-TR-2013-006, Science and Technology Facilities Council (2013)
3. Benson, A.R., Gleich, D.F., Demmel, J.: Direct QR factorizations for tall-and-skinny matrices in MapReduce architectures. In: Proceedings of the IEEE International Conference on Big Data, pp. 264–272, October 2013
4. Diaconis, P., Forrester, P.J.: Hurwitz and the origins of random matrix theory in mathematics. Random Matrices Theory Appl. **6**(1), 1730001 (2017)
5. Frerix, T., Bruna, J.: Approximating orthogonal matrices with effective Givens factorization. In: Proceedings of the 36th International Conference on Machine Learning (ICML), Long Beach, CA, 9–15 June 2019, pp. 1993–2001 (2019)
6. Genz, A.: Methods for generating random orthogonal matrices. In: Niederreiter, H., Spanier, J. (eds.) Monte-Carlo and Quasi-Monte Carlo Methods 1998, pp. 199–213. Springer, Heidelberg (2000). https://doi.org/10.1007/978-3-642-59657-5_13
7. Hofmann, M., Kontoghiorghes, E.J.: Pipeline Givens sequences for computing the QR decomposition on a EREW PRAM. Parallel Comput. **32**(3), 222–230 (2006)
8. Johnson, K.T., Hurson, A.R., Shirazi, B.: General-purpose systolic arrays. Computer **26**(11), 20–31 (1993)
9. Merchant, F., et al.: Efficient realization of Givens rotation through algorithm-architecture co-design for acceleration of QR factorization, March 2018. http://arxiv.org/abs/1803.05320
10. Mezzadri, F.: How to generate random matrices from the classical compact groups. Not. Am. Math. Soc. **54**(5), 592–604 (2007)
11. Modi, J.J., Clarke, M.R.B.: An alternative Givens ordering. Numer. Math. **43**(1), 83–90 (1984). https://doi.org/10.1007/BF01389639
12. Parfieniuk, M., Petrovsky, A.: Structurally orthogonal finite precision implementation of the eight point DCT. In: Proceedings of the IEEE International Conference on Acoustics, Speech, Signal Processing (ICASSP), Toulouse, France, 14–19 May 2006, vol. 3, pp. 936–939 (2006)
13. Parfieniuk, M., Park, S.Y.: Versatile quaternion multipliers based on distributed arithmetic. Circuits Syst. Signal Process. **37**(11), 4880–4906 (2018)
14. Pinchon, D., Siohan, P.: Angular parameterization of real paraunitary matrices. IEEE Signal Process. Lett. **15**, 353–356 (2008)
15. Pinheiro, J.C., Bates, D.M.: Unconstrained parametrizations for variance-covariance matrices. Stat. Comput. **6**(3), 289–296 (1996)
16. Sun, X., Bischof, C.: A basis-kernel representation of orthogonal matrices. SIAM J. Matrix Anal. Appl. **16**(4), 1184–1196 (1995)
17. Vaidyanathan, P.P., Doğanata, Z.: The role of lossless systems in modern digital signal processing: a tutorial. IEEE Trans. Educ. **32**(3), 181–197 (1989)

# Author Index

Almeida, Francisco II-134
Alrabeei, Salah II-287
Antkowiak, Michał II-312
Arioli, Mario I-104
Arrigoni, Viviana I-129
Augustynowicz, Paweł I-49

Bader, Michael I-311, II-59
Balis, Bartosz I-287
Banaszak, Michał II-333
Bartolini, Andrea II-159, II-181
Bashinov, Aleksei I-335
Bastrakov, Sergei I-335
Bečka, Martin I-555
Benoit, Anne I-211
Berlińska, Joanna II-230
Bielech, Maciej II-489
Bielecki, Wlodzimierz II-25
Blanco, Vicente II-134
Bokati, Laxman II-364
Bouvry, Pascal II-144
Brun, Emeric I-528
Brzostowski, Bartosz II-341
Byrski, Aleksander II-489

Cabaleiro Domínguez, Jose Carlos II-205
Cabrera, Alberto II-134
Cacciapuoti, Rosalba II-75
Calore, Enrico I-187, II-169
Calvin, Christophe I-528
Castellanos–Nieves, Dagoberto II-134
Castiglione, Aniello II-111
Çatalyürek, Ümit V. I-211
Cavazzoni, Carlo II-181
Cesarini, Daniele II-181
Chang, Tao I-528
Chen, Wei-yu I-385
Chien, Steven W. D. I-301
Claeys, Xavier I-141
Cobrnic, Mate I-199
Coti, Camille I-250
Cramer, Tim I-237
Czajkowski, Marcin I-359, I-421
Czarnul, Paweł II-123

D'Amore, Luisa II-75
Danelutto, Marco II-191
Danoy, Grégoire II-144
Davidović, Davor I-13
De Sensi, Daniele II-191
Demczuk, Miłosz I-261
Désérable, Dominique II-433
Di Luccio, Diana II-111
Dmitruk, Beata I-93
Dourvas, Nikolaos II-445
Dragic, Leon I-199
Drozdowski, Maciej I-224
Duepmeier, Clemens I-432
Duspara, Alen I-199

Efimenko, Evgeny I-335
Ernst, Dominik I-505

Fernández Pena, Tomás II-205
Fernández Rivera, Francisco II-205
Focht, Erich I-237
Fuentes, Joel I-385
Funika, Włodzimierz I-467

Gabbana, Alessandro I-187, II-169
García Lorenzo, Oscar II-205
Garza, Arturo I-385
Gąsior, Jakub II-433
Gębicka, Karolina II-333
Georgoudas, Ioakeim G. II-445
Gepner, Pawel I-370
Gergel, Victor I-174
Gokieli, Maria II-277
Gonoskov, Arkady I-335
Gosek, Łukasz II-478
Grochal, Lukasz I-272

Hagenmeyer, Veit I-432
Hager, Georg I-505
Haglauer, Monika II-312
Halbiniak, Kamil I-370
Halver, Rene II-35
Hladík, Milan II-374
Hoffmann, Rolf II-433

Hoffmann, Tomasz   II-407
Hössinger, Andreas   I-348

Inoue, Takahiro   I-493
Ito, Yasuaki   I-396, I-493, II-46

Jakob, Wilfried   I-432
Jankowska, Malgorzata A.   II-384, II-407
Jokar, Mahmood   II-287
Jurczuk, Krzysztof   I-359, I-421

Kannan, Ramakrishnan   I-543
Karbowiak, Łukasz   I-479
Kasagi, Akihiko   II-46
Khalloof, Hatem   I-432
Kieffer, Emmanuel   II-144
Kirik, Ekaterina   II-457
Kitowski, Jacek   I-261
Kjelgaard Mikkelsen, Carl Christian   I-58,
    I-70, I-82
Klinkenberg, Jannis   II-59
Klusáček, Dalibor   II-217
Kobayashi, Hironobu   I-396
Kondratyuk, Nikolay   I-324
Koperek, Paweł   I-467
Korch, Matthias   I-3
Kosheleva, Olga   II-364
Kosmynin, Boris   I-237
Kosta, Sokol   II-111
Kotlarska, Anna   II-333
Kovac, Mario   I-199
Kozinov, Evgeny   I-174
Kravčenko, Michal   I-141
Kreinovich, Vladik   II-364
Krenz, Lukas   I-311
Kretowski, Marek   I-359, I-421
Kruse, Carola   I-104
Krzywaniak, Adam   II-123
Krzywicka, Agata   II-333
Krzyżanowski, Piotr   II-267
Kubica, Bartłomiej Jacek   II-364, II-395,
    II-418
Kucharski, Łukasz   II-312
Kulawik, Adam   I-370
Kurek, Jarosław   II-418
Kuznetsov, Evgeny   I-324
Kwedlo, Wojciech   I-457

Lapegna, Marco   II-101
Laso Rodríguez, Rubén   II-205

Łasoń, Aneta   II-333
Lewandowski, Krzysztof   II-333
Lirkov, Ivan   II-93
Logunov, Mikhail   I-324
Lorenzo del Castillo, Juan Ángel   II-205
Low, Tze Meng   I-162
Łubowicz, Michał   I-457
Lueh, Guei-yuan   I-385
Lyutenko, Vladyslav   I-261

Malawski, Maciej   I-287
Malony, Allen D.   I-250
Malyshev, Andrey   II-457
Mamica, Maria   II-467
Manstetten, Paul   I-348
Maratea, Antonio   II-111
Marciniak, Andrzej   II-384, II-407
Marcinkowski, Leszek   II-245, II-287
Markidis, Stefano   I-301
Marowka, Ami   II-13
Marszałkowski, Jędrzej M.   I-224
Massini, Annalisa   I-129
Meinke, Jan H.   II-35
Mele, Valeria   II-75, II-101
Merta, Michal   I-141
Meyerov, Iosif   I-335
Michałek, Przemysław   II-478
Mitsopoulou, Martha   II-445
Mlinaric, Hrvoje   I-199
Montella, Raffaele   II-111
Morawiecki, Piotr   II-489
Mukunoki, Daichi   I-516
Müller, Matthias S.   I-237, II-59
Muras, Fryderyk   II-478
Murray, Charles D.   I-25
Myllykoski, Mirko   I-58, I-70

Nakano, Koji   I-396, I-493, II-46
Nikolskiy, Vsevolod   I-324
Nishimura, Takahiro   II-46
Ntallis, Nikolaos   II-301

Ogita, Takeshi   I-516
Okša, Gabriel   I-555
Orłowski, Arkadiusz Janusz   II-395
Orzechowski, Michal   I-287
Ostheimer, Phil   I-432
Ozaki, Katsuhisa   I-516
Özkaya, M. Yusuf   I-211

Paciorek, Mateusz  II-489
Palkowski, Marek  II-25
Panov, Alexander  I-335
Panova, Elena  I-335
Parfieniuk, Marek  I-567
Parikh, Devangi N.  I-162
Paszkiewicz, Andrzej  I-49
Pawlik, Krystian  I-287
Pawlik, Maciej  I-287
Pawłowski, Filip  I-38
Pęcak, Tomasz  II-489
Pękala, Agnieszka  II-467
Peng, Ivy B.  I-301
Pietroń, Marcin  II-374
Pilch, Marek  II-323, II-341
Piljic, Igor  I-199
Pinel, Frédéric  II-144
Popel, Egor  II-457
Popovici, Doru Thom  I-162

Quell, Michael  I-348
Quintana-Ortí, Enrique S.  I-13

Rahman, Talal  II-245
Rannabauer, Leonhard  I-311
Reinarz, Anne  I-117
Renc, Paweł  II-489
Revol, Nathalie  II-353
Riha, Lubomir  II-159
Rinaldi, Fabio  I-187
Romano, Diego  II-101
Römmer, Manoel  I-237
Rüde, Ulrich  I-104
Rutkowska, Danuta  I-445

Samfass, Philipp  II-59
Sao, Piyush  I-543
Sasak-Okoń, Anna  I-408
Scheichl, Robert  I-117
Scherson, Isaac D.  I-385
Schifano, Sebastiano Fabio  I-187, II-169
Schwarz, Angelika  I-82
Seelinger, Linus  I-117
Selberherr, Siegfried  I-348
Seredyński, Franciszek  II-433
Shahoud, Shadi  I-432
Singh, Gaurav  I-224
Sirakoulis, Georgios Ch.  II-445

Skalna, Iwona  II-374
Słota, Renata  I-261
Smyk, Adam  I-272
Sosonkina, Masha  I-104
Soysal, Mehmet  II-217
Spampinato, Daniele G.  I-162
Stegailov, Vladimir  I-324
Stpiczyński, Przemysław  I-93, II-3
Suita, Shunsuke  II-46
Surmin, Igor  I-335
Suter, Frédéric  II-217
Sutmann, Godehard  II-35
Szczepańczyk, Andrzej  II-277
Szustak, Lukasz  I-370

Tabaru, Tsuguchika  II-46
Takahashi, Daisuke  I-151
Tardieu, Nicolas  I-104
Terboven, Christian  II-59
Thies, Jonas  I-505
Tokura, Hiroki  I-493, II-46
Tripiccione, Raffaele  I-187, II-169
Trohidou, Kalliopi N.  II-301
Tudruj, Marek  I-272
Turek, Wojciech  II-489
Tytko, Karolina  II-467

Uçar, Bora  I-38

Valdman, Jan  II-256
Varrette, Sébastien  II-144
Vasilakaki, Marianna  II-301
Vitova, Tatýana  II-457
Volokitin, Valentin  I-335
Vysocky, Ondrej  II-159

Wąs, Jarosław  II-467, II-478, II-489
Weinbub, Josef  I-348
Weinzierl, Tobias  I-25
Wellein, Gerhard  I-505
Werner, Tim  I-3
Wiaderek, Krzysztof  I-445
Wiatr, Roman  I-261
Wojtkiewicz, Jacek  II-323, II-341

Yzelman, Albert-Jan  I-38

Zapletal, Jan  I-141

Printed in the United States
By Bookmasters